Solving the Pulses Crisis

by

Anil Singh

B. Gangwar

Pages	:	**382 pages, 25cm**
Year	:	**2013**
ISBN 10	:	**9381450481**
ISBN 13	:	**9789381450482**
Price	:	**INR 2000.00**

You can order @www.nipabooks.com

Anil Kumar Singh *formerly worked as Deputy Director General (NRM) in the Indian Council of Agricultural Research (ICAR).*

B. Gangwar: *Project Director, Project Directorate on Farming Systems Research, Modipuram, Meerut, Uttar Pradesh, India.*

India has achieved self sufficiency in food grain production in recent years with record production of 250 mt during 2011-12. However, the pulses production remained low and considered to be the major concern for researchers and development planners. Considering the much more importance in near future and to avoid pulses crisis situation, the present attempt was made to compile the available scientific information, so as to highlight the issues, technologies and strategies in the title of "Solving Pulses Crisis" in India. The publication is divided into two parts. The first part deals national issues, technologies and strategies while the second part deals with crop based issue and technologies.

The first part consists of 13 Chapters. The first three Chapters deals with pulses Related national issues, technologies and strategies including NEH region too. The Chapter IV deals with crop diversification involving pulses while Chapter V focused on pulses production under organic system. The issues related to legumes as a nutrient supplement in Chapter VI, tillage and crop establishment in Chapter VII water management in pulses in Chapter VIII and Integrated nutrient management in Chapter IX are discussed in detail. The aspects of weed and pest management are presented in Chapter X to XI, respectively. The specific issues related to post harvest, value addition are discussed in Chapter XII, while trade related policy issues are focused in Chapter XIII.

In part second, the crop issues, strategies and technologies are presented. Accordingly, Chapter XIV deals with pigeonpea while in Chapter XV issues related to greengram and black gram are discussed. The Chapter XVI to XIX deals with chickpea, lentil, field pea and lathyrus while in Chapter XX the issues technologies and prospects of Guar are discussed. In last Chapter XXI the issues and technologies related to arid legumes (mothbean, cowpea and horsegram) with special reference to arid areas are discussed. Hopefully, the publication will prove to be a reference and a way forward for solving pulses crisis in India and achieving the targets matching with food production strategies in years to come.

Contents

Foreword

Preface

Synonyms

Abbreviations

Part-I. Issues, Strategies and Technologies

Chapter 1. Solving the Pulses Crisis: A Road Map

Chapter 2. Technological Options for Increasing Pulses Production

Chapter 3. Based Issues and Technologies for Doubling Pulses Production in NEH Region

Chapter 4. Crop Diversification Through Pulses

Chapter 5. Pulses Production in Organic System

Chapter 6. Legume Inclusion Substitution in Cropping Systems

Chapter 7. Tillage Management and Crop Establishment Strategies

Chapter 8. Integrated Nutrient Management

Chapter 9. Integrated Water Management

Chapter 10. Weed Management

Chapter 11. Integrated Pest Management

Chapter 12. Post Harvest Losses of Pulses, Value Addition and Management

Chapter 13. Pulses Trade in India : Problems, Opportunities and Policy Issues

Part-II. Cropwise Issues and Technologies

Chapter 14. Pigeonpea

Chapter 15. Mungbean and Urdbean

Chapter 16. Chickpea

Chapter 17. Lentil Chapter 18. Fieldpea

Chapter 19. Lathyrus

Chapter 20. Guar

Chapter 21. French Bean

Chapter 22. Arid Pulses : Mothbean, Cowpea and Horsegram

Enhancing Pulses Production

Technologies and Strategies

Enhancing Pulses Production

Technologies and Strategies

Edited by :

Babooji Gangwar

Project Director
Project Directorate for Farming Systems Research
Modipuram, Meerut -250110
Uttar Pradesh

and

Anil Kumar Singh

Vice Chancellor
Rajmata Vijayaraje Sciendia Krishi Vishvwa Vidhyalaya
Gwalior - 474002, Madhya Pradesh

NEW INDIA PUBLISHING AGENCY

New Delhi – 110 034

Citation : Gangwar, B. and Singh, A.K. (Eds.). 2014. Enhancing Pulses Production: Technologies and Strategies. New India Publishing Agency, New Delhi; .... p. 1-545

NEW INDIA PUBLISHING AGENCY
101, Vikas Surya Plaza, CU Block, LSC Market
Pitam Pura, New Delhi 110 034, India
Phone: + 91 (11)27 34 17 17 Fax: + 91(11) 27 34 16 16
Email: info@nipabooks.com
Web: www.nipabooks.com

Feedback at feedbacks@nipabooks.com

ISBN : 978-93-81450-77-2

Composed, Designed and Printed at Jai Bharat Printing Press, Delhi

शरद पवार
SHARAD PAWAR

कृषि एवं खाद्य प्रसंस्करण उद्योग
मंत्री
भारत सरकार
MINISTER OF AGRICULTURE &
FOOD PROCESSING INDUSTRIES
GOVERNMENT OF INDIA

11th February, 2014

Foreword

Indian Agriculture has made tremendous progress in feeding the second largest population in the world. The agricultural production system can take credit that during the period 1950-51 to 2011-12, the food grain production increased five times, the horticulture and milk production six times, fish production nine times and egg production twenty seven times. A creditable record production of pulses of more than 18 million tonnes was achieved in 2010-11 as a consequence of efforts of the Ministry of Agriculture, Line Departments, State Agricultural Universities and Indian Council of Agricultural Research. However, when the production of pulses is examined, the enhancement in production has been only 2.2 times during this period increasing from 8.4 million tonnes in 1950-51 to 18.2 million tonnes in 2011-12.

The production target for 2025 has been set as 30 million tonnes which is in itself a challenging task considering the fact that pulses are cultivated in rainfed areas under various abiotic and biotic conditions with a low level of inputs. Technological developments that have taken place over the years related to improve varieties and agronomic management have indicated that it is feasible to meet the production targets and ensure nutrition security to the vast

Office : Room No. 120, Krishi Bhawan, New Delhi-110 001 Tel.: 23383370, 23782691 Fax : 23384129
Resi. : 6, Janpath, New Delhi-110 011 (India) Tel. : 011-23018870, 23018619 Fax : 011-23018609 E-mail : sharadpawar.sp@gmail.com

population and improve soil health by inclusion of pulses in the cropping and farming systems prevalent in the country.

I am happy that efforts have been made here to document the information at the national level relevant for adoption in different states in the form of this publication entitled "Enhancing Pulses Production: Technologies and Stratégies" for the use of extension agencies, researches and planners. This publication will certainly be of immense use to all the stake holders concerned with pulses production in India.

(Sharad Pawar)

Office : Room No. 120, Krishi Bhawan, New Delhi-110 001 Tel.: 23383370, 23782691 Fax : 23384129 Resi. : 6, Janpath, New Delhi-110 011 (India) Tel. : 011-23018870, 23018619 Fax : 011-23018609 E-mail : sharadpawar.sp@gmail.com

Preface

As a result of concerted multi-agency efforts focussing on R&D and technology with effective policy and institutional mechanism, India could harvest a record production of foodgrains of about 250 million tonnes during 2011-12. However, low production/ productivity and almost stagnant production of pulses remained a matter of concern for the researchers and planners of the country. Backed up with higher support prices and technological backstopping by developmental agencies, pulses production could also be achieved to a record levels of 18.24 million tonnes in 2010-11 and 17.28 million tonnes in 2011-12. In spite of the record production, the country still imported 3.5 million tonnes of pulses (2010-11) costing about Rs.7,000 crores in foreign exchange, being the biggest importer with close to 20 percent share of the world. Whenever availability of pulses fell short of domestic demand, there has been a spurt in prices causing widespread concerns among the planners and policy makers because pulses are considered a major source of proteins and play an important role in improving food and nutritional security particularly for the poor and small and marginal farmers in India. To alleviate protein-energy malnutrition in the country, a minimum of 50 gm of pulses per capita per day is required in addition to other sources of proteins such as cereals, milk, meat, egg, etc. To make up this shortfall, about 27 million tonnes of pulses are required by 2015 and 32 million tonnes by 2030. To meet this target, the overall pulses production must increase at the rate of around 4 percent per annum. Though the latest data shows that pulses productivity has reached a record level of approximately 700 kg/ha, the present productivity growth is still not sufficient enough to meet the diversified pulses demand of the rising population. This seems a herculean task in the light of the past trends in pulses production. India's dependence on global market to meet the domestic demand for pulses would increase in future. Completely relying on imports would not be a good strategy for a country with a large vegetarian population. Therefore, a well thought strategy is urgently needed to meet future demand of pulses and avoid crises situations. The key elements of such a strategy should aim at providing technological backstopping for managing biotic and abiotic stresses and achieving a breakthrough in pulse productivity on sustainable basis alongwith remunerative price support for the farmers. It must be remembered that pulses are going to play crucial role not only for nutritional security but also for improving human- livestock- soil health and also for saving water resource.

Experiences from farming system research reveal that opting for specific cropping pattern for different regions and adopting improved agronomic practices provide us with substantial scope for incorporating pulses in the prevailing farming systems, and realising the target of pulses production is not difficult if location-specific strategies are implemented effectively. The content of the present book addresses all the relevant issues by dovetailing the region-specific strategy and options for solving pulses crises in the country. It is expected that the book would serve to sensitise the planners and policy makers as well as the local agricultural development officials to explore and implement region-specific strategy for enhancing pulses production by incorporating these in the prevailing farming systems by harnessing the available technological potential. Moreover, various technological options and strategies presented in this book would also be useful for all the stakeholders to realise the target of pulses production in near future.

B. Gangwar

Anil Kumar Singh

Synonyms

1. **Pigonpea/Redgram (*Cajanus cajan* (L. *Millsp.*):** *Arhar dal* (Bengali/ Hindi/Kashmiri), *Tuvara parippu* (Malayalam), *Thugare bele* (Kannad), *Tur dal* (Marathi), *Harada* (Oriya), *Tuvaram paruppu* (Tamil), *Tuver* (Gujarati).

2. **Greengram/Mungbean (*Vigna radiate* L. R. Wilczek):** *Mug* (Gujarati/ Bengali), *Hesare kalu* (Kannad), *Muang* (Kashmiri), *Moongi* (Punjabi), *Cheru pararu* (Malayalam), *Mung* (Hindi/Marathi), *Muga* (Oriya), *Pasipayir* (Tamil), *Pesalu* (Telugu).

3. **Blackgram/Urdbean (*Vigna mungo* L. Hepper):** *Mashkalair dal* (Bengali), *Uddina bele* (Kannad), *Kaha* (Kashmiri), *Mah-di-dal* (Punjabi), *Uzhunnu parippu* (Malayalam), *Urd dal* (Hindi), *Biri* (Oriya), *Ulutham paruppu* (Tamil), *Minapa pappu* (Telugu), *Uddachi dal* (Marathi), *Aalad* (Gujarati).

4. **Chick pea (*Cicer arietinum* L.):** *Chana* (Gujarati/Hindi), *Kadale* (Kannad), *Kadala* (Malayalam), *Chanu* (Kashmiri), *Buta* (Oriya), *Harbara* (Marathi), *Chole* (Punjabi), *Kothu kadalai* (Tamil), *Sanagalu* (Telugu), *Chola* (Bengali).

5. **Lentil (*Lens culinaris*):** *Masoor* (Bengali), *Masur dal* (Hindi/Gujarati/ Marathi), *Masur bele* (Kannad), *Masur parippu* (Malayalam), *Musur* (Kashmiri), *Masura* (Oriya), *Mysore paruppu* (Tamil), *Misur pappu* (Telugu).

6. **Field pea (*Pisum sativum* L.):** *Matar* (Bengali/Hindi), *Vatana* (Gujarati/ Marathi, *Batani* (Kannad/Telugu), *Pattani* (Malayalam/Tamil), *Matara* (Oriya), *Kara* (Kashmiri), *Kabli chole* (Punjabi).

7. **Lathyrus/Grass pea (*Lathyrus sativus* L.):** *Khesari* (Hindi/Bengali/Oriya), *Lakh* (Marathi), *Khesari parippu* (Tamil), *Lanka pappu* (Telugu).

8. **Cluster bean (*Cyamopsis tetragonoloba* L. Taub):** *Gaur* (Hindi/Bengali), *Gori Kayi* (Kannad), *Guwar* (Marathi), *Kothavarnkai* (Tamil).

9. **Mothbean/Dew gram (*Vigna aconitifolia* Jacq. Marechal):** *Moth* (Hindi), *Matki* (Marathi), *Kancha muga* (Oriya), *Thanni kaalu* (Kannad), *Bobbarlu* (Telugu), *Thatta payaru* (Tamil).

10. **Cowpea (*Vigna unguiculata* L. Walp.):** *Barbati* (Bengali), *Lobia* (Hindi), *Alasande* (Kannad), *Payteluguar* (Malayalam), *Chavli* (Marathi), *Chani* (Oriya), *Karamani* (Tamil), *Bobbarlu* (Telugu).

11. **Horsegram (*Macrotyloma uniflorum* L. Verdc.):** *Kulthi-kalai* (Bengali), *Kuleeth* (Gujarati/Marathi), *Hurule* (Kannad), *Muthira* (Malayalam), *Kulthi* (Hindi), *Kolatha* (Oriya), *Kollu* (Tamil), *Ulavalu* (Telugu).

12. **Rajmash/Common Bean (*Phaseolus vulgaris* L.):** *Rajma* (Hindi), *Rajmah* (Assamese), *Raajma* (Bengali, Gujarati, Kannad), *Rajama* (Oriya), *Rajma phalli* (Urdu), *Kidney Beans* (Tamil), *Van Payar* (Malayalam).

Contents

Foreward v
Preface vii
Synonyms ix

1. **Enhancing Pulses Production: The Way Forward** **1**
B. Gangwar, Harbir Singh and N. Ravisankar

2. **Andhra Pradesh** **21**
V. Raja, M. Venkaṭa Ramana and S. Sridevi

3. **Assam** **39**
A. Baishya, J.P. Hazarika and M.C. Kalita

4. **Bihar** **55**
R.P. Sharma

5. **Chhattisgarh** **79**
S.K. Sarawgi, Shrikant Chitale, Alok Tiwari, G.K. Shrivastava, H.C. Nanda and Sunil Nair

6. **Gujarat** **123**
B.S. Patel, R.R. Patel, S.M. Patel and I.C. Patel

7. **Haryana** **141**
Pawan Kumar, Manoj Kumar and S. K. Yadav

8. **Himachal Pradesh** **161**
SC Negi, SS Rana and Daisy Basandrai

9. **Islands** **201**
N. Ravisankar, B. Gangwar, T.P. Swarnam, S.K. Zamir Ahmed and N. Bommayasamy

10. **Jammu & Kashmir** **221**
Dileep Kachroo, Jyoti Kachroo, Mudasir Iqbal, N.P. Thakur and Ashok Gupta

11. **Jharkhand** 247

M.K. Singh, R. Thakur, R.R. Upasani and U.N. Verma

12. **Karnataka** 267

A.P. Vishwanath, M. Byre Gowda and T.K. Prabhakara Setty

13. **Kerala** 287

Kuruvilla Varughese, Jacob John, Rani, B. and Thomas Mathew

14. **Madhya Pradesh** 301

V.K. Shukla, S.K.Vishwakarma, Lokesh Dubey and H.S. Yadav

15. **Maharashtra** 325

A.G. Wani, W.N. Narkhede, B.V. Saoji and L.G. Panwar

16. **North-East States** 345

A.S. Panwar, S. Gupta and S.V. Ngachan

17. **Orissa** 365

L. M. Garnayak, B.S. Rath, B.K. Mahapatra, A.K. Mohanty and M. Mohanty,

18. **Punjab** 395

S.S. Walia and Roopeinder Singh

19. **Rajasthan** 429

Surendra Singh, O.P. Gill, G.S. Bhatnagar and Ramniwas

20. **Tamil Nadu** 469

K. Siddeswaran and P.M. Shanmugam

21. **Uttar Pradesh** 489

H.P.Tripathi, Alok Kumar and M.P.Yadav

22. **Uttarakhand** 505

AK Bharadwaj, Purushottam Kumar, Sumit Chaturvedi, DK Singh, A.P. Singh, Mohit Sharma and Vivek Patrak

23. **West Bengal** 527

Manabendra Ray and Kajal Sengupta

1

Enhancing Pulses Production: The Way Forward

B. Gangwar, Harbir Singh and N. Ravisankar
Project Directorate for Farming Systems Research, Modipuram, Meerut -250110

India has witnessed a phenomenal increase in food grains production since Independence, and achieved a record harvest of about 259 million tonnes in 2011-12. Similarly, there was an impressive record production of pulses : 18.1 and 17.2 million tonnes in 2010-11 and 2011-12, respectively. However, to meet the increasing demand of the continuously rising population, about 2 to 3 million tonnes of pulses have to be imported every year costing the country about Rs.3000 crores in foreign exchange. Despite the changes in the dietary habits, pulses in India are still considered an important source of proteins, particularly, for the poor. However, the per capita availability of pulses has been declining over time decreasing from 60.7 gm/day in 1951 to about 32 gm/day in 2010 mainly due to stagnant production (except in recent years), rising demand for consumption and other uses. Further, pulses play an even greater role in maintaining and enhancing soil fertility through nitrogen (N_2) fixation.

In spite of diverse roles of pulse crops in improving soil fertility, nutritional security and environmental sustainability, their production hovered around 13-15 million tonnes during last decade or so until 2010-11. Whenever pulses availability has fallen short of demand, there has been a spurt in prices causing widespread concern among the policy makers. The issue is what can be done to enhance the pulses production, and what type of challenges have to be faced if the country has to become self-reliant in pulses production. For realising the potential of pulses production, the focus of the strategy should be on the specific crops and regions with location-specific technologies to realise the production potential of each pulse crop in specific states/regions. Therefore, state-based issues, technological options and strategies are discussed in detail under the subsequent chapters. However, this chapter attempts to assess the production potential of pulses and suitable strategies to harness that potential at the national level.

Present Scenario

Pulses in India are grown under diverse agro-climate conditions. The long term trend in area, production and yield of pulses shows that though productivity of pulses has been fluctuating around 600 kg/ha, an increasing trend has been visible, and the latest data shows pulses productivity touching a record 697 kg/ha in 2010-11. However, the present productivity growth is not sufficient enough to meet the diversified pulses demand of the rising population. To alleviate protein-energy malnutrition in India a minimum of 50 gm of pulses per capita per day is required in addition to other sources of protein such as cereals, milk, meat, egg, etc. To make up this shortfall about 27 million tonnes of pulses are required by 2015 (TMOP, 2000). Also, an analysis carried out by the Indian Institute of Pulses Research, Kanpur indicated that the demand for pulses in 2030 would be 32 million tonnes (IIPR, 2011). Thus there is need to almost double the pulses production in next one decade.

Major Constraints in Production

A number of biotic and abiotic constraints affect pulses production in the country. Important among these are growing of pulses in marginal input environment, pest and disease incidence, micro-nutrient deficiency, drought and lack of marketing and post-harvest processing technologies. Few states like Andhra Pradesh, West Bengal, Chhattisgarh and Andaman & Nicobar Islands are facing the shorter length of growing period coupled with unseasonal rainfall. Yield gaps between achievable (on-station) and actual in farmers field (on-farm) are very high in pulses in different states owing to various location-specific constraints. The state-wise production constraints along with yield gaps are summarised in table 1. A close perusal of the table shows that there are vast opportunities to realise the yield gaps in different states which range from 100 kg/ha (Odisha) to more than 1000 kg/ha (Madhya Pradesh). The pulses productivity can be enhanced by around 548 kg/ha at national level by addressing the state-specific constraints (Table 1).

Strategies for Enhancing Production & Productivity

A. Area Expansion

The scope for bringing more area under cultivation of pulses is limited. However some areas remain fallow after cultivation of a single crop, and such fallow areas hold the key for increasing pulses production. For example, about

Table 1 : Production related constraints and yield gap in pulses

State	Major pulses	Area* (lakh ha)	Production* (lakh tonnes)	Production related constraints	Yield gap (kg ha^{-1})
A & N Islands	Blackgram, greengram	0.02 (2119)	0.01 (1154)	• Shorter length of growing period coupled with unseasonal rainfall • Lack of marketing and post harvest processing facilities	460
Andhra Pradesh	Chickpea, greengram, blackgram	17.71	14.48	• Pulses grown as intercrops are not fertilized • Excess moisture in the seedling stage and deficient moisture at flowering and seed formation stages • Heliothis and spodoptera problems in chickpea	683
Assam	Lentil, blackgram, greengram	1.18	0.73	• Pest and disease incidence due to humid weather • Improper nutrition of pulses	283
Bihar	Chickpea, lentil, fieldpea, pigeonpea, greengram, blackgram, rajmah, cowpea and lathyrus	6.41	5.15	• Inadequate supply of quality seeds of improved varieties and slow seed replacement rate. • Insufficient and unbalanced use of fertilizers and poor or zero management of salt affected soils. • Disruption of drainage system leading to water logging throughout the state.	800
Chattisgarh	Lathyrus, (grasspea), black gram, pigeonpea, chickpea, horsegram	12.50	7.22	• Continuous heavy rains at flowering (August) • Lack of irrigation at critical stages • Lack of short duration, synchronous, photo-thermo insensitive varieties of summer Greengram and blackgram	707

Contd.

	(kulthi), lentil and field pea				
Gujarat	Greengram, cowpea, blackgram, mothbean, clusterbean, pigeonpea, chickpea	8.0	6.0	• Marginalized cultivation due to intrusion of Bt cotton • Limitation in area expansion	540
Haryana	Chickpea, pigeonpea, greengram, lentil and blackgram	1.72	1.45	• Lack of suitable varieties for late sowing after sugarcane in irrigated areas • Inadequate use of fertilizers and plant protection measures	879
Himachal Pradesh	Pigeonpea, greengram, kulthi and lentil	0.23	0.12	• Generally sown as mixed or intercrop with cereals or side bunds on the rice or maize fields under poor agronomic practices and on marginal lands	879
Jammu & Kashmir	Chickpea, Lentil, fieldpea, blackgram, greengram and Rajmah	0.3	0.15	• Poor utilization of residual soil moisture • Pest and disease of pulses • Limited area	272
Jharkhand	Pigeonpea, blackgram, horsegram, greengram, chickpea, lentil, field pea and lathyrus	3.68	2.94	• Low productivity in *kharif* • Low area and high productivity in *rabi*	513

Contd.

Karnataka	Pigeonpea chickpea, greengram and blackgram	24.34	11.00	• Low varietal turnover with lower yield gains combined with low resilience to moisture stress and pest infestation have contributed to variable production	750
Kerala	Cowpea, greengram, blackgram, pigeonpea	0.04	0.03	• Grown as rice fallow crops with low production inputs	588
Maharashtra	Greengram, blackgram, pigeonpea, chickpea	30.92	16.61	• Lack of adoption of production technology. • Very little fertilizer application to these crops	446
NEH region	Blackgram, Cowpea	1.77	1.17	• Low soil pH • Low adoption of improved package of practices	204
Madhya Pradesh	Greengram, blackgram, pigeonpea, chickpea, lentil, field pea, lathyrus	43.77	32.76	• Lack of adoption of weed management practices and poor care due to lack of irrigation.	1006
Odisha	Greengram, blackgram, pigeonpea, horsegram	20.0	9.94	• Poor *Rhizobium* population in medium and low land paddy fields and non-availability of quality *Rhizobium* cultures. • Heavy infestation of weeds particularly in initial growth stages of 20 to 45 days after sowing. Yield reduction ranged between 25 to 50% in greengram, 30 to 50% in blackgram, 20 to 40% in pigeonpea, 20 to 30% in peas and lentil, 15 to 30% in cowpea and 15 to 25% in chickpea	100

Contd.

Punjab	Summer greengram, blackgram	0.22	0.19	• Irrational use of irrigation water • Micronutrient deficiency	500
Rajasthan	Chickpea, lentil, fieldpea, pigeonpea	36.73	12.85	• Cultivated on marginal & sub-marginal lands, which are generally poor in soil fertility and moisture retention capacity	750
Tamil Nadu	Blackgram, greengram, cowpea,	6.09	1.85	• 78% of pulse area in Tamil Nadu is under rainfed conditions and mostly grown as mixed crop, intercrop, rice fallow, bund crop	462
Uttar Pradesh	Pigeonpea, greengram , blackgram chickpea, lentil and peas	23.71	19.80	• Large area under rainfed conditions (88%) and moisture stress at reproductive phase of the crop and emerging deficiency of secondary and micro nutrients due to use of high analysis fertilizers	625
Uttarakhand	Urdbean, horsegram, rajma, and black soybean, lentil	0.63	0.50	• Grown as alternate crops on marginal lands after taking input intensive crops like paddy and wheat by most of the farmers	250
West Bengal	Chickpea, pigeonpea, greengram, blackgram, lentil, peas, lathyrus	2.00	1.58	• Delayed precipitation during *kharif* season, or excessive rainfall or prevalence of winter rains is another reason for decrease in area, production and yield as a whole	348
	All India	**241.97**	**146.53**	-	**548**

Source: Compiled by authors

12 million hectares of rice-fallows is available in India. At least half of the area can be conveniently brought under cultivation of pulses. Even assuming the present productivity levels, utilization of this area would result in additional production of 4 million tonnes per annum. Similarly, a vast area is acidic in nature and remains fallow during post rainy season. With soil amelioration, this area can be brought under pulse cultivation. Another way of area expansion can be through introduction of pulses in cereal-based cropping system, intensification of pulses and cultivation of pulses as summer/catch crop.

Intensification of pulses: The diversification strategies vary with available resources, local needs, social economic interactions and market. A large number of experiment carried out under All India Coordinated Research Project on Cropping Systems (AICRP_CS) and other programmes in different agro-climatic conditions have indicated the beneficial effect of legumes (grown as intercrops, catch crops, substitute crops, green fodder or green manure) on the production profitability and soil fertility.

Pulses as summer/catch crop: The most feasible way of including legumes in cereal based systems without decreasing land area of the cereal crops is to grow legume as catch crop. In RWCS, medium duration rice varieties followed by wheat varieties suited for normal sowing or late sowing in north-eastern plain did not leave any scope for growing a legume between rice harvest and wheat sowing. However, the period between wheat harvesting and rice planting can be utilized for growing short duration (60-70 days) summer legumes. Experiments at Masodha, Varanasi and Ludhiana indicated that inclusion of mung bean in RWCS enhances the total productivity and net returns as compared to RWCS alone (AICRP_CS report). In other studies, the yield of rice in rice-wheat system generally remained higher when preceded by moong bean (Singh and Sharma, 2002).

Pulses as intercrop: Intercropping is one of the important ways to increase the productivity and provide income stability under limited soil moisture conditions. Extensive studies carried out under AICRP-CS have helped in identification and standardization of several highly profitable inter-cropping systems. Some of the promising inter-cropping systems are maize+black gram at Palampur, Ranchi and Banswara, maize+cowpea at Karjat, sorghum+pigeonpea at Indore, pigeonpea+mungbean at Bichpuri and Hanumangarh, and rice+soybean at Kalyani and Jabalpur. Most of these inter-cropping systems have been evaluated on farmers' fields and found to be highly remunerative (15-20%) over the sole crops.

Introduction in Niche Areas: For identifying the suitable niche areas for enhancing pulses production in major pulse producing states, **relative spread index (RSI)** and **relative yield index (RYI)** were estimated. While RSI shows relative intensification of production in a particular state, RYI indicates the yield level in that state relative to the yield at the national level. While a combination of RSI (>100) and RYI (>125) would indicate most efficient pulses producing state, their lower values (<75) would imply that the area/state does not have much potential for pulse production (Table 2). An analysis of relative spread and yield index of pulses indicates some niche areas where intensification of pulses production holds good potential (Table 2).

Table 2 : Computed values of relative spread index and relative yield index of pulses in major producing states of India (10 years' average)

States	Relative Spread Index (RSI)	Relative Yield Index (RYI)
Andhra Pradesh	127.3	108.7
Bihar	63.5	129.0
Chhattisgarh	129.7	87.0
Gujarat	92.5	110.5
Haryana	22.8	124.3
Jharkhand	113.4	106.5
Karnataka	147.1	67.5
Madhya Pradesh	185.6	118.3
Maharashtra	132.5	97.7
Orissa	72.3	69.0
Rajasthan	141.3	66.9
Tamil Nadu	83.3	62.5
Uttar Pradesh	86.7	132.6
West Bengal	18.6	122.5

Source: Calculated by the authors.

A perusal of Table 2 clearly brings out the relatively higher efficiency of pulses production in Andhra Pradesh, Jharkhand and Madhya Pradesh. These are the areas where production potential can be enhanced further by adopting improved agronomic practices which would help bridge the gap between existing and potential yields. Though the states of Chhattisgarh, Karnataka, Maharashtra and Rajasthan also show higher intensity of area under pulses but these have lower RYI implying that more focus on increasing the productivity and realizing the yield gaps would substantially enhance the pulses production in these states. If additional area can be brought under pulses cultivation in Bihar, Gujarat, Haryana, Uttar Pradesh and West Bengal, it would result in comparatively higher contribution to total production as these states show higher level of relative yield index. These scenarios are summarized in Table 3.

Table 3 : Extent of relative spread index (RSI) and relative yield index (RYI) of pulses in major producing states of India

Relative Spread Index (RSI)	Relative Yield Index (RYI)	
	High	Low
High	Andhra Pradesh (127.3, 108.7), Jharkhand (113.4, 106.5), Madhya Pradesh (185.6, 118.3)	Chhattisgarh (129.7, 87.0) Karnataka (147.1, 67.5) Maharashtra (132.5, 97.7) Rajasthan (141.3, 66.9)
Low	Bihar (63.5, 129.0) Gujarat (92.5, 110.5) Haryana (22.8, 124.3) Uttar Pradesh (86.7, 132.6) West Bengal (18.6, 122.55)	Orissa (72.3, 69.0) Tamil Nadu (83.3, 62.5)

Figures in parentheses indicate relative spread index and relative yield index, respectively.

B. Adoption of Improved Technology

There is lot of scope to enhance pulse production and productivity even with the available technological portfolio. Adoption of improved/ recommended varieties, provision of quality seed, large scale demonstrations of improved agronomic practices particularly emphasising control of insect, pest and diseases and site-specific nutrient management will contribute significantly to realise the yield gaps and improve the pulses production in the country. Table 4 provides an elaborate state-wise strategy which has the potential to enhance pulses production up to 29 million tonnes with the present levels of technology. It implies that it is very much possible even to exceed the target (27 million tonnes) by 2 million tonnes set by the Technology Mission on Oilseeds & Pulses (TMOP) for the year 2015. The domestic demand for pulses is going to be around 32 million tonnes in 2030. Considering the average pulses productivity in the world (836 kg ha^{-1}), USA (1908 kg ha^{-1}), Canada (1804 kg ha^{-1}) and China (1752 kg ha^{-1}), productivity in India is lower (624 kg ha^{-1}). According to the estimates, besides bringing an additional area of 3 m ha and reducing post harvest losses, the productivity needs to be brought to 1361 kg ha^{-1} to meet the projected domestic requirement in 2030. If current technologies are utilized and technological gap between a farmer and on-station experiments are bridged, the All India productivity of pulses can be enhanced to 1172 kg ha^{-1} only. Thus, in addition to realising the yield gaps, major technological breakthroughs in pulses by way of increased resistance to biotic and abiotic constraints needs to be attempted by way of forming a breeding-biotechnology-agronomic-microbiology-plant protection consortia in different regions of the country to develop new strains of pulses, associated microbes and package of practices to achieve higher production potential from the marginal input based production system (rainfed) of pulses. The following section presents a state-wise overview of the action plan for bridging the yield gaps and the extent of additional production (Table 4).

Table 4 : Technological Options for Enhancing Pulses Production in India

State(s) /Crops	Recommended pulses varieties / hybrids	Special package for additional yield	Action plan for bridging the yield gap	Achievable yield(kg/ha)	Achievable production (lakh t)
A & N Islands Blackgram Greengram	Type 9 Pusa-105, Pusa-115 and ML-320	Seed treatment with 5 g each of *Trichoderma viridie* and *Pseudomonas*/kg of seeds and or 3 g of Thiram or Captan /kg of seeds before sowing + 20 kg ha^{-1} seed rate + 30 X10 cm spacing + 10:20:20 kg NPK ha^{-1} + 2 % DAP (20 g/litre of water) at 55 & 65 DAS	Large scale demonstrations of improved package of practices	1000	0.02 (2119 t)
Andhra Pradesh Blackgram Greengram Chickpea	Late sowing conditions: PBG 107 For intercropping: MGG–348(Bhadradri) LBEG7 (Lam sanaga)	Application of N:P:K: S @ 20-50-0-40 for chickpea and 20-50-0-0 for pigeonpea, greengram and blackgram	Region specific interventions on balanced fertilizer application for pulses and promotion of drought and flood tolerant varieties	1500	26.56
Assam Blackgram Greengram Lentil	T 9 T44 B 77	Incentive based promotion of pest and disease resistant varieties +15-35-0 $N-P_2O_5-K_2O$ kg/ha for blackgram and greengram	Large scale demonstration of pest and disease resistant varieties of pulses	900	1.06

Contd.

Bihar Chickpea Lentil Pigeonpea Greengram Balckgram	Rajendragram 1 BR-25 BR-65 Samrat T 9	Application of $ZnSo_4$ @ 25 kg/ha for all pulses at the interval of three years + Compulsory inoculation of rhizobium + draining creation in lowlying areas	Infrasturure development in the form of biodrainage in pulse regions and promotion of use of micronutrient Zn.	1694	10.86
Chattisgarh Pigeonpea Chickpea Lentil Lathyrus	Asha, Rajeev-lochan, Laxmi, Prabhat. UPAS 120JG 11, JG 74, JG 315, Vaibhav, Shweta, BG 256, Annegiri, ICCV 10, JG 16, KAK 2JL 3, K-75, DPL 62, IPL 81, Lens 4076 Ratan, Prateek, Mahateora, Pusa 24	Sowing of pulses in ridges and furrows or FIRB method to avoid waterlogging during flowering + low cost infield water harvesting techniques	Large scale demonstration of improved package of practice through farmers participatory approch	1250	15.62
Gujarat Pigeonpea Chickpea Greengram Blackgram Mothbean	Gujarat tur-100, Gujarat tur-1Dahod yellow, ICCC-4,K-851, GM-3T-9, TPU-4G MO-1	Application of balanced nitrogen and phosphorus with secondary and micro nutrients	Multiple cropping systems with pulses. Higher support prices will ensure competition with Bt cotton	1507	12.10
Haryana Chickpea Field pea	H 208 and C 235, Haryana Masar 1	Application of 15 kg and 40 kg ha^{-1}, of N and P to lentil +	Development of suitable varieties for	1600	2.75

Contd.

Lentil Pigeonpea Greengram	Aparna (HFP 4), Manak (H 77-216) MH 83-20 (Asha)	25 kg zinc sulfate ha^{-1} + 0.5 per cent zinc sulfate and 2.5 per cent urea spray twice at 10 to 14 days interval.	late sowing conditions and supply of balanced fertilizers		
Himachal Pradesh Blackgram Cowpea Rajmah Greengram Pigeonpea	Himachal Mash-1, Kailash (SRC-74) C 475 (Himachal Lobiya), DPM 8909 (Saketi 1) Sarita (ICPL 85010)	20 kg N and 60 kg P_2O_5/ ha along with 10 t of FYM ha^{-1} + weed control Adequate supply of balanced nutrients to pulses grown as mixed and intercropping systems	1699	0.39	
Jammu & Kashmir Chickpea Lentil Field pea Urdbean Mungbean Rajmah	C-235, L-9-12 T-163 Pant U-19 Ps-18, C-235	Sowing in lines of 30 cm apart by '*Kera*' method with 4-6 cm depth.	Promotion of "Kera" method of sowing	840	0.25
Jharkhand Pigeonpea Blackgram Lentil	ICPH2671, Birsa Arhar-1, U 19, Birsa Urd 1 Pant U-31, Shekhar-2 PL 639 L 4076, KLS 218, K75 under late sown conditions	40: 60: 40: 20 Kg NPKS / ha, Half N and full phosphorus, potash and sulphur should be applied as top dressing at 15 days after germination	Increasing the productivity in *kharif through* technology demonstrations	1200	4.42
Karnataka Pigeonpea	TTB-7 (150-200 days),	25-50-50 kg NPK/ha under	Improving the	1200	29.20

Contd.

	HYD-3C (160-210 days), BRG-1 (170-190 days), BRG-2 (150-170 days) and ICP-7035 (160-170 days).	irrigated while 13-25-25 kg NPK /ha in rainfed conditions as basal. At the time of last ploughing or at the time of sowing, 100 kg DAP has to be given as a basal dose. As a source of sulphur, 100 kg gypsum has to be given to supply applied to provide 20 Kg sulphur per hectare	extension activities of pulse programme as the state is having high potential for pulses but their yield levels are low and farmers participatory varietal selection is important		
Chickpea	Annegere-1 (95-100 days), JG-11(95-100 days), KAK-2 (80-85 days) and Vishal (100-110 days)				
Greengram	PS-16 (65-70 days), Pusa Baisaki (65-70 days) and PDO-84-178 (65-70 days)				
Blackgram	Karagav-3 (85-90), T-9 (75-80 days) and				
Horsegram	Rashmi (lbg-625) (70-75 days) KBH-1(BGM-10) (90-100 days) and PHG-9 (90-100 days)				
Kerala					
Cowpea	Bhagyalakshmy, Pusa Barsathi, Pusa Komal. CO-3 and Pournami (in summer rice fallows)	FYM @ 20 t/ha + Lime @250 kg/ha or dolomite 400 kg/ha +N @20 kg/ha + P_2O_5 30 kg/ha +K_2O 10 kg/ha	Development of cost effective storage and production packages are essential.	1110	0.04
Greengram	Philippines, Madiera, Pusa Baisakhi, NP-24				
Blackgram	T-9, Shyama, Sumanjana, CO-2, S-1,				

Contd.

Pigeonpea	TAU-2. TMV-1 and KM-2 SA 1				
Maharashtra Pigeonpea Greengram Blackgram Rajmah Cowpea Chickpea	 BDN-708 (Amol), BSMR-736 Kopergaon, BM-4 BDU-1, TAU-1 Varun Konkan Sadabahar, BDN – 9 -3	The optimum fertilizer dose to chickpea and pigeonpea is 25 kg N and 50 kg P_2O_5/ ha, for greengram and black gram 20 kg N + 40 kg P_2O_5/ ha, for horsegram and mothbean 12.5 kg N + 25 kg P_2O_5/ha + 20 kg/ha elemental S is recommended for all pulses.	Cluster based demonstration of production packages of all the pulses along with improved varieties	930	28.76
NEH region Blackgram Cowpea	 T-9, UG 203, B3-8-8, Pant U 19, UG 301, PDU 2, LU 4, JU 77-41 Amba, V 385, C 152, V 16, V 154	30: 60: 40 kg NPK + 5 tonnes of FYM + lime application	Promotion of pulses in non traditional areas of north eastern hills	883	1.56
Madhya Pradesh Pigeonpea Blackgram Greengram Gram Lentil Peas	 Gwalior-3, Khargone-2, JA- 3 Gwalior-2, Khargone-3 J- 45, Khargone- 1 Gwalior – 2, JG – B2 JL-1, JM-1	Weed control with recommenced herbicides and irrigation at critical stages of crop	Growing of pulses after harvest of rice is more profitable. Farmers can harvest the rain water in community making ponds and irrigate the pulse crops for improving yield.	1580	69.15

Contd.

Odisha Greengram Blackgram Pigeonpea Horsegram	PDM- 54, HUM-1 Sarala (B 12-4, T-9 UPAS-120, Durga (ICPL-84031 Urmi (DS-2-2), Madhu	Recommended seed rate + recommended population + balanced nutrition along with rhizobium inoculation in seed and soil	Pest and disease tolerant varieties coupled with rhizobium inoculation.	600	12.0
Punjab Greengram Blackgram	SML-668, PAU 911 Mash 114	Combine harvesting: When 80 per cent of pods are mature then apply Gramoxone (paraquat) @ 2 L/ha in 375-500 litres of water so that leaves and stem dries up in 5-6 days. Then the moongbean can be harvested with combine harvester to save the crop from untimely rains and save labour	Replacement of 5 per cent rice-wheat area (2.6 million ha) by rice-greengram or rice-wheat-green gram system	1320	0.29
Rajasthan Chickpea Lentil Fieldpea Pigeonpea	RSG-44 LH-84-8 (Sapana) Type-163 T-21	In rain-fed conditions, 15-20 kg N and 30-40 kg P_2O_5 per hectare, in irrigated condition with normal sowing 20 kg N, 40 kg P_2O_5 and 20 kg K_2O, and in irrigated late sowing conditions 30 kg N, 40 kg P_2O_5 and 20 kg K_2O as a basal dose lead to better production. Basal application of 25 kg zinc	Large scale promotion of drought resistant varieties having higher production	1083	39.78

Contd.

		sulphate is quite effective for good production. Foliar spray of $ZnSO_4$ @ 0.5% or seed coating with zinc is also very much effective in correcting zinc deficiency			
Tamil Nadu Blackgram Greengram Cowpea	 Co 5, ADT-3, ADT-5, Vamban-1 KM-2, Co 4, Co-6 Co2, Co-6	20 kg sulphur /hashould be applied in the form of gypsum. Soil application of 25 kg $ZnSO_4$/ha under irrigated conditions.	Creation of community based irrigation facilities and sufficient nutrient application in intercropping systems	765	4.66
Uttar Pradesh Gram Pea Lentil Rajmah Pigeonpea Greengram	 DCP-92-3, KGD-1168, Awrodhi, Pusa-256, KWR-108, Rachna, Malviya pea-15, Sapana (KPMR-144-1), Narendra Masoor-1, Pant masoor-5, DPL-62, L-4076, K-75, HUL-57, Shekhar-2, Shekhar-3, Arun Bahar, Amar, Narendra Arhar-1, Type 44, Pant moong-3, Narendra moong-1,	Specific rhyzobium culture 1 packet (200 g) /10 kg seed and soil inoculation of PSB + 20 kg N + 60 kg P_2O_5 + 20 kg K_2O + 20 kg S/ha	The raised bed planting in the areas prone to water logging or in poorly drained soils. Integrated nutrient management including secondary and micronutrients.	1250	29.64

Contd.

Blackgram	Pant moong-4 Narendra Urd-1, Pant U-30				
Uttarakhand Blackgram Lentil Pea	Pant urd 31 Pant lentil 4, Pant pea 13,	Sulphur (S) deficiency is widespread in areas under pulse crops. Among these, chickpea, fieldpea and pigeonpea require 40 kg $Sha^{-1,}$ while lentil, greengram and black gram need 20 kg S/ ha. Application of S in pulse crops increased yield in the range of 176-592 kg/ ha	Promotion of use of secondary and micro nutrients	1043	0.66
West Bengal Chickpea Pigeonpea Greengram Blackgram	Mahamaya-1, Mahamaya-2, Anuradha, Bahar, NDA-1, Pusa-9, Azad, Sonali (B-1), Panna (B-105), PDM-54, Uttara, Kalindi (B-76), T-9	Chickpea requires N: P_2O_5: K_2O @ 20: 40: 40 kg ha^{-1} along with 8-10 tonnes of FYM ha^{-1}. Under deficient conditions, Borax spray @ 0.2% at 30 and 50 DAS may increase the yield. If there is any deficiency of molybdenum, spraying of ammonium molibdate @ 0.05% may also increase the seed yield of chickpea	Varietal development for late sowing and contingency measures for various moisture conditions. Promotion of secondary and micro nutrients	850	1.70
All India	-	- -	**1172 (mean)**	**291.47**	

Source: Compiled by authors

Policy Perspectives

There are a number of key economic, political and technological factors that are reshaping pulse production scenario in India. Keeping in view the stagnant production of pulses, the government has adopted two-pronged strategy for enhancing pulses production and meeting domestic requirements. While incentives are given in the form of assured minimum support prices (MSP) for increasing production, the import of pulses are freely allowed for meeting overall domestic requirements. However, the mechanism of price support has not been very effective in the past as the prices declared by the Government always remained lower than the ruling prices. A quick look on the MSP for pulses during the period 2000-01 to 2007-08, shows that the MSP were raised only marginally (maximum 11.8%) and in some years for few crops it remained even constant. Therefore, it is not surprising that pulses production also stagnated during this period and it hovered around 14 million tonnes. For the first time in 2008-09, MSP for greengram, black gram and pigeonpea were enhanced substantially (29-48%), and that played major role in achieving a record pulse production (17-18 million tonne) in recent years.

Recognizing the importance of pulses in household livelihoods and nutritional security, the Government has set up a National Mission on Pulses to further bolster domestic production in the country. For technological back up, many ICAR research institutes as well as various State Agricultural Universities (SAUs) are engaged on R&D in pulses in the country. Particularly, the Indian Institute of Pulses Research (IIPR), Kanpur with a major focus on basic and applied research in pulses such as chickpea, pigeonpea, *moong* bean, *urad* bean, lentil, etc. has developed a host of high yielding varieties in respect of these crops. The R&D efforts of the ICAR and SAUs have paved the way for Green Revolution in pulses as a group of thirty-one Indian scientists have recently decoded the genome of pigeonpea (*Arhar*), the second most important pulse crop of India. In addition, the Government of India has launched several programmes including National Food Security Mission (NFSM), Agricultural Technology Management Agency (ATMA), Rastriya Krishi Vikas Yojana (RKVY) and intensification of pulses and oilseeds in 60,000 villages for enhancing their production in the country. It was because of proper policy focus and emphasis on development activities that NFSM target of increasing pulses production by 2 million tonnes was achieved even before the planned period. However, many more concerted efforts and effective policy prescriptions would be required to enhance the pulses production to meet future demand. Specifically, the following issues need attention:

- Price incentives in the form of MSP should take into account the market signals. In other words, the level of price support should not be less than the prevailing market prices.

- Local level seed production must be encouraged to ensure quality seed at the right time.
- Providing support for low cost storage structures and developing market linkages is one of the most effective ways to attract farmers toward growing more pulses.
- Special focus should be given to the most efficient zones of pulses as identified on the basis of relative spread and yield indices (RSI & RYI). For example, efforts for increasing area in the states with higher RYI but low spread will result in higher production with minimum efforts.
- Higher R&D investment should be made to address biotic as well abiotic stress in pulses production, because pulses are grown in marginal environments and are susceptible to pest and disease attacks.

Conclusion

In spite of diverse roles of pulses in improving soil fertility, nutritional security and environmental sustainability, their production remained low during last two decades or so. The major issue is what can be done to enhance the pulses production in the country to a level where we could become self-sufficient for meeting the demand for pulses for which there is need to almost double the pulses production in the next decade. A number of biotic and abiotic constraints affect the doubling of the pulses production in the country. For realising the potential, the focus of the strategy should be on the specific crops and regions with location-specific technologies and modern agronomic practices for given pulse crop in the specific states/regions. By focusing on an elaborate state-wise strategy, it is possible to enhance pulses production upto 29 million tonnes even with the present levels of technology. Further, bringing rice fallows under pulses cultivation would result in an additional 4 million tonnes of production, thereby enhancing the total production up to 33 million tonnes with the help of harnessing the full production potential of the given technology portfolio. All these efforts would be successful only if there are proper policy incentives backed by adequate marketing and post-harvest facilities. Thus, there is no doubt that a proper mix of policy initiatives, efforts for bridging the technology gap through rapid dissemination, path breaking technologies with capability to tolerate harsh environment, price incentives, infrastructure and institutional support will result in enhancing the pulses production to the level which is more than sufficient to meet our domestic demand and to avert any supply crisis in the near future.

References

Annual Reports (different years), All India Coordinated Research Project (AICRP) on Cropping System (CS), PDFSR, Modipuram.

IIPR (Indian Institute of Pulses Research), (2011). Vision 2030, IIPR, Kanpur.

Singh VK and Sharma BB (2002). Economic evaluation of rice based cropping systems in foothills of Himalayas. Indian Journal of Agronomy, 47(1), 12-19.

TMOP (Technology Mission on Oilseeds and Pulses), (2000). Expert Committee Report on Pulses, DARE, New Delhi.

❑❑❑

2

Andhra Pradesh

V. Raja, M. Venkata Ramana and S. Sridevi
AICRP on Integrated Farming Systems, Diamond Jubilee Block, College of Agriculture, ANGRAU, Rajendranagar, Hyderabad-500030

Pulses, grown on marginal and poor soils under low input management with very low yields, are the integral part of dryland farming systems and cover 12.8% of the total cropped area in Andhra Pradesh. There are no pulse based cropping systems in the state and they are only component crops of cereals or commercial based cropping systems as intercrops or catch crops in between two main crops or as sequence crops after exhaustive cereal or commercial crops raised on residual fertility and receding soil moisture. The important pulse crops grown are Chickpea, Pigeonpea (Redgram), Mung bean (Greengram) and Urd bean (Blackgram) which contribute to 97.7% of total pulse production in the state. Pigeonpea is mostly grown as intercrop under rainfed conditions with crops like sorghum, maize, castor, cotton etc. It is also grown as sole crop and at times, pigeonpea is intercropped with greengram/ blackgram. Greengram is mainly grown during *kharif*, as a preceding crop to rice while blackgram is raised in *rabi*/summer, in rice fallows. In the recent years, because of remunerative market price, availability of improved cultivars and crop production technologies, chickpea area has gone up in the state in black soils replacing sorghum and commercial crops like tobacco, cotton etc. Generally, it is grown as sole crop on *Vertisols* on residual soil moisture during *rabi* season and in medium black soils with one or two irrigations.

Trends of Pulses Production

Total pulses area in Andhra Pradesh has marginally increased from 16.32 lakh ha in 1990-91 to 17.71 lakh ha in 2008-09 with a corresponding production figures of 6.95 lakh tonnes and 14.48 lakh tonnes, indicating that productivity of pulse crops in the state has almost doubled. During 2007-08, the pulse area in the state was 21.13 lakh ha with a production of 16.97 lakh tonnes (Table 1).

Table 1 : Trends in Area, Production and Productivity of total pulses in Andhra Pradesh

Year	Area (lakh ha)	Production (lakh tonnes)	Productivity (kg / ha)	National Productivity (kg/ha)
1990-1991	16.32	6.95	425	578
1991-1992	16.44	7.92	480	533
1992-1993	15.89	7.39	465	573
1993-1994	15.64	6.77	432	598
1994-1995	16.02	6.73	420	610
1995-1996	16.13	7.71	477	552
1996-1997	16.39	8.47	516	635
1997-1998	15.65	5.16	330	572
1998-1999	15.87	8.27	521	634
1999-2000	16.47	8.01	486	635
2000-2001	19.02	10.53	554	544
2001-2002	19.19	11.38	593	607
2002-2003	21.00	10.63	507	543
2003-2004	21.85	12.39	567	635
2004-2005	18.03	10.17	564	577
2005-2006	17.82	13.75	772	598
2006-2007	19.84	13.47	680	680
2007-2008	21.13	16.97	803	792
2008-2009	17.71	14.48	818	–
2009-2010	19.32	14.29	740	978
2010-2011	20.64	14.03	679	1053

A distinct change in pulses area and production was seen from the dawn of twenty first century. The area has gone up, primarily due to the adoption of large scale cultivation of chickpea. The area under chickpea was 0.89 lakh ha during 1990-91 and it reached to 5.84 lakh ha by 2010-11 (Table 2). Many factors favoured the cultivation of chickpea. Most of the newly developed cultivars are faring well even in areas with mild winter. The black soils that are left fallow during *kharif*, are effectively utilized for chickpea cultivation. The other factors that favoured the expansion in area of chickpea include, mechanization, higher and stable yields with low input levels and remunerative market prices. Presently, the mean yield of chickpea is more than 1.0 t/ha and very high yields (1.6 t/ha) were realized during 2005-06. These values are much above the national average (0.85 t/ha).

Table 2 : Trends in Area, Production and Productivity of Chickpea in Andhra Pradesh

Year	Area (lakh ha)	Production (lakh tonnes)	Productivity (kg/ha)
1990-1991	0.89	0.57	643
1991-1992	0.64	0.47	733
1992-1993	0.60	0.28	469
1993-1994	0.89	0.65	730
1994-1995	1.68	1.36	808
1995-1996	1.13	0.71	630
1996-1997	1.06	0.90	850
1997-1998	1.47	0.59	398
1998-1999	1.46	1.30	891
1999-2000	1.63	0.95	583
2000-2001	2.01	2.29	1138
2001-2002	2.85	3.63	1274
2002-2003	3.90	3.82	979
2003-2004	4.22	4.57	1081
2004-2005	3.41	3.45	1012
2005-2006	3.94	6.27	1590
2006-2007	6.02	6.53	1087
2007-2008	6.30	9.12	1448
2008-2009	6.07	8.57	1413
2009-2010	7.30	9.38	1285
2010-2011	5.84	7.19	1233

Pigeonpea, part of the staple food and an important *kharif* crop, has witnessed wide fluctuations in area and production (Table 3). The area in the last two decades fluctuated between 2.99 lakh ha (1994-95) and 6.38 lakh ha (2010-11). The productivity also varied very widely, ranging between 177 kg/ha (1997-98) and 651 kg/ha (2007-08). The performance of pigeonpea is highly dependent on rainfall, its distribution and moisture storage capacity of the soils. Apart from the soil and climatic factors, the biotic factors that have a greater influence on productivity, are the occurrence of pod borer (Heliothis) and wilt diseases (Fusarium and Macrofomina wilts).

Other important *kharif* pulse crop is greengram. It is also subjected to wide fluctuations and a negative trend is noticed in its cultivation in the recent past. Greengram area reached a peak of 6.81 lakh ha during 2003-04 and then onwards, it declined reaching a low of 3.02 lakh ha by 2009-10. For most of the years, the yields ranged between 350 and 450 kg/ha and the yield barriers in this crop are not broken (Table 4). Yellow Mosaic Virus (YMV) is a serious threat to the cultivation of greengram.

Table 3 : Trends in Area, Production and Productivity of Pigeonpea in Andhra Pradesh

Year	Area (lakh ha)	Production (lakh tonnes)	Productivity (kg / ha)
1990-1991	3.46	0.74	214
1991-1992	3.39	1.19	352
1992-1993	3.13	0.75	238
1993-1994	3.07	1.04	340
1994-1995	2.99	0.99	330
1995-1996	3.28	1.02	310
1996-1997	3.61	1.38	382
1997-1998	3.19	0.57	177
1998-1999	3.94	1.70	433
1999-2000	4.32	1.55	358
2000-2001	5.13	2.18	427
2001-2002	4.20	1.88	448
2002-2003	4.32	1.50	348
2003-2004	5.21	2.18	418
2004-2005	4.80	2.19	456
2005-2006	4.94	3.01	608
2006-2007	4.01	1.61	401
2007-2008	4.63	3.02	651
2008-2009	4.43	2.02	455
2009-2010	5.42	2.79	515
2010-2011	6.38	2.65	416

Table 4 : Trends in Area, Production and Productivity of Greengram in Andhra Pradesh

Year	Area (lakh ha)	Production (lakh tonnes)	Productivity (kg / ha)
1990-1991	4.93	1.36	277
1991-1992	5.15	2.19	426
1992-1993	5.19	2.26	436
1993-1994	5.17	2.18	423
1994-1995	5.01	1.65	330
1995-1996	4.64	1.91	411
1996-1997	4.91	2.30	469

Contd.

1997-1998	4.55	1.34	295
1998-1999	4.70	2.17	462
1999-2000	4.59	2.11	460
2000-2001	5.20	1.84	354
2001-2002	4.76	1.55	326
2002-2003	4.73	1.18	250
2003-2004	6.81	3.12	459
2004-2005	4.54	1.60	351
2005-2006	3.85	1.65	429
2006-2007	3.94	1.44	364
2007-2008	4.38	1.92	439
2008-2009	3.20	1.36	426
2009-2010	3.02	0.52	174
2010-2011	3.78	1.66	439

Blackgram, an important ingredient of breakfast foods, is primarily cultivated in the rice fallows in the coastal districts of A.P. and during *kharif* and summer seasons in other parts of the state. The mean yields of blackgram are better than those of greengram, but fluctuations in area are witnessed. The rice-fallow blackgram is haunted by YMV and the parasitic weed, Cuscuta. During the last 2-3 years period, most of these coastal district farmers switched over to zero-till maize in the rice-fallows, causing drastic reduction in the area of blackgram (Table 5). Other pulses like cowpea, lab-lab, rajmash, horsegram are cultivated to a smaller extent in certain localized pockets.

Table 5 : Trends in Area, Production and Productivity of Blackgram in Andhra Pradesh

Year	Area (lakh ha)	Production (lakh tonnes)	Productivity (kg / ha)
1990-1991	5.39	3.66	681
1991-1992	5.63	3.41	606
1992-1993	5.31	3.48	656
1993-1994	4.78	2.41	504
1994-1995	4.65	2.19	472
1995-1996	5.45	3.52	647
1996-1997	5.24	3.27	625
1997-1998	5.02	2.16	429
1998-1999	4.30	2.62	608
1999-2000	4.61	2.95	640
2000-2001	5.55	3.90	704

Contd.

2001-2002	6.27	3.83	611
2002-2003	6.95	3.76	541
2003-2004	4.38	2.00	456
2004-2005	4.27	2.59	605
2005-2006	4.21	2.50	593
2006-2007	4.98	3.50	703
2007-2008	5.03	2.50	496
2008-2009	3.38	2.20	656
2009-2010	4.39	1.51	344
2010-2011	4.64	2.53	546

Cropping Systems

Most of the *kharif* pulses are grown either as intercrops or as sole crops. In sequence to cereals or commercials crops. These systems differ from region to region based on the main crop grown and the agro-climatic conditions.

Redgram

Redgram is grown as sole crop in certain districts of Telangana region (mainly Ranga Reddy district) and also in some parts of north coastal districts. It is cultivated in all other districts (its cultivation is spread over the entire state) mostly as intercrop. In northern Telangana region, it is intercropped with cotton, soybean, sorghum, maize and greengram. In the scarce rainfall region, it is intercropped with groundnut. In hilly areas and high altitude zone, it is grown as intercrop with upland rice, sesamum, niger and greengram. In some of the Rayalaseema districts like Kurnool, it is intercropped with sorghum and sunflower. Because of the increasing demand for redgram dal and fluctuating productivity, the state government has initiated a special drive to encourage the cultivation of redgram during the *rabi* season. Sufficient technology for *rabi* redgram has been generated by the State Agricultural University and this has helped to extend the area in all the possible agro-climatic situations. During 2009-10, *rabi* redgram fared well in many parts of the state. In areas with very fair distribution of rainfall, a sequence system of greengram-redgram is encouraged. A unique intercropping system with redgram is existing in parts of adjoining district Maharashtra. Redgram is sown in wide rows (2.1 m) and intercropped with soybean during *kharif*. After the harvest of soybean, the inter spaces are tilled and any of the different *rabi* crops (bengalgram, coriander, sorghum, safflower, blackgram) are sown in the standing crop of redgram. Sprinkler irrigation is adapted for the *rabi* crops and redgram.

Greengram

Greengram during *kharif* is grown either as sole crop or intercropped with redgram. In certain areas, greengram is grown as intercrop in cotton. In certain parts of Telangana and the adjoining coastal districts, this crop is sown with the onset of monsoon and is followed by *maghi* jowar (special season sorghum). Such practice of sequential cropping with early season greengram is seen with crops like tobacco and maize, where the second crop is planted or sown late in the season after the greengram. In the rice growing areas, where transplanting of rice is delayed due to late release of water in the canal commands, it is advocated to grow greengram early in the season, and depending on the availability of water for puddling, the greengram pods are picked and haulms are ploughed into the soil, or, the entire crop is ploughed in as green manure crop. Greengram is also grown as a summer pulse after the *kharif* crops like cotton and maize. A three crop sequence of rice-zero-till maize/sunflower-greengram system is now being encouraged in the rice tracts.

Blackgram

Blackgram or urdbean, is grown to a limited extent during *kharif*, that too as intercrop either with cotton or redgram. It is predominantly grown as relay crop with rice. Prior to the harvest of rice, blackgram seed is broadcasted in the field. After the harvest of rice, it is raised with residual soil moisture. All the practices for its establishment, growth and development are well established. However, its area is dwindling due to YMV and Cuscuta problems and blackgram is being fast replaced with zero-till maize. Now it has become an important summer pulse in the irrigated areas after crops like maize, sunflower and cotton. It is an excellent intercrop in sugarcane planted during February.

Chickpea

Chickpea is the most potential pulse crop in A.P. In districts like Prakasam (rain shadow areas) land is kept fallow during *kharif* and is cropped with chickpea during September. Chickpea is now cultivated to a large extent in the tank beds after the cessation of rains. In Rayalaseema and parts of Telangana, chickpea is grown in sequence to many *kharif* crops-soybean, maize, sorghum and sunflower. Intercropping of chickpea and coriander (6:1) is seen in some areas.

The various cropping systems involving pulses being followed in Andhra Pradesh are as follows :

- **Intercropping systems under rain fed conditions**

  Cotton + greengram (1:3/1:7), Cotton + blackgram (1:3), Cotton +

Redgram (1:3/ 1:5), Redgram + blackgram (1:7), Redgram + Soybean (1:7), Redgram + Sesamum (1:7), Redgram + Dry paddy (1:7), Redgram + Castor (5:1 / 6:1 / 1:1), Redgram + Maize (1:3), Redgram + greengram (1:10), Redgram : Sorghum (1:2, 1:5, 1:8, 1:10, 1:12), Redgram : Groundnut (1:7), Greengram : Redgram : Groundnut (1:3:5)

- **Double / Triple cropping systems under irrigated conditions**

  Rice-Blackgram, Sesamum- Blackgram, Greengram-Redgram, Blackgram-Redgram, Sesamum-Redgram, Greengram-Tobacco, Greengram-Rice, Greengram-Maize, Greengram-Vegetables, Greengram-Maize-Vegetables, Maize-Groundnut-Greengram/ Sesamum, Greengram-Maize-Greengram, Greengram-Groundnut-Sesamum, Greengram-Blackgram-Soybean, Redgram-Sorghum-Dry Rice, Blackgram/ Turmeric / Sorghum / Maize-Chickpea, Redgram / Fallow-Jowar / Safflower / Chickpea,

Yield Gaps

The estimated pulse requirement in Andhra Prdesh is around 30 lakh tonnes as against the present production of 14.03 lakh tones. Further, yield levels of pulses realized in the state are lower than their potential yields in experimental plots. This shows that there is a scope to narrow down these gaps substantially through better management approaches. There is a potential to harvest around 1.0 t/ha of pigeonpea under intercropped situation and 1.5-2.0 t/ha when grown as a sole crop as against the state average yield of 0.45 t/ha. Similarly, there exists a wide gap between the potential yield of greengram and blackgram over their average figures in the state and there is a good scope to double the yields of these crops through better management practices. Despite higher yield levels realized with chickpea in the state (1.4-1.5 t/ha), there is a gap of around 0.5 t/ha when compared to its potential yield. This gap can easily be bridged with proper management of this crop especially through IPM and other best agronomic management practices.

Major Constraints and Technologies

The crop wise constraints and technologies for different zones of Andhra Pradesh are given in table - 6.

Table 6 : Cropwise constraints and technologies for Andhra Pradesh.

Zone	Base crops	Constraints	Technologies
Red gram			
Northern Telangana Zone (NTZ)	Cotton, soybean, Sorghum, maize and greengram	1. Base crop Fertilized, intercrop of Redgram not fertilized	1. Proportionate dose of RDF to be applied to red gram inter crop
Scare rainfall zone Hilly and high attitude zone	Ground nut Upland Rice sesamum, Niger and greengram	2. After base crop harvest, inter crop of red gram not taken care	2. Care of red gram crop specially during flowering and pod formation stage needs to be taken for pests and diseases
Kurnool	Sorghum and sunflower	3. Stubbles left over after base crop harvest rejuvenate and exhausts available soil moisture and nutrients without productive use, which could have been utilized by redgram crop.	3. Base crop should be harvested at physiological maturity and left over stubbles should be ploughed as there will be sufficient moisture in the soil which prevents unproductive use of residual moisture and could be beneficially used by red gram The soil mulch created helps in better insitu water soil in filtration in case of un seasonal rain and moisture conservation by preventing evaporation during rest of the periods
Adilabad	*Kharif* - Soybean Chickpea *Rabi* – coriander, sorghum, sunflower black gram	4. The crop in the vegetative stage suffers from excess water due to heavy rains while at flowering and pod formation suffers from moisture stress due to withdrawal of monsoon	4. Raising crop on ridge and furrow protects the crop from excess water in the vegetative stage. Supplemental irrigations at flowering and pod formation through drip or mobile sprinklers from rain harvested water protects the crop from moisture stress at critical stages
Greengram			
Central Telangana Zone (CTZ)	With on set of monsoon before *Maghi* jowar	1. Good seed for the respective zone not available and crop not fertilized	1. Suitable seed multiplied and made available. Seed treatment with rhizobium and INM practices to be followed
Krishna Zone NTZ and CTZ	Tobacco Maize	2. Plant protection not taken care	2. Seed treatment with insecticides, fungicides and IPM to be followed
All nine zones	Delay in planting *Kharif* rice due to late release of water	3. Subjected to intermittent long dryspells	3. Irrigation through mobile sprinklers to be provided whenever there

Contd.

	in from canals, tanks etc.		are dry spells of more than 15 days
		4. Crop hurriedly harvested to grow the sequence main crop	4. Crop should be harvested at physiological maturity or at green pod stage
NTZ , CTZ and Krishna	*Rabi*/ summer greengram sown after kharif cotton & maize	1. Crop not fertilized after exhaustive crops	1. RDF must be applied as per soil test
Scarce rain fall zone	As inter crop in cotton	2. Irrigation scheduled as per water availability at less frequent intervals in early stages and at long intervals during terminal stages	2. Mobile sprinklers should be used as per moisture need of the crop and not on availability of water.
Godavari	As pre-*kharif* crop to rice or in rice fallows in *rabi*	3. Yellow vein mosaic and cuscuta problem	3. New varieties to be identified and crop rotation to be followed.
		4. No Weed control	4. Pre-emergence application of pendimethaline and one hand weeding
Black gram			
NTZ	As inter crop in cotton	1. Intercrop of black gram not fertilized	1. Proportionate RDF to be applied through INM mode to blackgram
CTZ	As inter crop in cotton	2. Plant protection not taken care	2. Location and pest specific IPM to be followed
Krishna zone	As inter crop in cotton	3. Crop subjected to long dry spells which cotton withstands but not pulse	3. When dry spells lasts more than 15 days, irrigation should be given through mobile sprinklers
Scarce rainfall zone	As inter crop in cotton		
Southern Telangana Zone (STZ)	As inter crop in Red gram		
Godavari zone	As relay crop after *kharif* Rice	1. Yellow vein mosaic	1 Resistant varieties to be identified
		2. Cuscuta problem	2. Crop rotation
Krishna	As relay crop after *kharif* Rice	3. Weed problem	3. Premergence application of pendimethaline and hand weeding
NTZ CTZ	Cotton and maize After *Kharif* Cotton and maize	4. Excess moisture in the seedling stage and deficient moisture at flowering and seed seed formation stages.	4. Laying drainage channels at 10m intervals and irrigation through mobile sprinklers at critical stages
Scarce rainfall	Cotton and sunflower		
zone	inter crop in planted sugarcane		
North coastal Zone	As inter crop in sugar cane	Fertilizer scheduled to sugarcane and not black gram	Proportionate RDF should be applied to black gram.

Contd.

Godavari and Krishna zone	As inter crop in sugar cane	Irrigation scheduling based on sugarcane crop condition	Irrigation scheduling should also be based on black gram crop condition
Southern zone and NTZ	As inter crop in sugar cane	No plant protection measure	Pest specific and prophylactic plant protection measures should be undertaken
Chickpea			
NTZ	*Kharif* maize and soybean	1. Delay in harvest of kharif preceding crops leading to delay in sowing of chickpea	1. Summer land preparation, post harvest poloughing and harvesting kharif crop at physiological muturity.
CTZ	*Kharif* maize	2. Long dry spells during flowering subsequently prolong duration of soybean leading to late sowing of chickpea	2. Early duration varieties of soybean to be developed and hormones to be sprayed to hasten flowering in soybean
Scarce rainfall zone	*Kharif* Sorghum and sunflower	3. Dry root rot and wilt	3. Seed and soil treatment and crop and rotation
Krishna zone	Fallow	4. Heliothis and spodoptera problems	4. Pest specific IPM measures
		5. Terminal stress	5. Scheduling irrigation at flower initiation and pod formation through mobile sprinkler system

The improved varieties of pulses recommended for Andhra Pradesh given in table - 7.

Table 7 : The recommended improved pulse crop varieties for Andhra Pradesh.

S. No.	Variety	Seasons	Duration (days)	Yield (kg/ha)	Specific features
	Blackgram				
1	LBG 17 (Krishnaiah)	*Rabi* & Rice fallows	80-85	1800-2000	Resistant to powdery mildew and bold shining seed
2	LBG 402 (Prabhava)	*Rabi* & Rice fallows	90-95	2000-2200	Resistant to wilt, bold seed, tall growth habit, smothers the weeds in rice-fallows
3	LBG 20 (Teja)	*Kharif, Rabi* & Summer	70-75	1400-1600	Resistant to YMV
4	LBG 611 (Kotha Buttaminumu)	*Rabi* & Rice fallows	90-95	2000-2300	Resistant to wilt
5	LBG 22 (Teja)	*Rabi* & Rice fallows	70-75	2000-2300	Resistant to wilt
6	LBG 648	*Rabi* & Rice fallows	90-95	2000-2300	Multiple disease resistant
7	LBG 623	*Kharif, Rabi* & Summer	75-80	1500-1700	High yield, photo insensitive, bold shining seed

Contd.

8	LBG 685	*Rabi* & Rice fallows	85-90	2000-2200	High yield, resistant to wilt, main stem bearing
9	LBG 645	*Rabi* & Rice fallows	85-90	2000-2400	Bold seed, shining, resistant to wilt
10	LBG 709	*Rabi* & Rice fallows	80-85	1500-1700	Photo-sensitive, resistant to wilt and tolerant to YMV, Shining seed, hairy pods
11	LBG 752	*Kharif, Rabi* & Summer	70-75	1700-2000	Photo-insensitive, resistant to YMV and wilt, shining seed
12	WBG-26	*Kharif*	70-75	1000-1200	YMV resistant
13	PBG-107	*Rabi* & Rice fallows	80-85	1700-2000	Suitable for late sown conditions
14	PBG - 1	*Kharif*	70-75	1000-1200	Non shiny with hairy pods
15	MBG 207	*Kharif, Rabi* & Summer	75-80	1300-1400	Photo-insensitive variety, tolerant to stress and YMV
	Greengram				
1	LGG 127 (Kondaveedu)	*Kharif, Rabi* & Summer	60-65	1000-1200	Early maturing dull seeded variety
2	LGG 407 (Lam 407)	*Kharif, Rabi* & Summer	65-70	1300-1500	Resistant to YMV, ABLS
3	LGG 450 (Pushkara)	*Kharif, Rabi* & Summer	65-70	1300-1500	Resistant to pre-harvest sprouting
4	LGG 410 (Lam 410)	*Kharif, Rabi* & Summer	65-70	1300-1500	Resistant to leaf spot diseases and suitable for rice fallows
5	LGG 460 (Lam 460)	*Kharif, Rabi* & Summer	60-65	1300-1500	Resistant to YMV, top bearing with more pods per cluster
6	TM 96-2	*Kharif, Rabi* & Summer	60-65	1300-1500	Bold seed, resistant to powdery mildew and suitable for rice fallows.
7	MGG 347	*Kharif, Rabi* & Summer	60-65	1000-1300	Synchronous maturity and suitable for machine harvesting
8	WGG - 2	*Kharif, Rabi* & Summer	65-70	1300-1500	Resistant to ABLS
9	WGG - 37	*Kharif, Rabi* & Summer	60-65	1800-2000	YMV resistant
10	MGG - 348 (Bhadradri)	*Kharif* & *Rabi*	65-70	1200-1500	Suitable for intercropping
11	MGG 295	*Kharif, Rabi* & Summer	60-65	1000-1200	Resistant to ABLS, top bearing with more pods per cluster
	Bengalgram				
1	Jyothi	*Rabi*	100-110	1500-1800	Profuse branching from the base, desi type

Contd.

2	ICCV 2 (Swetha)	*Rabi*	80-85	1500-1800	Resistant to wilt, early maturing, kabuli type
3	ICCV 37 (Kranti)	*Rabi*	100-105	2000-2200	Resistant to wilt, desi type
4	LBEG 7 (Lam sanaga)	*Rabi*	90-95	2000-2200	Resistant to wilt, bold seeded, high yielding, Kabuli type
	Redgram				
1	LRG 30 (Palnadu)	*Kharif, Rabi*	170-180 (K) 120-130(R)	2000-2200	Regeneration capacity and high yield
2	ICPL 332 (Abhaya)	*Kharif*	160-165	1800-2000	Moderately resistant to pod borer
3	ICPL 85063 (Lakshmi)	*Kharif, Rabi*	160-170	1800-2000	Suitable for *rabi* and bold seed
4	LRG 38 (Ranga bold)	*Kharif, Rabi*	170 (K) 120-130(R)	2000-2200	High yielding, less susceptible to pod borer
5	LRG 41 (Lam Kandi)	*Kharif, Rabi*	180 (K) 120-130(R)	2000-2200	Tolerant to Helicoverpa pod borer
6	WRG – 27	*Kharif, Rabi*	180 (K) 120-130(R)	1800-2000	Resistant to pod borer
7	ICPL 84031 (Durga)	*Kharif*	115-125	1000-1200	High yielding, short duration
8	PRG – 100	*Kharif, Rabi*	140-150 (K) 120-130 (R)	1500-1800	Medium duration, tolerant to wilt
9	PRG – 158	*Kharif, Rabi*	150-155 120-130 (R)	1500-1800	Medium duration, resistant to Fusarium wilt
10	MRG 66	*Kharif*	190-200	1800-2000	Resistant to Macrophomina wilt
11	MRG 1004	*Kharif, Rabi*	180-190 (K) 120-130 (R)	2000-2200	Resistant to Macrophomina wilt, improved over MRG 66

National Varieties

Greengram	–	ML 267, Pusa 105, PDM 54, K 851, PS 16, TM 96-2
Blackgram	–	T 9, PDU 3, Pant U-30, PU 31, TAU-1, Co5
Redgram	–	C 11, ICP 8863, ICPL 87119, BDN 2, BSMR 736
Chickpea	–	BDN 9-3, ICCV 10 (Bharathi), Phule G 95311 (Vihar), KAK 2, JG 11, Vijay, Vishal, Co 3

Production technologies

The cropwise production technologies for different pulses are given in table - 8.

Table 8 : The Crop -wise Production technologies are as follows.

Particulars	Chickpea	Pigeonpea	Greengram	Blackgram
Soils	Well drained medium to deep black soils.Saline soils are not suitable	All soils with good drainage facility. Saline soils are not suitable	All soils with good drainage facility. Saline soils are not suitable	All soils with good drainage facility. Saline soils are not suitable
Land preparation	One deep ploughing followed by two harrowings and perfect leveling	One deep ploughing followed by two harrowings and perfect leveling	One deep ploughing followed by two harrowings and perfect leveling	One deep ploughing followed by two harrowings and perfect leveling
Seed rate (kg/ha)	50-60	*Kharif* : 5-8 *Rabi* : 15-20	*Kharif* : 15-18 *Rabi* /Rice fallows/Summer : 25-30	*Kharif / Rabi*: 15-18 Summer : 25-30Rice fallows : 40-50
Sowing management				
a. Seed treatment	Seed treatment with captan/thiram @ 3 g/kg or *Trichoderma* @ 4 g/kg seed followed by *Rhizobium* inoculation	Seed treatment with captan/thiram @ 3 g/kg or *Trichoderma* @ 4 g/kg seed followed by *Rhizobium* inoculation	Seed treatment with carbosulfan @ 30 g/kg seed followed by captan/thiram @ 3 g/kg followed by *Rhizobium* inoculation	
b. Sowing time	October-November	*Kharif* : June-July *Rabi* September to first fortnight of October	*Kharif* : June-July; *Rabi* : October Rice fallows : Nov-Dec ; Summer : Feb-March	
c. Spacing (cm)	30 x 10	*Kharif* : 150-180 X 20 *Rabi* 45-60 x 10 Closer spacing with delayed sowings beyond July.	30 x 10	30 x 10
Fertilizer application				
a. FYM (t/ha)	10	10	5	5
b. N :P:K:S(kg/ha)	20-50-0-40	20-50-0-0	20-50-0-0	20-50-0-0
Weed Control & inter-culture	Pre plant incorporation of fluchloralin @ 1 -1.5	Pre plant incorporation of fluchloralin @ 1 -1.5 kg	Pre-emergence application of pendimethalin /alchlor @ 1-1.5 kg a.i/ha followed by inter culture at 30DAS.	

Contd.

	kg a.i/ha or Pre-emergence application of pendimethalin @ 1-1.5 kg a.i/ha followed by inter culture at 30DAS	a.i/ha or Pre-emergence application of pendimethalin @ 1-1.5 kg a.i/ha followed by inter culture at 30DAS	Post emergence application of Fenoxoprop ethyl @ 60 g a.i/ha in rice fallows at 15-20 DAS to control *Echinocloa*	
Water management	Generally grown on residual moisture. However, two light irrigations, one before flowering and one at yield formation stage.	Generally the crop is grown as rain fed crop. However, two light irrigations at flowering and at pod development stage will boost up the yields.For *rabi* crop 4-5 irrigations are required.	Generally the crop is grown as rain fed crop. However two light irrigations, one each at flowering and at pod development stage will boost up the yields.For *rabi* and summer crop 3-4 irrigations are required.	
Plant protection	IPM to be followed for *Helicoverpa* control.For managing wilts and dry root rot diseases grow resistant varieties like ICCV 37/ ICCV 2/ICCV 10	IPM to be followed for *Helicoverpa* control. For managing Fusarium wilts grow resistant varieties like ICPL 87119/ICPL 85063/ PRG 158 while for Macro-phomina wilt grow MRG 66/ MRG 1004	**Powdery mildew:** Two sprays at 10-15 days interval of Carbendazim @ 1 g/l or Thiphanate methyl @ 1 g/l. Leaf **curl & YMV** : Control of vectors **White fly** : Triazophos or Monocrotophos @ 2 ml/l **Thrips :** Acephate 1.5 g/l or Monocrotophos @ 2 ml/l **Spodeptera :** Poison baiting with monocrotophos/carbaryl Spray NPV 200LE during evening times. **Maruca :** Chlorpyriphos @ 2.5 ml /l +Dichlorovos 1 ml/l Quinolphos @ 2.5 ml /l +Dichlorovos 1 ml/l	
Sequence cropping systems	Fallow-Chickpea Greengram- Chickpea	Black/Greengram - Redgram	Rice-greengram Greengram - Tobacco Rice/ Maize/Vegetables	Rice – Blackgram Blackgram-Redgram
Intercropping systems	Chickpea + coriander (6 :1)	Redgram + Groundnut (1:7) Redgram + Maize /Jowar - (1:2) Redgram + Greengram /blackgram /Soybean (1:4-5)	Greengram + cotton (3:1) Redgram + Greengram (1:4-5) Cotton +greengram (1:3)	Blackgram + cotton (3:1) Redgram + Blackgram (1:4-5) Cotton + Blackgram (1:3)
Nonconventional seasons and areas	The crop may be grown in rice fallows under zero till conditions provided the land is vacated by October end.	Crop has a good potential as early *rabi* crop under contingency situations of late onset of monsoon.	They have good potential as pre *kharif* crops under contingency situation of late release of water in canals	

Strategies Specific to Pulse Crops

The pulse cultivation in A.P. is dependent either on rainfall during *kharif,* or stored soil moisture during *rabi.* Pulses are raised as component of cereal/ commercial cropping systems as inter crop or sequence crop preceding or succeeding base crop. Hence importance is not given to pulse crop in terms of package for high productivity and production. Long term strategy is to introduce pulse in irrigated cropping system under pulse based cropping systems making it a base crop in cropping system instead as component crop raised on residual fertility or as catch crop. Short term strategy is to provide all package to the component crops of pulses in all important cropping systems whether it is intercropping or sequence cropping to increase production and productivity. The specific development strategies for each of the pulse crops in A. P. are detailed here under include research as well as extension measures

The per capita availability of pulses is around 40 g as against the requirement of 85 g per day as per WHO standards. Though pulses are part of food grains, their production and productivity did not improve much as in the case of cereals like rice, wheat and maize

Keeping the pulse crops free from moisture related stress at critical periods (drought or excess moisture) will have a definite impact and pave the way for increased production. Introduction of pulses into irrigated cropping systems is the first option in this direction.

Indeterminate flowering habit of *kharif* pulse crops is helpful at times. But, the flower to pod conversion ratio is less because of this genetic factor. Development of new plant types in the pulse crops with improved flowering behavior together with mechanization will result in large benefits for farmers including mechanization.

Multiple resistance to biotic and abiotic stresses is very much desired. Breeding programmes should aim at development of resistance or tolerance to Helicoverpa and wilts in redgram, to YMV in blackgram and greengram and to dry root rot and wilt in chickpea. The specific developmental strategies for each of the four pulse crops in A.P. are detailed here under, that include researchable items as well as extension measures.

Greengram

- To develop new varieties with synchronous maturity, resistance to pre-harvest sprouting, YMV, ABLS and Powdery mildew, which could fit as intercrop and sequence crop.
- To demonstrate the available production technology on greengram in Telangana region where the area under this crop is largest, but the yields are poor.

- To popularize greengram as preceding crop to rice in tankfed and tail end areas in canal commands. A massive campaign is essential for this programme.
- To introduce greengram as intercrop in cotton on large scale to make the cotton system more sustainable.
- To popularize Rice-maize-greengram sequence in Godavari delta.

Blackgram

- To raise it as intercrop in the early stages of orchard crops.
- To develop multiple disease resistant varieties.
- To extend *rabi* blackgram under irrigated conditions as an alternative to *rabi* rice under wells and canal commands.
- To popularize *rabi* blackgram in place of tobacco.
- To demonstrate effectively, the plant protection measures against rust and Cercospora leaf spot and Maruca pod borer, especially in rice fallows.
- To popularize the photo-insensitive varieties like LBG 752 in rice-fallows under delayed sowings (Dec - Jan).
- To popularize it as summer pulse intercrop specially in January and February planted sugarcane.
- To popularize Rice-Maize-blackgram sequence.

Redgram

- To develop short statured compact types with 150-155 days duration which could fit as inter crops
- To effectively disseminate IPM practices in a big way for the management of Helicoverpa
- To popularize *rabi* redgram in south coastal districts of A.P. in place of Sorghum and Tobacco
- To encourage cultivation of redgram as fringe crop to meet local demands
- To develop slightly early maturing varieties (150 days) with resistance to wilt and tolerance to Helicoverpa pod borer and escape terminal moisture stress.
- To popularize it as refugee crop in cotton as the flowering time of both crops coincide and this can reduce the development of resistance in bollworm.

- To popularize it as border crop around commercial crop of chillies, turmeric and cotton.

Chickpea

- To develop extra short duration (75-80 days) varieties with resistance to wilt and dry root rot diseases
- To popularize varieties like JG 11, JAKI 9218, JG 130, KAK 2, and Phule G 95311 (Vihar) to sustain the growth rate

Non-Traditional Crops and Cropping Systems

Rajmash

This crop has good potential where *kharif* crops are harvested by November first week. Because of multiple utility (green pod stage use as vegetable and later as pulse) could be raised in areas where irrigation facilities extend only till February. Also can be raised as soil restorer after exhaustive *kharif* crops.

Cowpea

Though it is a traditional crop in some tracts it has good potential even as soil restorer crop after exhaustive commercial crops of Bt cotton and turmeric. It has multiple utility as legume fodder, vegetable and pulse.

References

Anonymous 2008. An outline of Agricultural Situation in AP 2008-09. Directorate of Economics and Statistical, Government of Andhra Pradesh, Hyderabad.

Anonymous-2007-09. Annual Reports of Acharya N.G. Ranga Agricultural Univeristy, Rajendranagar, Hyderabad- 30.

Anonymous 2004-2008. Status Reports of different agro-climatic zones. Acharya N. G. Ranga Agricultural University, Rajendranagar, Hyderabad- 30 Andhra Pradesh.

❑❑❑

3

Assam

A. Baishya, J.P. Hazarika and M.C. Kalita
All India Coordinated Research Project on Integrated Farming System, AAU, Jorhat-785013

Assam is surrounded by Bhutan and Arunachal Pradesh on the North; Nagaland and Manipur on the East; Meghalaya and Mizoram on the South and Bangladesh, Tripura and West Bengal on the West. The total land area of Assam is 78,523 sq. km and it accounts for 2.4% of total geographical area of the country. Demographically, Assam is the most populous state in the North East India. Economically, out of eight North East states, Assam is the most developed one, but compared to other states like Maharastra, Tamilnadu, Gujrat, it lags far behind in almost all economic spheres.

The state of Assam comprising of 25 plain districts and 2 hill districts, falls under high potential zone as far as agricultural productivity is concerned. Assam is basically an agrarian state with more than 70 percent of its population (2.66 crore as per 2001 census) involved in agriculture either fully or partially. Out of the total geographical area of 78.43 lakh hectares, almost one-third (30.4 percent) is under cultivation covering an area of 23.86 lakh hectares (excluding tea area). The gross cropped area is about 36.37 lakh hectares. The cropping intensity is 152.4 percent. Rice, which is presently grown mainly during *kharif* season, dominates the agriculture scenario in Assam, of which winter rice (*Sali*) occupies an area of around 17.49 lakh hectares, Autumn rice (*Ahu*) is another 4.64 lakh hectares and summer rice (*Boro*) grown in an area of 3.27 lakh hectares. The other food grain crops like wheat (0.70 lakh hectares), pulses (1.8 lakh hectares) and maize etc (0.20 lakh hectares) are also grown. The important commercial crops are oilseeds (3.39 lakh hectares), sugarcane (0.25 lakh hectares) and Jute (0.68 lakh hectares).

Although regular floods, droughts, heavy population pressure on land and infrastructural weakness are impediments to growth, yet the farmers have started to increase production through the use of improved technologies.

Pulses, wheat and oilseeds (Mustard) continue to elude the efforts in achieving self-sufficiency. At present, the state produces 0.73 lakh metric tonnes of pulses, 1.90 lakh metric tonnes of oilseeds and 0.78 lakh metric tonnes wheat only. But in the Horticulture sector, the state is self-sufficient in vegetables, fruits and spices (ginger). The area covered under fruit crops is 1.11 lakh hectares producing 13.35 lakh metric tonnes, vegetables in 2.10 lakh hectares producing 28.37 lakh metric tonnes and spices in 0.84 lakh hectares producing 1.91 lakh metric tonnes.

Trends of Pulse Production

The major pulses grown in Assam are presented in Table 1 which shows that the maximum area (0.40 lakh ha) and highest production 0.24 metric tonnes were observed in Black gram crop whereas highest productivity of 725 kg/ha was observed in Arhar crop only.

Table 1 : Area (lakh ha), production (lakh mt) and average yield (kg/ha) of crops in Assam

Crop/ Trends	2000 -01	2001 -02	2002 -03	2003 -04	2004 -05	2005 -06	2006 -07	2007 -08	2008 -09
Arhar									
Area	0.07	0.07	0.07	0.07	0.07	0.06	0.07	0.06	0.05
Prodn.	0.05	0.05	0.05	0.05	0.05	0.05	0.05	0.04	0.04
Av. Yield	707	708	708	707	705	700	714	718	725
Peas									
Area	0.25	0.24	0.23	0.25	0.22	0.18	0.19	0.22	0.21
Prodn.	0.16	0.14	0.14	0.16	0.13	0.11	0.12	0.14	0.15
Av. Yield	582	587	593	582	599	633	612	613	700
Lentil									
Area	0.20	0.21	0.20	0.20	0.20	0.19	0.20	0.22	0.22
Prodn.	0.11	0.11	0.11	0.11	0.11	0.10	0.11	0.11	0.13
Av. Yield	527	526	541	527	542	532	527	471	581
Black gram									
Area	0.37	0.42	0.40	0.37	0.36	0.35	0.37	0.38	0.40
Prodn.	0.21	0.23	0.20	0.21	0.21	0.17	0.19	0.20	0.24
Av. Yield	568	556	498	568	575	490	514	531	610
Greengram									
Area	0.07	0.06	0.07	0.07	0.07	0.08	0.07	0.08	0.06
Prodn.	0.03	0.04	0.03	0.03	0.03	0.03	0.03	0.04	0.03
Av. Yield	457	456	458	457	470	456	460	460	527
Gram									
Area	0.03	0.02	0.02	0.03	0.02	0.02	0.02	0.02	0.02
Prodn.	0.01	0.01	0.01	0.01	0.01	0.01	0.01	0.01	0.01
Av. Yield	482	526	504	482	507	513	511	505	548
Total Pulses									
Area	1.23	1.30	1.23	1.26	1.15	1.07	1.14	1.17	1.18
Prodn.	0.68	0.72	0.67	0.69	0.66	0.57	0.62	0.64	0.73
Av. Yield	556	556	539	546	568	533	542	544	617

Efficient Cropping Systems

In Assam rice-rice, rice-toria and rice-vegetable sequences are mainly practiced. However, following sequences are found to be economically viable and can be followed by the farmers (Table-2).

Table 2 : Economics of different rice based sequences involving pulses

Cropping sequence	Gross return (x10^3 Rs/ha)	Total variable cost (x10^3 Rs/ha)	Return over variable cost (x10^3 Rs/ha)	B : C ratio
Winter rice-Rajmah	46.53	16.79	29.75	2.77
Winter rice-Pea	30.53	14.40	16.13	2.12
Winter rice-Toria-Greengram	94.19	17.63	76.56	5.34
Greengram-Toria-Lady's finger	88.24	28.42	59.81	3.07
Winter rice-Rajmah-Autumn rice	74.96	24.02	50.94	3.12
Winter rice-Pea-Autumn rice	55.49	21.78	33.71	2.55
Winter rice-Lentil-Autumn rice	47.41	24.86	22.55	1.91
Winter rice-Cauliflower-Black gram	101.11	38.35	62.76	1.64
Winter rice-Toria-Black gram	51.08	22.10	28.99	2.31
Winter rice-Oat-Greengram	52.89	20.29	32.60	2.61
Winter rice-Cabbage-Greengram	335.39	49.71	285.68	5.75
Winter rice-Knolkhol-French bean-Greengram	146.74	51.69	950.45	1.84
Winter rice-Radish-French bean-Greengram	140.91	50.50	90.41	1.79

Yield Gap

On an average Assam has a pulse yield of 617 kg/ha which is lower than National average (659 kg/ha). In India, the highest yield of pulse was observed in Haryana (824 kg/ha) whereas globally, the highest yield of 4920 kg/ha was obtained in Ireland. The gap between expected and yield achieved are presented in Table-3.

Table 3 : Gap between yield expected and yield achieved (kg/ha) [2008-09]

Pulse crop	Yield Expected	Yield Achieved
Arhar	775	725
Peas	740	700
Lentil	610	581
Black gram	640	610
Greengram	540	527
Gram	575	548

Annual exponential growth of production of total pulses during the period 2004-05 to 2008-09 was 3.23%.

Recommended Varieties

The improved varieties of pulse crops as recommended for the state of Assam are given in Table 4.

Table 4 : Improved varieties of Pulse crops for Assam.

Varieties	Duration (days)	Grain yield (q/ha)	Specific features/remark
Black gram			
T 9	80-90	10-12	Tolerant to YMV and Cercospora leaf spot
T 27	80-90	10-12	-do-
Pant U 19	80-95	10-12	-do-
T 122	80-90	10-12	Susceptible to YMV and Cercospora leaf spot
Saonia mah (SB 121)	70-80	10-12	Recommended for central Brahmaputra Valley Zone only.
UG 157	75-85	10-13	Tolerant to YMV and Cercospora leaf spot
JU 78	75-85	10-13	-do-
Greengram			
T 44	60-70	10-12	Susceptible to YMV and Cercospora leaf spot
Kopergaon	60-70	8-9	-do-
K 851	60-70	10-14	Tolerant to YMV and Cercospora leaf spot
ML 56	65-70	10-14	-do-
PIMS 1	70-80	10-13	-do-
ML 131	65-70	10-14	Recommended for North Bank plains Zone only.
AAU 34	65-75	10-12	Resistant to Cercospora leaf spot and susceptible to YMV
AAU 39	65-75	10-12	-do-

Contd.

Pratap (SG-1)	60-70	10-14	Tolerant to Cercospora leaf spot and susceptible to YMV
SG 21-5	60-70	10-14	Resistant to Cercospora leaf spot and YMV
Lentil			
B-77	110-120	8-12	Erect, Branched, 25-30 cm tall Seeds medium sized, 1000 seeds wt 12-15 g
L-9-12	120-130	8-12	Erect, Branched, 20-30 cm tall Seeds medium sized, 1000 seeds wt 15-19 g
PL 406	120-125	10-12	Branched, Semi spreading, 35-40 cm tall Seeds medium sized, 1000 seeds wt 17-19 g
Pl 81-4	120-125	11-13	Branched, Semi spreading, 35-40 cm tall Seeds medium sized, 1000 seeds wt 15-17 g
Pea			
T-163	110-120	10-11	Medium tall (100-120cm), double podded variety Pods 7-8cm long, light green, 5-6 seeds per pod. 1000 seeds wt. 150-200g. Susceptible to Powdery mildew and Fusarium root rot
Bonville	100-120	10-12	Medium tall (100-120cm), double podded variety used as vegetables Seeds wrinkled, light greenish to white. Pods light green, straight, 8-9cm long with 5-6 seeds per pod. Susceptible to powdery mildew and Fusarium root rot.
HUP-2	115-125	13-15	Tall (130-140cm), Semi spreading and semi leafless 4-5 seeds per pod, seeds white, 1000 seeds wt 200-250 g. Resistant to powdery mildew and downy mildew.
KFPD 1	100-105	18-20	Plants are dwarf (50-60 cm) 5-6 seeds per pod. Seeds white and bold. 1000 seeds wt 250-300g. Susceptible to powdery mildew but escapes the disease because of earliness.
Rachna	120-125	10-12	Tall (135-138 cm)Almost similar to T-163, tolerant to powdery mildew.
New variety			
FP-255 (Pant-14) hill zone.	108	11.49	Moderately resistant to rust most suitable for
Rajmash			
HUR-301	95-105	20.0	Resistant to wilt, leaf crinkle and BCMY recommended for all zones
HUR-203	95-105	20.0	Resistant to major diseases. Recommended for all zones
Uday	100-120	15-20	Promising variety selected by IIPR, Kanpur. Erect plant type, branched, determinate, plant height 35-50 cm , pods: 10-12 cm long,3-4 seed/pod and 15-20 pods/plant.

Contd.

Lathyrus			
Ratan(Biol 212)	110-120	10-12	Negligible ODAP content (<0.1%), suitable for utera cultivation. Recommended for all zones (except Hill Zone and Barak Valley zone).
Rice bean			
RBL 4	88-93	19-21	Resistant to YMV and Cercospora leaf spot
RBL 5	90-95	19-21	-do-
Beziamah	90-95	20-22	-do-
Soybean (varieties for *rabi* season)			
JS-2	85-90	20-22 (irrigated)	erect
Moti	85-100	11-12 (rainfed)	-do-
Pusa Soybean-1	90-100	-do-	-do-
Soybean (varieties for *kharif* season)			
Bragg	120	25	C,L,B
JS 2	90-95	25	N,U,L,H
PK 71-21 (Alankar)	115-120	25	U,H
Ankur	120-125	25	U,L,H
UPSM 19 (Silajit)	120-125	26	U,H
# DS 73-16 (Pusa Soybean)	100-106	25	U,C,H
# PK 327+	105-110	25	N,U,B,H
# PK 271+	110-115	26	N,U,H
# PK 262+	110-115	27	U,B,H
Pk 308	115-120	25	U
Moti	95-100	27	N,U,C,H

* Refer to agro-climatic zone; C=Central Brahmaputra Valley Zone; L= Lower Brahmaputra Valley Zone; B= Barak Brahmaputra Valley Zone; N= North Bank Plains Zone; U= Upper Brahmaputra Valley Zone; H= Hill Zone

Pigeonpea

It is cultivated in a very small area in the state. Till now only one variety named T-21 has been recommended for all the six zones of the state. This variety is a semi spreading, medium tall (170-185 cm in height) variety with 170-200 days duration.

Cowpea

RC 19 : Recommended for all zones. Plant type erect and height is 80-120cm. It bears 6-12 pods per plant and pods are 10-12cm long. Average 100 seed weight is 10g. Average yield 10-14 Q/ha.

TVX 644-02E : Recommended for all zones. Plant type erect and height is 100-120cm. Pods are long containing 10-14 seeds per pod. Seeds are creamy in colour and bold. Average 100 seed weight is 15g. Average yield 12-15 Q/ha.

Chickpea

This is considered as a minor pulse crop in the state and is cultivated around 2000 ha producing around 1000 kg/ha. At present no variety is recommended for the state but farmers are cultivating non descript variety at this moment. Chickpea is yet to be included in the package of practices by Assam Agricultural University, Jorhat.

Land Preparation

Greengram/ Black gram : The land is to be ploughed 2-3 times followed by leveling. The stubbles are to be removed. Surface drains should be provided to facilitate quick removal of excess water from the field.

Lentil : The field is to be ploughed 3-4 times to obtain a good tilth.

Pea : Peas generally require reasonably deeply worked soils hence; the field should be ploughed 3-4 times to obtain a good tilth.

Rajmah : Land should be prepared thoroughly to a fine tilth. Avoid sowing in highly moist soil.

Pigeonpea : The land is to be ploughed 3-5 times followed by harrowing and removing weeds.

Rice bean : Fine tilth is necessary for cultivation of rice bean. The land is to be ploughed 2-3 times followed by leveling. The weeds should be removed completely. Surface drains should be provided to facilitate quick removal of excess water from the field.

Cowpea (G) : The land is to be ploughed 2-3 times followed by leveling.

Lathyrus : The field is to be ploughed 3-4 times to obtain a good tilth.

Soybean : Soybean should be grown on well pulverized bed. Land should be prepared by giving deep ploughing followed by laddering. There should be provision of surface drains with gentle slope for good drainage.

Seed Rate

The cropwise optimum seed rates are as follows.

Crop	Seed Rate (s) Kg/ha	Remark
Black gram		
Line sowing	25.00	–
Broadcasting	27.00	–
Greengram		
Line sowing	20-25	For AAU 34 and AAU 39
	15-18	For other varieties
Broadcasting	25	For AAU 34 and AAU 39
	20-22	For other varieties.
Lentil	30	–
Pea	50	For T-163.
Bonneville	60	–
	65	For HUP-2
	77	KFPD-1
	30-40	For riverine tract.
Pigeonpea	12-15	–
Rajmash	75	
Cowpea	30-35	For grain
Rice bean	20-27	RBL 4 and RBL 5
Beziamah	15-20	–
Lathyrus	40	
Soybean	60	For branched type soybean varieties (viz., Bragg, Ankur, Alankar, Silajit, PK 271 and PK 262)
	75	For less branched type soybean (viz., JS 2, Moti, and Pusa Soybean)

Sowing Management

Black gram/ Greengram : Before sowing seeds should be inoculated with rhizobium; any one culture of Shillongoni, Ahatguri, Kamalabari or other suitable strains should be used. Seeds are to be moistened carefully to avoid excessive wetting. Rhizobium culture is to be mixed with seeds @ 50 g/kg seed so that a thin layer of inoculum is deposited on the seed coat. Treated seeds are to be dried under shade. Inoculated seeds should not be exposed to sun.

Seeds are sown with a spacing of 30 cm x 10 cm (i.e, 30 cm row to row and 10 cm plant to plant).

Lentil : Seeds are to be moistened with clear water avoiding excess wetting and *Rhizobium* culture @ 50 g/kg seed or a prescribed rate given on the packet of the culture be mixed thoroughly and dried in shade before sowing.

The seeds are to be sown in line at a spacing of 25 cm between rows and 5-7 cm seed to seed.

Pea : Strains PAU from IARI or any other suitable strains of Rhizobium culture are to be used @ 50 g/kg seed.

Seed treatment with carbendazim or benomyl @ 2 g/kg of seed or captan or thiram @ 3 g/kg of seed is recommended.

Seeds are to be sown with a spacing of 30 cm between rows and 10 cm between plants.

Rajmah : Seeds need to be treated with Captan @ 3 g/kg of seeds or mancozeb @ 2.5 kg/kg of seeds before sowing.

Seeds are sown with a row spacing 30 cm and plant spacing of 20 cm; however, for North Bank Plain Zone of Assam row to row spacing of 40 cm and plant to plant spacing of 30 cm is recommended.

Pigeonpea : Seeds are sown with a spacing of 75 cm between rows and 30 cm between plants.

Cowpea (G) : A spacing of 30cm between rows and 10-15 cm between plants is to be given.

Rice bean : Sowing is performed with row to row spacing of 30 cm and plant to plant spacing of 10 cm.

Lathyrus : Sowing is done with a spacing of 30 cm x 10 cm.

Soybean : For inoculation seed should be moistened with clear water avoiding excessive wetting. *Rhizobium* culture should be mixed with seeds @ 15 g/kg so that a thin coat of inoculum is deposited on each seed. The inoculated seeds should not be exposed to the sun.

Seeds are sown in lines 45 cm apart at a distance of 5 to 10 cm between seeds. Depth of seeding should not exceed 4 cm. The depth can further be reduced to 2- 3 cm if moisture control in the soil is high. The rows should be covered with thin layer of paddy straw for better germination.

Fertilizer Application Including Micro-Nutrients

Black gram/Greengram

Compost or FYM to be applied @ 1t/ha along with 15-35-0 N-P_2O_5-K_2O kg/ha for non inoculated crop; however, when seeds are inoculated with *Rhizobium* culture then 1 t FYM or compost + 10-35-0 N-P_2O_5-K_2O kg/ha needs to be applied.

For Upper Brahmaputra Valley Zone (UBVZ), Lower Brahmaputra Valley Zone (LBVZ) and North Bank Plains Zone (NBPZ) K_2O @ 10kg /ha is recommended in addition to normal dose of N and P. For Central Brahmaputra Valley Zone (CBVZ), 15 kg K_2O/ha is recommended as basal dressing. Diammonium Phosphate (DAP) @ 75kg/ha can be applied in lieu of Urea and SSP in non-inoculated crop. NPK may be applied in the form of mixed fertilizers.

Lentil : FYM or compost to be applied @ 4-5 t/ha along with the fertilizer as mentioned in case of Greengram/ Black gram both for inoculated as well as non-inoculated crop; however, in this case 10 kg K_2O/ha is not recommended for CBVZ, LBVZ and NBPZ but 15 kg K_2O/ha is recommended CBVZ.

Pea **:** Fertilizer to be applied @ 4-5 t/ha compost or FYM+20-46-0 N-P_2O_5-K_2O kg/ha for non-inoculated crop and @ 4-5t/ha compost or FYM+10-46-0 N-P_2O_5-K_2O kg/ha when seeds are inoculated with *Rhizobium* culture.

For UBVZ, LBVZ and NBPZ, 10 kg K_2O/ha is recommended in addition to normal dose of NPK. In NBPZ 10 kg/ ha Borax is also recommended.

DAP @ 100 kg/ha may be applied in lieu of Urea and SSP in non-inoculated crop. NPK may also be applied in the form of mixed fertilizer.

Pigeonpea : Fertilizer to be applied @ 15-40-0 N-P_2O_5-K_2O kg/ha + 4-5 t FYM/Compost for non-inoculated crop whereas for *Rhizobium* inoculated crop 10-40-0 N-P_2O_5-K_2O kg/ha to be applied along with 4-5 t FYM/Compost

Rajmah **:** 30-40-20 N-P_2O_5-K_2O kg/ha to be applied along with 10 t FYM/ Compost/ha.

Half of the N and full doses of FYM, P_2O5 and K_2O needs to be applied as basal dressing and the remaining half of N to be top dressed 30 days after sowing.

Since no viable nodulation takes place, the variety 'Uday' respond to high nitrogen fertilizer (upto 120 kg/ha). For better yield of Rajmah, a higher fertilizer dose of 60:80:40 kg NPK/ha is recommended for North Bank Plains Zone of Assam.

Rice bean : 4-5t/ha compost or FYM to be applied along with 20-40-0 N-P_2O_5-K_2O kg/ha.

Cowpea (G) : 15-35-0 N-P_2O_5-K_2O kg/ha to be applied along with 4-5t compost or FYM.

Lathyrus : Compost @ 1 t/ha to be applied along with 15-35-15 N-P_2O_5-K_2O kg/ha in case seeds are sown without the inoculation of Rhizobium culture, however, in case seeds inoculated with *Rhizobium* culture, 10-35-15 N-P_2O_5-K_2O kg/ha are to be applied. DAP @ 75 kg/ha may also be applied in lieu of Urea and SSP in non-inoculated crop. NPK may also be applied in the form of mixed fertilizer.

2% Urea solution as Foliar spray at branching and pod filling stage is also recommended.

Soybean : 20-60-40 N-P_2O_5-K_2O kg/ha to be applied.

Potash is required for healthy development of seeds. Foliar application of MOP is recommended in soils with low potash content. Fertilizer should be placed in furrows at a depth of 5-7 cm below the seeds and should be covered with a layer of 3-4 cm soil and make it level with the field hoe. In case placement is not possible, the fertilizer should evenly broadcast and incorporated in the soil.

Weed Control and Interculture

Black gram/ Greengram/ Lentil/ Pigeonpea/ Cowpea (G)/ Rice bean/ Lathyrus : One weeding at 20-25 days after sowing is to be done.

Pea : Fluchloralin 45% @ 1.5 kg a.i/ha as pre-emergence spray one day before sowing be incorporated in soil with light hoeing.

Rajmah : First weeding should be given 20-25 days after emergence and subsequent weeding as and when necessary. Earthing up should follow immediately after weeding.

Soybean : Two manual weedings at 15 and 35 days after sowing need to be carried out. A light earthing up may be given at the time of second weeding to keep the plants erect.

Water Management

Black gram/ Greengram : No irrigation is recommended; however draining out of excess water is very much necessary. Raised beds of around 6-8 cm height and 2 m wide were found to be suitable for proper germination as well as growth and development for the Summer Black gram/ Summer greengram crop because it saves the plants from temporary water logging.

Lentil : If the weather is very dry then, one light irrigation to be provided at pod filling stage.

Pea : Under very dry condition, one irrigation should be provided at 40-50 days after sowing.

Rajmah : Two irrigations one at flowering (40 days after sowing) and other at early pod formation stage (65-70 days after sowing) are needed for proper growth and pod filling. Water logging should be avoided.

Soybean : Two irrigations one at flowering and the other at pod formation stage are needed for proper growth and pod filling.

Under rainfed culture, soil moisture can be conserved by application of straw mulch over the rows after sowing. This also suppresses the weed growth.

Plant Protection Measures

The schedule of plant protection measures for important Pulse crops are given in Table 5.

Table 5 : Schedule of plant protection measures for important Pulse crops

Diseases/ Insect pests	Fungicide/ Pesticide	Dose/ha	Spraying interval/ mode of application
Black gram/ Greengram diseases			
Leaf spot	Cu-oxychloride or carbendazim	1.8-2.0 kg in 600-700 liters of water (0.3%)300-500 g in 600-700 liters of water(0.05%)	7-10 days12-15 days
Blight	Carbendazim	300-500 g in 600-700 liters of water (0.05%)	12-15 days
Powdery mildew	Wettable sulpher	1-1.4 kg/ha in 500-700 liters of water (0.2%)	As soon as disease appears
Black gram/ Greengram Insect pests			
Aphids, Jassids	Malathian 50 EC	1-1.5 lit in 500-700 liters of water (0.10-0.15%)	Repeat spray if necessary
Flea beetle	Malathian 50 EC	1-1.5 lit in 500-700 liters of water; repeat spray if necessary or 20 kg (0.10-0.15%)	Repeat dusting if necessary
White fly (to control YMV)	Dimethoate 30 EC or Endosulfan 35EC	2 ml/ lit of water (0.06-0.07%)	1st spray 10 days after germination then at 15 days interval
Hairy caterpiller	Endosulfan 35EC	2 ml/ lit of water (0.07%)	do-
Lentil diseases			
Wilt	Carbendazim or captan or	2 g/kg of seed 3 g/kg of seed	Seed treatment. -do-

Contd.

	Thirum carbendazim		
Wet rot or white blight	Carbendazim	Foliar spray of 0.05% (0.3 to 0.35 kg mixed with 600-700 liter of water/ha) or Ziram: 0.30% (1.8 to 2.0 liter. Mixed with 600-700 liter of water/ha) or Mancozeb: 0.2% (1.12-1.14 mixed with 600-700 liter of water/ha)	Infected plants are to be collected along with sclerotia & burnt.
Pea (Diseases)			
Wilt	Carbendazim or Benomyl or Captan	2 g/kg Seed treatment 3 g/kg of seeds	-do-
	Thiram		
	Carbendazim or Benomyl	0.1%	Drenched in soil in localized patches
Rust	Tridemorph	500-700 ml in 500-700 liter of water (0.1%)	As soon as disease appears
	Mancozeb	1.0 to 3.5 kg in 500-700 liter of water (0.2%)	-do-
Powdery mildew	Wettable sulphur	2.5 to 1.4 kg @ 500-700 ml in 500-700 liter of water (0.5%)	-do-
	Tridemorph	500-700 ml in 500-700 liter of water (0.1%)	-do-
	Dinocap	500-700 g in 500-700 liter of water (0.1%)	-do-
Pea (Insect pests)			
Pod borer	Dichlorovos 100 EC	0.05%	Foliar spray of the insecticides 0.5 ml mixed with per liter of water
Leaf miner, Aphid	Malathion 50 EC	0.10%	Foliar spray of the insecticides 2 ml mixed with per liter of water
Lathyrus (Diseases)			
Wilt	Carbendazim or Thiram Carbendazim	2 g/kg of seed3 g/kg of seed0.1% solution	Seed treatment Drenching the soil in localized patches
Lathyrus (Insect pests)			
Aphids and pod borers	Malathion 50 EC	2 ml/lit of water	As per need
Aphids and pod borers	Dimethoate 30 EC	2 ml/lit of water	As and when need arises

Rajmah : To control aphids, jassids and beetles, spray Dimethoate or Malathion @ 1-1.5 litre in 800-1000 litre of water/ha.

Pigeonpea

To control wilt disease in pigeonpea, rogueing and burning of disease plants partly helps in controlling the disease.

For the control of pod borer, apion borer, aphids, jassids, flea beetle and leaf roller- spray Malathion 50 EC @ 1.0-1.5 litre or Fenitrothin @ 1.0 to 1.5 litre mixed in 500-700 litres of water/ha at an interval of 15 days is recommended. For control of flea beetle, leaf roller, pod bugs and jassids insecticides should be applied if felt necessary.

Cowpea (G)

No serious disease has so far been observed. However, for control of insect pests the spray of Malathion 50 EC @ 1-1.5 liters mixed in 500-700 litres of water is to be sprayed in one hectare against insect pests like-aphids, jassids, flea beetles, pod borers and leaf roller.

Rice bean

The insect pests like Jassids/ Flea beetle/ Hairy caterpillar cause major concern. The spraying of Malathion 50 EC @ 1.0-1.5 litre/ha in 500-700 litre of water or dusting with Malathion 5% dust @ 20-25 kg/ha is recommended to controls these pests.

Soybean

To control major diseases, soybean seeds should be treated with Thiram @ 3 g/kg of seed before sowing. Soybean is susceptible to yellow mosaic virus. In order to stop spread of the vector of Yellow Mosaic in field, spray Dimethoate 30 EC @ 1 to 1.5 liter in 800-1000 liter of water/ha. *Rhizoctonia solani* induced dry root rot in soybean can be effectively managed by seed treatment with Trichoderma spp.

Spraying of Dichlorovos 100 EC @ 0.5 ml/litre or Chloropyriphos 20 EC @ 1 ml/lit is recommended against hairy caterpillar, semilooper, leaf roller and flea beetle.

Protection Against Storage Pests

The pulses seeds properly dried should be mixed thoroughly with black pepper seed powder @ 3 g/kg of seed for protection against bruchid infestation during storage. Treated seeds should be kept in poly bags with outer covering of gunny bags. Moreover, treatment of seeds with Thiram @ 3 g/kg of seed is absolutely necessary when stored for seed purposes.

Pulse based sequential cropping : The sequential cropping can be pursued with better economic viability.

For Medium land situation : Summer greengram/ Black gram/ Cowpea (V)-Winter rice- Vegetables/ Oil seeds/ Pulses/ Fodder crops

For upland situation :

Kharif Greengram/ Black gram- Toria- Lady's finger

Pigeonpea- vegetables

Intercropping of Pigeonpea

Intercropping of Pigeonpea with Greengram/ Black gram and Sesame for the hill slope (20-25%) either in single/ double rows of Pigeonpea is recommended. The crops are to be sown in lines across the slope. A basal dose of 15 kg N and 40 kg P_2O_5/ha are to be applied. To get optimum yield from both main and intercrop, the crops should be sown within last week of July to first of August.

Relay Cropping of Lathyrus

Broadcasting of Lathyrus seeds needs to be done @ 50 kg/ha 15 to 20 days before the harvest of winter rice when the soil is in moist condition. Before sowing, seeds should be soaked in water for 6 hrs.

Relay Cropping of Pea

Pea seeds at 25-50% higher seed rate are to be broadcast in the standing crop of rice about 15-20 days before harvest provided the soil is in moist condition. In the medium land situation of Hill Zone of Assam, 90 kg/ha seed is recommended. Pea seeds are to be soaked overnight before sowing.

Cultivation of Pulses in Non-conventional Seasons/ Areas

In Assam, pulses are generally cultivated either in *rabi* season or as late *kharif* season, hence Summer season pulse crop will be considered as crops grown in non- conventional season. Also, Pulses are cultivated in upland situation but research information generated under Cropping System Programme of Assam Agricultural University clearly favours the possibility of growing pulse crop also in medium land situation during *rabi* as well as during summer season. Hence, pulses can be accommodated after harvest of medium duration variety of winter rice in medium land situation which would not only increase cropping intensity and farmer's income but also improve soil fertility.

Future Thrust

The following areas may be considered as future lines of action on pulses:

- Pulse area expansion incorporating the AAU recommended varieties with concurrent research thrust for developing/ testing new varieties;
- Development of agro-ecosystem specific pulse based cropping system;
- Promotion of pulse crops in the identified areas for organic agriculture since its tap root system facilitates soil micro manipulation thus ensuring both pulse production enhancement and organic agriculture;
- Testing and development of safety protocols for transgenic pulse crops providing resistance to pests and diseases;
- Organize large-scale demonstration of pulse production technology involving KVKs and other programmes of the university.

Acknowledgement

We thankfully acknowledged the advice and inspiration received from Hon'ble Vice Chancellor, Assam Agricultural University, Jorhat in preparing the manuscript. We are also thankful to Dr. N.N. Sharma, Director of Research (Agri.) and Dr. K.K. Pathak, Sr. Scientist, Directorate of Research (Agri.) for providing some official information, which helped in preparing the manuscript.

References

GOA (2009): Statistical Handbook of Assam, 2008

GOA (2009): Package of Practice for *Kharif* crops of Assam. Prepared jointly by Assam Agricultural University and Department of Agriculture, Assam

GOA (2009): Package of Practice for *Rabi* crops of Assam. Prepared jointly by Assam Agricultural University and Department of Agriculture, Assam

GOA (2010): Economic Survey of Assam, 2008-09

GOA (2010): Pulse crop statistics, Directorate of Agriculture, Assam

❑❑❑

4

Bihar

R.P. Sharma
Bihar Agricultural University, Sabour, Bhagalpur

Pulses are considered to be the most effective means of overcoming protein and micronutrient malnutrition since they are excellent sources of dietary protein (19-25%), besides being rich in important minerals and vitamins (Ali and Kumar, 2004). In Bihar, people are mostly vegetarian in dietary habits and depend largely on pulses as the source of dietary protein. But production of pulses has not kept pace with the population growth and per capita availability of pulses has thus, declined from 61 g/ day in 1950-51 to less than 28.0 g/ day in 2001 as against 70 g/ day the standards prescribed by ICMR. Moreover, in Bihar the contribution of cereals in total protein consumption is more than 73.0% which is not a healthy trend as cereal proteins lack many essential amino acids that are present in the pulses from which only 10.6% of total protein intake come in the diet of rural population. This alarming situation, perhaps, has arisen largely due to the quest for food security rather than nutritional security, being the prime national agenda. According to an estimate, more than 70% of gross cropped area in the state is occupied by paddy and wheat whereas, only 9.0% of it is shared by the pulses. As a result, production of pulses is lagging far behind the domestic requirement. Pulses contribute even less than 5.0% in the total food basket. Due to the mismatch of supply and demand of pulses (i.e. shortage of supply of pulses), the prices of pulses increased exorbitantly over the years and have gone beyond the reach of common mass in the state. To meet the high demand for pulses, India has been importing a large quantity in the recent years spending a lot of foreign exchange. The high dependence on the global market for essential source of protein for the vegetarian population in the country is a matter of serious concern in the national perspective. The requirement of pulses will continue to increase in future mainly due to ever increasing population and preference for pulses as the cheapest source of dietary protein. Enhancing pulses production is, thus, essentially required to make the state a prime contributor for ensuring.Golden Revolution in the country (Chaturvedi and Ali, 2002).

Besides being the excellent source of dietary proteins, mineral and vitamins, pulses also serve as valuable animal feeds and cheaper fuels in rural areas.

Over and above, cereal-cereal cropping systems particularly in irrigated areas appear to have exerted adverse effect on the soil health and the sustainability of such cropping system is at stake. Due to continuous rice-wheat (cereal-cereal) sequential cropping, the organic matter content in soil declined to less than 0.2% and the fertilizer use efficiency reduced from 50.25 kg in 1969-70 to 11.52 kg grains/kg of NPK in 1999-2000. The ground water reserve started depleting, deficiency of micronutrients increased, the expression of yield potential and vulnerability to diseases and pests increased causing a big threat to Indian Agriculture. Pulse crops, on the other hand, play a vital role in maintaining and sustaining soil fertility through biological nitrogen fixation and addition of huge amount of organic matter in the soil (Singh *et al.*, 2009). Of the total fixed nitrogen, about 20-65% undergoes mineralization and accrued an economy ranging from 20-60 kg N/ha in succeeding non-legumes. Hence, the diversification of cropping system involving pulses has become essential for imparting sustainability in the contemporary agriculture. Considering the importance of pulses towards house hold food as well as nutritional security, and their role in maintaining soil fertility and sustainability of cropping systems, there is a dire need to boost total pulses production through area expansion and yield enhancement both at the state as well as national level.

Promotion of legume based cropping systems would certainly meet the requirement of crop diversification for sustainable production and also ensure the nutritional security for growing population in the state. There exists ample scope for area expansion as well as yield enhancement of different pulse crops, resulting in increased pulses production and generating the employment opportunities via seed production, processing, marketing etc. Economic condition of pulses growers is expected to improve due to increase in net return ranging from Rs. 10,000 to Rs. 30,000/ha depending upon the type of crop.

Bihar has a unique feature of agro-climatic conditions suited for growing a wide variety of pulses in different seasons and regions (Kisan Diary, 2010). The diversity and seasonal distribution of pulses in Bihar is as follows.

Rabi (Winter)	:	Chickpea, lentil, peas, lathyrus, rajmash
Summer	:	Mung bean, urd bean, cow pea
Kharif	:	Pigeonpea, urd bean, horse gram, cow pea
Pre-*rabi*	:	Pigeonpea
Autumn	:	Mung bean, urd bean
Spring	:	Mung bean

Trends of Pulse Production

Area under major pulse crops over the years as indicated in Table-1 showed somewhat a mixed scenario in terms of a steep decline in pigeonpea, chickpea and black gram, a steady increase in spring/summer green gram, slight fluctuation but stable in lentil and almost stagnant position in case of peas.

Table 1 : Trends in area (lakh ha), production (lakh t) and productivity (kg/ha) of major pulse crops in Bihar (1981-2005).

Crop	Year	Area (Lakh ha)	Production (Lakh tonnes)	Productivity
Pigeonpea (*Kharif* + Pre-*rabi*)	1981-85	6.77	6.77	1176
	1991-95	6.34	6.34	1305
	2001-05	7.98	7.98	1350
Chickpea	1981-85	16.18	13.36	825
	1991-95	12.31	12.6	1024
	2001-05	8.57	8.59	1001
Green gram (Spring+Summer)	1981-85	16.95	7.41	413
	1991-95	19.21	11.38	596
	2001-05	20.52	13.88	676
Black gram(*Kharif*+Summer)	1981-85	11.38	5.12	450
	1991-95	8.59	4.58	533
	2001-05	7.21	3.92	544
Lentil	1981-85	16.46	11.51	699
	1991-95	17.45	14.65	840
	2001-05	17.47	15.00	859
Field peas	1981-85	2.64	1.65	624
	1991-95	2.71	1.91	703
	2001-05	2.39	2.23	932

The productivity, on the other hand, of individual pulse crops has shown an increase consistently in pigeonpea, green gram, lentil but only a minor fluctuation in chickpea and a slight upward trend in black gram. However, productivity level of all these crops witnessed improvement over the years as compared to that in 1981-85.

The productivity of individual pulse crop has either increased or attained somewhat stability over the years, definitely due to adoption of improved varieties, production as well as protection technologies. Yet, there is ample scope of further improvement in productivity of individual pulse crop. For increasing total pulse production in the state, there is need to give greater emphasis on increasing area besides adoption of improved production and protection technologies (Agricultural Technology Module, 2010).

Decline in Area

The reasons for decline in area of major pulse crops in Bihar are as follows.

Pigeonpea

- Increasing problems of water logging.
- More area diverted to rice in *kharif* season.
- Smaller land holdings & longer crop maturity.
- Labour intensive threshing operations coinciding with pre-monsoon rainfall.

Chickpea

- Insect pest & disease problems. Diversion of land to wheat in irrigated areas.
- Unsecured harvest of crop in isolated pockets due to social factors.

Black gram

- Increasing problems of water logging in *kharif* season.
- Labour intensive harvesting and threshing operations in summer season.

Constraints of Pulses Production

- Inadequate supply of quality seeds of improved varieties and slow seed replacement rate.
- Insufficient and unbalanced use of fertilizers and poor or zero management of salt affected soils.
- Inadequate supply and poor adoption of quality bio-fertilizers based on efficient strains of *Rhizobia* and PSB.
- Post adoption of IPM and IDM.
- Labour intensive operations *viz;* harvesting and threshing of crops like pigeonpea, green gram and black gram.
- Post harvest problems-enormous losses in pulse grains due to faulty storage practices and processing.
- Poor and undeveloped infrastructural facilities like covered threshing floor, road, godown etc.
- Disruption of drainage systems leading to water logging throughout the state.
- Slow awareness of developed technologies.
- Fluctuating prices of pulses.

It became obvious from the (Table - 2) comparative statement that average productivity level of all the six major pulse crops maintained a better position in Bihar as compared to national average of the corresponding pulse crop during the period under report. Of these crops, pigeonpea and green gram recorded relatively high productivity. Agro-climatic conditions of this state, thus, are quite favorable for pulses cultivation.

Table 2 : Comparative pulses productivity level in Bihar Vs. India (2001-05)

Crop	Productivity (Kg/ha)		
	Bihar (A)	India (B)	Difference (A-B)
Pigeonpea	1350	667	683
Chickpea	1001	796	205
Lentil	859	665	194
Green gram	676	366	310
Black gram	544	420	124
Field pea	932	901	31

Existing cropping systems : The existing cropping systems in different zones of Bihar are given in table 3.

Table 3 : The existing cropping systems under irrigated and rainfed conditions of Bihar.

Zone	Rainfed			Irrigated		
	Kharif	*Rabi*	Summer	*Kharif*	*Rabi*	Summer
I.North West	Paddy	Chickpea	-	Early Paddy	Maize	-
Gangetic	Paddy	Wheat	-	Maize	Potato	Green gram
plains	Paddy	Paira Lentil	-	Maize/ Early Paddy	Rai	Green gram
	Paddy	S/cane	S/Cane	Paddy	Wheat	-
	Fallow	Tobacco	-	Paddy Maize	S/cane	Sugarcane
				Early Paddy	Potato	Onion
					Pulses	Green gram
II.North	Jute	Pulses	-	Paddy	Wheat	-
East	Jute	Wheat	-	Paddy	Wheat	Green gram
Gangetic	Paddy	Chickpea	-	Paddy	Paddy	-
plains	Paddy	Oilseed	-	Jute	Mustard	Maize
	Maize	Pigeonpea	-	Early Paddy	Potato	Groundnut
				JutePaddy	Maize	Onion
				JuteJuteMaize	Potato	Green gram
					Potato	
					Wheat	
					Vegetables	

Contd.

III.South Gangetic plains	Maize	G/nut	-	Paddy	Wheat	-
	Maize	Pigeonpea	-	Paddy	Wheat	Green gram
	Paddy	PairaLentil	-	Paddy	Maize	-
	Maize	Oilseeds	-	Paddy	Potato	Onion
	Fallow	Chickpea	-	Maize	Vegetables	Vegetables
	Fallow	Lentil	-	Paddy	Berseem	Maize + Cowpea
	Black gram	Wheat	-	Paddy	Potato	Sunflower
				Maize	Oilseeds	Sunflower

(Agricultural Technology Module, 2010)

Yield Gaps

The yield gap analysis (Table - 4) shows that there are very wide yield gaps between average and potential yield in Bihar. The productivity & production can be doubled with the use of available technologies.

Table 4 : Yield gaps in different pulses grown in Bihar (Actual yield data of 2008-09)

Crops	Yield potential (kg/ha)	Average actual yields (kg/ha)	Yield gap (kg/ha)
Chickpea	2000-2200	960	1040-1240
Lentil	2000-2500	800	1200-1700
Field pea	1200-1500	920	280-580
Pigeonpea	2000-2500	950	1050-1550
Green gram	1200-1500	540	660-960
Black gram	1000-1200	800	200-400
Rajmas	2000-2500	885	1115-1615
Cowpea	1000-1200	740	260-460
Lathyrus	1500	610	890

(Kisan Diary, 2010)

Improved Varieties of Pulse Crops

Chickpea : Rajendra gram-1 (1500-1800), Pusa 256 (2000-2500), RAU-52 (2200-2500), Uday or KPG-59 (2000-2200), Pusa-372 (1500-2000), KWR-108 (2000-2200), SG-2 (2000-2200), Pusa Kabuli 1003 (1200-1400), HK Kabuli 94-134 (1200-1500).

Lentil : BR 25 (1400-1500), Pant L 406 (1800-2000), DPL-15 (2250), Arun or Pant L 77-12 (2000-2200), DPL-62 (2500), HUL-57 (2200), KLS- 218 (2200), Malika or L 75 (1800-2000), Pant L 639 (1800-2000).

Field Pea : Rachna (2000-2100), Malviya matar-2 or HUDP-02 (2100), Malviya matar-15 or Hudp-15 (2000-2200), Aparna or HFP-4 (2000-2500), Harbhajan (1200-1300), DDR-23 or Pusa Prabhat (1200-1500).

Pigeonpea : BR 65 (1600-1800), Bahar (2500-3000), Pusa 9 (2000-2500), Narendra Arhar -1or NDA-1 (1800-2000), MAL -13 (2000-2200), Sharad (1800-2000)

Greengram : Samrat (100-1200), Pusa Vishal (1200-1500), Meha or IPM 99-125 (1200-1400), SML 668 (1500-1800), HUM 16 (1500-1600), PDM 54 (1000-1200), T 44 (800-1000)

Black gram : T 9 (800-1000), Azad or Urd 1 (1000), PDU 1 (1300-1400), Pant U 31 (1000), KU 300 (1000), Pant Urd 19 (1000-1200), Naveen (1000-1200)

Cowpea : Pusa komal, Pusa rituraj, Pusa dofasali, Arka garima, Narendra lobia, Pusa barsati, Kashi kanchan, Pusa 1, Pusa 3, Pusa phalguni (1000-1200)

Rajmash : PDR 14 or Uday (2000-2500), Amber or IPR96-4 (1500-2000), Malviya rajma 15 (1600-2000)

Lathyrus : BIO L 212 (1500), Pusa 24, LSD 3, LSD 6

Figures in brackets indicate yield potential (kg/ha) of the varieties.

Cropwise Production Technology

Chickpea

Land preparation : One deep ploughing followed by two harrowing and planking.

Seed rate (kg/ha) : Bold seeded -90-100; Small seeded – 70-75.

Sowing time : Mid Oct. for *Tal* areas; Mid. Oct - Mid. Nov. for *Diara* areas; Mid Nov. - Mid Dec. for whole Bihar; Mid. Oct - Mid. Nov. for *kabuli chana* variety

Sowing method : Line sowing

Spacing : 30 cm x 10 cm

Fertilizer application including micro nutrients : The schedule is as follows.

- Organic : FYM - 50 q/ha or Vermicompost- 25q/ha before sowing
- Macro : NPKS - 20 : 40 :20 : 20 kg /ha at sowing
- Micro : Zink sulphate- 25 kg/ha (After every 3 yrs) at sowing
- Bio-fertilizers : Rhizobium culture-5 packets (each of 200g)/ha and PSB - 5 Packets (each of 200g)/ha for seed inoculation.

Weed control & inter culture : One hand weeding at 25-30 days after sowing (DAS) or pre-emergence application of Pendimethalin or Alachlor @ 0.75 kg a.i./ha followed by one hand weeding at 40-45 DAS (Agricultural Technology Module, 2010).

Water management : One light irrigation in case of severe drought at pod development stage.

Plant protection measures : The plant protection mesures to control the important insect-pests and disease are discussed below.

Crop stage	Important insects/diseases	Technology for Management
Insect-pests		
Seedling stage	Cutworm *(Agrotis* spp.)	Pre sowing seed treatment with chlorpyriphos 20 EC at 8.0 ml/kg seeds or , Pre-sowing insecticidal soil treatment with chlorphyriphos (1.5 D) at 20-25 kg/ha in endemic areas only.
Flowering and poding stage	Pod borer *(H. armigera.)*	Timely sowing of recommended varieties, intercropping with coriander (4:1) / linseed (4:1)/ barley (4:2) / mustard (6:2), use of trap plants viz., Africaln gaint, marigold or sunflower or Vicia, spraying of NSKE (5.0%) at 10 days interval.
Storage	Pulse beetles *(Callosobruchus* spp.)	Store the thoroughly dried seeds (8-10 % moisture) in scientifically improved storage structure, after seed coating with Neem or Mahua oil (1.0 %) w/v for seed purpose or with mustard oil (1.0%) w/v for grain purpose for upto six months.
Diseases		
Seedling stage to flowering and grain filling stage	Wilt C.o.: *Fusarium oxyspsorum f.* sp. *ciceri*	• Deep summer ploughing • Seed Treatment with carbendazim 2 g or carbendazim + Thiram (1:2) @ 3g or, *Trichoderma viride 6 g or Trichoderma viride* 6g + Vitavax 1 g / kg seed • Intercropping with linseed (1:2) or mixed cropping with linseed (1:1) • Grow tolerant varieties like Pusa 256 and BG 372
flowering to poding stage	Target leaf spot C.o. : *Stemphylium sarciniforme*	• Seed treatment with carbendazim @ 2 g/kg seed. • Two Foliar spray of carbendazim + Mancozeb (companion or saff) @ 0.2% or 2g/lit. water.

(Agricultural Technology Module, 2010)

Sequential Cropping : ***Kharif*** fallow - chick pea (in *'tal'* lands) ; paddy - chick pea ; maize - chick pea ; jute - gram

Inter Cropping/mixed Cropping : Chick pea + wheat / barley / linseed / rapeseed /mustard

Lentil

Land preparation : One deep ploughing followed by 2-3 harrowing and planking are adequate to prepare good seedbed.

Seed rate (kg/ha) : Bold seeded - 40-50 (Timely), 55 - 65 (Late)

: Small seeded - 30-35 (Timely), 45 - 55 (Late)

Sowing time : 15 Oct. - 15 Nov (Timely); 15 Nov - 25 Dec. (Late)

Sowing method : Line sowing

Spacing : 25 cm x 10 cm

Fertilizer application including micro nutrients : The recommended schedule of fertilizer are as follows.

- Organic : 20 q/ha Vermicompost before sowing
- Macro : NPKS - 20 : 40 :20 : 20 kg /ha at sowing
- Micro : Zink sulphate- 25 kg/ha at sowing (for soil deficient in Zn, especially in North Bihar)
- Bio-fertilizers : Rhizobium culture-5 packets (each of 200g)/ha before sowing as seed inoculation

Weed control & inter culture : One hand weeding at 25-30 days after sowing (DAS) or pre-emergence application of Fluchloraline or Pendimethalin @ 1.0 kg a.i./ha followed by one hand weeding at 45-50 DAS.

Water management : One light irrigation at the time of flowering (45 DAS) and another light irrigation at the time of pod filling stage (80 DAS) in case of severe drought.

Sequential Cropping : *Kharif* fallow - lentil (in '*tal*' lands); paddy - chick pea ; maize - chick pea; jute -lentil.

Inter Cropping/mixed Cropping : Lentil + barley / linseed / rapeseed and mustard,

Lentil + sugarcane (inter cropping).

Field Pea

Land preparation : One deep ploughing followed by 2-3 harrowing and plankings are adequate to prepare good seedbed.

Seed rate (kg/ha) : 75-80; 95-100 (for Harvajan and Pusa Prabhat varieties)

Sowing time : 15 Oct. - 15 Nov; 15 Oct. - 30 Nov. for Harvajan and Pusa Prabhat varieties.

Sowing method : Line sowing

Spacing : 30 cm x 10 cm; 20 cm x 05 cm (for Harvajan and Pusa Prabhat varieties)

Fertilizer application including micro nutrients: The recommended management schedule is as follows.

- Macro : 30 N + 40 P + 20 K + 20 S kg /ha as basal application
- Micro : $ZnSO_4$ - 25 kg/ha (Zn deficient soils of at sowing North Bihar after every 3 years) at the time of sowing.
- Bio-fertilizers : Rhizobium culture-5 packets (each of 200g)/ha before sowing as seed inoculation.

Weed control & inter culture : Weed competition may be minimized by giving one hand weeding at 25-30 days after sowing (DAS) or pre-emergence application of Pendimethalin or Alachlor @ 1.0 kg a.i./ha .

Plant Protection Measures : The recommended packages for contral of insect-pests and diseases are detailed below.

Crop stage	Important insects/diseases	Technology for Management
Seedling stage	Cutworms *(Agrotis* spp.)	Pre sowing seed treatment with chlorphyriphos 20 EC at 8.0 ml/kg of seeds or Pre-sowing insecticidal soil treatment with chlorphyriphos (1.5D) at 20-25 kg/ha in endemic areas only.
	Stem fly *(Ophiomyia phaseoli Tryon)*	Delayed sowing between 5th and 15th November intercropping with Oat (2:1) in timely sowing , Pre-sowing seed treatment with carbofuran 3G (15 g/kg seed)
Storage	Pulse beetles *(Callosobnechus* spp.)	Store the thoroughly dried seeds (8-10% moisture) in scientifically improved storage structure, after seed coating with Neem or Mahua oil (1.0%) w/v for seed purpose or with mustard oil (1.0%) w/v for grain purpose for upto six months.
Diseases		
Flowering and grain filling stage	Powdery mildew C.o.: *Erysiphe pisi*	• Foliar spray of sulfex @ 2.5 g/lit. of water or karathane @ 0.1% on first appearance of symptom and second spray after 10 days.
Flowering and grain filling stage	Rust C.o.: *Uromyces fabae*	• Field sanitation • Foliar spray of Mancozeb @ 2 g/lit. of water as soon as the disease is detected and second spray at 10 days interval.

Water management : Two irrigations one at branching and second at pod initiation stage.

Sequential Cropping : Maize - field pea; paddy- field pea

Inter Cropping/mixed Cropping : Field pea + chick pea / wheat / barley / oat / linseed / rapeseed / mustard,

Field pea + sugarcane (inter cropping)

Pigeonpea

Land preparation : One deep ploughing followed by two harrowing and planking

Seed rate (kg/ha) : 18-20; 40-45 (for Sharad and Pusa 9 varieties sown in pre- *rabi* season)

Sowing time : 1 July. - 30 July; 25 Aug. - 15 Sept. (for pre- *rabi* sowing of Sharad varieties).

Sowing method : Line sowing

Spacing : 60 cm x 20 cm; 30 cm x 15 cm (for pre- *rabi* sowing).

Fertilizer application including micro nutrients : The recommended schedule of nutrient application is given below.

- Organic : Apply FYM-50 q/ha / Vermicompost- 25 q/ha before sowing
- Macro : N:P:K:S - 20:40:20:20 kg /ha at sowing
- Micro : $ZnSO_4$ - 25 kg/ha (After every 3 years) at sowing
- Bio-fertilizers : Rhizobium culture-5 packets (each of 200g)/ha before sowing as seed inoculation.

Weed control & inter culture : Pendimethalin @ 0.75 kg a.i./ha (Pre-emergence) followed by one hand weeding (45 DAS)

Water management : In case of severe drought one light irrigation at pod development stage preferably in light textured soil (Kisan Diary, 2010).

Plant Protection Measures : The recommended plant projection measures for control of insect-pests and diseases are as follows.

Crop stage	Important insects/diseases	Technology for Management
Insect-pests		
Flowering and grain filling stage	Pod fly *(Melanagromyra obtuse Malloch)*	Use of recommended varieties, intercropping with maize (2:1) or turmeric (2:2), foliar spray of endosulfan 35 EC (2.0 l/ha) at 50% flowering followed by another one with dimethoate 30 EC (1.0 lit. /ha) at 50% podding stage.
	Pod borer *(Helicoverpa armigera Hub.)*	
Storage	Pulse beetles *(Callosobruchus* spp.)	Store the thoroughly dried seeds (8-10% moisture) in scientifically improved storage structure, after seed coating with Neem or Mahua oil (1.0 %)w/v for seed purpose or with mustard oil (1.0%) w/v for grain purpose for upto six months.
Diseases		
Seedling stage to flowering and grain filling stage	Wilt C.o.: *Fusarium* udum	• Hot summer ploughing and removal of stubbles • Use of tolerant varieties- Narendra Arhar-1, Sharad • Seed Treatment with carbendazim 2g/ kg or Trichoderma 6g/kg or Trichoderma 6g + Vitavax 1 g/kg of seed. • Mixed or intercropping with sorghum (1:1 ratio). • Avoiding monocropping. • Timely sowing and normal spacing. • Three year crop rotation with sorghum or non host crops..
Vegetative stage to flowering stage	Sterility mosaic C. o. sterility mosaic virus	• Discourage ratooning of pigeonpea. • Grow resistant/tolerant variety-Bihar, Pusa-9 and MAL-13. • Foliar spray of Kelthane, Metasystox or Dimethoate @ 0.1 % or 1ml/litre of water on first appearance of symptom.
Flowering stage	Alternaria leaf spot *Alternaria tenuisima*	• Seed treatment with Thiram @ 2.5 g/kg of seed. • Foliar spray of Iprodion (Rovral) or Mancozeb (Indofil M-45) @ 0.2% or 2 g / litre of water on first appearance of symptom.

(Agricultural Technology Module, 2010)

Sequential Cropping : Short duration pigeonpea fits well in cropping systems viz.: Pigeonpea-wheat; Pigeonpea-late potato; Pigeonpea-lentil; Pigeonpea-sugarcane; Pigeonpea-wheat-moong.

Inter Cropping/mixed Cropping : Pigeonpea + maize / black gram / greengram / cowpea / groundnut/sorghum.

Intercropping with blackgram (Pant U-30), moong and cowpea (Pusa do phasali) gives about 4-7 quintals per hectare additional yield without affecting yield of pigeonpea.

Greengram

Land preparation : One deep ploughing with soil turning plough followed by 2-3 harrowing and planking for good seedbed.

Seed rate (kg/ha) : Bold seeded (such as Pusa Vishal, SML-668) - 30-35; Small seeded (like Samrat) 20-25.

Sowing time : 10-25 March (spring season); 25 March- 10 April (summer season)

Sowing method : Line sowing behind the plough or with the help of seed drill.

Spacing : 30 cm x 08 cm (spring) and 25 cm x 08 cm (summer)

Fertilizer application including micro nutrients : The recommended schedule of nutrient application is as follows.

- Macro : 15-20 kg N, 35-40 kg P/ha basal application
- Bio-fertilizers : Rhizobium culture -5 packets/ha as seed inoculation.

Weed control & inter culture : One hand weeding at 25-30 days after sowing (DAS) or pre-emergence application of Pendimethalin or Alachlor @ 0.75 kg a.i./ha followed by one hand weeding at 40-45 DAS.

Water management : One irrigation at 20-25 DAS in addition to pre-sowing irrigation for summer season crop.

Plant Protection Measures : The recommended plant protection schedule for important insect-pests and diseases of greengram are as follows.

Crop stage	Important insects/diseases	Technology for Management
Insect-pests		
Vegetative stage	White fly (*Bemisia tabaci G.*), Leaf hopper (*Empoasca kerri Pruthi*), Bihar hairy caterpillar (*Spilosoma oblique W.)*	Pre sowing seed treatment with dimethoate 30 EC @ 8.0 ml/kg of seeds, collection and destruction of egg mass or gregarious larvae of Bihar hairy caterpillar.
Flowering and pod formation stage	Bean thrips *(Megalurothrips distalis Karny)* Blue butter fly *(Lampides boeticus L.)*	Spraying of endosulfan (0.07 %) or dimethoate (0.03%) or malathion (0.04%) at bud formation stage.
Storage	Pulse beetles *(Callosobnechus* spp.)	Store the thoroughly dried seeds (8-10% moisture) in scientifically improved storage structure, after seed coating with Neem or Mahua oil (1.0%) w/v for seed purpose or with mustard oil (1.0%) w/v for grain purpose for upto six months.
Diseases		
Vegetative stage to flowering and grain filling stage	Yellow masaic C.o.: Mungbean yellow mosaic virus	• Removal and destruction of affected plants as soon as noticed in field. • Use of tolerant varieties viz., SML 668, Pusa bold-1, Samrat, TMB 37, Meha, Pusa Vishal, HUM-16
Vegetable stage to grain filling stage	Cercospora leaf spot C.o.: *Cercospora cruenta*	• Destruction of infected crop debris. • Seed treatment with carbendazim 2g, or Thiram 2.5 g / kg of seed. • Foliar spray of carbendazim @ 0.1% on first appearance of the disease and subsequently at 12 days interval depending on disease severity.
Flowering to grain filling stage	Powdery mildew C.o.: *Erysiphe polygoni*	• Two foliar spray of karathene or carbendazim @ 0.1% or sulfex @ 0.3% on first appearance.

Sequential Cropping : Maize-wheat-greengram; Potato-wheat-greengram; Greengram-wheat; Greengram-potato.

Inter Cropping/mixed Cropping : Green gram + maize/ sorghum/ pigeonpea during *kharif* season. Green gram + spring planted sugarcane (inter cropping). From this inter cropping system, an additional grain yield of 5-6 quintals per hectare may be obtained without any adverse effect on the performance of the main crop sugarcane. Sugarcane is planted at distance of 90 centimetre from row to row. Two rows of moong 30 centimetre apart in the centre of sugarcane rows leaving 30 centimetre distance between cane and moong rows are sown with a seed rate of 7-8 kg per hectare.

Black Gram

Land preparation : One deep ploughing followed by 2-3 harrowing and planking to make the field level.

Seed rate (kg/ha) : 30 - 35 (spring season); 15-20 (*kharif* and post- *kharif* season); 20 - 25 (summer season).

Sowing time : Whole March (spring season); 15 Mar. - 10 Apr. (summer season); Middle of July (*Kharif* season); 15 Aug. – 15 Sept. (Post-*kharif* or pre *rabi* season in *diara* areas).

Sowing method : Line sowing behind the plough or with the help of seed drill. Broadcasting in zero tilled condition in diara areas just after recede of flood water.

Spacing : 30 cm x 10 cm (spring) and 25 cm x 05 cm (summer)

Fertilizer application including micro nutrients : The recommended schedule of nutrient managment for Blackgram is Bihar is as follows.

- Macro : 15-20 N + 35-40 P + 20 K + 20 S kg/ha as basal application.
- Bio-fertilizers : Rhizobium culture 5 packets/ha as seed treatment. (Agricultural Technology Module, 2010).

Weed control & inter culture : One hand weeding at 20-25 days after sowing (DAS) or pre-emergence application of Pendimethalin @ 1.0 kg a.i./ha.

Plant Protection Measures : The recommended plant protection measures for central of insect pest and diseases of blackgram as follows.

Crop stage	Important insects/diseases	Technology for Management
Vegetative stage	White fly (*Bemisia tabaci G.*), Leaf hopper (*Empoasca kerrif Pruthi*), Bihar hairy caterpillar (*Spilosoma oblique W.*)	Pre sowing seed treatment with dimethoate 30 EC @ 8.0 ml/kg seeds, collection and destruction of egg mass, or gregarious larvae of Bihar hairy caterpillar.
Flowering and pod formation stage	Bean thrips (*Megalurothrips distalis Karny)* Blue butter fly (*Lampides boeticus L.)*	Spraying of dimethoate (0.03%) or malathion (0.04%) at bud formation stage.
Storage	Pulse beetles (*Callosobnechus spp.)*	Store the thoroughly dried seeds (8-10% moisture) in scientifically improved storage structure, after seed coating with Neem or Mahua oil (1.0 %) w/v for seed purpose or with mustard oil (1.0%) w/v for grain purpose for upto six months.

Contd.

Diseases		
Vegetative stage to flowering and grain filling stage	Yellow mosaic complex Mungbean yellow mosaic virus	• Removal and destruction of affected plants as soon as noticed in field. • Use of tolerant varieties (Plant Urd 19, Pant Urd 30, Navin, Shekhar). • Foliar spray of Metasystox or Dimethoate @ 0.1% on first appearance of disease and subsequently at 10 days interval.
Vegetable stage to grain filling stage	Cercospora leaf spot Complex *Cercospora cruenta*	• Destruction of infected crop debris. • Seed treatment with carbendazim 2g, or Thiram 2.5 g/ kg of seed. • Foliar spray of carbendazim @ 0.1% on first appearance of the disease and subsequently at 12 days interval depending on disease severity.
Flowering to grain filling stage	Powdery mildew Complex *Erysiphe polygoni*	• Two foliar spray of karathene or carbendazim @ 0.1% or sulfex @ 0.3% on first appearance of symptom.

(Agricultural Technology Module, 2010)

Water management : One irrigation at 20-25 DAS in addition to pre-sowing irrigation for summer season crop.

Sequential Cropping : Maize-potato-black gram; Maize-toria-black gram; Maize-wheat- black gram; Rice-wheat- black gram; black gram-wheat- black gram.

Inter Cropping/mixed Cropping : Black gram + maize/ sorghum during *Kharif* season. The black gram is profitably grown as intercrop in widely spaced crops like pigeonpea and sugarcane. One to two rows of black gram can be planted between two rows of pigeonpea planted 75 centimeter apart. Urd bean yields 4 to 7 quintals of grain without affecting the growth and yield of arhar. Similarly intercropping of urd bean can be done with spring planted sugarcane. Planting of two rows of urd bean between two rows of sugarcane had given an additional urd yield of 5-7 quintals per hectare without any adverse effect on the performance of cane crop (Kisan Diary, 2010).

Rajmash

Land preparation : One deep ploughing followed by 2-3 harrowing and planking are adequate to prepare good seedbed.

Seed rate (kg/ha) : 90-100

Sowing time : 25 Oct-20 Nov.

Sowing method : Line sowing behind the plough or with the help of seed drill.

Spacing : 30 cm x 10 cm

Fertilizer application including micro nutrients : The recommended nutrient schedule for Rajmash in Bihar is as follows.

- Organic : Apply FYM 100 q/ha before sowing
- Macro : 90-100 N + 60 P + 30 K + 20S kg/ha. Apply½ N + full P, K and S/ha at sowing time and ¼ N after first irrigation as well as ¼ N at pod initiation stage.
- Micro : $ZnSO_4$ @ 25 kg/ha (Zn deficient soil of North Bihar after every three years). Apply at final land preparation before planking.

Weed control & inter culture : Pre- emergence application of Pendimethalin @ 0.75-1.0 kg a.i./ha followed by one hand weeding at 45 DAS or one hand weeding at 30-35 DAS after first irrigation proved most critical.

Water management : Two irrigations at 25 DAS i.e., during branching and second at 50-60 days after sowing i.e., flowering or pod initiation stage.

Plant Protection Measures : Rajmash is generally free from insect damage. Stored grain pests which pose threat to pulses generally do not harbour seeds of Rajmash. Among the diseases white mold *(Sclerotinia sclerotiorum)* affects the crop at flowering grain filling stage. To control the disease treat the seed with Thiophanate methyl (Roko) or Carbendazim 2g/kg seed. Two foliar spray of carbendazim @ 1 g/litre of water beginning from initiation of first symptom and 2nd spray at 10 days interval. Prefer to grow tolerant varieties PDR-14 (Udai) and Amber.

Sequential Cropping : Paddy- rajmash; maize- rajmash ;

Inter Cropping/mixed Cropping : Rajmash + potato/ linseed / rabi maize. Intercropping of 2 rows of rajmash in between 2 rows of sugarcane is also recommended.

Cow pea

Land preparation : One deep ploughing followed by 2-3 harrowing and planking.

Seed rate (kg/ha) : 20 – 25; 35-45 (for fodder crops).

Sowing time : Middle of June- end of July (*kharif*); March-April (spring and summer).

Sowing method : Line sowing behind the plough or with the help of seed drill.

Spacing : 35 to 45 cm x 8 to10 cm (*kharif*); 25 to 30 cm x 8 to10 cm (spring and summer)

Fertilizer application : Apply 15-20 N + 40-50 P + 20-25 K kg/ha as basal application. Seed treatment with Rhizobium culture 5 packets/ha is also recommended.

Weed control & inter culture : One hand weeding 30 days after sowing and /or one inter cultivation. As chemical control pre-planting spraying of Fluchloralin (Baslin) @ 0.75 kg a.i./ha or Pendimethalin @ 0.75 kg a.i. / ha (Pre - emergence).

Water management: No irrigation but good drainage during rainy season. In case of failure of monsoon one to two irrigation irrigation at critical moisture stress period like flowering and pod development 5 to 6 irrigations are given to spring/summer crops at an interval of 10 to 15 days. (Pan *et al.,* 2010).

Plant Protection Measures : The recommended plant protection measures for control of insect port and diseases of cowpea are as follows.

Crop stage	Important insects/diseases	Technology for Management
Insect-pests		
Vegetative stage	Jassids (Empoasca kerri.), Leaf hopper (*Empoasca kerrif Pruthi*), Bihar hairy caterpillar (*Spilosoma oblique W.)*	Pre sowing seed treatment with dimethoate 30 EC @ 8.0 ml/kg seeds, collection and destruction of egg mass, or gregarious larvae of Bihar hairy caterpillar.Spray 0.1% metasystox or 0.03% monocrotophos against Jassids and Leaf hopper
Flowering and pod formation stage	Bean thrips *(Megalurothrips distalis Karny)* Pod borer (*Helicoverpa armigera*), Aphids(*Aphis craccivora*)	Spraying of systemic insecticide like endosulfan (0.07%) or monocrotophos (0.04%) or malathion (0.05%) at flowering and pod filling stage.
Diseases		
Seedling stage to flowering and grain filling stage	Wilt Complex: *Fusarium oxyporum*Bacterial blight Complex : *Xanthomonas vignicola*	• Hot summer ploughing and removal of stubbles • Grow resistant variety. • Use healthy seed from disease free field. • Seed Treatment with carbendazim 2g/ kg or Trichoderma 6g/kg or Trichoderma 6g + Vitavax 1 g/kg of seed. • Foliar spraying of 0.2% Blitox. • Timely sowing and normal spacing. • Three year crop rotation with sorghum or non host crops..
Vegetative stage to flowering stage	Cowpea Mosaic (virus transmitted by aphids)	• Use healthy seed from healthy crops. • Grow resistant variety. • For controlling aphids, do foliar spray of any systemic fungicide like Metasystox (0.1%) or monocrotophos (0.4%).
Flowering stage	Powdery mildew Complex : *Erysiphae polygoni*	• Two foliar spray of karathene or carbendazim @ 0.1% or sulfex @ 0.3% on first appearance of symptom.

Sequential Cropping : Maize-wheat-cowpea; Rice-wheat-cowpea; Cowpea-wheat-cheena; Maize-toria-wheat-cowpea; Sorghum + cowpea (Fodder)- berseem (Fodder)- maize + cowpea (Fodder); Maize/Sorghum (Fodder)- berseem (Fodder)- maize + cowpea (Fodder)

Inter Cropping/mixed Cropping : Maize + cowpea (summer); Sorghum + cowpea (*kharif*)

Lathyrus

Land preparation : One deep ploughing followed by 1-2 harrowing and planking. Paira cropping of lathyrus on standing paddy field does not require any tillage operation.

Seed rate (kg/ha) : 40-50; 80 - 100 (paira cropping)

Sowing time : It depends on cessation of monsoon rains and maturity of rice crops. Seeds of lathyrus are broadcasted 4-7 days before harvest of rice.

Sowing method : Broadcasting seeds in standing rice crop without tillage operation. Line sowing will facilitate interculturing operation.

Spacing : 30cm x 10 cm (Line sowing), Broadcasting in rice field.

Fertilizer application : Apply 10 kg N + 40 kg P_2O_5 / ha at sowing.

Weed control & inter culture : One hand weeding at 25-30 days after sowing (DAS).Removal of weeds like *Vicia sativa* may reduce the possibility of lathyrism.

Water management : Lathyrus is generally grown under relay cropping with rice under rainfed condition. However, one light irrigation at 60-70 days crop stage under high moisture stress condition has proved highly productive.

Sequential Cropping : *Kharif* fallow- lathyrus ; paddy- lathyrus.

Inter Cropping/mixed Cropping : Lentil + barley/ linseed/ rapeseed and mustard and Lentil+sugarcane (inter cropping).

Plant Protection Measures : The schedule of plant protection measures for control of insect-pest and diseses of Lathyrus are discussed below.

Crop stage	Important insects/diseases	Technology for Management
Insect-pests		
Vegetative stage	Bihar hairy caterpillar (*Spilosoma oblique W.)*	Pre sowing seed treatment with dimethoate 30 EC @ 8.0 ml/kg seeds, collection and destruction of egg mass, or gregarious larvae of Bihar hairy caterpillar.
Flowering and pod formation stage	Aphids(*Aphis craccivora*)	Spraying of systemic insecticide like endosulfan (0.07%) or monocrotophos (0.04%) or malathion (0.05%) at flowering and pod filling stage.
Diseases		
Seedling stage and flowering stage	Wilt Complex : *Fusarium oxyspsorum*	• Hot summer ploughing and field sanitation • Seed treatment with carbendazim + Thiram (1:2) @ 3 g/kg seed or Trichoderma @ 6 g / kg of seed. • Crop rotation with Paddy, Sorghum and Maize.
Flowering and grain filling stage	Rust Complex *Uromyces* fabae	• Field sanitation • Foliar spray of Propiconazole (Tilt) 1.5 ml or Mancozeb 2 g/litre of water. Repeat the spray after 10 days.
Flowering stage	Powdery mildew Complex : *Erysiphae polygoni*	• Two foliar spray of karathene or carbendazim @ 0.1% or sulfex @ 0.3% on first appearance of symptom.

Pulses in Niche Areas

The cultivation of pulses in non conventional seasons/areas are discussed below wherein pulses offers potential to increase area and production in Bihar.

Crop	Cropping System/situations	Region/Zone
Pigeonpea	• Inter cropping with turmeric in *Kharif* season.	Zone I & II
	• Pre-*rabi* mono-cropping after harvest of *kharif* maize/finger millets, land vacated by paddy nursery, land available after receding of flood water.	North & South Bihar plain region
	• Intercropping of pre-*rabi* pigeonpea with yam bean and autumn planted sugarcane.	Zone I
	• Peripheral cropping on bunds of rice fields.	Zone I
Chickpea	• Relay cropping with medium/late maturing rice.	Most parts of North & South Bihar

Contd.

	• Inter/mixed cropping with coriander, barley, linseed etc in rain fed situation	North & Sough Bihar, more particularly in *Tal & Diara*. Zone-I
	• Pure /inter/mixed cropping after reclamation of salt affected soils.	
Lentil	• Adoption of suitable genotypes in non-traditional areas.	Zone-II
	• Inter/mixed cropping after reclamation of sodic soils.	Zone-I
Urd/Mungbean	• Inter/mixed cropping with sunflower, sesamum, sugarcane, Oat, etc. in spring/ summer seasons.	Whole Bihar
	• Cultivation of short duration mung bean in low lands after harvest of paddy	Zone-I
Rajmash (Non-traditional crop)	• Inter/mixed cropping with potato, winter maize etc.v	Zone-I & II
	• Mono-cropping after harvest of paddy under irrigated conditions.	Zone I & II

Bringing additional area under pulse crops : The Bihar state offers good opportunities to bring the additional area in pulse crops as detailed below.

Crop	Existing area ('000ha)	Additional area by 5th year ('000 ha)	Total ('000ha)
Pigeonpea	55	62(112.7)*	117
Chickpea	70	366(523.2)	436
Mungbean	200	639(319.5)	839
Urd bean	50	20(40.0)	72
Lentil	170	692(407.20)	862
Field peas	24	15(62.5)	39
Rajmash	1	45(250.0)	46
Total	**570**	**1839(322.6)**	**2409**

* Figures in brackets indicate (%) increase

Possibilities of productivity enhancement : The possibilities of increase in pulse production through adoption of improved package of practices in Bihar are as follows.

Crop	Extent of yield increase (%) due to improved package of practices over farmer's practice
Pigeonpea	39.4
Chickpea	48.6
Lentil	47.4
Field Peas	49.8
Rajmash	45.5
Mungbean	40.5

Taking new avenues/niches for pulse crops into consideration, it is expected that area under individual crop might be increased to the extent of 27.8% in urd bean to 523.2% in chickpea by the end of 10th year. Total area under pulses can be increased from 5.70 lakh hectare to 24.09 lakh hectare in the state. Thus, there is a vast scope for area expansion which would certainly contribute to jump in pulse production.

Adoption of improved crop varieties with technological back up, productivity of different pulse crops can be enhanced reasonably to the extent of 40-50% over the farmer's practices. It is estimated that even small gain of 100 kg/ha in yield can bring profound impact on total production of pulses in the state. According to conservative estimate, pulses production would increase to the tune of 6.5 lakh quintal/year with the adoption of improved package of practices even without area expansion.

Future Thrusts

- Promotion of legume based improved cropping system suited to agro-climatic situations prevailing in the different districts.
- Enhancement in the existing productivity level of different pulses crops through the transfer of improved pulses production technologies in tune with the agro-climatic situation of the project area.
- Improvement in productivity and development of production / protection techniques for minor pulses like faba bean and horse gram.
- Expansion in the area under pulses with greater emphasis on newer avenues/niches available in prevailing cropping pattern in the project area.
- Insurance for the sustainability of various cropping systems through the promotion of legume crop based diversification in the project area.
- Ensure greater house holds nutritional security to the people within and outside the target area for alleviation of malnutrition.

- Improvement in income of the farming community and creation of job opportunities for the unemployed youths.
- Optimal utilization of the locally available resources in the project area.
- Management of post-harvest problems related to storage and processing of pulses.

References

Agricultural Technology Module (2010). Rajendra Agricultural University, Pusa (Bihar). pp.1-211.

Ali Masood and Kumar, Shiv (2004). Pulses-amazing nutrient composition. The Hindu Survey of Indian Agriculture. pp : 39-43.

Chaturvedi, S.K. and Ali, Masood (2002). "Poor man's meat" needs fresh fill in. The Hindu Survey of Indian Agriculture. pp : 63-69.

Pan, R.S., Kumar Suresh, Pal Govind & Tirky, I. (2010). Economics and resource use efficiency of cowpea under late cultivation in Jharkhand. Agriculture Situation in India. LXVII. pp : 371-376.

❑❑❑

5

Chhattisgarh

S.K. Sarawgi, Shrikant Chitale, Alok Tiwari, G.K. Shrivastava, H.C. Nanda and Sunil Nair
All India Coordinated Research Project on Integrated Farming System, Department of Agronomy, Indira Gandhi Krishi Vishwavidyalaya, Raipur (CG) 492 006

Rice is the principal crop of the region and cultivated in 3.9 million hectares area with an average productivity of about 1.3 t ha^{-1}, which is very low. Irrigation resources are limited and confined to only 26 per cent and utilized for rice crop only. The region has rice based cropping system, in which lathyrus (grasspea), is grown as *utera* (relay crop) in 0.46 million hectares. However, its productivity is not more than 2 to 3 q ha^{-1}. Pulse crops are cultivated in an area of 0.44 m ha in *kharif* and 0.81 m ha in *rabi* season. The main pulse crops of the region are lathyrus, black gram, pigeonpea, chickpea, horsegram (kulthi), lentil and field pea. Area-wise, urd in *kharif* and lathyrus in *rabi* occupy the top position but as far as productivity is concerned, pigeonpea and chickpea are important. Pigeonpea and chickpea together can contribute a lot in increasing the pulses production in region. The average productivity of pigeonpea is 5.19 q ha^{-1}. It is also grown on rice bunds, which may give an additional production of 0.06 mt. Mung and urd are grown throughout the year in the region, however, powdery mildew and yellow mosaic virus diseases pose severe threat to these crops.

Chhattisgarh is divided into three agro-climatic zones *viz;* Chhattisgarh plains zone, Bastar plateau zone and northern hill zone, and comprises of 18 districts (Table 1). The region is endowed with abundant rainfall (1400-1600 mm), optimum solar radiation, favorable temperature, relatively well-conserved soils and surplus manpower. Inspite of this, the region lags behind the rest of the country in crop productivity and has highest concentration of poverty, population and small fragmented holdings. The state is spread over in 13.51 m ha of which 4.52 m ha is cultivable. Because of limited availability of irrigation in *rabi* season, only 25% area in under cultivation that too with *utera* (relay) cropping of lathyrus, chickpea and linseed.

Table 1 : Districts of Chhattisgarh in the various agro-climatic zones.

Agroclimatic zone	Districts
Chhattisgarh plain	Raipur, Durg, Rajnandgaon, Bilaspur, Dhamtari, Mahasamund, Korba, Raigarh, Kawardha, Jangir and Kanker
Bastar Plateau	Bastar, Narayanpur, Bijapur and Dantewada
Northern hills	Surguja, Jashpur and Koria

Cropping Intensity

The cropping intensity of the state was only 114 per cent in 1998-99 (Fig. 1). It was 100% in Bastar plateau and 109 per cent in Northern hills zones in the same year. The cropping intensity has been increasing gradually. It was 136 per cent in 2006-07 but is almost 140% now. The rate of increase in cropping intensity was higher in Chhattisgarh plains zone than Bastar plateau and Northern hills. It was noticed that *kharif* crops including vegetables in Chhattisgarh plains and increase in area under maize in Bastar plateau and Northern hills contributed significantly to increase in cropping intensity. The area under rice-chickpea in rainfed and rice-vegetable cropping system in irrigated conditions are also showing an increasing trend. In fact, the major irrigation source is canal and canal water is available when the reservoirs are filled with rainwater. Since canal water itself is rain dependent, drought plays an important role in reducing the cropping intensity of Chhattisgarh

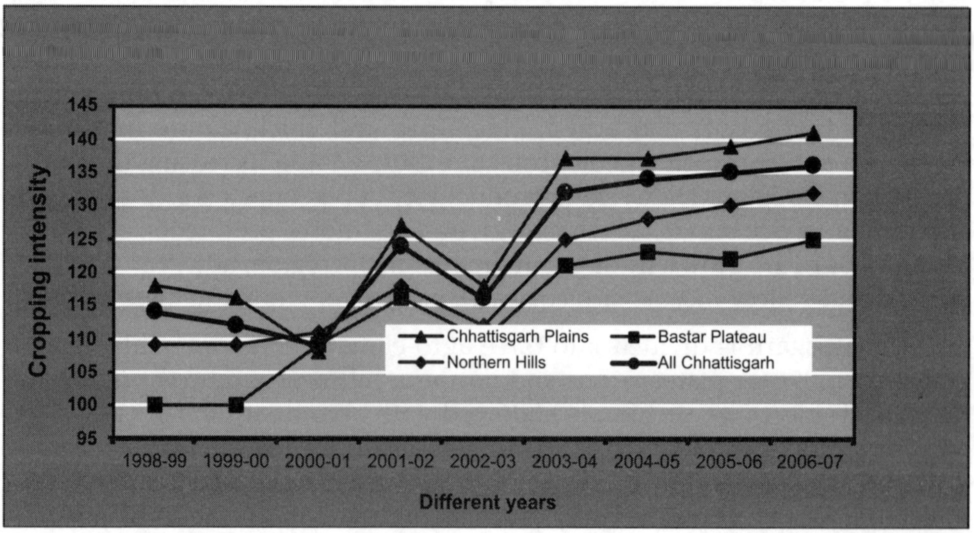

Fig. 1 : Cropping intensity % in three agro-climatic zones

Although, considerable progress has been made in increasing the productivity of pulse crops during the last decade, yet concerted efforts are required to keep pace with the demand of the ever growing population. After the establishment of Indira Gandhi Krishi Vishwavidyalaya, Raipur in 1987, there was an increase in the pulses research at Raipur. Besides, research work on developing varieties suitable for different agro-ecosystems, pulses research has also been extended to other thrust areas. The experiences gained through Front Line Demonstrations have also shown that the existing technology is capable of giving higher yield levels of different pulse crops. Attempts are being made to introduce Rajmash/Frenchbean in Chhattisgarh plains under intensive input management. It is a short duration early maturing crop and can successfully be grown after the harvest of early and medium duration rice and, therefore, short duration crops like black gram, cowpea can also be taken in summer season under irrigated conditions. In addition to this, there are good chances or opportunities of its exports as canned seeds and green pods as vegetable. This will definitely increase the economic status of the farming community of Chhattisgarh.

Crop diversification or crop shift is a new paradigm of sustainable agriculture. Crop diversification is not only a shift from traditional and less remunerative crop (or crops) to more remunerative crops but also a demand driven, need based, situation specific and national goal. The crop substitution and even variety substitution and addition of more crops in existing cropping systems have been the major approach of diversification in India. The nature of crop diversification has been mainly from low value coarse cereals to high value pulses and other food grain crops.

Trends of Pulse Production

In Chhattisgarh, unbunded upland *Vertisols* (*Bharri* soil) is most suitable for cultivation of pulses with good drainage conditions. Poor drainage affects the germination, crop growth and subsequently crop yield. Adopting various land configuration techniques like sowing on flat beds, ridges, opening of furrows etc. help to overcome this difficulty. In raised bed and furrow system, the furrows drain out excess water at the time of heavy rainfall. The productivity of pulse crops remains low i.e. 578 kg ha^{-1} due to which the availability of pulses has declined to 37 g/capita/day.

In Chhattisgarh, total pulses were grown in 12.5 m ha land during both the seasons in 2008-09 and with a production of 7.22 m tonnes and productivity of 578 kg ha^{-1}. The maximum area, production and productivity of different pulse crops were in Chhattisgarh plains, followed by Northern hill zone and Bastar plateau zone. During the last decade i.e. from 2000-01 to 2008-09, the total pulse production increased significantly from 2.67 m tonnes to an all time high of 7.22 m tonnes in 2008-09 (Table 2 and 3, Fig.2).

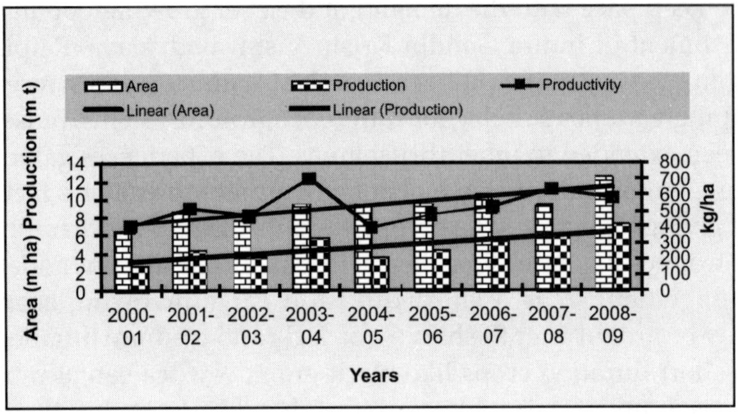

Fig. 2 : Trends in pulse area, production and productivity of pulses during the last decade.

Table 2 : Area, production and productivity of pulses in different district of Chhattisgarh (2008-09)

S. No.	Zones /	Area (000′ ha)		Production (000′ tonnes)		Productivity (kg/ha) District	
		Kharif	*Rabi*	*Kharif*	*Rabi*	*Kharif*	*Rabi*
	CG Plains						
1	Raipur	18.20	127.76	6.76	66.35	371	519
2	Mahasamund	23.05	14.62	9.56	6.75	415	462
3	Dhamtari	3.07	41.27	1.18	26.50	386	642
4	Durg	14.72	197.15	7.12	153.05	483	784
5	Rajnandgaon	47.82	74.07	20.50	49.13	429	656
6	Kawardha	19.21	78.09	7.81	55.51	407	711
7	Bilaspur	25.45	131.41	11.50	97.03	452	738
8	Janjgir	4.07	22.82	1.80	14.41	443	632
9	Korba	15.64	11.75	5.04	7.40	322	630
10	Raigarh	30.15	25.45	9.78	10.41	325	409
	Northern Hills						
11	Surguja	74.89	28.12	36.82	19.52	492	694
12	Jashpur	43.74	11.79	15.16	6.14	347	521
13	Korea	28.17	9.95s	11.84	4.71	420	473
	Bastar Plateau						
14	Jagdalpur	43.22	17.27	20.25	8.54	469	494
15	Narayanpur	7.49	0.54	3.51	0.40	469	738
16	Dantewada	11.24	4.87	6.26	1.29	557	265
17	Beejapur	1.45	1.84	0.61	0.43	422	235
18	Kanker	25.27	16.13	11.29	7.94	447	492
	Total for states	**436.85**	**813.70**	**186.80**	**535.52**	**428**	**658**

Table 3 : Area, production and productivity of pulses in Chhattisgarh during the last decade

Crops	Pulses in Chhattisgarh: Area (m ha), production (m tonnes) and productivity (kg^{-1} ha)									
	Years	**2000-01**	**2001-02**	**2002-03**	**2003-04**	**2004-05**	**2005-06**	**2006-07**	**2007-08**	**2008-09**
Pigeonpea, (*Cajanus cajan*)	Area	0.49	0.50	0.56	0.52	0.60	0.58	0.53	0.54	1.54
	Production	0.21	0.20	0.24	0.32	0.30	0.26	0.23	0.28	0.87
	Yield	429	402	433	602	511	448	426	519	570
Greengram (*Vigna radiata*)	Area	0.14	0.15	0.16	0.18	0.17	0.17	0.23	0.21	0.55
	Production	0.04	0.04	0.04	0.05	0.04	0.04	0.04	0.05	0.16
	Yield	248	288	244	265	240	256	176	233	286
Blackgram (*Vigna mungo*)	Area	1.22	1.19	1.13	1.21	1.20	1.20	1.65	1.56	2.02
	Production	0.34	0.36	0.29	0.35	0.33	0.34	0.38	0.41	0.70
	Yield	280	305	261	289	272	288	229	262	345
Horsegram (*Macrotyloma uniflorum*)	Area	0.65	0.61	0.58	0.56	0.56	0.54	0.36	0.34	0.82
	Production	0.19	0.21	0.14	0.18	0.16	0.17	0.10	0.14	0.29
	Yield	295	336	240	330	295	311	271	396	349
Chickpea (*Cicer arietinum*)	Area	1.40	1.70	1.76	2.05	2.11	2.32	3.11	3.05	2.98
	Production	0.72	1.25	1.13	1.97	1.14	1.63	2.29	2.63	2.97
	Yield	515	735	644	964	542	706	735	862	998
Lathyrus (*Lathyrus sativus*)	Area	2.51	4.17	3.30	4.61	4.28	4.27	5.73	4.81	3.89
	Production	1.09	2.31	1.70	2.79	1.58	1.94	2.81	2.81	2.01
	Yield	434	553	516	605	369	455	488	584	515
Peas (*Pisum sativum*)	Area	0.09	0.12	0.12	0.17	0.17	0.18	0.54	0.46	0.43
	Production	0.03	0.04	0.04	0.07	0.05	0.06	0.16	0.17	0.15
	Yield	360	370	344	376	319	352	302	362	360
Lentil (*Lens culinaris*)	Area	0.13	0.16	0.15	0.18	0.17	0.19	0.27	0.26	0.22
	Production	0.03	0.05	0.05	0.06	0.04	0.06	0.08	0.08	0.072
	Yield	270	327	306	344	255	324	286	333	330
Total Pulses	Area	6.69	8.67	7.82	9.55	9.32	9.50	10.50	9.35	12.50
	Production	2.68	4.48	3.66	5.81	3.68	4.53	5.53	5.92	7.22
	Yield	400	517	468	608	395	477	527	633	578

**Source:* Directorate of Agriculture, Govt. of Chhattisgarh

A total of 1.75 lakh hectare area (main crop and pigeonpea on bunds) is under pigeonpea cultivation during *kharif* season with an average productivity of 570 kg ha^{-1}. The area under pigeonpea main crop has increased from 0.49 m ha in the year 2000-01 to 1.54 m ha in 2008-09 with a corresponding increase in productivity from 429 kg ha^{-1} in 2000-01to 570 kg ha^{-1} during 2008-09 implying a more than 30 per cent boost.

Chickpea is cultivated throughout the state and mostly grown in heavy soils (Vertisols) particularly in Durg, Rajnandgaon, Kabirdham, Bilaspur, and Raipur districts. Chickpea area, production and productivity also increased tremendously during the last decade. Chickpea was grown in 1.40 m ha in 2000-01 with only 0.72 m tonnes production. There was almost a 200 per cent increase in area (2.98 m ha) and production with (2.97 m tonnes) during 2008-09. Its productivity also enhanced from just 515 kg ha^{-1} to 998 kg ha^{-1} in 2008-09. However, its productivity is still low due to the sensitivity of the crop to various biotic and abiotic stresses in rice based farming systems particularly the moisture stress.

Greengram occupies an area of 0.55 m ha with a production of 0.16 m tonnes and productivity of just 236 kg ha^{-1} during *rabi* 2008-09. However, during *kharif* season, the area, production and productivity are 0.35 m ha, 0.11 m t and 315 kg ha^{-1}, respectively. No sizeable increase in area and production has been witnessed since the last ten years. The major growing districts for greengram in the state are : Mahasamund (15.8%), Durg (12.7%), Dantewara (11.7%), Kanker (11.1%) and Rajnandgaon (10.3%). Farmers prefer to grow greengram under upland unbunded conditions locally known as *Bharri* land (Vertisol).

In Chhattisgarh, the area under lathyrus had increased from 2.51 m ha to a maximum of 5.73 m ha in 2006-07 but declined subsequently. It was grown in an area of 3.89 m ha with an annual production of 2.01 m tonnes and productivity of 515 kg ha^{-1} during 2008-09. It is commonly referred as *khesari, tiora, lakh* and *lakhadi* or grasspea. The cultivation of lathyrus is done throughout the whole of Chhattisgarh under relay cropping (*utera*).

Lentil is a winter loving crop and restricted in northern districts of the state but is now getting popularity in plains due to introduction of new improved varieties. Its area had increased from a meagre 0.13 m ha to a high of 0.27 m ha during 2006-07 but declined to 0.22 m ha during 2008-09. Production increased from just 0.03 m tonnes to 0.07 m tonnes with a productivity of 330 kg ha^{-1} during 2008-09.

Significant increase in area and production of peas was observed as it increased from 0.09 m ha and 0.03 m tonnes in 2000-01 to an all time high of 0.54 m ha and 0.16 m tonnes in 2006-07, respectively, a marginal slight decline to 0.43 m ha and 0.15 m tonnes in 2008-09. However, the productivity remained almost unchanged and ranged between 300 to 360 kg ha^{-1}.

Horsegram shows an increasing trend in area and production. Its area and production increased from 0.65 m ha and 0.19 m tonnes in 2000-01 to 0.82 m ha and 0.29 m tonnes during 2008-09, respectively, with an average productivity of 349 kg/ha.

Predominant Cropping Systems

A cropping system adopted in particular area is the mirror of agricultural development of the state or a particular area. It not only illustrates the current land use but also reflects the change in land use pattern over the time particularly in irrigated and rainfed areas. Since, Chhattisgarh has a great potential for diversified agriculture a suitable cropping system will not only increase the income of farmers but also provide year round employment which will reduce the migration problem during lean periods. The major cropping systems followed in Chhattisgarh are listed below.

Mono Cropped Rainfed Systems in Medium to Light Soils : Rice- fallow, kodo/kutki - fallow and Pigeonpea-fallow, Blackgram - fallow.

Double Cropped Rainfed System in Heavy Soils : Rice-mostly lathyrus and partly linseed *utera,* Rice-chickpea, Maize-rapeseed, Rice-linseed and Rice-wheat (partly under rainfed).

Double Cropped Irrigated Systems : Rice-wheat, Rice-rice, Rice-chickpea (partly under irrigated) and Rice-mustard.

Emerging Cropping Systems : Soybean-wheat, Rice-lentil/ Lathyrus/Pea, Rice-Chickpea and Blackgram/Greengram-wheat, Rice-rice/vegetable/Hybrid Sunflower.

Analysis of Change in Crop Coverage and Crop-Ranking

The shift in crops and crop-ranking towards more remunerative crops based on available are indicative of scientific efforts made for technology generation suited to farming community. The districts-wise change in areas of major crops and shift in crop ranking were analyzed. Cropping system analysis not only illustrated the current land use but also reflected the change in land use pattern over time (Gangwar and Ram, 2005). Considering its importance, studies on crops coverage were analyzed in pre and post scenario of Chhattisgarh and are given in Table 4 and 5.

Table 4 : Change in area (000 ha) under different crops in Chhattisgarh.

Agro climatic zone	98-99	05-06	98-99	05-06	98-99	05-06	98-99	05-06
Kharif crops	**Rice**		***Kharif* Maize**		**Pigeonpea**		***Kharif* vegetables**	
Chhattisgarh plains	2792.60	2407.81 (-13.8)	20.40	42.90 (110.3)	25.30	86.31 (241.1)	NA	124.39
Bastar plateau	464.00	464.14 (-0.0)	20.90	65.81 (219.1)	2.00	13.73 (586.5)	NA	16.86
Northern Hills	551.20	517.10 (-6.2)	51.20	65.59 (28.1)	13.70	64.07 (367.7)	NA	21.32
Chhattisgarh	3807.80	3389.05 (-11.0)	92.50	174.30 (88.4)	41.00	164.11 (300.2)	NA	162.57
***Rabi* crops**	**Chickpea**		**Wheat**		**Rapeseed/ mustard**		**Vegetables**	
Chhattisgarh plains	166.30	296.22 (78.1)	63.00	111.49 (77.0)	12.80	57.17 (346.6)	27.75	98.81 (256.0)
Bastar plateau	1.20	2.09 (74.2)	0.60	1.83 (205.0)	8.20	20.63 (151.6)	1.95	8.63 (342.56)
Northern Hills	5.40	22.24 (411.8)	22.70	51.97 (128.9)	30.40	66.65 (119.2)	5.24	26.17 (399.4)
Chhattisgarh	172.90	320.55 (85.3)	86.30	165.29 (91.5)	51.40	144.45 (181.0)	34.94	133.61 (282.3)

Figures in parentheses indicate percent increase/decrease in area in 2005-06 over 1998-99.

Table 5 : District wise crop coverage ranking in Chhattisgarh

District	*Kharif* crops				*Rabi* crops		
	First		Second	Third	First	Second	Third
Raipur	1998-99	Rice	Kodo	Blackgram	Lathyrus	Chickpea	Wheat
	2005-06	Rice	Veg.	Til	Lathyrus	Chickpea	Wheat
Mahasamund	1998-99	Rice	Blackgram	Groundnut	Lathyrus	Wheat	Linseed
	2005-06	Rice	Blackgram	Veg.	Lathyrus	Rice	Veg.
Dhamtari	1998-99	Rice	Blackgram	Kodo	Lathyrus	Linseed	Chickpea
	2005-06	Rice	Pigeonpea	Veg.	Summer Rice	Lathyrus	Chickpea
Durg	1998-99	Rice	Kodo	Soybean	Lathyrus	Chickpea	Linseed
	2005-06	Rice	Soybean	Veg.	Lathyrus	Chickpea	Linseed
Rajnandgaon	1998-99	Rice	Kodo	Soybean	Lathyrus	Linseed	Chickpea
	2005-06	Rice	Soybean	Pigeonpea	Chickpea	Lathyrus	Linseed
Kabirdham	1998-99	Rice	Kodo	Pigeonpea	Chickpea	Lathyrus	Linseed
	2005-06	Rice	Soybean	Kodo	Chickpea	Lathyrus	Wheat

Contd.

Bilaspur	1998-99	Rice	Kodo	Maize	Lathyrus	Chickpea	Wheat
	2005-06	Rice	Veg.	Maize	Lathyrus	Chickpea	Wheat
Janjgir	1998-99	Rice	Blackgram	Pigeonpea	Lathyrus	Linseed	Wheat
	2005-06	Rice	Veg.	Pigeonpea	Lathyrus	Mustard	Linseed
Korba	1998-99	Rice	Kulthi	blackgram	Mustard	Lathyrus	Linseed
	2005-06	Rice	Veg.	blackgram	Lathyrus	Veg.	Mustard
Raigarh	1998-99	Rice	Blackgram	Groundnut	Lathyrus	Wheat	Chickpea,
	2005-06	Rice	Blackgram	Groundnut	Veg.	Summer Rice	Lathyrus
Kanker	1998-99	Rice	Kodo	Kulthi	Linseed	Toria	Lathyrus
	2005-06	Rice	Veg.	Maize	Toria	Linseed	Chickpea
Jagdalpur	1998-99	Rice	Kodo	Soybean	Toria	Chickpea,	Linseed
	2005-06	Rice	Maize	Blackgram	Toria	Veg.	Linseed
Dantewada	1998-99	Rice	Kodo	Kulthi	Toria	Kulthi	-
	2005-06	Rice	Maize	Kodo	Toria	Veg.	Lathy.
Jashpur	1998-99	Rice	Nizer	Blackgram	Toria	Chickpea,	Wheat
	2005-06	Rice	Blackgram	Pigeonpea	Toria	Veg.	Chickpea
Sarguja	1998-99	Rice	Maize	Niger	Mustard	Wheat	Linseed
	2005-06	Rice	Maize	Pigeonpea	Mustard	Wheat	Chickpea
Koria	1998-99	Rice	Maize	Kodo	Wheat	Mustard	Linseed
	2005-06	Rice	Maize	Pigeonpea	Mustard	Wheat	Veg.
Chhattigarh	1998-99	Rice	Kodo	Blackgram	Lathyrus	Chickpea	Linseed
	2005-06	Rice	Blackgram	Maize	Lathyrus	Chickpea	Wheat

The crop coverage ranking is an indicator of types of crops grown during *kharif* and *rabi* and how the cropping systems are changing. It was observed that rice was the major crop in all the districts. Next to rice was kodo-kutki millet in 13 districts, blackgram in two districts and niger in one district during *kharif* 1998-99 in Chhattisgarh. The cultivation of kodo-kutki, Blackgram and Niger was essentially a mono cropping system, which has changed significantly in the post Chhattisgarh period i.e. 2005-06. The second ranked mono cropped kodo-kutki millet was replaced by blackgram, soybean, vegetables and maize.

In the light soils of Mahasamund, Raigarh and Jashpur, blackgram continues to be cultivated. During *rabi*, Lathyrus was replaced by summer rice in Dhamtari, chickpea in Rajnandgaon and vegetables in Raigarh district. The first crop ranking of toria/mustard in Bastar plateau and Northern hills zone remained unchanged in 2005-06 also when compared to 1998-99. The irrigation facilities in different districts played an important role in shifting of crop ranking from 5th to 4th, 4th to 3rd and 3rd to 2nd position.

The analysis clearly indicated that there is shift in crops during *kharif* and *rabi* with more remunerative crops under rainfed and irrigated conditions. The increase in area under particular crop and cropping system also suggested the formulation of planning for development of market and post processing infrastructure in a particular district. For example increase in area under soybean, sugarcane, *kharif* and *rabi* vegetables, sunflower, maize, summer rice etc. need special attention for marketing strategies and infrastructure.

Potential crops : Considering the tremendous potential that the state of Chattisgarh has for agricultural development, the possible potential crops have been identified, for different situations in Chhattisgarh, they are given below.

Light unbunded soils (rainfed) : Maize, ragi, Blackgram, greengram, horsegram, grasspea, Sesame, groundnut, niger, castor, Colocasia, gemikand, thikhur, cassava, sweet potato, Turmeric, ginger and Vegetables.

Medium to heavy unbunded soils (rainfed) : Kodo, ragi, maize, Blackgram, pigeonpea, pigeonpea +soybean, Soybean, groundnut, sunflower Colocasia, gemikand, thikhur, cassava, sweet potato, Turmeric, ginger, Cotton, and vegetables during *kharif.*

In rabi : Wheat, Chickpea, Lathyrus, blackgram, Greengram, pea, lentil, cowpea, Sunflower, toria, linseed, Methi, coriander, karayat and Vegetables can be grown.

Cultivation on rice bunds : Pigeonpea, sesamum, blackgram, cowpea, beans, Colocasia, Lemon grass, Dinanath, sudan, napier, M.P. Chari and Fruit/ forestry trees

Bunded Rainfed (Medium to Heavy Soil)

Kharif : Rice and Pigeonpea on bunds, Soybean, Blackgram, Greengram, Pigeonpea and vegetables (removal of bunds and providing drainage for large scale cultivation)

Rabi : Wheat, Chickpea, pea, lentil, lathyrus, Safflower, linseed, mustard, Methi, Coriander and Karayat.

Irrigated Conditions (Varied Soil Types)

Kharif : Rice, maize, Blackgram, Greengram, Pigeonpea, Soybean, sunflower, castor, groundnut, All *kharif* vegetables, sugarcane, fodder, medicinal and aromatic plants.

Rabi : Wheat, maize, Mustard, sunflower, linseed, hybrid castor, groundnut, safflower, Pea, greengram, rajma, chickpea, lathyrus, cowpea, lentil, Onion, potato, tomato, brinjal, chilli, cucurbits cauliflower, cabbage and variety of *rabi* vegetables, Berseem, Lucerne grass, Sugarcane, Medicinal plants, Spices

Yield Gaps

Pulses remain unattractive to farmers. The widening gap between production and demand is reflected in the decrease in per capita consumption of pulses. Between 1972-73 and 2004-05 pulses consumption in rural areas reduced from 4.3 per cent of the food intake to 3.1 per cent; in urban areas it slipped from 3.4 per cent to 2.1 per cent, according to a Planning Commission report. The rural protein consumption came down by over 8 per cent during the same period, while urban consumption remained the same. According to Indian Council of Medical Research, 65 g of pulses is essential for an adult every day. However, people were actually consuming 40 g day^{-1}, which has now reduced to 30g day $^{-1}$ due to high prices.

Pulses have not been promoted like wheat and rice, though they are equally important for the Indian diet. Unfortunately, adequate attention has not given to research and technological interventions in pulses as in the case of cash crops. With the development of the seed-fertilizer-irrigation technologies for rice, wheat and maize, fertile lands were diverted to these crops and pulses were relegated to marginal pieces of land cultivated by marginal farmers. As irrigation improved in the past 10 years, pulses in Chhattisgarh, particularly in core pulse areas like North-West Chhattisgarh where the right climatic conditions for pulses exist, were replaced by wheat, paddy and sugarcane. This also affected the yield.

Only 10% of pulse growers use breeder/certified seeds. Even, these seed material is provided by Research Institutes and agencies like the National Seeds Corporation only after a few cycles. The disadvantages and risks involved in growing pulses are not compensated by the minimum support price (MSP) for pulses (pigeonpea), the MSP of Rs 2,3000 q^{-1} is higher than the MSP of Rs.1,000 q^{-1} for paddy. But while paddy yields about 3,000 kg ha^{-1}, pulses have a national average yield of 600 kg/ha. Even when market prices of pigeonpea were ruling above Rs. 100 kg^{-1}, the MSP was Rs. 3000 q^{-1}.

Nearly 400 improved varieties of pulses have been released for cultivation since the inception of the Coordinated Pulse Improvement Programme in 1967, but only 124 varieties are in the production chain and only a dozen are popular among farmers. In India, most of the research on pulses is confined to the Indian Institute Pulses Research (IIPR), and the AICRP on Pulses as well as the International Crops Research Institute for Semi Arid Tropics (ICRISAT).Very little biodiversity exists in case of pulses to develop desirable characteristics like high yield and resistance to pests. But even with available varieties and technologies, pulse production can be increased by at least 30 per cent. There is definitely a lot of scope for strengthening research, but even the available technology is not reaching the farmers. Private seed companies have also kept away from pulses because developing and distributing seeds are not economical.

Except pigeonpea, pulses do not have hybrids. Since pulses have been pushed to marginal farms, seed production has remained restricted to public-sector research organizations.

Bridging the Yield Gaps

In pulses, the yield potential being expressed consistently in research stations and demonstration plots through the first generation research efforts are yet to be realized on the farmers' fields. Substantial but untapped yield potential, locked up in already existing technologies, can bring substantial improvements in pulse production by narrowing the gaps. An aggressive approach towards transferring the technologies generated at research stations coupled with developmental policies to ensure timely availability of critical inputs is urgently needed for bridging the yield gaps.

In addition to the yield gaps, inter-regional and inter-seasonal disparities in yield realisation also exist. While very large area (37%) under various pulse crops fall under low productivity with lower than national average, mainly in Andhra Pradesh, Chhattisgarh, Karnataka, Orissa, Tamil Nadu and Rajasthan, there are states like Bihar, Madhya Pradesh, Punjab, Uttar Pradesh and West Bengal harvesting more than the national average. The major factors contributing to low yields are forced maturity due to rising temperature and terminal drought besides disease-pest syndrome, choice of inappropriate varieties, use of poor quality seeds and non-adoption of recommended cultural practices. In the low productivity regions, yield can substantially be raised by alleviating biophysical constraints like lack of water and unfavorable temperature through better management or by matching technologies to withstand the constraints without much reduction in yield. Even a modest gain of 100 kg ha^{-1} from 10 million ha low productivity area under pulses would have a far-reaching impact on the pulse supply in the country.

Major Constraints in Pulse Production in Chhattisgarh

- **Agro-ecological constraints**

  Unfavorable weather conditions

  - High rainfall (1200-1600 mm) especially for *kharif* mung, urd and pigeonpea
  - Continuous heavy rains at flowering (August)
  - Short span of winter and sudden rise of temperature in the month of February, disfavors to *rabi* crops
  - Susceptibility to environmental fluctuations

- **Abnormal soil conditions**
    - *Bhata* soils (*Entisols*) are gravelly with low water holding capacity and acidic in nature.
    - Unbunded black soils (*Vertisol*) have poor drainage.
    - Field preparation after paddy harvest in *Vertisol* is difficult.
    - Initial soil moisture during *rabi* sowing is poor.
    - Cultivation on marginal and wastelands.
- **Seed constraints**
    - Inadequate availability of good quality seed of improved varieties.
    - Faulty system of indenting and seed production.
    - High price.
    - Untimely supply
- **Varietal constraints**
    - Lack of high yielding, disease resistant varieties.
    - Lack of short duration, synchronous, photo-thermo insensitive varieties of summer Greengram and blackgram.
    - Narrow adaptability of cultivars.
    - Poor response to inputs specially high doses of fertilizer and irrigation.
    - Indeterminate growth habit.
    - Non-availability of genetic resistance particularly against yellow mosaic virus and pod borers in mung and urd; thrips in grass pea and lentil.
- **Biological constraints**
    - Highly prone to insect, pests and diseases.
    - Practically inadequate plant protection measures are adopted by farmers.
    - Severe problems of Diseases and Insects:

Mung & urd	:	Powdery mildew & yellow mosaic virus.
Grasspea	:	Powdery mildew & downy mildew
Lentil	:	Fusarium wilt, powdery mildew
Mung and urd	:	Hairy caterpillar (*Dycritia*) thrips, aphids and pod borer, whitely, jassids.
Lathyrus	:	Thrips, aphids, pod borers

- **Agronomical constraints**
  - Lack of irrigation
  - Heavy weed population especially in *kharif* season
  - Lack of post emergence herbicides.
  - Insufficient/Negligible use of organic and inorganic fertilizers
  - Sulphur deficiency in Rajnandgaon and Kawardha districts
  - Non-availability of granular single super phosphate
  - Water logging and soil moisture stress cause seedling mortality
  - *Utera* cultivation of Lathyrus
- **Constraints in use of cultures**
  - Less awareness about *Rhizobium* and phosphorus solubilizing bacteria (PSB) culture
  - Inadequate availability of efficient strains of bio-fertilizers.
  - Improper storage facility.
  - Unawareness of techniques of application.
- **Extension and training constraints.**
  - Development and extension workers are less exposed to pulse production technology.
  - Existence of communication gap.
  - Lack of strong linkages with developmental personnel for careful planning and implementation of the pulses productions programmes.
  - Lack of infrastructure (manpower, vehicles etc.) for transfer of technology.
  - More focus needed on training of farmers including farm women.
- **In adequate research infrastructure**
  - Basic research activities through biotechnology
  - On farm research
- **Socio-economic constraints**
  - Subsidiary status in farming system
  - Notions and beliefs which determine farmers response
  - Pulses are not major crops and have low production.

- **Post-harvest technology**
  - Improper post harvest technologies
  - Non-availability of suitable machinery for field preparation, harvesting and threshing.
  - Losses during storage
  - Losses during processing

Relay Cropping (*Utera* in Rice Fields)

Relay *(utera)* cropping refers to seeding/ planting of succeeding crop after flowering or at physiological maturity stage and before the harvest of the standing crop. It is analogus to a relay where one crop hands baton (land) to the next crop in quick succession. Farmers having assured irrigation facilities can take to relay cropping systems to augment their income. Such systems are not beyond the ingenuity of farmers who can adopt them keeping in view the local conditions. Deep-rooted pulses can be sown in cereals like wheat and rice with broadcast method a few days before harvest. The former crops tap nutrients and moisture from the lower layers while the latter from the upper layers without impairing soil health. The *utera/paira* method is a kind of relay cropping which is quite common in the states of Chhattisgarh and West Bengal where lentil, lathyrus, linseed seeds are broadcast in the standing crop of rice about 2-3 weeks before its harvest. This practice enables better and earlier establishment of crop seedlings because of an adequate availability of soil moisture which otherwise tends to be lost quickly once the rice is harvested. During *rabi*, lathyrus, urdbean and mungbean are sown with *utera* cultivation in standing rice. This practice is not only location-specific but also system-specific. This type of cultivation does not permit agronomic manipulations such as tillage, weeding, fertilizer and irrigation application. The cost of cultivation is low and net returns are high. The pulse crop production in rice-fallows is influenced by the duration of the rice variety. There are certain other constraints in *utera* system such as: (i) Varieties of rice are of long duration and photosensitive, (ii) Seedling mortality causes inadequate plant population, (iii) Rare use of fertilizers, (iv) Weeds infestation due to lack of post-emergence herbicides, and (v) Moisture stress at terminal phase. In the case of *utera* method, the net returns were three times higher over the sowing done after proper seedbed preparation. The higher yields of chickpea were possible due to timely sowing. Moreover, *utera* method requires lesser input cost.

Benefits of *Utera* System

- Judicious use of rainwater under rain fed conditions
- One of these best ways for taking double crop under rainfed conditions.

- Helps in increasing cropping intensity
- Farmers are kept engaged in *rabi* season
- Pulses and oilseeds are made available to farmers
- Availability of fodder for animals
- Adds to soil fertility
- Economic returns to farmers

Constraints of Low Production under *Utera* System

- Non-availability of good quality seed
- Low plant population per metre square
- Lack of proper soil moisture and nutrients.
- Poor weed control
- Untimely sowing of *utera* crops
- Stubble height of rice
- Seed treatment with fungicide and cultures are not followed
- Non adoption of plant protection measures
- Cattle grazing in *rabi* season
- Rainfall immediately after sowing of *utera* crop
- Accumulated water in the field after sowing of *utera* crop

Agronomic Points to be Considered in *Utera* System

- Early, medium and late duration rice should be planted as per availability of water in the field
- *Dorsa* (Alfisol) and *Kanhar* (vertisol) soils are suitable
- *Utera* crops perform better after *biasi* or drilled rice
- Lathyrus varieties like Ratan/Prateek/Mahateora should be sown @ 90 kg seed/ha with seed treatment with fungicide and *Rhizobium*, PSB culture
- *Utera* crops should be sown in the month of October.
- Water should be drained 12 hrs after sowing of improved varieties of lathyrus
- In absence of proper soil moisture, seed should be broadcasted after 6 hrs soaking of the seeds in water.

- Stubble height of 20-25 cm was found to be suitable for lathyrus as utera crop
- Besides recommended dose of fertilizers to rice, additional 20 kg P_2O_5 to rice gives better *utera* crop
- 2% urea spray, twice at 15 days interval, at the time of flowering
- Use of *Tauchi* hoe in transplanted rice once or twice helps *utera* crop
- Under *biasi* system, the field should be free from algae as it affects germination of *utera* crop
- *Utera* sowing should be done in the afternoon
- In absence of proper soil moisture in the field, *lakh* (lathyrus) perform better
- Water should be drained immediately after sowing, mung, urd or mustard as *utera* crop.
- Adoption of weed control and plant protection measures.

Precautions for Proper Germination of *Utera* Crops

Different *utera* crops require different soil water conditions for proper germination

Lathyrus

- For proper germination water should be drained 12-24 hrs after broadcasting the seeds.
- Draining water within 2 days leads to better germination

Linseed, Urd and Mung

- More water in the field in the time of sowing is not beneficial
- Seed rotting occurs in case of more water in the field, so these crops are sown just after draining water. Only a thin layer of water should be in the field so that seeds are not completely submerged. This condition generally occurs within a day after draining the water.

Lentil, Chickpea and Wheat

- More water at the time of sowing is harmful and leads to seed rotting, on the contrary, less water also leads to poor germination.
- Care should be taken that level of water in the field at the time of sowing should be such that it can be drained within 12 hrs of sowing
- Just after sowing, field should be open at different points in such a way that water is drained within 12 hrs of sowing.

- The portion of the field where draining of water is delayed, some extra seed should be broadcasted.

Rye-Mustard and Safflower

- Before sowing the seed should be soaked in water. Rye-Mustard for 12 hrs and safflower for 24 to 36 hrs.
- Seed should be dried in shade 2-3 hrs before sowing.
- Seed are light, so by soaking in water seeds become heavier and do not float in water
- Water should be drained immediately after sowing.

Research Work done on *Utera* in Chhattisgarh

Lathyrus

***Rhizobium*, Seed Rate and Nitrogen :** The results of field experiments in different parts of Chhattisgarh reported that use of 40 kg seed rate ha^{-1} of *Lathyrus* produced almost similar yield to 60 kg ha^{-1}, but 80 kg ha^{-1} proved inferior to either 40 or 60 kg seed ha^{-1}. Treatment with *Rhizobium* resulted significantly higher seed yield than no *Rhizobium*. Application of 20 and 30 kg N ha^{-1} proved at par and gave significantly higher seed yield than 0 and 10 kg N ha^{-1}. Economical returns also followed the same trend.

Sowing Technique : Data on sowing technique of *utera* crop (Lathyrus) on standing rice in district Mahasamund for more than two years revealed that during 2001-02, sowing of succeeding crop (Lathyrus) with minimum tillage after harvest of rice resulted in highest seed yield of Lathyrus and net returns from *utera* crop. Whereas, during 2002-2003, the highest seed yield of Lathyrus and net return from *utera* crop was noted under sowing of *utera* crop (Lathyrus) two weeks after flowering of rice. The highest total net return (Rs.15,663 ha^{-1}) was noted under rice- sowing of *utera* crop (Lathyrus) one week after flowering of rice.

In Rajnandgaon district, this experiment was conducted during second year (2002-03) only and the result indicated that sowing of *utera* crop (Lathyrus), 2 weeks after flowering recorded the highest seed yield of Lathyrus and net returns of *utera* crop (Lathyrus). The highest total net return (Rs.12,127 ha^{-1}) was also recorded under same system i.e. rice-sowing of *utera* crop (Lathyrus) two weeks after flowering of rice.

Productivity and Economic Viability : Productivity and economic viability of improved rice based *utera* cropping system was assessed at both Mahasamund and Rajnandgaon districts for two years. The data regarding comparison of farmer's practice of growing rice based *utera* vs. improved practices revealed that there was significant difference in grain yield of rice and Lathyrus at both Mahasamund and Rajnandgaon districts. The improved package of practice followed in rice gave 21% and 27% higher grain yield of rice over farmers' package of practice for rice at Mahasamund and Rajnandgaon districts, respectively. Similarly, improved *utera* cropping resulted 33% and 30% higher seed yield of Lathyrus over farmers' practice of *utera* cropping at Mahasamund and Rajnandgaon districts, respectively.

As regards to total net returns, improved *utera* cropping proved remunerative with net returns of Rs. 13,488 ha^{-1} and Rs. 18,749 ha^{-1} in comparison to traditional system with net return of 8,728 ha^{-1} and Rs. 11,418 ha^{-1} at Mahasamund and Rajnandgaon districts, respectively.

Adaptability of Legume Crops under *Utera* System : Performances of Lathyrus, chickpea and lentil were attractive among other pulses under *utera* system. Lathyrus gave maximum grain yield followed by chickpea and lentil. However, on the basis of economics, chickpea was more remunerative followed by lathyrus and lentil.

Suitable Varieties : Performance of different genotypes of pulse crops under *utera* system revealed that Pusa 24 (Grass pea), JG-315 (Chickpea) and RL-1 (Lentil) gave better yield as compared to other genotypes. On the basis of net return, JG-315 variety of chickpea was best followed by Pusa-24 variety of Grass pea. The Second alternative in order of merit was JG-74 variety of chickpea and RLK-1098 variety of Grass pea.

Rainfed Pulses

Performance of chickpea after early rice was best followed by lathyrus and linseed in relation to grain yield and chickpea equivalent seed yield. Rotovating twice with power tiller to a depth of 2-4 cm resulted in a better performance in relation to seed yield and chickpea equivalent seed yield as compared to conventional tillage with bullocks and tractor drawn cultivator.

Crop wise Production Technologies

Pigeonpea (*Cajanus cajan*)

Pigeonpea, commonly known as red gram and *arhar,* is a very ancient crop. India stands first in production of pigeonpea in the world and after chickpea; it is the second most important pulse crop in India. It is a rich source of protein and accounts for a major share of the protein requirement of the vegetarian population. The outer covering of its seed together with part of the kernel provides a valuable feed for milch cattle. The husk of pods and leaves obtained during threshing constitute a valuable cattle feed. Woody parts of the plant are used for fuel. It is a legume crop and, consequently, possesses valuable properties as restorer of nitrogen to the soil. In Chhattisgarh, it is mainly grown in Rajnandgaon, Kabirdham, Raipur, Durg, Surguja, Koriya, Bastar and Bilaspur districts. Pigeonpea is cultivated in 0.75 lakh hectare land of Chhattisgarh. Rice area under wide bunds (2.5 lakh ha) can be brought under pigeonpea having 1.5 t ha^{-1} productivity, which would add to the pulse production of the state by 3.75 lakh tonnes . (Table - 6 & Fig. 3).

Table 6 : Area, Production and Productivity of Pigeonpea in Chhattisgarh during *kharif* 2008

S.No.	Zones/ Districts	Area (000' ha)	Production (000' ton)	Productivity (kg/ha)	Suitable varieties
	CG Plains				
1	Raipur	8.96	3.45	385	
2	Mahasamund	2.50	1.00	399	
3	Dhamtari	2.62	1.00	380	
4	Durg	10.70	5.91	552	Asha, Rajeev-lochan, Laxmi, Prabhat. UPAS 120.
5	Rajnandgaon	21.32	12.17	571	
6	Kawardha	9.95	4.67	469	
7	Bilaspur	11.50	7.46	649	
8	Janjgir	1.71	1.12	656	
9	Korba	5.80	2.64	455	
10	Raigarh	4.60	2.37	516	
	Northern Hills				
11	Surguja	38.71	23.69	612	
12	Jashpur	9.12	5.44	596	Asha, Rajeev-lochan, Bahar, UPAS 120.
13	Korea	13.00	6.42	494	
	Bastar Plateau				
14	Jagdalpur	3.48	3.41	980	
15	Narayanpur	1.04	0.78	750	
16	Dantewada	6.51	4.52	695	Asha, Rajeev-lochan, No. 1048, Manak, UPAS 120.
17	Beejapur	0.21	0.16	742	
18	Kanker	2.36	1.62	682	
	Total for State		**154.09**	**87.83**	

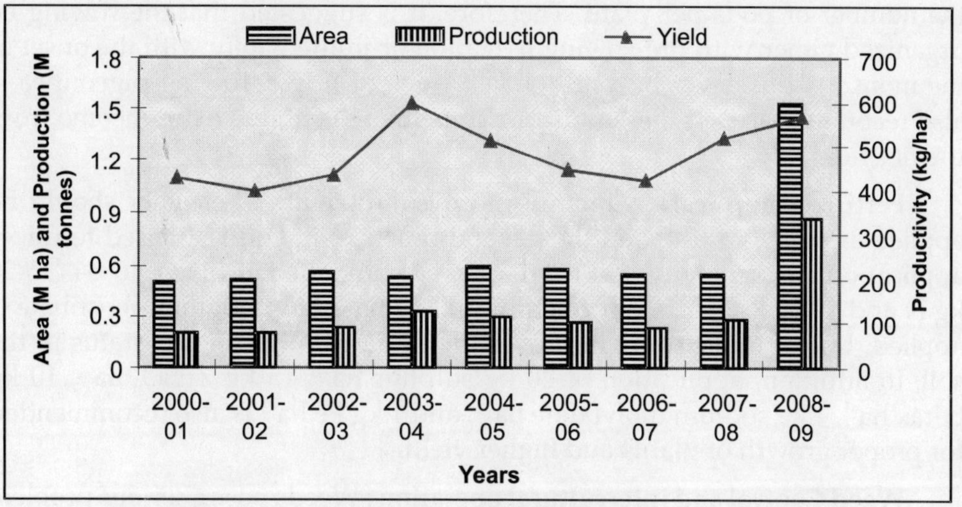

Fig. 3 : Area, Production and Productivity of pigeonpea in Chhattisgarh.

Major constraints in pegion pea cultivation : Lack of high yielding and hybrid varieties. Non availability of quality seeds of improved varieties. Occurrence of biotic stresses like sterility, insects etc. and Slow awareness of improved technologies.

Climatic requirement : Warm and humid climate is beneficial for better growth of pigeonpea. Dry and sunny days are essential for flowering and pod formation. This condition prevails in some districts of Chhattisgarh and, therefore, it is a potential crop for the state under upland heavy soil conditions.

Land preparation : A weed-free and firm seedbed ensures better plant stand and early vigour to the crop. This is achieved by opening the soil through soil-turning plough or disc harrow, followed by the cross harrowing or ploughing with *desi* plough on or before the onset of monsoon. Finally the seedbed should be planked and levelled. It is desirable to slope the land to provide excellent drainage. Sowing pigeonpea on ridges also allows it to tolerate excess moisture.

Seed rate : The seed rate of good quality of pigeonpea is 15 to 20 kg ha^{-1}. The row to row and plant to plant spacings for early pigeonpea and pre *rabi* crops are 45 cm and 15 cm, respectively, whereas for late pigeonpea these are 60 X 20 cm.

Sowing management : The optimum time of sowing of pigeonpea is between 25th June to 15th July. In rainfed areas sowing is done with the onset of the monsoon in *kharif*. However, sowing from the first week of June to the second week of July is considered best. Generally, delay in sowing beyond 15 July not only reduces the plant height and the total dry matter but also decreases the number of pods per plant. Therefore, it is suggested that the sowing be organized either with pre-sowing irrigation or immediately with the onset of the monsoon but never beyond the last week of July. Growing pigeonpea is also recommended on the bunds of rice fields which give extra income from the rice field.

Fertilizer application including micro-nutrients : Fertilizer should be applied after soil testing and the general practices of soil test based fertilizer application are commonly followed. The recommended fertilizer dose is 15-20 kg N and 40-60 kg ha^{-1}. Conveniently, 100 kg ha^{-1} diammonium phosphate is applied. Use of PSB culture improves the available phosphorus status in the soil. In addition, application of 20 kg sulphur ha^{-1}, 15 kg $ZnSO_4$ ha^{-1}, 10 kg borax ha^{-1}, 1 kg sodium molybdate ha^{-1}, and 1 kg Fe ha^{-1} is also recommended for proper growth of plants and higher yield.

Weed Control and intercultural operation : Weeds pose a serious problem during rainy season. As the early growth of pigeonpea is very slow, first 60 days is very critical in the life cycle of the plants. Two mechanical/hand weedings, one at 25-30 days and another at 45-60 days after sowing give excellent weed control. The pre-emergence application of alachlor @ of 3 liters ha^{-1} kills the germinating seedlings of weeds and keeps the field free from the weeds for the first 50 days. In case alachlor is not available, use fluchloralin 1 kg a.i. per hectare in 800-1000 liters of water as pre-planting spray. Pendimethalin or metolachlor @ 1-1.5 kg ha^{-1} dissolved in 500-700 liter water may be sprayed before sowing or within 72 hrs of sowing and/or germination. It should be well incorporated in the soil before sowing.

Water management : Being a deep-rooted crop, it can tolerate drought. For the crop planted in June, one or two pre-monsoon irrigations should be given as per requirement. After the start of monsoon, there is no need of irrigation but in case of prolonged drought during the reproductive period of growth, one or two irrigations may be needed. A pre-requisite for the success of arhar is proper drainage. Ridge planting is effective in areas where sub-surface drainage is poor. This provides enough aeration for the roots during the period of excess rainfall. During rainy season, water should not stand anywhere in the field.

Plant protection measures : Pigeonpea crop is attacked by many insect pests and diseases. Important pests which damage the crop are the pod-borer, pod-fly, plume-moth and hairy-caterpillar. Of the diseases, wilt, powdery-mildew and sterility-mosaic virus cause serious losses to the crop. Pod-borer (*Heliothis armigera*) causes considerable damage to pigeonpea, especially the early-maturing varieties. Spraying insecticides like Malathion (0.05%) provides good protection. Pod-fly (*Melanagromyza obtusa*) is a very serious pest of pigeonpea throughout the country. It causes extensive damage to the crop. The hairy-caterpillars are voracious feeders and defoliate the plants causing heavy losses to the crop. Dusting of chlorpyrifos 20 EC 40% @ 15 can control the pest incidence. To manage diseases follow the following strategies.

Vascular-wilt : Wilt caused by the fungus *fusarium udum* Bulter var. *cajani* Padwick, is characterized by gradual wilting of plants. Addition of manganese (100-200) to the infested soil prevents infection. Sanitary measures like rouging and burring of plants reduce disease incidence. As the fungus is soil-borne, rotation of crop for 4 to 5 years effectively reduces pathogen population and the extent of damage.

Stem-blight : Blight caused by the fungus *Phytophthora drechsleri* var. *cajani,* produces irregular to circular grey spots surrounded by brown lesions on the stem. Poor drainage and stagnation of rainwater in the fields during the monsoon need to be avoided. Application of potash fertilizer (K_2So_4) @ 25-50 kg ha^{-1} reduces the disease incidence.

Powdery-mildew : Sulphur dusting @ 30 kg/ha checks the spread of disease.

Sterility-mosaic : Mites can be controlled by spraying with 0.1% Metasystox. Removal of old and volunteer plants before sowing of crop would help in restricting the population of mites.

Intercropping System : Planting of castor and tall sorghum or maize varieties on borders conserves natural enemies, *viz.* predators of birds, etc. To facilitate spraying, short-stature crops, e.g. cowpea, Greengram, blackgram, fodder soybean, etc. may be grown in 1 m wide land after every 8-10 rows of pigeonpea. Soybean, black gram, mung, cowpea and groundnut are very common intercrops. Being deep rooted; the crop is very well suited for mixed cropping and intercropping with the shallow rooted ones. Intercropping with urd (Type-9), mungbean (Pusa baisakhi) and cowpea (Pusa do phasali) gives about 4-7 quintals per hectare additional yield without affecting yield of pigeonpea.

Chickpea (*Cicer Arietinum*)

Chickpea in the most important pulses crop of Chhattisgarh and occupies 2.98 m hectares with a production of 2.97 m tonnes. The average productivity of chickpea is 998 kg/ha. It is mainly grown in Durg, Rajnanadgaon, Kabirdham. Raipur and Bilaspur districts. (Table 7 & Fig. 4).

Table 7 : Area, Production and Productivity of Chickpea in Chhattisgarh (2008-09)

S.No.	Zones/ Districts	Area (000' ha)	Production (000' ton)	Productivity (kg/ha)	Suitable varieties
	CG Plains				
1	Raipur	18.60	18.69	1005	
2	Mahasamund	0.67	0.69	1032	
3	Dhamtari	8.85	9.99	1129	
4	Durg	90.87	107.68	1185	
5	Rajnandgaon	37.34	35.81	959	JG 11, JG 74, JG 315, Vaibhav, Shweta, BG 256, Annegiri, ICCV 10, JG 16, KAK 2
6	Kawardha	66.98	51.41	768	
7	Bilaspur	39.10	39.93	1021	
8	Janjgir	0.62	0.70	1132	
9	Korba	1.54	1.27	822	
10	Raigarh	3.00	3.16	1054	
	Northern Hills				
11	Surguja	16.41	14.77	900	
12	Jashpur	3.70	3.58	967	JG 16, JG 74, JG 315. KAK 2
13	Korea	3.60	2.40	666	
	Bastar Plateau				
14	Jagdalpur	2.54	2.79	1100	
15	Narayanpur	0.32	0.32	990	
16	Dantewada	0.20	0.20	990	Annegiri, ICCV 10, JG 74, JG 315, Vaibhav, Shweta, BG 256,
17	Beejapur	0.05	0.05	950	
18	Kanker	3.18	3.54	1113	
	Total for State	**297.57**	**296.97**	**998**	

The major constraints are :

- Non availability of quality seeds of improved varieties.
- Lack of suitable farm implements for field preparation and seeding particularly in rice-fallow system.
- High moisture stress at seeding and moisture stress with high temperature at grain filling.
- Incidence of disease like *Fusarium* wilt, dry root rot, stunt and pests like *Helicoverpa.*
- Lack of suitable drought and heat tolerant varieties suitable for rice based rainfed cropping.

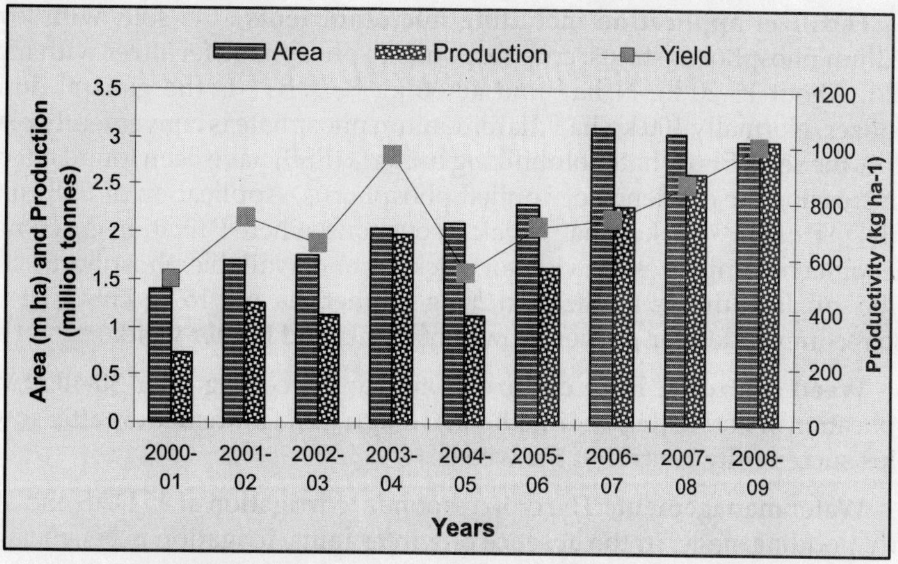

Fig. 4 : Area, production and productivity of chickpea in Chhattisgarh

Land preparation : Chickpea is highly sensitive to soil aeration. This imposes a restriction for its cultivation on heavy soils and calls for special care in seedbed preparation. A rough seedbed is required for gram. In case, it is taken after a *kharif* fallow, it would be desirable to go for a deep ploughing during the monsoon, as the same would help in larger conservation of rainwater in the soil profile for subsequent use by this crop. Very fine and compact seedbed is not good for chickpea. It requires a loose and well-aerated seedbed.

Seed rate : The crop is sown in rows, about 30 cm apart. Seeds are placed 7-10 cm deep in soil; the rate depends on seed size, which can vary from 60 to 100 kg ha^{-1}.

Sowing management : Rice is the dominant crop in Chhattisgarh state during *kharif* season and immediately after harvesting of rice, chickpea is sown. Second fortnight of October to second fortnight of November is the best time for gram sowing. Delay beyond this period results in a conspicuous reduction in yield. The crop may be sown by seed-drill or local plough at a row spacing of 30 cm. The seed should be placed 5-7 cm deep because the shallow sown crop is more liable to be damaged by wilt. Seed should be treated with 0.25 per cent Thiram before sowing followed by *Rhizobium* culture.

- *Chickpea under rainfed condition*: Suitable row spacing for planting of chickpea under rainfed condition is 30 cm x 10 cm spacing which produced more yield than by 20 x 10 or 45 x 10 cm.
- *Chickpea varieties for late planting*: Under late sown condition, chickpea variety RG-9218 performed best followed by BG-256, JG-315 and JG-74.

Fertilizer application including micro-nutrients : On soils with low to medium phosphorus status, crop responds to phosphatic fertilizer with higher yield. About 15-20 kg N ha^{-1} and 40-60 kg P_2O_5 ha^{-1} is the general dose of fertilizer. Normally 100 kg ha^{-1} diammonium phosphate is conveniently placed below the seed. Phosphate solubilizing bacteria (PSB) have been found effective in increasing the efficiency of applied phosphorus. Application of PSB culture with DAP or SSP @ 40 kg P ha^{-1} in calcareous soils, where P fixation is a common phenomenon, improves the yield of chickpea and available phosphorus status of the soil. In addition, application 20 kg sulpher/ha and 15 kg $ZnSO_4$ ha^{-1}, are also recommended for proper growth of plants and higher yield.

Weed control & inter culture : One hand-weeding after 30-40 DAS or application of herbicide pendimethalin @ 1 kg a.i./ha immediately after sowing helps successfully control of the weeds.

Water management : The crop responds to irrigation at 45 DAS and at the early podding stage. In the absence of winter rains, irrigation is beneficial.

Plant protection measures : Pod-borer is the major pest. It requires monitoring at weekly intervals at flowering and podding stages for eggs and larvae. Sex-pheromone traps (5 traps ha^{-1}) should be put in January-March. Economic threshold limit may be 1 larva/5 to 10 plants.

Mechanical practices : Collect and destroy the grown-up larvae.

Biological control : Spray nuclear polyhedrosis virus (NPV) @ 250 larval equivalents (LE)/ha on noticing the first-or second-instar larvae or eggs of *Helicoverpa armigera*. Apply 3 sprays at weekly intervals

Chemical control : Selective pesticides, *viz.* Monocrotophos (0.04%), and extract of neem-seed kernel (5%) or other neem-based formulations, like Nimbecidine, Repelin, Achook, Neemguard, etc., should be used as soon as economic threshold limit is crossed.

Wilt and root rot : Seed should be treated with fungicides, viz. Carbendazim+Thiram (1 g+2 g kg^{-1}) at sowing. Seed-dressing with bioagent Gliocladium virens @ 0.5 g kg^{-1} or Trichoderma viride @ 4 g kg^{-1} seed+fungicide Carboxin (Vitavex) @ 1 g kg^{-1} seed should be followed.

Ascochyta blight : Seed treatment with Hexacap (3 g kg^{-1}), Carbendazim (2 g kg^{-1} seed) or Calixin-M (Tridemorph 2 g kg^{-1} seed) is effective. Foliar sprays of Calixin-M (Thridemorph 0.1%) or Hexacap (0.02%) or Carbendazim (0.1%) may be adopted.

Intercropping system : Deep summer ploughing, clean cultivation, timely sowing (up to mid of November), and intercropping with linseed, mustard, wheat, coriander etc. minimizes damage by pod-borers and nematodes and provide good yield of main and intercrops.

Lentil (*Lens Culinaris*)

Lentil has a great promise as a pulse crop in the state as it gives higher price in market. It may serve as a good alternative to Lathyrus under *utera* cultivation. At present, the lentil occupies 28,000 ha with production of 16,300 tonnes and productivity of 734 kg ha^{-1} with positive growth rates. Generally it is a temperate season crop, but can be successfully grown in Kawardha, Rajnandgaon, Bilaspur, Koria, Durg and Dhamtari districts of the state (Table 8 & Fig. 5). A large area under rice fallows can be brought under large seeded lentil which will increase production and may also attract export.

Table 8 : Area, Production and Productivity of Lentil in Chhattisgarh during *kharif* 2008-09

S.No.	Zones/ Districts	Area (000′ ha)	Production (000′ ton)	Productivity (kg/ha)	Suitable varieties
	CG Plains				
1	Raipur	4.60	1.50	325	
2	Mahasamund	0.02	0.01	390	
3	Dhamtari	1.22	0.49	398	
4	Durg	4.70	1.69	359	JL 3, K-75, DPL 62, IPL 81, Lens 4076
5	Rajnandgaon	1.73	0.55	319	
6	Kawardha	1.40	0.38	271	
7	Bilaspur	1.74	0.55	316	
8	Janjgir	0.31	0.11	342	
9	Korba	0.35	0.12	347	
10	Raigarh	0.50	0.16	325	
	Northern Hills				
11	Surguja	3.04	1.01	331	
12	Jashpur	0.80	0.24	297	JL 1, JL 3, K-75, DPL 62, IPL 81
13	Korea	1.00	0.29	285	
	Bastar Plateau				
14	Jagdalpur	0.17	0.05	311	
15	Narayanpur	0.02	0.01	300	
16	Dantewada	0.02	0.01	290	
17	Beejapur	0.00	0.00	0	
18	Kanker	0.20	0.06	323	
	Total for State	21.82	7.20	330	

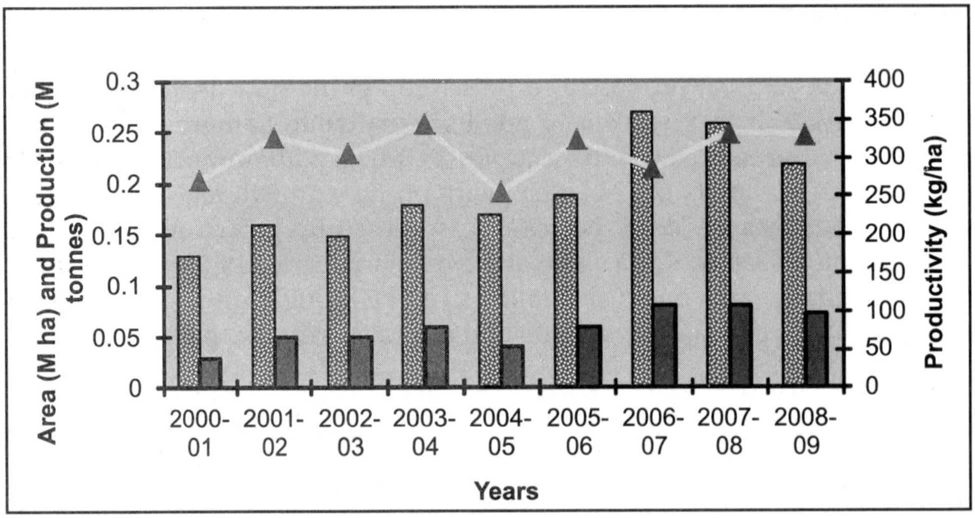

Fig. 5 : Area Production and Productivity of lentil in Chhatisgarh over the years.

The major constraints in lentil production are :

- Heavy infestation of wilt in local varieties. Therefore growing large seeded wilt resistant varieties such as JL 3, IPL 81, JLS 1 and L 4076.

  Foliar spray of 2% urea at flowering and pod formation stages is recommended
- Slow replacement of existing varieties by new improved varieties responsive to fertilizers.
- Occurrence of abiotic stresses like drought.
- Slow transmission of available technology.

Soil requirement and land preparation : Medium to heavy type of soil like *Alfisols* and *Vertisols,* are suitable for lentil cultivation. The soil should be made friable and weed-free so that seeding could be done at a uniform depth. On heavy soils, one deep ploughing followed by two to three cross harrowings should be sufficient.

Seed rate : The optimum seed rate for normal sown crop is 30-40 kg ha^{-1}. Seed rate should be increased to 50-60 kg ha^{-1} in case of late sowing.

Sowing management : Timely planting is the key factor for full realization of the yield potential of improved varieties of lentil. Middle of October is the most suitable time for sowing lentil. Delay in planting causes reduction in yield but the magnitude of reduction is large after 15th November. The reduction in yield could be minimized up to a certain extent by relatively closer spacing and use of higher seed rate. Sowing should normally be done in rows 30 cm apart.

This could be done either by using a ferti-seed drill or by seeding behind *desi* (country) plough.

Fertilizer application including micro-nutrients : Generally, lentil is grown without fertilizers and manures. In soils deficient in nitrogen and phosphorus, quite significant responses to application of starter nitrogen (20-25 kg ha^{-1}) and moderate level of phosphorus (50-60 kg ha^{-1}) have been obtained. Soil testing, therefore, becomes important to assess fertility status of the soil. If soil test is not possible, apply 20-30 kg Nitrogen, 50-60 kg P_2O_5 ha^{-1} in medium to low fertile soils as basal dressing. The deficiency can be rectified by spraying a solution of 0.5 per cent zinc sulphate and 0.25 per cent lime at the appearance of early symptoms. On soils where zinc deficiency is well established, apply about 25 kg zinc sulphate ha^{-1} before final discing of the field so that zinc sulphate is mixed well in the soil.

Weed control & inter culture : Two manual weedings, one at 25-30 days and another 45-50 DAS should be done. Weedicide like Basalin and Tribunil can be used safely for weed control. Basalin 0.75 kg a.i. ha^{-1} in 800-1000 litres of water as pre-planting spray may be used as an effective herbicide.

Water management : First irrigation should be given at 45 days of planting and second, if needed, at pod filling stage. Additional irrigations may affect the crop performance adversely.

Plant protection measures : The major disease of lentil is rust in Chhattisgarh plains and wilt in Northern hills. Use of resistant varieties is helpful in controlling the disease. Pests are not a major problem in lentil. In some years aphid population rises due to rise in temperature during January or February, and causes heavy damage. Spray of Monocrotophos (0.04%) is effective for aphid control.

Intercropping system : It is intercropped with barley and mustard. It is also grown as an intercrop in autumn-planted sugarcane. In north-eastern plains, it is also grown as *utera* crop after rice. The seeds are broadcast in standing crop of rice just before harvest.

Lathyrus (*Lathyrus Sativus*)

Lathyrus (grasspea) is an important pulse crop in the state that occupies 4.28 lakh ha with an annual production of 1.58 lakh tonnes. In the last five years, both area and production showed positive growth rates (area from 2.51 lakh ha in 2000-01 to as high as 5.5 lakh ha in 2006-07 and production from 1.09 lakh tonnes to 2.9 lakh tonnes). The productivity fluctuated between 369 and 605 kg ha^{-1} (Table 9 & Fig.6). *Lathyrus* is grown solely as rainfed crop under *utera* cultivation with poor management practices. The ban on trade of crop continues to be a major stumbling block in expansion of area under the crop, despite its unique ability to perform better even under severe moisture stress conditions. Following technological interventions are recommended:

- Growing high-yielding varieties with low beta n-oxalyl amino alanine (BOAA) content such as 'Ratan' and 'Prateek'.
- Foliar application of 2% urea at flowering and pod-development stage.
- Seed priming and broadcasting of seed 5 to 10 days before the harvest of rice.

Table 9 : Area, Production and productivity of Lathyrus in Chhattisgarh (2008-09)

S.No.	Zones/ Districts	Area (000' ha)	Production (000' ton)	Productivity (kg/ha)	Suitable varieties
1	Raipur	85.45	41.36	484	
2	Mahasamund	12.06	5.51	457	
3	Dhamtari	26.30	14.44	549	
4	Durg	95.83	42.64	445	Ratan, Prateek, Mahateora, Pusa 24
5	Rajnandgaon	25.63	9.97	389	
6	Kawardha	9.15	3.56	389	
7	Bilaspur	85.47	54.87	642	
8	Janjgir	20.49	13.15	642	
9	Korba	9.02	5.79	642	
10	Raigarh	6.02	3.14	522	
	Northern Hills				
11	Surguja	1.30	0.59	456	Ratan, Prateek, Mahateora, Pusa 24
12	Jashpur	3.78	1.78	470	
13	Korea	1.20	0.58	486	
	Bastar Plateau				
14	Jagdalpur	2.67	1.42	530	
15	Narayanpur	0.03	0.02	500	
16	Dantewada	0.00	0.00	0	Ratan, Prateek, Mahateora, Pusa 24
17	Beejapur	0.00	0.00	0	
18	Kanker	5.03	1.74	346	
	Total for State	**389.43**	**200.56**	**515**	

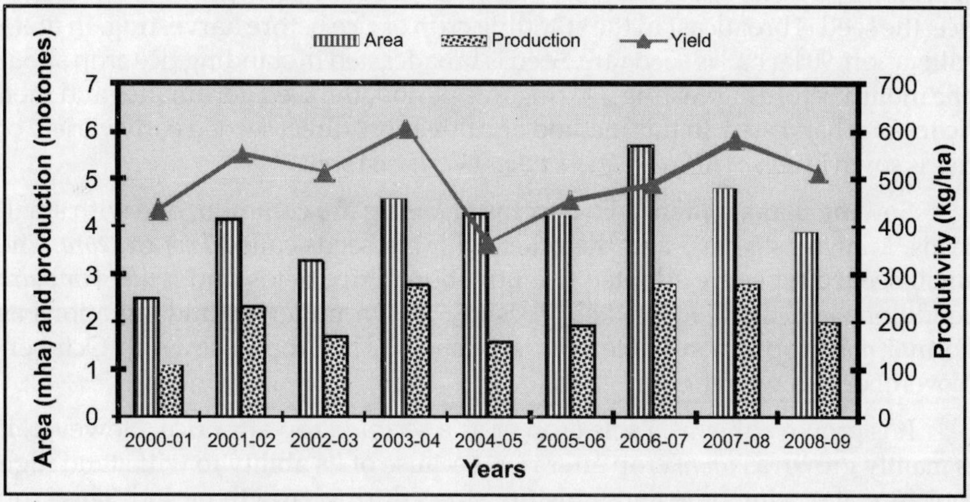

Fig. 6 : Area, production and productivity of lathyrus in Chhattisgarh (2000-01 to 2008-09)

In Chhattisgarh, lathyrus is a traditionally popular drought-tolerant crop grown for food and feed because of its easy cultivation, its relative resistance to drought, floods, and salinity, insect attacks and its good yield of tasty protein-rice seeds with about 30 g per 100 g edible seeds. Despite its tolerance to drought, lathyrus is not affected by excessive rainfall and can be grown on land subject to flooding. It has unique ability to sustain under harsh conditions of excessive moisture/standing water at sowing and moisture stress stages of crop growth. Hence, it is grown as relay/*utera* crop with rice. When other crops fail due to adverse conditions, it may be the only available food source for the weaker section of the population and sometimes as survival food in times of drought induced famine. In sprite of all its quality, lathyrus has remained neglected on account of its neurotoxin principle B-N-oxalyl-L--1, diaminopropionic acid (ODAP syn- BOAA) which is considered to be a probable causative factor of neuro lathyrism in human beings.

The major constraints in lathyrus production are :

1. Lack of High yielding varieties with low ODAP content
2. Non-availability of quality seed of improved varieties,
3. Lack of thrips and downy mildew resistant varieties and
4. Lack of dual purpose high yielding varieties for solo and *utera* (relay) cropping coupled with low ODAP.

Land preparation : The crop is sown in conserved moisture after rains. Two or three ploughings are given soon after harvesting of *kharif* rice.

Seed rate : The seed is broadcast or sown in rows using a seed rate of 50-75 kg ha^{-1}. The recommended row-to-row distance is 30 cm. Before sowing, the seed is treated by 3 g Thirum or Bavistin @ 1.5 g kg^{-1} of seed. When grown as *utera* after rice, the seed is broadcast in the standing crop of rice before harvesting. In utera cultivation, 90 kg ha^{-1} is seed rate. Seed is broadcasted in standing rice crop about one month before harvesting. During this period, the seed germinates and then rice crop is harvested. In this method of cultivation, direct seeded early variety of rice is sown in *dorsa* (*Alfisols*) and *kanhar* (*Vertisols*) soils.

Sowing management : Two forms of *khesari* are common, one with small seeds, known as *lakhodi* and the other with big seeds called *lakh* or *teora*. The small-seeded types are adapted to cultivation in rice fields under *utera* or *paira* conditions and the bold-seeded types are grown under upland situations as normal *rabi* crops, in suitable crop sequences. The crop is sown in October-November as a winter crop.

It is grown either as a sole crop or as a second crop after rice. However, it is mainly grown as *utera* crop after rice, because of its ability to withstand high moisture at sowing time and moisture stress during growth period. This crop is unique in tolerance to adverse environmental conditions, tolerating not only drought but also water logging.

Fertilizer application including micro-nutrients : Fertilizers @ 20 kg N and 40 kg P_2O_5 ha^{-1} are recommended; 20 kg sulphur ha^{-1} or 100 kg diammonium phosphate ha^{-1} may also be applied.

Weed control & inter culture : Hand-weeding 30-40 DAS is beneficial.

Water management : First irrigation should be given at 40 days of planting and second, if needed, after flowering stage. Additional irrigations may affect the crop performance adversely.

Plant protection measures : Of all the diseases, rust, downy mildew and powdery mildew causes substantial losses to the crop. Two or three sprays of Dithane at 8 days interval reduce the rust infection considerably. Spraying with a copper fungicide prevents incidence of downy mildew. Sulphur dusting @ 30 kg ha^{-1} of spraying with wet table sulphur @ 5 kg ha^{-1} checks the spread of powdery mildew. Aphids are also serious pests of this crop. Spraying with nicotine sulphate (40% at 1:600 dilution in water by volume) or Malathion 0.03% checks the infestation of aphids.

Greengram (*Vigna radiata*)

In Chhattisgarh, greengram (mungbean) occupies 16.59 thousand hectares and contributes to 6.06 thousand tonnes in pulse production with productivity of 365 kg ha^{-1} (Table 10 and Fig.7). The grains (whole or split) are used as *dal* or to make flour. The straw and husk are used as fodder for cattle. The germinated grains are also used as sprouts. Unlike other pulses, it does not produce heaviness or flatulence.

Table 10 : Area, Production and Productivity of greengram in Chhattisgarh (during *kharif* & *rabi* (2008-09)

S. No.	Zones/ Districts	Area (000' ha)		Production (000' ton)		Productivity (kg/ha)		Suitable varieties
		Kharif	*Rabi*	*Kharif*	*Rabi*	*Kharif*	*Rabi*	
	CG Plains							
1	Raipur	2.16	8.20	0.58	1.42	269	173	
2	Mahasamund	5.60	0.49	1.72	0.09	307	178	
3	Dhamtari	0.10	0.81	0.03	0.16	293	203	
4	Durg	0.45	0.72	0.13	0.14	294	197	
5	Rajnandgaon	6.75	0.99	1.59	0.22	235	218	BM 4, HUM 1, HUM 12, HUM 16, Pusa Vishal, K 851
6	Kawardha	0.48	0.03	0.11	0.01	237	256	
7	Bilaspur	3.85	0.31	1.10	0.08	287	249	
8	Janjgir	0.60	0.22	0.18	0.06	30	258	
9	Korba	1.50	0.04	0.39	0.01	257	254	
10	Raigarh	2.15	5.60	0.81	1.74	379	311	
	Northern Hills							
11	Surguja	1.18	0.13	0.29	0.04	246	314	
12	Jashpur	1.07	0.10	0.40	0.03	372	311	Pusa Vishal, HUM1, HUM 12, K 851
13	Korea	1.02	0.05	0.26	0.01	257	256	
	Bastar Plateau							
14	Jagdalpur	0.80	0.13	0.37	0.04	462	346	
15	Narayanpur	0.13	0.03	0.06	0.01	463	350	
16	Dantewada	1.61	0.78	0.64	0.23	399	290	HUM1, HUM 12, K 851, TARM 1.
17	Beejapur	0.38	0.61	0.15	0.18	401	290	
18	Kanker	5.12	1.12	2.19	0.34	427	302	
	Total for State	**34.95**	**20.36**	**11.01**	**4.80**	**315**	**236**	

The major constraints are :

- Breakdown of resistance against YMV.
- Lack of High yielding, medium maturing varieties suitable for delayed sowing.
- Lack of *Maruca* resistant varieties and varieties with synchronous maturity suitable for *kharif* planting.

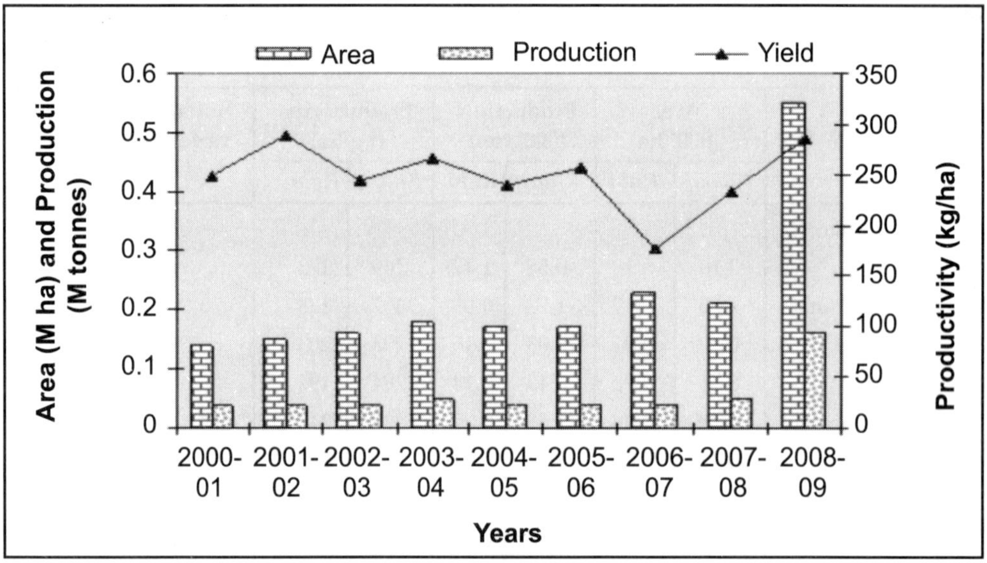

Fig. 7 : Area, production and productivity of Greengram in Chhattisgarh (2000-01 to 2008-09)

Land preparation : Field is prepared by giving two or three cross harrowinges followed by planking. The field should be well leveled and free from weeds. For summer season crop, give a pre-irrigation immediately after harvesting of *rabi* crop. When the field comes in condition, prepare it by giving two to three ploughings by local plough or harrowings. Each ploughing/ harrowing should be followed by planking to the field level and minimize the loss of moisture by evaporation from the soil surface.

Seed rate : During *kharif* season, 12-15 kg seed ha^{-1} should be sown in rows 45 cm apart while during *rabi* and summer seasons 20 kg seed ha^{-1} should be sown in rows 30 cm apart. Sowing can be done behind the local plough or with the help of seed drill. Before sowing seed should be treated with thiram or captan at the rate of 2.5 g kg^{-1} of seed. It is also desirable to treat the seed with suitable *Rhizobium* culture as well, if the crop is being taken in the field for the first time or after a long period.

Intercropping Systems : In *kharif*, intercropping with maize, sesame, pigeonpea and cotton is common.

Sowing management : The sowing time is the first week of March for the spring Greengram, last week of March to first fortnight of April for the summer crop. In *kharif*, seed is sown with the onset of monsoon, preferably first week of June to mid-July. In the event of the late onset of monsoon, it can be sown by July end. In *rabi*, seed can be sown from September to December or January, depending on the time of rice harvest and availability of irrigation.

Kharif **:** In *kharif* season, sowing is done with the onset of monsoon in the second fortnight of July.

Rabi **:** It should be sown in October-November.

Summer : Moong is grown in summer season too under irrigated conditions. It is sown after the harvest of sugarcane, wheat, potato etc., and sowing is done from middle of March to middle of April.

Fertilizer application including micro-nutrients : Apply 15-20 kg nitrogen and 40 to 50 kg P_2O_5 ha^{-1} at the time of sowing. If available, drill 100 kg of diammonium phosphate to meet the requirement of both nitrogen and phosphorus. The fertilizer should be applied by drilling either at the time of sowing or just before sowing in such a way that they are placed about 5-7 cm below the seed. In addition 20 kg sulphur ha^{-1}, 0.5 kg manganese ha^{-1}, 4 kg zinc ha^{-1}, 1 kg copper ha^{-1}, 0.6 kg boron ha^{-1}, 0.1 kg molybdenum ha^{-1} and 1 kg iron ha^{-1} improve the yield considerably.

Weed control & inter culture : The control of weed during 30-40 DAS is very beneficial. The young crop is given 1 hand weeding or hoeing. Use of weedicide pendimethalin or alachlor @ 1.0-1.5 kg a.i./ha immediately after sowing is also useful.

Water management : For rainy season crop, irrigation is not needed but drainage is very important. There should be adequate drainage in the field because this crop is sensitive to water logging conditions. For *rabi* and summer crops, five to six irrigations may be given. The field should be irrigated when soil appears to be deficient in moisture. During summer, because of high temperature and low relative humidity, more irrigation is needed as compared to *kharif* crop. The first irrigation should be given about 20-25 DAS. The subsequent irrigations should be given at an interval of 12-15 days. No irrigation should be given when the crop is in full bloom stage.

Blackgram (*Vigna mungo*)

Blackgram or urd differs from other pulses in its peculiarity of attaining a some what mucilaginous pasty character when ground up with water, giving additional body to the mass. It is also fried to serve as a savory dish. Urd dal is also used in the preparation of *'halwa'* and *'imarti'*. It is used as a nutritive fodder especially for milch cattle. It is also used as a green manuring crop. Urd plant possesses a deep root system which binds soil particles and thus prevents soil erosion. The details of area, production and productivity are given in (Table 11 & Fig.8).

The major constraints in blackgram productivity are :

1. Lack of high yielding, medium maturing varieties suitable for delayed sowing.

2. Breakdown of resistance against YMV.
3. Lack of varieties with sympodial plant habits and suitable for *kharif, rabi* and summer planting.

Table 11 : Area, Production and Productivity of Urdbean in Chhattisgarh (*kharif* & *rabi* 2008-09)

S. No.	Districts/ Zones	Area (000' ha)		Production (000' ton)		Productivity (kg/ha)		Recommended varieties
		Kharif	*Rabi*	*Kharif*	*Rabi*	*Kharif*	*Rabi*	
	CG Plains							
1	Raipur	5.87	1.20	2.31	0.28	391	230	
2	Mahasamund	14.95	0.01	6.84	0.00	457	234	
3	Dhamtari	0.20	1.17	0.10	0.32	442	275	
4	Durg	3.37	1.31	1.00	0.32	298	246	KU 96-1, Barkha, TU 94-2, TAU 2, Pant U 30.
5	Rajnandgaon	19.33	3.46	6.60	0.83	341	241	
6	Kawardha	8.50	0.03	3.03	0.01	355	323	
7	Bilaspur	7.10	0.24	1.90	0.06	267	233	
8	Janjgir	1.76	0.23	0.50	0.05	282	238	
9	Korba	7.60	0.06	1.79	0.01	235	181	
10	Raigarh	20.00	1.12	5.44	0.20	272	183	
	Northern Hills							
11	Surguja	24.80	0.16	9.50	0.04	383	260	
12	Jashpur	30.30	0.00	8.24	0.00	272	0	TU 94-2, TAU 2, Pant U 30
13	Korea	9.40	0.10	3.60	0.03	383	255	
	Bastar plateau							
14	Jagdalpur	23.10	0.21	10.10	0.06	437	285	
15	Narayanpur	3.47	0.00	1.52	0.00	438	0	
16	Dantewada	2.81	0.60	0.99	0.16	354	265	Pant U 4, TU 94-2, Barkha
17	Beejapur	0.40	0.27	0.15	0.07	358	255	
18	Kanker	9.03	0.70	3.59	0.27	398	390	
	Total for State	**191.99**	**10.87**	**67.20**	**2.72**	**350**	**250**	

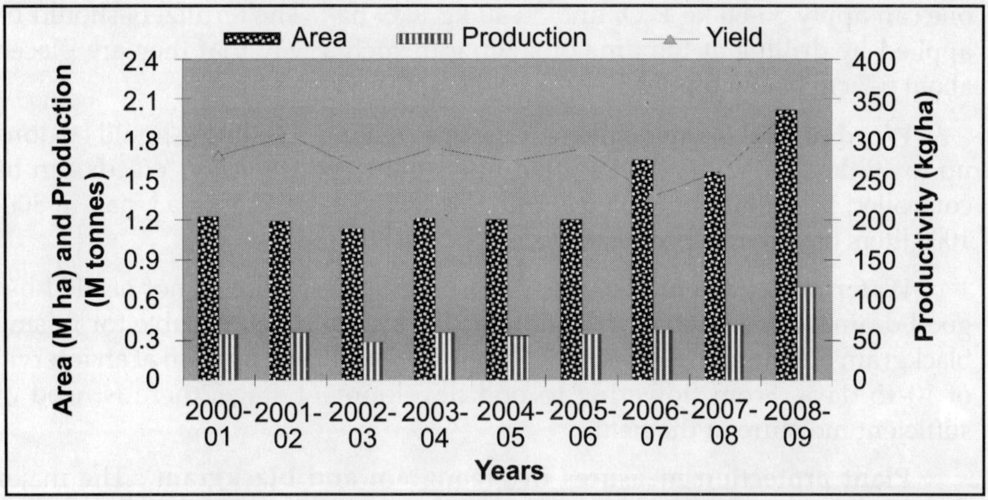

Fig. 8 : Area, production and productivity of Blackgram in Chhattisgarh (2000-01 to 2008-09)

Land preparation : Field is prepared by one ploughing with mouldboard plough and followed by two to three harrowings followed by planking. The field should be well leveled and free from weeds. For summer crop, after harvest of *rabi* crops, give two to three harrowings followed by planking. To ensure good germination, a pre-sowing irrigation should be given.

Seed Rate : During *kharif* season 12-15 kg seed is sufficient for sowing in one hectare. The crop should be sown in furrows opened at a distance of 45 cm.

In summer : About 20-25 kg seed is required for sowing in one-hectare area. Sowing should be done in furrows opened at a distance of 20-25 cm. Seed drill could be used for this purpose. Before sowing, seed should be treated with Thiram at the rate of 2.5 g kg^{-1} of seed. Seed should also be inoculated with suitable *Rhizobium* culture, if urd is being taken for the first time in the field or after a long duration.

Sowing management : In *kharif* season sowing is done with the onset of monsoon in later part of June or early part of July.

In summer : The sowing could be done form the third week of February to first week of April. Sowing done thereafter, yields low due to rains at the time of maturity. Urdbean is planted under flat bed system where the plant population is very often reduced due to accumulation of water in the field. By adopting raised bed system, water stagnation can be avoided in field.

Fertilizer application including micro-nutrients : Being a leguminous crop, blackgram needs a small quantity of nitrogen for early growth period on those soils which are poor in organic matter. Such soils should get about 15-20 kg nitrogen ha^{-1} as a starter dose. However, phosphatic and potassic fertilizers should be applied as per soil test values. If soil test facilities are not available, one can apply 50-60 kg P_2O_5 and 30-40 kg K_2O ha^{-1}. The fertilizers should be applied by drilling at the time of sowing in such a way that they are placed about 5-7 cm below the seed.

Weed control & inter culture : One or two hand weedings should be done up to 40 days of sowing depending upon the weed intensity. Weeds can be controlled by the use of chemical too. Use fluchloralin @ 1 kg a.i. ha^{-1} in 800-1000 liters of water as pre-planting.

Water management : For rainy season crop, irrigation is not needed but good drainage is essential. Irrigation facilities should be available for raising blackgram during summer season. The crop should get irrigation at an interval of 10-15 days. From flowering to pod development stage, there is need of sufficient moisture in the field

Plant protection measures in greengram and blackgram : The major diseases of Greengram and blackgram are yellow mosaic virus, Cercospora leaf spot, powdery mildew and Macrophomina blight. The integrated pest management strategy consists of cultural, mechanical, biological and chemical control measures.

Cultural practices : Deep ploughing, early sowing and timely irrigation are good for crop health. In disease prone areas, resistant varieties should be used.

Mechanical practices : Hairy caterpillars have a definite egg-laying pattern in masses. The first stage of larvae remains restricted to leaves of plants where eggs were laid. Collect and destroy them.

Biological control : Natural enemies and birds reduce the pest population.

Chemical control : For control of thrips, stem fly, jassids and whitefly, apply phorate or Carbofuran granules in soil before sowing @ 1 kg a.i. ha^{-1}. Seed treatment with phorate or Disulfotone @ 2% weight basis is also effective. Spraying with Monocrotophos (0.05%) or Quinalphos (0.05%) controls the leaf defoliators and hairy caterpillars. Dusting with 5% Malathion @ 35-40 kg ha^{-1} is also effective against hairy caterpillars.

Intercropping Systems : In *kharif*, intercropping with maize, sesame, pigeonpea and cotton is common.

Horsegram (*Macrotyloma uniflorum*)

Horsegram or *kulthi* is extensively cultivated in dry areas of Chhattisgarh. It is also drought tolerant and hardy. It is also grown to provide feed and fodder for cattle and horses. Horse gram possesses medicinal properties also. In Chhattisgarh, horsegram occupies 32.13 thousand hectares with a production of 13.49 thousand tonnes. The average productivity is 420 kg ha^{-1}. (Table 12 & Fig. 9).

The major constraints are :

- Lack of high yielding varieties suitable for Maize based cropping sequence and semi-*rabi* cultivation.
- Lack of bushy plant types coupled with responsiveness to fertilizers.

Table 12 : Area, Production and Productivity of horsegram in Chhattisgarh during *kharif* & *rabi* 2008-09

S. No.	Districts/ Zones	Area (000′ ha)		Production (000′ ton)		Productivity (kg/ha)		Recommended varieties
		Kharif	*Rabi*	*Kharif*	*Rabi*	*Kharif*	*Rabi*	
	CG Plains							
1	Raipur	1.21	2.41	0.42	0.65	347	268	
2	Mahasamund	0.00	0.25	0.00	0.07	400	268	
3	Dhamtari	0.15	0.15	0.06	0.05	393	316	
4	Durg	0.20	0.47	0.07	0.13	362	283	
5	Rajnandgaon	0.42	4.23	0.14	1.17	326	277	VZM 7, Deobhog local
6	Kawardha	0.28	0.00	0.00	0.00	0	0	
7	Bilaspur	3.00	0.50	1.03	0.14	344	270	
8	Janjgir	0.00	0.00	0.00	0.00	374	0	
9	Korba	0.74	0.00	0.23	0.00	313	0	
10	Raigarh	3.40	3.51	1.16	0.74	340	210	
	Northern Hills							
11	Surguja	10.20	0.00	3.35	0.00	328	0	
12	Jashpur	3.25	0.00	1.09	0.00	334	0	AK 21, BK 1
13	Korea	4.75	0.00	1.56	0.00	328	0	
	Bastar Plateau							
14	Jagdalpur	15.84	9.77	6.37	3.63	402	372	
15	Narayanpur	2.85	0.00	1.15	0.00	404	0	
16	Dantewada	0.31	3.02	0.10	0.69	333	228	Bastar White and Bastar Black, BK 1, VZM 7, AK 21
17	Beejapur	0.46	0.58	0.15	0.13	335	228	
18	Kanker	8.76	1.31	3.89	0.47	444	360	
	Total for State	**55.82**	**26.20**	**20.77**	**7.86**	**372**	**300**	

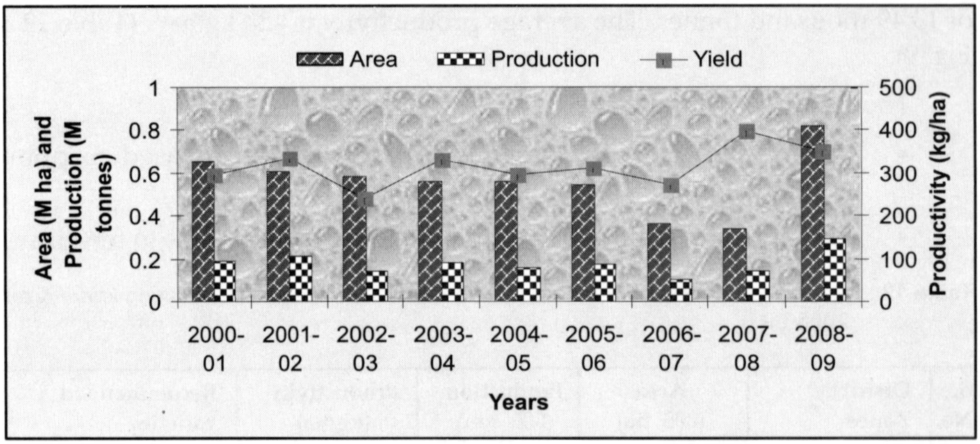

Fig. 9 : Area, production and productivity of horsegram in Chhattisgarh (2000-01 to 2008-09)

Land preparation : The crop needs minimum field operations. Only 1 or 2 ploughings followed by planking create the desired seedbed.

Seed rate : The seed is generally broadcast. It is also sown in rows with the help of seed drill. When drill-sown, the spacing between rows is about 30 cm. The seed is covered immediately after sowing with a blade harrow. The normal seed rate is 25 kg ha^{-1} for grain and 40 kg ha^{-1} for fodder.

Sowing management : The main season for sowing horse gram is late August to November. As a fodder crop it is sown during June to August. In fields, where groundnut and pigeonpea are grown, horse gram is sown in between rows of pigeonpea after the harvest of the groundnut crop.

Fertilizer application including micro-nutrients : The crop normally receives no fertilizers. A basal dose of 10 kg N and 25 kg P_2O_5 ha^{-1} promotes good crop growth and increases the yield of grain and fodder.

Weed control & inter culture : One hoeing or weeding at an early stage is beneficial.

Water management : First irrigation should be given at 45 days of planting and second, if needed, at pod filling stage. Additional irrigations may affect the crop performance adversely.

Plant protection measures : Pod-borers cause damage to the crop. In the early stage of crop growth, grasshoppers, leaf-eating caterpillars and aphids also infest the crop. These pests can also be controlled by spraying of malathion 0.07%.

Intercropping Systems : It is grown as a pure crop as well as intercrop with pigeonpea, sesame or niger.

Field pea (*Pisum sativum*)

It is a high-yielding pulse crop. Unripe pods are used as green vegetable and dry seeds are used for *dal* purpose and *chat* after boiling. Other parts are used for feeding cattle. It is cultivated on 0.42 lake hectare area particularly in Northern hills of Chhattisgarh (Table 13 & Fig. 10). Field pea is getting popularity in the state as it offers high remuneration if comes to fruiting early in the season as then it can be sold at a premium price in local market.

The major constraints in pea cultivation are : Slow replacement of existing varieties by new improved tall and dwarf varieties responsive to fertilizers., Susceptibility to powdery mildew, wilt and rust., lack of dual purpose varieties (Seed and green pod) and lack of resistant varieties to *Helicoverpa.*

Table 13 : Area, Production and Productivity of Fieldpea in Chhattisgarh in 2008-09

S. No.	Districts/ Zones	Area (000′ ha)	Production (000′ ton)	Productivity (kg/ha)	Recommended varieties
	CG Plains				
1	Raipur	7.30	2.47	338	
2	Mahasamund	1.12	0.38	340	
3	Dhamtari	2.77	1.05	380	
4	Durg	1.25	0.44	349	
5	Rajnandgaon	1.49	0.58	390	Ambika, Paras, Subhra, Rachna, JP 885, Aparna (HFP 4), KPMR 400
6	Kawardha	0.50	0.14	276	
7	Bilaspur	4.05	1.41	348	
8	Janjgir	0.91	0.34	371	
9	Korba	0.74	0.20	271	
10	Raigarh	3.50	1.26	361	
	Northern Hills				
11	Surguja	7.08	3.07	434	Ambika, Rachna, JP 885, Aparna (HFP 4), KPMR 400
12	Jashpur	1.48	0.52	351	
13	Korea	4.00	1.41	352	
	Bastar Plateau				
14	Jagdalpur	1.78	0.53	300	
15	Narayanpur	0.14	0.05	360	
16	Dantewada	0.04	0.01	359	Ambika, Paras, Rachna, JP 885, Aparna (HFP 4), KPMR 400
17	Beejapur	0.02	0.01	325	
18	Kanker	4.59	1.51	330	
	Total for State	**42.76**	**15.39**	**360**	

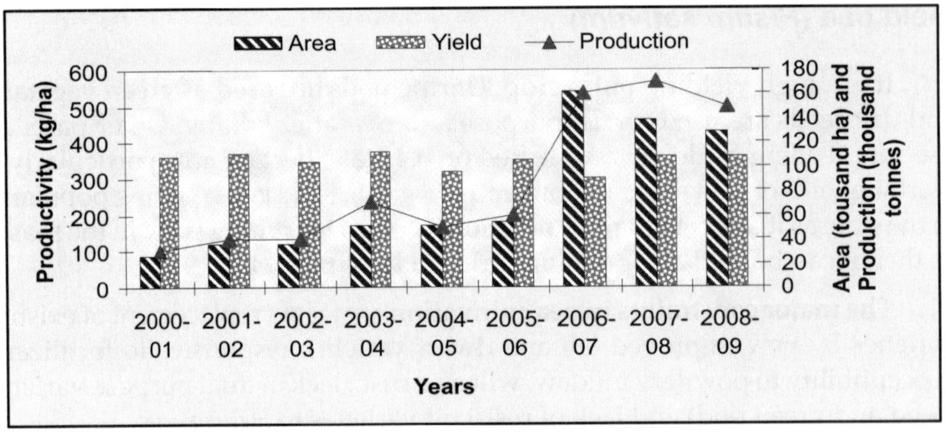

Fig. 10 : Area, production and productivity of Fieldpea in Chhattisgarh

Land preparation : Field pea needs relatively better seed-bed. After harvest of *kharif* crops, the land should be ploughed deep with soil-turing plough and then prepared with 2-3 harrowings or ploughings, followed by planking. In rice fields, efforts must be made to break clods and attain friable structure. Optimum moisture must be ensured for proper germination.

Seed rate : Second fortnight of October is the optimum time of sowing of field peas in north Indian conditions. Sowing after October results in drastic yield reduction. Seeds are sown at the rate of 60-80 kg ha^{-1} in rows 30 cm apart. Before sowing, seeds should be treated with Thiram at the rate of 0.25 per cent. In case of early maturing dwarf varieties, the crop should be sown in rows 20 cm apart and seed @ 100-125 kg ha^{-1}. In late maturing and taller varieties, row spacing of about 30 cm is considered to be optimum. The seed rate should be reduced to 75-80 kg ha^{-1}.

Water management : In the absence of winter rains, irrigations are to be given at 40-45 DAS and pod formation stages. In the absence of winter rains, 1 or 2 irrigations would be required; the first irrigation at the preflowering stage and the second, at the pod formation stage are required

Plant protection measures : Field-pea is attacked by a variety of insect pests, like leaf-minor, semi-looper, aphids, thrips, pod-borer and leaf-eating caterpillar; and diseases, like powdery-mildew, wilt, foot-rot, downy-mildew, etc.

Insect Pests

Leaf-minor [*Phytomyza horticola* Gour.(=atricornis)], Semi-looper (*Plusia orichalceu* Fb.) : Spraying with 10% chlorpyrifos as soon as the symptoms appear controls the leaf-minor. In case of severe infestations, 0.25% Methy1 Parathion spray is recommended.

Aphids (*Aphis craccivora Koch.*, A. *gossypii* Gl.) : Spraying the crop with nicotine sulphate (40%) at 1:600 dilutions in water by volume or malathion 0.03% reduces the incidence. About 650-850 liters of spray solution and 25-40 kg of dust are required per hectare.

Thrips [*Caliothrips indicus* (Bagnall)] : For the control of thrips, the crop may be sprayed with 1,250 g DDT (50 WP) in 650 liters of water per hectare and the spraying may be repeated after 2 weeks, if necessary.

Pod-borer (*Etiella zinckenella Treit*). Dusting BHC 10% or DDT 10% @ 20-25 kg ha^{-1} or spraying with a mixture of 0.05% Diazinon and 0.25% DDT twice during pod-formation stage, minimizes the incidence of pod-borer.

Leaf-eating caterpillar [*Spodoptera exigua* (Hubner), *Diachrysia orichalcea* (fabricius)]. Dusting of BHC @ 10% or DDT 10% @ 20-25 kg ha^{-1} is very effective in controlling the pest.

Diseases

Powdery-mildew (*Erysiphe polygoni* DC.). Seed treatment with 0.5% Benomy1 gives protection to the seedlings up to 60-65 days after emergence. Four sprays of Sulfex (Wet table sulphur) @ 3 kg/ha in 1,000 liters of water at intervals of 7 days starting immediately at the appearance of disease, controls powdery mildew effectively.

Wilt and foot rot. Treatment of seeds with organo-mercurial compounds, like Ceresan, Captan or Bavistin @ 2.5 to 3.0 g ha^{-1} of seed, is recommended.

Downy mildew : The disease is favoured by a temperature of 15^0C and heavy dew, and can be controlled by spraying with a copper fungicide as a preventive measure.

Intercropping system : Field pea is more adapted than chickpea crop for late sown conditions as well as high input conditions. With these stipulations, it could replace chickpea in crop sequences and intercrops especially as pea-rice, pea-cotton and pea-summer moongbean.

Future Thrust

- The land remains fallow after harvest of *kharif* and can be used for growing pulses, grain legumes and food legumes. Since paddy grows in flooded conditions, the soil has adequate moisture even after irrigation has been stopped. Even if the moisture content is less, the legumes being drought-resistant and deep-rooted can take-in the required moisture from the soil. Farmers can harvest water in community farm ponds and irrigate the pulses during moisture stress.
- To make pulses profitable, the cost should be brought down by developing short-duration and pest-resistant varieties.

- The cost can also be brought down by eliminating middle men in supply of certified seeds.
- Development of integrated pest management and integrated nutrient management techniques which will stabilize yields at levels 20 to 60 per cent higher than normal. Pulses are also extremely sensitive to heat and cold. Developing tolerance to abiotic stresses like excessive moisture condition, heat and extreme cold in pulses, may be the thrust areas for future research.
- Identification of varieties, which are suitable for double cropping system even under rainfed conditions.
- Pulses are rich in protein, so they are an easy target for pests. They are also slow growing crops as they need a lot of heat energy to break down the protein molecules required for growth. Therefore, development of varieties having resistance to biotic stresses such as pod borer and *Fusarium* wilt should a priority area.
- Rice area under wide bunds (2.5 lakh ha) can be brought under pigeonpea (cv. *Asha*) having 1.5 tonnes / ha productivity, which would add to the pulse production of the state by 3.75 lakh tonnes .

References

Ahlawat, I.P.S. and Shivakumar, B.G. (2002). Cropping systems research in pigeonpea. In Pulses for Sustainable Agriculture and Nutritional Security. Proceedings of National Symposium (Eds M. Ali, S.K. Chaturvedi and S.N. Gurha), Indian Society of Pulses Research and Development. pp. 23-32. Kanpur.

Ali, M. and Kumar,S. (2009) Problems and prospects of pulses research in India. Indian farming 50 (8), 4-13

Ali, Masood and Kumar Shiv (2005). Current status and Prospects of pulse production (Eds) Singh G, Sekhon, H.S.and Kolar, J.S. In Pulses.

Ali, M. (2001). Research focus on pulses in 21st century. In Souvenir of National Symposium on Pulses for Sustainable Agriculture and Nutritional Security, Kanpur: Indian Society of Pulses Research and Development. pp 24-30.

Ali, M. and Mishra, J.P. (2000). Nutrient management in pulses and pulse based cropping systems. Fertilizer News 45 (4), 57-69.

Ali, M., Singh, N.B. and Kumar, S. (2002). Crop diversification through pulses for nutritional security in new millennium. In proceedings of the International Symposium on Food, Nutrition and Economic Security through Diversification in Sugarcane Production and Processing Systems, pp. 117-123, Lucknow: ASTI/IISR.

Annual Report (2007-08) AICRP on MULLARP

Status Report (2001-05) of AICRP on MULLARP.

Jeswani, L.M. and Baldev, B. (1990). Advances in Pulse Production Technology. ICAR, New Delhi

Rabi faslon ki krishi karya mala. (2009). Extension bulletin, IGKV Publication.

Saxena, K.B. (2002). Heterosis breeding in pulses-problems & prospects. In Proceedings of National Symposium on Pulses for Sustainable Agriculture & Nutritional security. (Eds) Kanpur: Indian Society for Pulses Research & Development. pp. 61-72.

❑❑❑

6

Gujarat

B.S. Patel, R.R. Patel, S.M. Patel and I.C. Patel
AICRP on Integrated Farming Systems, Sardarkrushinagar Dantiwada Agricultural University, Sardarkrushinagar - 385 506 (Gujarat)

Gujarat grows more than half a dozen pulse crops on 7.84 lakh hectares area with an annual production of 6.09 lakh tonnes and a productivity of 777 kg/ha. Amongst these, pigeonpea, chickpea, greengram, blackgram, cowpea, mothbean and clusterbean are the important pulse crops grown in the state. Mothbean, greengram and clusterbean are the predominant pulse crops grown in arid region of North Gujarat, though other pulses are also grown on a sizable area. Despite this distinction of being the largest producer, the availability of pulses is far from satisfactory which is evident from the fact that per capita availability has not kept pace with the population increase and has declined to merely 28 g/capita/day as against the mandated minimum and optimum quantity of 80 and 104 g/capita/day, respectively, as per WHO standards. As a result, with 11 per cent of the world's import, it is also the largest importer of pulses. In Gujarat, the condition is still dismal (23 g/capita/day), though there are ample chances to improve the situation by inclusion of short duration pulse crops in cropping systems. Pulses are the cheapest source of protein and minerals (Ca, Fe, Cu & Mo) and compliment cereal proteins excellently particularly for lysine. Rajmash, though it is an unconventional pulse crop grown in North Gujarat, has the yield potential of around 35 q/ha. Similarly, fieldpea is also a unconventional potential crop having yielding ability of around 40 q/ha. Hence, these new crops should be introduced in Gujarat as catch crops. Moreover, the farmer's should be encouraged to cultivate recommended high yielding varieties of each and every pulse crop, through transfer of technology.

Trends of Pulses Production

Gujarat has diversified agriculture with good entrepreneur business-minded farming community with excellent policy support from the government. Gujaratis as a community, are god fearing and predominantly vegetarian. Agriculture, in general, hovers around water availability and off late water resources have improved with the linking of rivers and ambitious water recharging programmes like *Sujalam Sufalam*, construction of check dams as well as irrigation water availability from Sardar Sarovar Project etc. Still pulses that comprise about 9 per cent of the total cultivated area are predominently grown as rain fed crop in *kharif* season. Gujarat is 7th in acreage and 8th in production. Productivity hovers around 7 to 8 q/ha that is more than the national average but far less than USA and China that harvest double the yield around 15 q/ha. The area under pulses is around 0.8 m ha. The area has shown an increase particularly after 1980s and increase in production that at present is around 0.6 million MT, is primarily attributed to increase in area. Despite increase in area, the pulses have been marginalized to very poor growing situations. The major reason is Bt cotton cultivation that has replaced pulses. The major concern is to reduce cereals to pulses ratio that has come down to 15:1 from 5:1 in 1950s. Pigeonpea followed by mungbean, chickpea, urdbean and mothbean are the major crops though cowpea, Indian bean and horsegram are also cultivated in some pockets. Clusterbean is cultivated in around 0.2 m ha and Gujarat is the only state that has both sizable acreage and productivity of the crop. Limitation in the expansion of agricultural land suggests that increase in gross cropped area can come from multiple cropping. The state has well developed pulses and clusterbean industries. Marketing is through APMCs.

Existing Cropping Systems

In traditional cropping systems practiced in Gujarat, there is yield stability at lower level but adaptable to favourable and adverse agro-climatic conditions prevailing over period of time. These cropping systems are adopted by the farmers considering the market price, family and fodder requirement of the milch and draft animals. The existing cropping systems followed in Gujarat by the farmers are as under.

North Gujarat

- Pearl millet-Mustard
- Cluster bean-Wheat
- Castor-Fallow
- Cotton-Fallow

Middle Gujarat

- Pulses-Mustard/Wheat
- Tobacco-Fallow
- Paddy-Wheat
- Cotton-Fallow

South Gujarat

- Paddy-Summer Paddy
- Paddy-Summer ground nut
- Sugarcane-Sugarcane ratoon (2 Years)- Banana- vegtables (2 Years)

Saurashtra

- Groundnut-Wheat
- Groundnut-Onion/Garlic/Coriander
- Cotton alone
- Groundnut-Pigeonpea/Castor relay cropping

The above mentioned different cropping systems are widely adopted in different parts of Gujarat because of different prevailing favorable agro-climatic conditions and socio-economic situation of the farmers but as stated earlier, the yield level is low. Therefore, it is the need of the day to standardize and develop adaptable alternative efficient cropping systems in existing agro-climatic conditions with high yield potential, system productivity, system profitability, water profitability and sustainability under available natural resources at the farmer's disposal.

Efficient Alternative Cropping Systems

The major soil groups of Gujarat range from sandy to clayey and, hence due to this variability, crops are as diverse as pearlmillet- low water requiring crop to paddy- a very high water requiring crop. Therefore, as far as land capability classification is concerned, the majority of soil comes under land capability class IInd and III rd which require moderate management practices for taking good production. Therefore, alternative cropping sequences identified in crop diversification/intensification experiments which are suitable as per land capability classification and resource position of the farmers of Gujarat state are as follows.

The efficient alternative cropping systems recommended for North, Middle and South Gujarat as well as Saurashtra region of Gujarat are as under:

North Gujarat

- Groundnut-Potato-Summer Pearl millet
- Castor-Summer Greengram
- Castor-Summer fodder Sorghum
- Greengram/Cowpea-Mustard- summer Pearl millet
- Pearl millet-Mustard + Lucerne

Middle Gujarat

- Tobacco-Summer Pearl millet
- Paddy- Wheat/ *Rabi* Castor
- Cotton – Summer Pearl millet/ Summer Greengram
- Paddy- Lucerne fodder- Lucerne seed production after two cuts

South Gujarat

- Paddy- *Rabi* Castor
- Paddy- Summer groundnut/Greengram
- Paddy- Sorghum- Sorghum Ratoon
- Paddy- Fennugreek vegetable/Onion- Summer Okra Vegetable

Saurashtra

- Groundnut-Wheat
- Groundnut-Garlic/Coriander
- Groundnut-Onion-Summer Greengram
- Groundnut-Pigeonpea/Castor relay cropping

These cropping sequences, if adopted by the farmers will be very helpful for increasing productivity, soil fertility status, nutrient profitability, water profitability, system productivity and profitability, employment generation etc.

Yield Gaps

The yield gap of different crops was computed based on the yield of a crop in efficient cropping system/research station and average yield of Gujarat State as depicted in Table 1.

Table 1 : Yield gap of important pulse crops grown in Gujarat

Crop	Maximum yield(kg ha^{-1})	Average yield (kg ha^{-1})	Yield gap (kg ha^{-1})	Yield gap (%)
Greengram	1417	453	964	213
Cowpea	1261	427	834	195
Blackgram	1322	671	651	97
Mothbean	1310	232	1078	465
Clusterbean	879	350	529	151
Pigeonpea	2111	987	1124	114
Chickpea	2250	659	1591	244
Field pea	2266	New introduction	–	–
Rajmash	2716	New introduction	–	–

Source : Final forecast report of area, production and productivity, Directorate of Agriculture, Gujarat state (2008-09)

Yield gap of 964, 834, 651, 1078, 529, 1124 and 1591 kg/ha was observed in greengram, cowpea, blackgram, mothbean, clusterbean, pigeonpea and chickpea, respectively, in Gujarat state. If yield gap is considered in percentage terms, the gap ranged from 97 per cent in blackgram to as high as 465% in mothbean. This indicated that there is more scope of yield improvement in mothbean followed by chickpea and greengram.

Improved Varieties of Pulse Crops

Improved varieties of pulse crops with their characteristics are given in Table 2.

Table 2 : Improved varieties of pulse crops with their characteristics

Pulse crop/ Varieties	Days of maturity	Av. yield (kg/ha)	Remarks
Pigeonpea			
Gujarat tur-100	145-155	1500-2000	Bushy, determinate and white seeded.
Gujarat tur-1	125-130	5000-6000 green pods	Spreading, nondeterminate, white seeded, recommended for vegetable as green pods.
Gujarat tur-101	130-140	2000-2500	Spreading, nondeterminate, white seeded.
Banas	135-145	2100-2300	Spreading, nondeterminate, white seeded.
Gujarat tur hybrid-1	135-145	2800	Semi-spreading, nondeterminate, white seeded, 87% dal recovery.
Chickpea			
Dahod yellow	100-105	2000	For rain fed condition, susceptible to wilt.
ICCC-4	115-120	2800	For irrigated condition, resistant to wilt.

Contd.

Gujarat Chickpea-1	100-105	2500	For irrigated condition, resistant to wilt, 30% higher yield over Dahod yellow.
Gujarat Chickpea-2	90-95	2500	For rain fed condition, specially for Bhal region.
Greengram			
K-851	65-70	1200-1400	Erect type, for *kharif* and summer, synchronized maturity, resistant to YMV.
GM-3	60-65	1100-1200	Spreading type, for summer season, green lustrous seed, resistance to YMV.
GM-4	61-68	1200-1500	Erect type for *kharif* and summer, synchronized maturity, green bold seeded.
Blackgram			
T-9	70-75	1000-1200	Semi erect, pods in bunch, blackish seeds.
TPU-4	75-80	1200-1600	Pods in bunch
Gujarat Urd-1	75-80	1500-2000	Greenish black seeds, more fuzz on pods, 24% higher yield than T-9.
Mothbean			
GMO-1	80-85	500-700	Spreading type, dual purpose indeterminate, brown seeded.
GMO-2	65-70	800-1000	Semi spreading type, indeterminate, brown seeded.
Cowpea			
GC-3	70-75	800-1000	Erect type, indeterminate, seed color light white, for *kharif*.
GC-4 color	67-70	900-1100	Spreading type, indeterminate, bold white seed, for *kharif* and summer, dual seed and vegetable purpose.
GC-5 brown,	65-70	1000-1300	Erect type, indeterminate, seed color light for *kharif* and summer, suitable for inter cropping, resistant to moisture stress.
Clusterbean			
Gujarat guar-1	100-110	800-1000	Spreading type, indeterminate, creamy white seed, medium resistant to bacterial blight, high gum per cent, synchronized maturity.
Gujarat guar-2	100-105	1000-1200	Spreading type, indeterminate, attractive pink seeds color, synchronized maturity, 31% gum content, 17% higher yield than GG-1, resistant to bacterial blight and root rot.
Rajmash			
HUR-137	105-110	2000	Purple red seed color, synchronized maturity.
Gujarat Rajma-1	98-105	2500	Purple red seed color, synchronized maturity.
Field pea			
DDR-23 (Extra early)	65-70	1100	Extra early suited as pure as well as intercrop.
HFP-4	105-110	2000	Medium height without vine.
Indianbean			
Kadva val 125-36	150	600-700	Grown after paddy in conserved moisture.
Gujarat val-1	135-140	900-1000	Grown after paddy in conserved moisture

Crop-wise Production Technologies

The crop wise production technologies are highlighted in Tables 3 to 6.

Table 3 : Production technologies for short duration *kharif* season pulse crops

Cultural operations	Greengram	Cowpea	Black gram
Land preparation	Pulses needs fairly deep, well drained soil which should be free from salt. Land is prepared by 2-3 cross harrowing followed by planking. The seedbed must be free from stubbles and weeds. Apply 5 tonnes FYM/ ha before land preparation for better moisture conservation and microbial activity.		
Seed rate	18 kg/ha	15-20 kg/ha	18 kg/ha
Sowing management			
• **Seed treatment**	Seed treatment with Carbendazim, *Rhizobium* and PSB	Seed treatment with Carbendazim, *Rhizobium* and PSB	Seed treatment with Carbendazim, *Rhizobium* and PSB
• **Time**	3rd week of June to 2nd week of July	3rd week of June to 2nd week of July	3rd week of June to 2nd week of July
• **Spacing**	30 cm x 10 cm	45 cm x 20 cm	30 cm x 10 cm
• **Gap filling**	10 DAS	10 DAS	10 DAS
• **Thinning**	15-20 DAS	15-20 DAS	15-20 DAS
Fertilizer application	20-40-0 NPK kg/ha	20-40-0 NPK kg/ha	20-40-0 NPK kg/ha
Weed control and interculturing	1 kg/ha Pendimethalin as PE. Interculturing at 20 and 40 DAS	0.75 kg/ha Pendimethalin as PE. Interculturing at 20 and 40 DAS	1 kg/ha Pendimethalin as PE. Interculturing at 20 and 40 DAS
Water management	Life saving irrigation if dry spell is prolonged		
Plant Protection measures	For sucking pest, spray 0.03% dimethiate. For the control of pod borer, spray metasystox 0.03%. To control powdery mildew disease spray Carbendanzim 50WP @ 0.07 % twice at 15 days interval after appearing disease.	For sucking pest, spray 0.03% dimethiate. For the control of pod borer, spray Metasystox @ 0.03% To control powdery mildew disease spray Carbendanzim 50WP @ 0.07 % twice at 15 days interval after appearing disease.	For sucking pest, spray 0.03% dimethiate. For the control of pod borer, spray metasystox @ 0.03%. To control powdery mildew disease spray Carbendanzim 50WP @ 0.07 % twice at 15 days interval after appearing disease.

Table 4 : Production technologies for short and long duration *kharif* season crops

Cultural operations	Mothbean	Clusterbean (Long duration *kharif*)	Pigeonpea
Land preparation	Crops need fairly deep, well drained soil which should be free from salt. Land is prepared by 2-3 cross harrowing followed by planking. The seedbed must be free from stubbles and weeds. Apply 5 tonnes FYM/ha before land preparation for better moisture conservation and microbial activity.		
Seed rate	15-20 kg/ha	15-20 kg/ha	15-20 kg/ha
Sowing management			
• **Seed treatment**	Rhizobium and PSB	Seeds soaking in 250 ppm streptocycline for 15 minutes, *Rhizobium* and PSB.	Rhizobium and PSB
• **Time**	3rd week of June to 2nd week of July	3rd week of July to 2nd week of August	On set of monsoon (3rd week of June to 2nd week of July)
• **Spacing**	45cm x 10cm	45cm x 10cm	60 cm x 20cm (early), 90 cm x 20cm (mid late)
• **Gap filling**	10 DAS	10 DAS	10 DAS
• **Thinning**	15-20 DAS	15-20 DAS	15-20 DAS
Fertilizer application	20-40-0 NPK kg/ha	20-40-0 NPK kg/ha	25-50-0 NPK kg/ha
Weed control and interculturing	Interculturing at 20 and 40 DAS	1 kg/ha Pendimethalin as PE. Interculturing at 20 and 40 DAS	1 kg/ha Pendimethalin as preemergence. Interculturing at 20 and 40 DAS
Water management	Life saving irrigation if dry spell is prolonged		
Plant protection measures	For sucking pest, spray 0.03% dimethiate. For the control of pod borer, spray metasystox @ 0.03%	For sucking pest, spray 0.03% dimethiate. For the control of bacterial blight, spray streptocyclin.	For sucking pest, spray 0.03% dimethiate. For the control of pod borer, endosulfan @ 0.07%

Table 5 : Crop-wise production Technologies for *rabi* season crops

Cultural operations	Rajmash	Chickpea	Field pea	Indian bean
Land preparation	Land is prepared by 2-3 cross harrowing followed by planking. The seed bed must be free from stubbles and weeds.			
Seed rate	120-140 kg/ha (Rainfed)	60 kg/ha (Irrigated)	50-60 kg/ha	40 kg/ha
Sowing management				
• **Seed treatment**	Bavistin/Thirum 2-3 g/kg, *Rhizobium* (Raj 5) and PSB each 30 g/kg	Bavistin/ Thirum 2-3 g /kg, *Rhizobium* (IC-86) and PSB each 30 g /kg	*Rhizobium* and PSB each 30 g/kg	*Rhizobium* and PSB each 30 g/kg
• **Time**	2nd week of November	3rd week of Oct. (Rainfed), to 1st week of Nov. (Irrigated)	2nd week of Nov. to 3rd week of Nov.	Last week of October
• **Spacing**	30 cm x 10 cm	30 cm (Rainfed) to 45 cm (Irrigated)	30 cm to 45 cm	45 cm to 60 cm
• **Gap filling**	10 DAS	10 DAS	10 DAS	-
• **Thinning**	15-20 DAS	15-20 DAS	15-20 DAS	-
Fertilizer application	120-60-0 NPK kg/ha	20-40-0 NPK kg/ha, 25 kg/ha ZnSO4 in NG	20-40-0 NPK kg/ha	20-40-0 NPK kg/ha
Weed control and interculturing	1.0 kg/ha Pendimetheline as PE. Interculturing at 20 and 40 DAS	1.0 kg/ha Pendimethalin as PE. Interculturing at 20 and 40 DAS	1.0 kg/ha Pendimetheline as PE. Interculturing at 20 and 40 DAS	HW and IC at 30 DAS
Water management	Pre sowing irrigation 20 DAS , 35 DAS45 DAS, 55 DAS65 DAS	Pre sowing irrigation 30 DAS, 45 DAS, 65 DAS	Pre sowing irrigation20 DAS, 35 DAS,50 DAS, 65 DAS	Branching (25-30 DAS), Flowering (50-55 DAS), Pod development (70-75 DAS)
Plant Protection measures	For control of sucking pest, spray 0.03% dimethiate.	For control of sucking pest, spray 0.03% dimethiate. For control of pod borer and green larvae spray Dimethoate 0.03%.	For control of sucking pest, spray 0.03% dimethiate.	For control of sucking pest, spray 0.03% dimethiate. For control of pod borer and green larvae spray Dimethoate 0.03%.

Table 6 : Crop-wise production Technologies for summer season crops

Cultural operations	Greengram	Cowpea	Clusterbean
Land preparation	Land is prepared by 2-3 cross harrowing followed by planking. The seedbed must be free from stubbles and weeds.		
Seed rate	20-22 kg/ha	20-22 kg/ha	15-18 kg/ha
Sowing management			
• **Seed treatment**	*Rhizobium* and PSB	*Rhizobium* and PSB	Seed soaking in 250ppm solution of streptocycline, *Rhizobium* and PSB.
• **Time**	3rd week of February to 2nd week of March	3rd week of February to 2nd week of March	2nd week of February to 1st week of March
• **Spacing**	30 cm x 10 cm	45 cm x 15 cm	45 cm x 15 cm
• **Gap filling**	10 DAS	10 DAS	10 DAS
• **Thinning**	15-20 DAS	15-20 DAS	15-20 DAS
Fertilizer application	20-40-0 NPK kg/ha	20-40-0 NPK kg/ha	20-40-0 NPK kg/ha
Weed control and interculturing	1 kg/ha Pendimethalin as pre-emergence. Interculturing at 20 and 30 DAS	0.75 kg/ha Pendimethalin as pre-emergence. Interculturing at 20 and 30 DAS	1 kg/ha Pendimethalin as pre-emergence. Interculturing at 20 and 40 DAS
Water management	Pre sowing irrigation 20 DAS 35 DAS 45 DAS 55 DAS	Pre sowing irrigation 20 DAS 35 DAS 45 DAS 55 DAS	Pre sowing irrigation 25 DAS 40 DAS 55 DAS 70 DAS
Plant protection measures	For control of sucking pest, spray 0.03% Dimethoate, Metasystox or Dimecron. To control pod borer, Dimethoate (0.03%) should be sprayed at 50 per cent flowering.	For control of sucking pest, spray 0.03% Dimethoate. To control pod borer, Dimethoate (0.03%) should be sprayed at 50% flowering.	For control of sucking pest, spray 0.03% Dimethoate.

Pulses Based Sequential Cropping

The cropping system is a sub system of the farming system. System means a set up of elements/components that are interrelated and interact among themselves. The objective of any cropping system is efficient utilization of all resources *viz.*, land, water and solar radiation, maintaining stability in production and obtaining higher net returns. The produce obtained per unit resource in a unit time is higher.

Space concept relates to increasing the intensity in rain fed areas by way of building up crops/other agricultural allied components in vertical dimension, the land equivalent area can be increased. Thus, by making use of these time and space dimensions either in irrigated or in rain fed areas within specified time span, say a year and unit area of land, productivity is sought to be increased by repeated and/or intensified cropping.

Considering the above facts, the following technologies were developed for optimum utilization of natural resources to increase the productivity in different zones of Gujarat.

North Gujarat

Castor-green Gram : Farmers of North Gujarat Agro-climatic zone growing crops in light textured soil and assured irrigation facility are advised to follow castor-summer greengram with recommended doses of fertilizers for getting higher yield and net returns.

Chickpea-pearlmillet : The *kharif* pearlmillet growing farmers of north Gujarat agro-climatic zone are advised to follow the cropping system of *kharif* pearlmillet after *rabi* chickpea and the *kharif* pearlmillet crop should be fertilized with 90 kg N/ha.

Mustard Based Cropping Sequences : Farmers of north Gujarat region growing mustard (Gujarat Mustard -1) with assured irrigation throughout the year, are advised to adopt any of the following cropping sequences.

Kharif	***Rabi***	**Summer**
Clusterbean (Guj-1)	Mustard (GM-1)	Pearlmillet (GHB-235)
Greengram (K-851)	Mustard (GM-1)	Pearlmillet (GHB-235)

Profitable Crop Sequence (Crop Rotation) in Relation to Castor Wilt : Under north Gujarat Agro-climatic region, the most profitable crop sequence in relation to castor wilt disease is pearlmillet - mustard, clusterbean - wheat and sorghum – cumin. Castor (GCH-4) should not be sown in the same field every year but sown after every three years to avoid wilt disease infestation.

Groundnut as a Preceding *Kharif* Crop to *Rustica Tobacco* : Farmers of Vijapur area of north Gujarat zone are advised to take either clusterbean or sesame or groundnut as a preceding *kharif* crop to *Rustica tobacco* to get higher monetary returns.

Cropping System Diversification and/or Intensification : The farmers of north Gujarat agro-climatic zone having assured irrigation facility, are advised to adopt cowpea/greengram - mustard - pearlmillet or castor- fodder sorghum/greengram (summer) instead of existing cropping sequences *viz.*, pearlmillet- wheat and clusterbean-wheat for getting higher yield and net realization.

Middle Gujarat

Broad bed and furrow system for Cotton - Gram Cropping System : Farmers of Bhal and Coastal zone are advised to adopt raised bed (15 cm above surface) and furrow (15 cm bellow surface) bed system (6 m width) of cultivation for getting higher benefit in terms of substantial yield of *kharif* and *rabi* crops. They are advised to grow cotton on raised bed during *kharif* and wheat or gram on sunken beds during *rabi* for securing higher income.

South Gujarat

Paddy-Gram Cropping System : Farmers of South Gujarat in the heavy rainfall zone growing *rabi* crops on residual moisture after *kharif* paddy, are advised to grow gram without fertilizers. *Kharif* paddy, of course, should be fertilized with 100 per cent of the recommended doses (120-30-00 NPK kg/ ha).

Paddy-Chickpea Cropping System and Fertilizer Management : Farmers are recommended to adopt Paddy-Gram sequence in clayey soils of South Gujarat heavy rainfall zone. They should apply only 100 kg N/ha to Paddy var. IR-22 instead of the recommended dose of 120 kg N/ha.

Chickpea-Sorghum Cropping System and Fertilizer Management : The farmers of South Gujarat Heavy Rainfall zone growing *kharif* sorghum are recommended to follow Chickpea- Sorghum sequence. They are further advised to fertilize sorghum with 60 kg N/ha so as to obtain higher profit.

Paddy-Chickpea Cropping System Fertilizer Management : The farmers of South Gujarat agro-climatic zone following paddy-chickpea crop sequence, are advised to fertilize each crop with 60 kg P_2O_5/ha to secure the highest total economic return.

Marginal farmers, however, may reduce the phosphorus dose to chickpea crop as 30 kg P_2O_5/ha.

Paddy-Gram and Paddy-Groundnut Cropping System : The paddy growing farmers of South Gujarat heavy rainfall zone are advised that under conditions of constraints of irrigation water they should adopt paddy-gram system in *rabi* instead of paddy-wheat. Similarly, during summer they should adopt paddy-groundnut system instead of paddy-sorghum. Further, the paddy yield after gram and groundnut will be higher by 25 and 30 per cent, respectively than those after wheat and sorghum.

Paddy-Vegetable Fenugreek - Summer Okra/ Paddy- Onion-Vegetable Cowpea Cropping System : The farmers of South Gujarat agro-climatic zone having assured irrigation facilities are advised to adopt paddy-vegetable fenugreek- okra and paddy-onion- vegetable cow pea instead of existing cropping sequence of paddy-wheat for getting higher yield and net realization.

Saurashtra Region

Groundnut-Wheat-Greengram cropping system : For south Saurashtra zone, Groundnut-Wheat-Greengram cropping system has been found to be the best viable groundnut based cropping sequence.

Groundnut– Onion- Greengram Cropping System : The Farmers of south Saurashtra Agro-climatic zone having assured irrigation facilities are advised to adopt Groundnut (*kharif*)- Onion (*rabi*)- Greengram (summer) cropping sequence as crop diversification for obtaining higher yield and net realization.

Inter Cropping/mixed Cropping Systems

North Gujarat

Inter Cropping in Castor with Greengram under Irrigated Conditions: The farmers of north Gujarat Agro-Climatic zone growing irrigated castor (GCH- 5) at wider spacing (120 cm x 60 cm) are advised to intercrop with greengram (GM- 4) or sesamum (GT-1) in 1:1 row ratio to achieve about 30% and 24% higher yield and 34% and 27% more net returns, respectively, over sole castor.

Agri. Horti System in class- IV Sandy Soils of North West Zone : The farmers of North West agro climatic zone growing. *Ber* var. Gola (6.0 m x 6.0 m) on sandy soil are advised to take intercrop of greengram for getting higher monetary returns (33%) as compared to sole crop of *ber* and to over come the risk involved in dry farming.

Mixed Cropping in Pearl Millet with Mothbean : The farmers of north – West Agro - Climatic Zone growing pearlmillet, are advised to adopt mixed cropping of pearlmillet with mothbean in proportion of 3.0 kg seed of pearlmillet and 4.0 kg seed of mothbean/ha for getting higher yield and gross returns under dry farming conditions.

Chickpea-Mustard Intercropping : Farmers of north Gujarat are advised to adopt inter cropping of mustard in chickpea (1:5) in light soil areas so as to have more stable returns.

Castor-groundnut/Greengram Intercropping : Although sole castor has been found to give higher net returns in north Gujarat zone (Sardarkrushinagar and Talod), the farmers are advised to intercrop it with groundnut or greengram in 1:1 ratio as a safe guard against the monsoon's erratic behaviour.

Ber-greengram/Fodder sorghum : Farmers of north Gujarat zone growing ber (10 m x 6 m) on light textured soils are advised to take intercrop of either greengram or sorghum (fodder) for getting 27 and 16% higher net return as compared to that obtained by growing ber as sole crop.

Tree- Crop Intercropping : Residual effect of tree species *viz., Acacia tortilis, Delbergia sissoo, Albazzia lebback* and *Prosopis juliflora* (Israel) on succeeding crops (castor, pearlmillet, greengram and clusterbean) grown on sandy soils of north Gujarat agro-climatic zone was found economically beneficial while improving the fertility and productivity of the soils. The average increase in gross income over control was 52, 46, 40 and 22% due to residual effect of the respective tree species. Eucalyptus grown in sandy soils of north Gujarat agro-climatic zone improved the soil productivity. Percent increase in the gross income was 65% for greengram, 39% for castor, 28% for pearlmillet and 28% for clusterbean under irrigated conditions. This shows that 26% increased profit can be obtained from field crops grown in eucalyptus.

Pearlmillet-Clusterbean Intercropping : The farmers of north Gujarat agro-climatic zone growing pearlmillet on light textured soil in *kharif* season are advised to follow intercropping (2:2) with clusterbean variety Malosan or HG- 75 for insuring the income under risk prone dry land agriculture. To overcome the risk involved in dry land agriculture under light textured soils of north Gujarat agro-climatic zone, the farmers are advised to take pearlmillet+clusterbean (2:1) intercropping during *kharif*. It gives a higher net profit than that of sole pearlmillet. In the low rainfall areas of north Gujarat agro-climatic zone having light textured soils, the farmers are advised to take intercropping of pearlmillet +clusterbean (2:1) at an inter row spacing of 45 cm. They should apply only 50 % of the recommended doses of N to the crops (pearlmillet 37.5 kg N/ha and clusterbean 10 kg N/ha). However, P_2O_5 @ 25 kg/ha should be applied to pearlmillet as a common dose to both the crops.

Cotton-green gram intercropping : Farmers of north Gujarat zone cultivating cotton (Hy- 8) are advised to intercrop it with greengram (1:2) to get 19 % more income.

Mixed cropping of greengram and pigeonpea with pearlmillet : Cultivators of north Gujarat Agro-climatic zone are advised to grow pearlmillet (100% recommended seed rate) as mixed crop with greengram (20% recommended seed rate). The mixed crop of greengram and pigeonpea with pearlmillet recorded 40% and 26% higher returns, respectively, as compared to the sole crop of pearlmillet.

Pearlmillet + Greengram + Pigeonpea mixed cropping : Cultivators of north Gujarat Agro-climatic zone are advised to grow pearlmillet (100% recommended seed rate) as mixed crop with greengram (20% recommended seed rate). The mixed crop of greengram and pigeonpea with pearlmillet gives 40% and 26% higher returns, respectively, as compared to the sole crop of pearlmillet.

Relay Cropping

Greengram-castor relay cropping : Farmers of North Gujarat Agro-climatic zone growing greengram on light textured soils under rainfed conditions, are advised to inter relay castor crop after first picking in order to make efficient use of available soil moisture and get higher returns. Castor should be line sown after every two rows of greengram.

Middle Gujarat

Pigeonpea + Maize intercropping : Farmers adopting pigeonpea + maize intercropping system (2:1) in paired row method are advised to fertilize maize crop with 100 kg N and 50 kg P_2O_5 /ha and not apply any fertilizer to pigeonpea.

Maize + Mid late Pigeonpea intercropping : Farmers are advised to intercrop mid late pigeonpea (BDN-2) in 1:1 ratio with maize (one row of maize in between two rows of pigeonpea planted at 120 cm) for getting higher grain yield as well as monetary realization.

Pigeonpea + Blackgram intercropping : Farmers are advised to intercrop pigeonpea (BDN-2) with Blackgram in 1:1 ratio (one row of blackgram in between two rows of pigeonpea planted at 90 cm) for getting higher seed yield as well as monetary realization.

Maize+ Pigeonpea intercropping : The farmers of Panchmahal district of middle Gujarat agro-climatic zone growing maize (Farm Sameri), are advised to grow pigeonpea variety BDN-2 as a intercrop in the ratio of 1:1 at a distance of 45 cm between two rows with the application of N and P @ 80 kg and 40 kg per ha, respectively. The entire dose of P_2O_5 and 50% of N should be applied as a basal dose. Remaining 50% of N should be applied one month after sowing.

Cotton +Greengram or Blackgram inter cropping : It is recommended that the farmers of Bhal region should adopt intercropping of greengram and blackgram in a paired row system of planting cotton for getting higher additional realization.

South Gujarat

Cotton + Soybean or Blackgram or Greengram intercropping : For securing higher economic returns, farmers of south Gujarat zone growing irrigated cotton G.Cot Hy-6 at a distance of 120 cm between the rows are advised to intercrop one row of Soybean (G-1) or blackgram (Zande wal) or greengram (GM-2) in between the rows.

Cotton or Pigeonpea + Blackgram or Greengram intercropping : For securing higher profit, farmers of south Gujarat zone growing rainfed cotton G.Cot.-11 or Pigeonpea at a distance of 120 cm between the rows, are advised to intercrop it with Blackgram (Zande wal) or Greengram (GM-2) in between the rows.

Saurashtra Region

Pearlmillet + Pigeonpea intercropping : Farmers are advised to apply recommended doses to the base crop pearlmillet (80 kg N and 40 kg P_2O_5 / ha) while intercropping with pigeonpea for getting maximum monetary returns. The marginal farmers, however, may apply 40 or 60 kg N/ha to the intercropping system. Phosphorus dose, however, remains unchanged.

Sorghum + Pigeonpea intercropping : Farmers are advised to adopt intercropping system of Sorghum + Pigeonpea in 1:1 row ratio. They should apply recommended doses of fertilizers (90 kg N/ha and 30 kg P_2O_5 /ha) to the base crop for getting higher equivalent yield of sorghum and maximum monetary returns.

Cultivation of Pulses in Non-Conventional Season/Area

Rajmash, though it is an unconventional pulse crop grown in North Gujarat, has yield potential of around 35 q/ha. Similarly, Fieldpea is also a potential unconventional crop having yielding ability of around 40 q/ha. Hence, these new crops should be introduced in Gujarat as catch crops. Growing of short duration (about 60-70 days) summer pulses like greengram and cowpea instead of growing summer pearlmillet saves precious irrigation water and increases soil fertility.

Future Thrusts

It is an era of integration and globalization in which technological advancements have linked and integrated every aspect of life. Along with this, the impact of global warming is also being realized. Therefore, along with other crop production aspects, impact of global warming should also be taken into consideration for deciding the future course of action for realizing sustainable productivity of the pulses crops for feeding the teeming millions. The important future action points for sustaining productivity of the pulse crops in different cropping sequences of Gujarat are as follows:

- Technology generation and development with respect to weed, water and nutrient management must be based on systems approach rather than an individual crop for getting higher productivity of the system.
- Currently, a lot of varieties of diverse crops under cultivation are available which need different agronomic package of practices. Therefore, future research should focus on development of variety specific agronomic packages rather than crop specific for getting higher resource use efficiency.
- There is a great diversity in response of crops with respect to nutrients depending upon fertility status of the soil. Therefore, in order to achieve high nutrient use efficiency, crop and variety specific response equations based on targeted yield must be developed.
- All over the world, health awareness is increasing and developed countries are framing stringent regulations with respect to quality parameters of agricultural products. Therefore, in order to promote or maintain export in these countries, it is necessary to meet the quality standards of various parameters.
- Social audit of research and extension system must be strengthened.

Acknowledgements

Authors are highly grateful to Director of Research and Dean P.G. Studies, Dean (Agriculture), Professor and Head, Department of Agronomy, C.P. College of Agriculture, S.D.A.U., Sardarkrushinagar-385 506 (Gujarat), Research scientist (Main Pulses Research Station), Research scientists (pulses, castor and mustard, potato and crop improvement) and taff of AICRP on Integrated Farming Systems for rendering help in preparation of this chapter.

References

Acharya, S and Patel, J.B. (2010). Strategies of research on pulses and vision 2025, Directorate of Research, SDAU, Sardarkrushinagar-385 506.

Bharodia, P.S. (1995). Krushi Geeta, Director of Extension, GAU, Sardarkrushinagar-385 506.

Directorate of Agriculture (2008-09). Final forecast report of area, production and productivity, Gujarat state.

Patel, B.S., Patel, B.T., Patel, R.M., Patel, A.M., Prajapati, J.N. and Patel, P.K. (2007). Cropping Systems Research in North Gujarat: Technical Bulletin No.07/01 published by Directorate of Research, SDAU, Sardarkrushinagar-385 506 (Gujarat).

Annual Report (2000-01, 2001-02, 2002-03, 2003-04, 2004-05, 2005-06 and 2006-07) AICRP on Cropping Systems, SDAU, Sardarkrushinagar.

Panda, S.C. (2005). Agronomy : Agrobios (India),Jodhpur.

Panda, S.C. (2004). Cropping and Farming Systems: Agrobios (India), Jodhpur.

Research recommendations Bulletin (1980 to 2009), Director of Research, Gujarat State Agricultural University.

Singh, S.S., Gupta, P. and Gupta, A.K. (1996). Hand book of Agriculture Sciences. Published by Kalyani Publishers Ludhiana.

❑❑❑

7

Haryana

Pawan Kumar, Manoj Kumar and S. K. Yadav
Department of Agronomy, CCS Haryana Agricultural University, Hisar - 125 004 Haryana.

Viable cropping systems involving new rewarding crops infuse new opportunities and challenges as well as demonstrate the potential for land productivity with higher growth and economic achievements. As far as agricultural production is concerned Haryana is a surplus agricultural state and is one of the largest food supplier to the national kitty. The cropping patterns have undergone drastic changes over the years with a major shift in favour of cereals like paddy and wheat. Consequently, the existing cropping system *viz.* paddy-wheat, has started showing its adverse effect on soil health and consequent decline in crop productivity. Diversification of existing system by involving pulses, vegetable, fodder and other high value crops, is considered to be the potential alternative to address many present day problems. For example, under assured irrigation, pearlmillet-potato-moongbean gave almost 65% higher yield than the existing cotton-wheat cropping system because the leguminous crop in summer gave higher economic yield and benefit to pearlmillet crop. The package of practices with focus on pulses crops grown in Haryana state are discussed here.

In Haryana, Chickpea is the major food legume crop and predominantly cultivated in moisture stress prone areas of south and western agro-ecological zones. Despite the development of genetically improved varieties and improved production technology, the area under chickpea cultivation in the state has drastically decreased from 1543000(1960-61) to 123000 hectares (2008-09) and fluctuates considerably over the years, primarily depending upon rainfall patterns in the traditionally chickpea growing regions of the state (Table-1). The drastic decrease in areas under chickpea cultivation is mainly attributed to the introduction of high yielding mustard and cereal crop varieties, the process of secondary salinization caused by increased irrigation facilities in traditionally chickpea growing areas and governmental policies in favour of oilseed and cereal production. These factors further relegated chickpea cultivation to marginal lands with low inputs conditions in the arid and semi arid agro-ecological regions of the state.

Table 1 : Area, production and productivity of chickpea in Haryana

Year	Area(000 ha)	Production(000 tones)	Productivity(kg/ha)
1960-61	1543	1274	0826
1965-66	0868	0385	0444
1970-71	1163	0789	0678
1975-76	1108	0909	0820
1980-81	0746	0466	0625
1985-86	0761	0625	0821
1990-91	0650	0469	0722
1995-96	0377	0381	1011
2000-01	0080	0050	0675
2008-09	0123	0128	1041

The successful introduction of chickpea in the prevailing cropping systems in the state of Haryana depends upon its increased tolerance against abiotic stresses particularly salt, moisture and temperature stresses. The introduction of chickpea for value addition in autumn planted sugarcane intercropping in north-eastern and central zone or mustard-chickpea based intercropping system in south-western zone of the state can open up a new avenue for increasing the areas under chickpea cultivation.

Pigeonpea : Cultivation in Haryana started in seventies and the introduction of short duration varieties developed farmers' interest in this crop. The area under this crop increased upto 1990 and remained fluctuating thereafter (Table 2).

Table 2 : Area, Production and productivity of pigeonpea in Haryana

Year	Area(000 ha)	Production(000 tones)	Productivity(kg/ha)
1979	6.6	6.1	925
1985	21.4	15.5	725
1990	51.3	49.6	967
1995	20.5	16.1	785
2000	41.1	46.6	1134
2004	31.0	32.2	1039

The area, productivity and production of MULLaRP crops in the state has generally decreased except the area under mungbean which is showing an increasing trend in the recent past (Table 3). All these crops except lentil are grown in rainfed areas and therefore, their area and yield depend on rainfall.

Yield gaps : The yield productivity gaps in major pulses grown in Haryana varies from 189 to 323% (Table 4.)

Table 3 : Area (000 ha), production (000 tonnes) and productivity (kg/ha) of different pulses in Haryana

Year	Mash			Mungbean			Lentil			Others (Peas, Moth-bean, Lobiya etc)			Total Pulses		
	A	P	Y	A	P	Y	A	P	Y	A	P	Y	A	P	Y
1980-81	12.5	7.6	606	4.1	2.9	713	27.4	13.1	478	28.9	23.9	-	794.8	502.5	-
1985-86	5.8	3.8	692	4.2	3.1	727	26.1	13.8	529	49.3	40.9	-	846.3	686.6	-
1990-91	3.4	1.3	376	10.7	5.8	483	18.3	10.8	587	60.3	54.8	-	742.0	541.7	-
1995-96	2.3	1.1	474	7.0	3.1	444	10.4	7.2	692	53.4	58.3	-	449.8	450.7	-
2000-01	1.5	0.3	148	5.3	1.2	226	6.5	5.2	798	19.2	13.1	-	157.0	99.8	-
2004-05	2.3	0.9	341	17.2	6.8	393	6.4	5.0	793	39.5	61.1	-	172.4	144.8	-
2007-08	-	-	-	21.8	9.6	439	-	-	-	-	-	-	-	-	-

Table 4 : Yield gaps in pulses production

Crop	Actual yield (kg/ha)	Achievable yield (kg/ha)	Gap	
			Productivity (kg/ha)	Per cent
Chickpea	1041	2250	1209	216
Pigeonpea	1039	1900	861	183
Mungbean	393	1250	857	318
Lentil	793	1500	707	189
Mash	341	1100	759	323

Ecological Zones of Haryana

Haryana is divided into 2 zones *viz.,* eastern zone and western zone.

The Eastern zone : It comprises of Faridabad, Gurgaon, Stayamevpuram, Sonepat, Karnal, Panipat, Kurukshetra, Ambala, Panchkula, Kaithal and Jind districts and part of Rohtak district. The soils of this zone are heavy and rainfall is high. The important pulses grown are mungbean and lentil.

The Western zone : It comprises of Sirsa, Fatehabed, Hisar, Bhiwani, Mahendragarh, Rewari, Jhajjar and part of Rohtak. The soil of this zone is light and rainfall is low. The important pulses grown are chickpea, pigeonpea and mungbean.

Existing Major Cropping Systems of Haryana

Eastern zone: Paddy-wheat, Paddy-wheat-mungbean/green manuring/ green fodder, Paddy-potato onion/ vegetables, Groundnut-wheat, Sugarcane-fallow and Paddy-lentil

Western zone: Cotton-wheat, Bajra-mustard, Bajra-chickpea, Mungbean-wheat/mustard, Sorghum-chickpea, Guar-mustard/wheat and Fallow-gram/mustard/barley

Major Constraints in Pulse Production

Chickpea

- Wide gap between potential yield and realized yield due to non - adoption of improved varieties and improved production technologies.
- Lack of varieties and farm machinery for mechanized harvesting.
- Non- availability of seed of improved varieties at the proper time.
- Non-responsiveness of the released varieties to inputs particularly fertilizer and irrigation
- To regain area under chickpea in Haryana, input responsive varieties which can compete with mustard/wheat are required.
- Lack of suitable varieties for late sowing after sugarcane in irrigated areas.
- Non-availability of quality *rhizobium* culture to the farmers.
- Cultivation of chickpea on poor, marginal and problematic soils under moisture stress (rainfed) conditions.
- Uncertainty of moisture at sowing time in rainfed areas greatly affects the area under chickpea.
- Inadequate use of fertilizers and plant protection measures.
- Poor and substandard quality of insecticides/herbicides.
- Lack of extension of the technology like ridge planting and irrigation to which chickpea is highly responsive.

- Price fluctuation and large price gap between declared minimum support price (MSP) and retail price in the market i.e. farmers interests are not being properly protected by the govt. policies as being done under the cereal MSP regimes.
- Marketing and procurement problems.

Pigeonpea

- Wide gap between potential yield and realized yield due to non adoption of improved varieties and improved production technology.
- Cultivation of pigeonpea on neglected, marginal and problematic land.
- Inadequate use of fertilizers and plant protection measures.
- Price fluctuation and large price gap between declared minimum support price (MSP) and retail price in the market i.e. farmers interests are not being properly protected by the govt. policies as being done under the cereal MSP regimes.
- Marketing and procurement problems.
- Pod-borer and blue bull are the main constraints in the production of pigeonpea.
- Terminal moisture stress affects the yield seriously.
- Longer duration and low productivity of crop is another constraint.
- Lack of resistant varieties against *Helicoverpa*, pod fly and *Maruca*.
- Lack of spread of the technology of intercropping of short duration pulses like mungbean, urdbean, cowpea *etc.*, in medium duration pigeonpea.

Mungbean, Urdbean, Cowpea and Mothbean

- Pre-harvest sprouting of grains in *kharif* season.
- Lack of varieties with bold seeds and synchronized maturity.
- Lack of varieties with extra early maturity of mungbean for summer cultivation after harvest of wheat.
- Lack of short duration mungbean and urdbean varieties.
- Lack of varieties for multiple disease resistance like powdery mildew, yellow mosaic virus and *Cercospora* leaf spot.
- Lack of varieties resistant to pests such as *Helicoverpa,* thrips, *Maruca etc.*

- Lack of short duration, high yielding varieties of mothbean for shallow soils.

Peas, Lentil and Rajmash

- Susceptibility of existing varieties to rust, wilt, root rots and viral diseases.
- Poor biological and grain yield.
- Lack of rajmash varieties with wider adaptability suitable for *kharif,* spring and summer cultivation.

Recommended Varieties of Pulse Crops

The recommended varities of pulses for Haryana are as follows:

Chickpea

H 208 : Suitable for rainfed, tolerant to wilt, average yield 20 q ha

C 235 : Suitable for irrigated areas, tolerant to blight but susceptible to wilt, Average yield 20 q ha^{-1}.

Haryana Chana No. 1 : Suitable for rainfed, irrigated and late sown conditions, early maturing, less attack of pod borer and tolerant to wilt, average yield 20-25 q ha^{-1}.

Haryana Chana No. 3 : Suitable for irrigated areas and should be sown in 1st week of November, resistant to wilt, blight and other diseases, average yield 20-25 q ha^{-1}.

Haryana Chana No. 5 : Suitable for full Haryana except dry land areas, tolerant to wilt and root rot, average yield 21-25 q ha^{-1}.

Haryana Kabuli No. 1 : Suitable for full Haryana except dry land areas, Comparatively less attack of wilt, average yield 20-25 q ha^{-1}.

Haryana Kabuli No. 2 : Suitable for full Haryana except dry land areas, resistant to main diseases of gram, average yield 17.5-20 q ha^{-1}.

Gora Hisari : Suitable for irrigated areas of western Haryana, less prone to damage of pod borer, average yield 20 q ha^{-1}.

Lentil

Haryana Masar 1 : Suitable for whole Haryana, matures in 140 days, resistant to insects and diseases, average yield 16.25 to 17.5 q ha^{-1}.

Sapna : Suitable for irrigated areas, matures in 140 days, resistant to pod borer and diseases, average yield 15 q ha^{-1}.

Garima : Suitable for irrigated areas for timely sowing, matures in 135 days, average yield 15 q ha^{-1}.

Field pea

Aparna (HFP 4) : Dwarf leafless variety, recommended for irrigated areas of Haryana, fairly tolerant to powdery mildew, leaf minor has no effect on yield, does not lodge, escapes from frost, average yield 25 q ha^{-1}.

Jayanti (HFP 8712) : Suitable for irrigated areas, it is leafless dwarf variety, resistant to powdery mildew, less attack of blight and wilt, duration 125-130 days, average yield 27.5 q ha^{-1}.

Uttra (HFP 8909) : Suitable for irrigated areas, suitable for NW India including Haryana, leafless dwarf variety, can tolerate powdery mildew and blight, duration 131 days, resistant to lodging, average yield 25-30 q ha^{-1}.

Harial (HFP 9907-B) : Tall growing, matures in 128 days, resistant to powdery mildew, root rot, blight and virus, average yield is 21 q ha^{-1}.

Type 163 : Suitable for irrigated areas, matures in 150 days, dual purpose crop, average yield 20 q ha^{-1}.

Pigeonpea

Manak (H 77-216) : Even suitable for late sowing (1st fortnight of July), medium in height, indeterminate growth habit, matures in 120-130 days, average yield 19 q ha^{-1}.

UPAS 120 : Matures in 130-140 days, indeterminate growth habit, droops at maturity, pods well distributed on the branch, average yield 15 q ha^{-1}.

Type 21 : Long duration variety which matures in 150-160 days, tall and highly branched, indeterminate in flowering, only late sown wheat can be sown after its harvest, should be planted before mid-June, average yield 17.5-20 q ha^{-1}.

Paras (H 82-1) : May be sown up to mid-July, mid-tall in height and matures in 135-140 days, bears more branches and pods, average yield 17.5-20 q ha^{-1}.

Greengram

MH 83-20 (Asha) : Suitable in irrigated areas as summer and *kharif* crop, tall growth habit, resistant to yellow mosaic, synchronous in maturity, matures in 75-80 days, average yield is 12-13.5 q ha^{-1}.

K 851 : Suitable in irrigated areas during summer season, semi-spreading, synchronous in maturity, matures in 65-70 days, average yield is 8-10 q ha^{-1}.

Type 44 (Pusa Baisakhi) : Early maturing and suitable as summer crop, matures in 65 days, average yield is 6.25-8 q ha^{-1}

MH 96-1 (Muskan) : Suitable in irrigated areas as summer and *kharif* crop, resistant to yellow mosaic, synchronous in maturity, harvest in time to avoid shattering, average yield is 10 q ha^{-1} (summer) and 12.5-15 q ha^{-1} (*kharif*).

Black gram

T 9 : Recommended for general cultivation, early maturing and synchronous in maturity, average yield is 12-12.5 q ha^{-1}.

Cowpea

FS 68 : Suitable for multiple cropping system and fits well after wheat or barley, can be sown in july also, matures in 65-70 days, average yield 8 q ha^{-1}.

Rajmash

Him 1: Suitable for irrigated area, dwarf and straight, matures in 80-85 days, average yield is 10-12.5 q ha^{-1}.

Jawala : Suitable for irrigated area, dwarf and straight, matures in 85-90 days, average yield is 7.5-10 q ha^{-1}.

VL 63 : Suitable for irrigated area, dwarf and straight, matures in 85-90 days, average yield is 12.5-15 q ha^{-1}.

Production Technologies

Chickpea

Chickpea has special importance in western Haryana and 88 per cent area falls in western districts.

Land preparation : Very fertile soil is not required for chickpea cultivation. Loose and aerated soil is best for its cultivation. A deep ploughing of 22 cm enhances chickpea productivity and this ploughing may be given in July or August by disc/mould bold plough. It facilitates the movement of roots and soil moisture conservation.

Seed rate : The optimum seed rate for *desi* gram is 37.5 to 45.0 kg ha^{-1} but for Haryana Chana No. 3 and Gaurav optimum required seed rate is 75 to 80 kg ha^{-1}. The seed rate for *kabuli* gram is 80 kg ha^{-1} and for Gora Hisari is 62 to 65 kg ha^{-1}. The seed rate for late sown Haryana Chana No. 1 is 50-55 kg ha^{-1}.

Sowing management : The optimum time for chickpea cultivation is mid-October, however, it can be sown from mid-October to 30th October. *Kabuli* gram may be sown in the last week of October. In irrigated areas the optimum time for Haryana Chana No. 1 and Gora Hisari is from second week of November to 3rd week of November. Haryana Chana No. 1 may be sown upto mid December in irrigated areas. Early sowing leads to wilt disease and vigorous growth.

Rhizobia culture : To harvest good yield of chickpea, the seed may be treated with rhizobia culture.

Method of seed treatment : Dissolve 150 gm of *"gur"* or *"sakkar"* in 5 cups of water. Mix seeds for one hectare with this solution. Then open one hectare Rhizobium culture packet and mix with the seeds thoroughly. Allow the seeds to dry under shade and then sow as usual. This practice should be followed every year.

Sowing method : In well moisturized soil the sowing of chickpea may be made by seed drill or pora at 30 cm row spacing and in light to medium soils, where moisture is limited, the row spacing of 45 cm may be kept. Generally the moisture is low in the fields where it is sown after bajra and therefore it may be sown at a spacing of 45 cm under such situations. This crop can also be sown in paired rows also with spacing of 30 cm between two rows and 60 cm between two pairs. It facilitates inter-cultivation, more aeration and better light interception and thereby, helps in increasing yield of chickpea and traditional method of sowing.

Intercropping systems : Cross sowing of mustard in chickpea is an accepted technology. In the main crop of chickpea, mustard is sown crossly in the ratio of 6:1 or 8:1.

Fertilizer management : The recommended fertilizer nitrogen and phosphorus dose for chickpea is 15 kg ha^{-1} and 40 kg ha^{-1}, respectively. The fertilizer may be applied along with last ploughing. In irrigated areas 25 kg zinc sulfate ha^{-1} may also be applied.

Weed management : One to two hoeings depending on weed infestation may be carried out.

Water management : In general, gram is cultivated under dry land situations. However, it gives good response to irrigation. One irrigation may be applied before flowering; 45 to 60 days after sowing. In sandy loam soil, if the crop has been sown on pre-sowing irrigation then first irrigation may be applied at pod development stage. Chickpea should not be irrigated when grown in paddy-chickpea rotation.

Sulphate dominated (SO_4=70 per cent or more) saline waters of electrical conductivity up-to 5000 μmhos cm^{-1} can be used as protective irrigation to gram on sandy loam well drained soils in areas, which receive at least 400 mm rainfall annually. However, protective irrigation with chloride rich waters (Cl=50 per cent and above) of electrical conductivity higher than 2000 μmhos cm^{-1} should not be given.

Plant Protection Measures

Insects and Pests

Termites : Termites cause damage from sowing to harvesting; however, it is more in light soils with low moisture content. It can be controlled by treating with any one of these insecticides i.e. monochrotophos 36SL 850 ml/ chlorpyriphos 20EC 1500 ml and dilute this with water so as to make the volume to 2 litres and spray it on one quintal seed. Mix the seed well. Keep the treated seed as such overnight before sowing. Care must be taken to remove crop stubbles before sowing. Avoid application of raw FYM in the field.

Cutworm : This insect cuts germinating seedling or the branches of the grown up plants. It may be treated with the spray of fenuvalrate 20EC 200 ml or cypermethrin 25EC 125 ml or decamethrin 2.8EC 375 ml in 250 litres of water ha^{-1} or with the dusting of fenuvalrate 0.4 per cent at 25 kg ha^{-1}.

Helicoverpa species : This insect is green in color and attacks leaves, flower buds and pods. It destroys the developing seeds in the pod. It can be controlled with quinalfos 25EC 1 lt. or carboryl 50WP 1 kg or monochrotophos 36SL 500 ml or fenuvalrate 20EC 200 ml or cypermethrin 10EC 312.5 ml or cypermethrin 25EC 125 ml or decamethrin 2.8EC 375 ml in 250 litres of water ha^{-1}. If required, 2nd spray may be made at 15 days interval. It can also be controlled with the dusting of quinalfos 1.5D/carboryl 5D at 25 kg ha^{-1}. If required, 2nd dusting may be made at 15 days interval. The control should be initiated when at least one insect is present in 1 meter length and the crop is at 50 per cent pod development stage.

Diseases

Wilt : It appears 3-6 weeks after sowing. The leaves droop down but remain green. A longitudinal knife cut of the stem reveals black discoloration of the xylem vessel. To control it, keep the field moist and do not sow the crop before 10th October. Grow resistant varities like H208, Haryana Chana No. 1, 3 and 5. The seed may be treated with bavistin 2.5 g kg^{-1} seed. The disease may also be controlled by treating the seed with Bioderma 4 g kg^{-1} seed + Vitavax 1g kg^{-1} seed.

Root Rot : It occurs at seeding stage when soils are moist or at adult stage during the crop growth period in irrigated areas. Dark brown lesions on stem near the soil level are noticed. The stem and root below the lesion rot with pinkish mycelial growth.

Stem Rot : The disease generally occurs due to overgrowth or excessive rainfall. Petioles and leaflets droop without discoloration. Turf of white mycelial stands girdle the collar region. Later appearance of white sclerotia in the mycelial mass is noticed.

Virus : The affected plants remain stunted and turn orange or brown in color. It is more conspicuous in *'Desi'* than in *Kabuli* types. A shallow vertical cut at the collar region invariably exhibits phloem browning. The crop can be protected from this disease by proper soil moisture conservation and sowing the crop after 10th of October.

Fungal disease : This disease occurs due to excessive rainfall and when temperature remains 18-20°C.

a) ***Ascochyta***: Initially the symptoms are characterized by the appearance of minute brown lesions on leaves, stems and pods. On the green pods, the lesions are usually circular with dark margins and have pyenidia arranged in concentric rings. On the stem and petiole, the lesions are brown, elongated (3-4 cm), bearing black dots, and often girdle the affected portion. As the disease progresses, patches can be noticed in the field and consequently, it spreads over the entire field.

b) ***Alternaria Blight***: Small grey circular spots appear on the leaflets. With the progression of disease development, these spots enlarge, becoming pale yellow. The plants droop and dry. Sporulation on the necrotic spots takes place on the stem. Acute damage is also caused when this occurs on the base of the plant, resulting in its weakening. Pods are heavily affected.

c) ***Grey mould***: The leaflets get discolored starting along the margin and at the terminal portion of the leaflet. Drooping of the terminal branch becomes conspicuous. The fungal growth can be seen on infected tissues on close examination. The infection spreads further to branches and main stem resulting in rotting.

Control Measures

- The gram sowing may be avoided in the area with previous record of blight incidence. The resistant varieties like Haryana Chana No. 3, C235 and Gaurav may be sown.

- Disease free seed should be sown.
- The seed may be treated with *Bavistin* @ 2.5 g kg^{-1} seed.
- Remove diseased plants and destroy them.

Chickpea Based Sequential Cropping : Maize-chickpea, Fallow-chickpea, Pearl millet-chickpea and Sorghum-chickpea.

Lentil

It is one of the important pulse crops of Haryana and mainly grown in paddy areas. It is also recommended for western Haryana, Narwana and Palwal. The average yield in the state is low.

Land preparation : Land should be ploughed 2 or 3 times and each ploughing should be followed by "*suhaga.*" The soil should be well pulverized as that results in good seed germination.

Seed rate : The optimum seed rate is 30-37 kg ha^{-1}. The seed rate may be kept at 45-50 kg ha^{-1} for late sowing and bold seeded varities.

Sowing time : Normally the crop should be sown in the month of November. However, the sowing can be done successfully up to 1st week of December but it results in 20 per cent lower yield.

***Rhizobia* culture :** To harvest good yield of lentil, the seed may be treated with rhizobia culture.

Method of seed treatment : Dissolve 150 gm of *gur* or *'sakkar'* in 5 cups of water. Mix seeds for one hectare with this solution. Then open one hectare Rhizobium culture packet and mix with the seeds thoroughly. Allow the seeds to dry under shade and then sow as usual. This practice should be followed every year.

Sowing method : The crop should be sown in lines 22.5 cm apart by *kera* method. It can also be sown by broadcast following paddy crop where shortage of time and wet field conditions do not permit good land preparation. Under late sown conditions the row spacing should be reduced to 18 cm.

Fertilizer management : The recommended fertilizer nitrogen and phosphorus dose for lentil is 15 kg and 40 kg ha^{-1}, respectively. The fertilizer may be applied along with last ploughing. In irrigated areas 25 kg zinc sulfate ha^{-1} may be applied. If zinc deficiency is observed in the standing crop a mixture of 0.5 per cent zinc sulfate and 2.5 per cent urea may be sprayed twice at 10 to 14 days interval.

Weed management : One to two hoeings may be carried out depending on weed infestation.

Water management : In general lentil is cultivated under dry land situations after paddy crop. However, when it is grown after groundnut in sandy loam soils, 2-3 irrigations are required.

Plant Protection Measures

Pod Borer : The most damaging pest of this crop is pod borer. This insect is green in color and attacks leaves, flower buds and pods. It destroys the developing seeds in the pod. It can be controlled with quinalfos 25EC 1 lt. or carboryl 50 WP 1 kg or monochrotophos 36 SL 500 ml or fenuvalrate 20 EC 200 ml or cypermethrin 10 EC 312.5 ml or cypermethrin 25 EC 125 ml or decamethrin 2.8EC 375 ml in 250 litres of water ha^{-1}. If required 2nd spray may be done at 15 days interval. It can also be controlled with the dusting of quinalfos 1.5D/carboryl 5D at 25 kg ha^{-1}. If required 2nd dusting may be made at 15 days interval. The control should be initiated when at least one insect is present in 1 meter length and the crop is at 50 per cent pod development stage. The insect is controlled well by the spray of insecticide at flowering stage.

Diseases

This crop is affected by lentil rust, phillody rust, leaf blight and root rot. However, only root rot appears to be causing serious damage. The research work is going on for the control of this disease.

Lentil Based Sequential Cropping

- Maize-lentil
- Paddy-lentil
- Cotton-lentil
- Groundnut-lentil

Fieldpea

It has been recommended for cultivation in Haryana in the recent past. It is a replacement for chickpea in irrigated and high water table areas. It can be successfully cultivated in western Haryana, Narwana and Palwal areas.

Land preparation : Land should be ploughed 2 or 3 times and each ploughing should be followed by *"suhaga"*.

Seed rate : The optimum seed rate for tall growing varieties is 75 kg ha^{-1} and for dwarf varieties is 90-100 kg ha^{-1}.

Sowing time : Optimum time of sowing is the first fortnight of November. Early sowing should be avoided because it suffers from stem fly which infests the plant at collar region and the plant dies. Sowing should be avoided after November as it results in drastic reduction in crop yield.

Rhizobia culture : To harvest good yield of field pea, the seed may be treated with rhizobia culture.

Method of seed treatment : Dissolve 150 gm of "*gur*" or "*sakkar*" in 5 cups of water. Mix seeds for one hectare with this solution. Then open one hectare *Rhizobium* culture packet and mix with the seeds thoroughly. Allow the seeds to dry under shade and then sow as usual. This practice should be followed every year.

Sowing method : Fieldpea can be sown by seed drill or pora or kera depending on soil moisture and tilth. Aparna variety should be sown in lines 20 cm apart while Uttra and Jayanti should be sown 25-30 cm apart. Tall growing varieties should be sown 30 cm apart.

Fertilizer management : The recommended fertilizer N and P dose for lentil is 20 kg and 40 kg ha^{-1}, respectively. The fertilizer may be applied before sowing or at sowing.

Weed management : One to two hoeings depending on weed infestation may be carried out by *khurpa/kasola.*

Water management : Sulphate dominated (SO_4=70 per cent or more) saline waters of electrical conductivity upto 5000 µmhos cm^{-1} can be used as protective irrigation to gram on sandy loam well drained soils in areas, which receive at least 400 mm rainfall annually. However, protective irrigation with chloride rich waters (Cl–50 per cent and above) of electrical conductivity higher than 2000 µmhos cm^{-1} should not be given.

Plant Protection Measures

Pea stem fly : It affects early sown crop at seedling stage. Infested plants become yellow, dry and ultimately die. To control it avoid early sowing.

Pea thrip : The nymph and the adults attack the young crops and damage by sucking cell sap from the leaves. For its control, spray cypermethrin 25EC 150 ml or endosulfan 25 EC 1250 ml in 625 litres of water ha^{-1}.

Pea leaf minor : During the month of December, January and February, the larvae of this insect make white tunnels in the leaves and cause serious damage. For its control, spray Dimethoate 30EC 1000 ml or methl dematon 25EC/ formathion 25EC 1250 ml in 625 litres of water ha^{-1}.

Pea pod borer : The larvae bore into the pods and eat developing grains. This insect is green in color and attacks leaves, flower buds and pods. It destroys the developing seeds in the pod. It can be controlled with quinalfos 25EC 1 litre or carboryl 50WP 1 kg or monochrotophos 36SL 500 ml or fenuvalrate 20EC 200 ml or cypermethrin 10EC 312.5 ml or cypermethrin 25EC 125 ml or decamethrin 2.8EC 375 ml in 250 litres of water ha^{-1}. If required, 2nd spray may be done after 15 days. It can also be controlled with the dusting of quinalfos 1.5D/carboryl 5D at 25 kg ha^{-1}. If required, 2nd dusting may be done after 15 days. The control should be initiated when at least one insect is present in 1 meter length and the crop is at 50 per cent pod development.

Diseases

Powdery mildew : Grow tolerant varieties like Aparna. Spray 0.3 per cent solution of wettable sulfur or 0.1 per cent benlate or bavistin or karathane. Repeat spray after one week.

Root rot and wilt : For the control of root rot and wilt, treat seed with *bavistin* @ 2 g kg^{-1} seed. Avoid early sowing.

Fieldpea Based Cropping System

- Maize-fieldpea

Pigeonpea

Now-a-days it is cultivated in all the districts of Haryana due to the introduction of short duration varieties.

Land preparation : Well-drained loam to light loam soil are more suitable. Land should be ploughed 2 or 3 times and each ploughing should be followed by *"suhaga"*. It will help in control of weeds and removing the clods in the field.

Seed rate : For pure crop, the seed should be sown in rows about 40 cm apart using 12.5 to 15 kg seed ha^{-1}, however, for intercropping of short duration varieties the row spacing should be 50 cm.

Sowing time : Under irrigated conditions where double cropping system is followed, the time of sowing for variety T21 is from mid-March to mid-June and for UPAS120 is from mid-March to 1st week of July. Manak and Paras varities should be sown from June to mid-July. Under rainfed conditions, all the recommended varities should be sown as early as possible after the on set of monsoon.

Rhizobia culture : To harvest good yield of pigoenpea, the seed may be treated with rhizobia culture.

Method of seed treatment : Dissolve 150 gm of *"gur"* or *"sakkar"* in 5 cups of water. Mix seeds for one hectare with this solution. Then open one hectare *Rhizobium* culture packet and mix with the seeds thoroughly. Allow the seeds to dry under shade and then sow as usual. This practice should be followed every year.

Sowing method : For pure crop the seed should be sown in rows about 40 cm apart; however, for intercropping of short duration varieties, the row spacing should be 50 cm.

Fertilizer management : The recommended fertilizer nitrogen and phosphorus dose for pigeonpea is 20 kg ha^{-1} and 40 kg ha^{-1}, respectively and should be applied at the time of sowing.

Weed management : Two hoeings at 25 and 40 days after sowing may be given for the control of weeds.

Water management : Irrigation, if possible, must be given once at commencement of flowering. Irrigate the summer crop as per the requirement of intercrop.

Plant Protection Measures

Insects

Pod borer : It damages the pigeonpea extensively. Spray the crop with quinalfos 25EC 1500 ml or monochrotophos 36SL 750 ml or cypermethrin 25EC 190 ml or fenuvalrate 20EC 300 ml or deltamethrin 2.8EC 540 ml in 750 litres of water ha^{-1}. If required, make 2nd spray after 15 days interval.

Pigeonpea Based Cropping System

- Pigeonpea-wheat

Intercropping : Pigeonpea when sown from middle of March to middle of April by keeping inter-row distance of 75 cm, 2 rows of mung variety K851/Muskan, Asha or cowpea variety HFC42-1 can be inter cropped between 2 rows of pigeonpea. These intercrops can be harvested in about 60-75 days. By this system of planting pigeonpea yield is not adversely affected and additional yield of intercrop is obtained. In pigeonpea when spaced 50 cm apart, one row of mung or cowpea can also be planted.

Greengram, Black gram and Cowpea

These are important *kharif* pulse crops but their average yield is low as they are grown in less fertile soils. Still their productivity can be increased by adopting best agronomic practices.

Land preparation : Well-drained loam to sandy loam soils are more suitable. Land should be ploughed twice and each ploughing should be followed by *"suhaga"*. The land should be well leveled and free from clods and weeds.

Seed rate : For mung and blackgram, the seed rate is 15 to 20 kg ha^{-1}. For summer mung, the seed rate may be kept 25 to 30 kg ha^{-1}. The optimum seed rate of cowpea variety FS 68 is 30 kg ha^{-1}.

Sowing time : The crop should be sown with the onset of monsoon in the 1st week of July. The optimum time for summer mung is the month of March.

***Rhizobia* culture :** To harvest good yield, the seed may be treated with *rhizobia* culture.

Method of seed treatment : Dissolve 150 gm of *"gur"* or *"sakkar"* in 5 cups of water. Mix seeds for one hectare with this solution. Then open one hectare *Rhizobium* culture packet and mix with the seeds thoroughly. Allow the seeds to dry under shade and then sow as usual. This practice should be followed every year.

Sowing method : For attaining higher productivity, proper spacing should be maintained. Too less or too large plant population leads to reduction in yield. For pure crop, the seed should be sown in rows 30 cm apart under irrigated conditions and 45 cm apart under rainfed conditions. It can be sown either by *"pora"* or *"kera"* method. For summer moong, line to line spacing of 20-25 cm may be maintained. If needed, plants may be thinned to maintain 10 cm plant to plant spacing.

Fertilizer management : The recommended fertilizer nitrogen and phosphorus dose for all these crops is 15-20 kg and 40 kg ha^{-1}, respectively, and it should be applied at the time of sowing.

Weed management : Two weedings are recommended to keep the fields free from weed infestation. First hoeing at 20-25 days after sowing and second hoeing at 30-40 days after sowing should be given.

Water management : Generally no irrigation is required by these crops. However, in case of failure of rainfall for a longer period, the crop should be irrigated. One irrigation in the end of September helps in increasing the production. For summer crop of mungbean, 1st irrigation may be applied 20 to 22 days after sowing and thereafter the irrigation may be made at 10 to 15 days interval as per requirement.

Plant protection measures

Insects

Hairy caterpillar, jassids, white flies and beetles damage these crops. For the control of hairy caterpillar, give deep ploughing after the harvest of *kharif* crops. Use light trap after rains. Avoid weeds around the fields. Destroy the eggs. Spray Monochrotophos 36SL 625 ml or Dichlorvas 76EC 500 ml quinalfos 25EC 1250 ml in 625 litres of water. For the control of jassid and white fly, spary, Malathion 50EC 1000 ml or Dimethoate 30EC 625 ml or Oxidameton methyl 25EC 625 ml or Formathion 25EC 625 ml in 625 litres of water. Repeat spray after 2 to 3 weeks.

Diseases

Cercospora leaf spot : The spots of brown red color appear on leaves, stalks and pods and reddish purple borders are formed. Spray the crop with blitox 50 or indofilM 45 @ 1.5 to 2 kg ha^{-1} in 500 litres of water.

Bacterial leaf spot : Symptoms appear as water soaked dots on lower side of leaf which remains small in size and the surrounding tissues become necrotic. Spray the crop with copper oxychloride 1.5 to 2 kg in 500 litres of water ha^{-1}.

Root rot : Affected plants show stunting and yellowing and roots begin to shred and rot leading to death of the plant. For its control, treat the seed with Thiram @ 4 g kg^{-1} seed. Follow crop rotation of 3 years.

Yellow mosaic : Affected plants show yellow and green areas on leaves. In Ambala and Karnal districts, this disease may affect 90 per cent of the crop. Grow tolerant variety Asha. White fly is the vector of this disease and therefore, control white fly with Dimethioate 30EC or Oxydematon methl 25EC or Formathion 25 EC 625 ml or malathion 50EC 1litre in 625 litres of water ha^{-1}. Rogue out the infected plants. Destroy weeds in and around the crop.

Crops Based Sequential Cropping System : Paddy-wheat-mungbean, Mungbean-wheat, Mungbean-mustard and Blackgram-barley

Intercropping : Sowing of cowpea in paired rows of pearlmillet at a distance of 30 cm has been successful over the years. Pearlmillet and mung sown in the ratio of 6:3 or 8:4 (strip method) has been successful in dry land areas.

Rajmash

It can be easily grown in north-western Haryana in irrigated areas.

Land preparation : Land should be ploughed 2 times and each ploughing should be followed by *"suhaga"*. The land should be well leveled and free from clods and weeds.

Seed rate : For rajmash, the seed rate of medium size variety is 100 kg ha^{-1} and for bold seeded variety, the seed rate is 120 kg ha^{-1}.

Sowing time : The crop should be sown between 10th to 20th September. Early sowing leads to death of seedling due to high temperature. Late sowings leads to poor development of pod and seed due to low temperature.

Rhizobia culture : To harvest good yield of rajmash, the seed may be treated with rhizobia culture.

Method of seed treatment : Dissolve 150 gm of *"gur"* or *"sakkar"* in 5 cups of water. Mix seeds for one hectare with this solution. Then open one hectare *Rhizobium* culture packet and mix with the seeds thoroughly. Allow the seeds to dry under shade and then sow as usual. This practice should be followed every year.

Sowing method : The rajmas seed should be sown in rows 30 cm apart. It can be sown either by *"pora"* or *"kera"* method. There should be sufficient moisture in the soil at sowing time.

Fertilizer management : The recommended fertilizer nitrogen and phosphorus dose for rajmash is 100 to 120 kg ha^{-1} and 45 kg ha^{-1}, respectively. The whole phosphorus and half dose of nitrogen should be applied at sowing time and remaining half dose of nitrogen may be top dressed at flowering stage.

Weed management : Two hoeings at 20-25 and 40-45 days after sowing may be carried out to control weeds effectively.

Water management : First irrigation should be applied at 15 to 20 days after sowing. Second and third irrigation may be applied at 15 to 20 days interval.

Efficient Alternating Cropping Systems

Existing	Alternative	Remark
Eastern zone		
Paddy-wheat	Paddy-lentil	Good scope for greengram in summer.
Paddy-wheat-summer paddy	Paddy-wheat-mungbean	
Western zone		
Pearlmillet-wheat	Pigeonpea-wheat Mungbean-wheat	The area under Pearlmillet-chickpea and Fallow-chickpea deserves to be increased.
Pearlmillet-mustard	Mungbean-mustard	
Fallow-mustard	Fallow-chickpea	There is scope of intercropping of mungbean with pearlmillet.

Future Thrust

- Chickpea area in the state can be increased subject to availability of a variety which responds well to high dose of fertilizers and irrigation and can be successfully grown in saline, high water table and irrigated areas.
- Lentil is generally grown after paddy in Haryana. As water is becoming a limiting factor in paddy-wheat cropping system; lentil can be a better substitute of wheat.

- As the area under chickpea is decreasing due to increase in irrigated area and high water table; fieldpea can be a good substitute of chickpea under such situations.
- Pigeonpea is grown in several districts of Haryana; however, its area is almost static for the last several years because of competition with cereals. The insect attack is severe in this crop. The growers interest in pigeonpea cultivation can be enhanced subject to dissemination of technology and better marketing facilities.
- Greengram, blackgram and cowpea has ample scope as summer crop as a replacement of *santhi* (Summer) paddy in north-east parts of Haryana. Blackgram cultivation is picking-up slightly in rainfed areas of Bhiwani district and around.

References

Package of Practices, Published by CCS HAU, Hisar. (2010)

Emerging Challenges to Haryana's Agricultural Economy vis-a-vis Diversification in Agriculture. Published by Directorate of Research, CCS HAU, Hisar. (2010)

Personal Communication from Pulses Section, Department of Plant Breeding, CCS HAU, Hisar.

❑❑❑

8

Himachal Pradesh

SC Negi, SS Rana and Daisy Basandrai
Department of Agronomy, Forages and Grassland Management, CSK Himachal Pradesh Krishi Vishvavidyalaya, Palampur 176 062, HP, India.

Pulses occupy an important place in hill agriculture. They are complementary to cereals in production and consumption being an important source of protein and nitrogen fixing ability which helps in improving soil health. Thus, cultivation of pulses has a positive relationship with our environment, human health and food security. Pulses produced in the hills are famous for their cooking quality, taste and flavour.

Himachal Pradesh has emerged as a model hill state in which agriculture is a dominant economic activity and major source of employment and income to the rural people. Being a hilly state, agriculture is practiced on about 10% of the geographical area. Out of which nearly 80% is rainfed. During the last three decades, the farming scenario has changed significantly. The area under fruits, vegetables and cereals has increased tremendously at the cost of pulses and oilseeds. There is, therefore, an urgent need to reverse this trend as pulse cultivation will not only increase soil health but save precious foreign exchange. The cropping pattern of Himachal Pradesh is dominated by cereal crops (84%), followed by fruits (6%) and vegetables (4%) (Gupta 2004). During the year 2006-07 the share of pulses in the state acerage was 3.00% (22.9 thousand ha). *Mash, kulthi,* chickpea and lentil with nearly 52, 12, 6 and 6% area, respectively, are the major pulses grown in the state. Mash, the major pulse crop, is the preferred dish of local people. Moreover, Mash grown in the hilly areas fetches premium price in the market due to its taste, cookability and aroma. Similarly, rajmash grown in temperate areas *viz*. Kinnaur, Lahaul and Spiti, Chamba and Mandi districts, is very delicious and in great demand. The total production of pulses in state during 2006-07 was 12.14 thousand tonnes with average pulse productivity of 530 kg/ha. Kangra was the largest producer of pulses followed by Mandi. Standard yields of gram, pigeonpea, mash, kulthi and lentil was 786, 236, 512, 456 and 541 kg/ha, respectively. There exists a wide variability in inter-district productivity. Productivity of chickpea and mash was higher in sub-tropical and temperate areas, respectively. In Himachal Pradesh, pulses are generally sown as mixed or intercrop with cereals or side bunds on the rice or maize fields under poor agronomic practices and on marginal lands (*rabi*) fallows. Hence, their productivity is very low.

Trends of Pulses Production

The pulses which were grown in about 71.4 thousand hectares in 1975-76, have dropped to 22.9 thousand hectares in 2006-07 but reached to 31.2 thousand hectares in 2011-12 (Table 1). The total pulse production was 31.5 thousand MT in 1975-76, 9.9 thousand MT in 2002-03 and 23.5 thousand MT in 2011-12. Mash, kulthi and rajmash are the *kharif* pulses whereas chickpea and lentil are the major *rabi* pulses. Although the productivity of pulses/ha has increased, the decline in area is a matter of major concern. Relative acerage under gram has reduced drastically from 30.4 thousand ha in 1975-76 to 0.76 thousand ha in 2011-12. Its production has also declined from 16.6 thousand MT in 1975-76 to 0.29 thousand MT in 2011-12.

Table 1 : Trends in area (000, ha), production (000, t) in Himachal Pradesh

Year	Gram		Other pulses		Total pulses	
	Area	Production	Area	Production	Area	Production
1975-76	30.4	16.6	41.0	14.9	71.4	31.5
1979-80	19.7	9.5	44.7	14.5	64.4	24.0
1989-90	5.8	1.7	34.7	9.2	40.5	10.9
1990-91	3.4	2.2	36.6	10.5	40.0	12.7
1994-95	3.1	1.3	33.4	14.0	36.5	22.8
2002-03	1.35	1.01	28.4	8.9	29.7	9.9
2003-04	1.3	1.21	27.5	9.8	28.8	11.01
2004-05	1.3	1.32	28	9.59	29.3	10.91
2005-06	1.34	0.72	26.17	18.91	27.51	19.63
2006-07	1.33	1.02	21.57	11.99	22.9	12.14
2007-08	1.45	1.37	32.42	34.59	33.87	35.96
2009-10	0.73	0.29	30.37	23.18	31.10	23.47
2010-11	0.68	0.37	29.66	20.29	30.34	20.66
2011-12	0.63	0.60	33.70	40.99	34.33	41.59
2011-12	0.67	0.66	31.58	30.12	32.25	30.78

Statistical Outlines of Himachal Pradesh, 1975-76 to 2011-12. *Important Statistics of Himachal Pradesh 2011-12.

Average yield of chickpea was approximately 5.46 q/ha in 1975-76 which increased to 9.45 q/ha in 2010-11. but dropped to 3.82 kg/ha in 2011-12. Average yield of other pulses was 3.63 in 1975-76 and 9.91 q/ha in 2011-12. It is evident that area under cultivation has decreased but the yields have increased which may be attributed to the cultivation of high yielding disease resistance varieties with improved agronomic practices.

Existing Cropping Systems

Wheat (363.4 thousand ha), maize (298.5 thousand ha) and rice (81.4 thousand ha) are the major crops covering 89% of the gross acreage in the state. Thus, maize-wheat and rice-wheat are the major cropping systems in the state. Barley is grown in an area of 24.3 thousand ha. The main barley based cropping systems are rice-barley, maize-barley, barley-peas and barley-potato. Rice/maize-oilseed/pulses/vegetables, oilseeds/vegetables/pulses – wheat, rice-berseem/oats/other fodder crops are followed to a lesser extent. The sugarcane based cropping systems are being followed in 2.6 thousand ha. Pea-buckwheat/sarson/turnip has been emerged as an important cropping system in Zone IV of the state. Rice-potato-potato has been emerging as a major cropping system in Una district of Himachal Pradesh. With increasing plantations of fruit crops, multi-tier cropping systems involving cereals/underutilized crops, pulses, oilseeds and potato have started dominating the horticulture based niches. Besides these, there are several niche based cropping systems being practiced by the farmers of Himachal Pradesh.

Zone-wise Efficient Cropping Systems

Recently, the State Agriculture University under the aegis of AICRP on Cropping Systems Research (now Farming Systems Research) came up with worthwhile cropping systems. By adopting these cropping systems, farmers of the state can strengthen their livelihood by increasing their income two-three folds as compared to the prevalent traditional cropping systems i.e. maize-wheat and rice wheat. The zone-wise efficient cropping systems are as follows.

Zone I. Sub-Mountain and Low Hill Zone Sub-Tropical Zone

Maize based cropping systems are Maize-*toria-gobhi sarson*, Maize-gram, Maize + soybean- wheat, Maize-potato-wheat, Maize + soybean- wheat + *sarson* and Maize-pea-potato.

Rice based cropping systems are Rice-onion, Rice- potato- French bean, Rice-potato- potato, Rice- radish- potato, Rice- toria-potato, Rice-wheat-french bean and Rice-potato-onion.

Zone II. Mid Hills Sub-Humid Zone

Maize based cropping systems are Maize-*toria*-potato, Maize-pea-potato, Maize+soybean-wheat, Maize-toria-potato-French bean, Maize-*Gobhi* sarson and Babycorn + asparagus bean/French bean-pea-summer squash.

Rice based cropping systems are Rice-pea-french bean, Rice-radish-potato, Rice-potato, Rice-*toria*-potato and Rice-pea-potato.

Vegetable based cropping systems are Okra- radish-onion, Okra-cauliflower-french bean.

Zone III. High Hills Temperate Wet Zone

Lavender + rajmash

Zone IV. High Hills Temperate Dry Zone

Maize-sarson, Potato-sarson, Pea-pea/buckwheat/turnip/sarson and *Kalazira* + rajmash/ buckwheat (relay)

The most remunerative cropping systems identified under situations of Zone 2 (600-1800 m altitude) can as such be adopted under Zone 1 (300-600 m altitude). In Zone 4 (Dry temperate zone, > 1800 m altitude) most parts remain covered with snow during winter and only single crop is generally possible. However, a few cropping systems with two crops (from April to mid October) have been found to be most promising in the zone.

Yield Gaps

Yield gap can be divided into two parts (Table 2). Yield gap I is the difference between the yield obtained on the experimental station and yield obtained in the on-farm experiments (on the basis of the highest yield during the last five years). In the present study, yield gap I worked out to be 59.1, 15.0 and 42.3% of the potential yield for rice, wheat and maize, respectively. For rajmash, *mash, kulthi,* gram and cowpea, the yield gap I was 43.4, 47.6, 34.3, 30.4 and 20%, respectively. This gap is due to several factors, which cannot be controlled by the farmers. Yield gap II is the gap between the maximum yield in on-farm experiments and the average farm yield/state average yield. For rice, wheat and maize, the yield gap II was very high, and of the order of 63.3, 77.6 and 55.3%, respectively, of the on-farm experimental potential yield. Yield gap II for rajmash, pigeonpea, mash, mung, kulthi, gram and lentil was 49.8, 38.5, 84.1, 16.7, 75.7, 56.9 and 79.3%, respectively. This gap is due to biological (biotic), soil, water (abiotic) and sociological constraints. Primary focus should be on yield gap II that can be managed to minimize the yield losses at farmers level.

Table 2 : Production potential and yield gaps in major crops

Crops	Yield (kg/ha)			Yield gap (%)	
	Experimental plot*	On farm	State average	I	II
Rice	9497	3887	1413	59.1	63.6
Wheat	5360	4557	1021	15.0	77.6
Maize	7139	4117	1839	42.3	55.3
Rajmash	3800	2150	1080	43.4	49.8
Pigeonpea	-	780	480	-	38.5
Mash	1899	1005	160	47.1	84.1
Mung	-	600	500	-	16.7
Kulthi	1065	1750	170	34.3	90.3
Gram	2155	1500	647	30.4	56.9
Lentil	-	1450	300	-	79.3
Cowpea	16.8	14	-	20.0	-

*Maximum yield during the last five years.

Yield gaps between actual and experimental rice yields under irrigation in major rice growing states of India indicate that with the exception of Tamilnadu (15%) and Punjab (22%), the yield gap was in the range of 35 to 75% (Siddiq 2000). If the gap is taken as an opportunity and efforts directed on priority for narrowing the same, achievements of targeted production growth would not be difficult. At national level, the gap in yield levels between national demonstration average and the national average is 0.6 t/ha in irrigated areas and 1.5 t/ha in rainfed areas.

Improved Varieties

Mungbean

Himachal Mash-1 (UPU00-31) has been released for Zone I of Himachal Pradesh. It has average grain yield of 14-16 q/ha. It is synchronous in maturity and matures in about 74-76 days. On farmers' fields, its mean yield ranged between 10-12 q/ha. It is having comparatively better resistance to mungbean yellow mosaic virus (MYMV) as compared to UG-218 and Pant U-19, the existing varieties. It is also resistant to leaf crinckle, anthracnose and powdery mildew and moderately susceptible to Cercospora leaf spot.

UG 218 : This variety was adopted for cultivation in sub mountain, low hill, sub tropical zone (Zone I). It matures in about 75-80 days and yields 11-12 q/ha. It is highly resistant to yellow mosaic virus and tolerant to leaf spots. It can also be cultivated as "zaid" crop in summer under irrigated conditions.

PANT U 19 : It was adopted for cultivation in Zone I of Himachal Pradesh. It is an early maturing variety and takes 85 days to mature. Its average yield is 8-10 q/ha. It is resistant to yellow mosaic virus but moderately susceptible to leaf spot diseases.

PDU 1 : It was released for Mid hills (Zone II) of Himachal Pradesh, especially for the rainfed areas in district Kullu. It is suitable for intercropping with maize and yields 10-12 q/ha. This variety is highly susceptible to mung bean yellow mosaic virus but due to low temperature this disease does not appear in Kullu area, hence, it is recommended for Kullu valley only.

KULLU 4 : It has tall plants with climbing habits. It matures in 100-105 days. Its average seed yield is 7-9 q/ha. It is susceptible to yellow mosaic virus. It is recommended for Kullu valley only.

T 9 : The variety was adopted for cultivation in Zone I of the state. It matures in 80-85 days and on an average yields 8-10 q/ha. It has become susceptible to mungbean yellow mosaic virus and leaf spots.

Rajmash

Kailash (SRC-74) : This variety was released for timely sown, irrigated and rainfed conditions for district Kinnaur. Its plants are determinate and mature in 124 days. It is resistant to most of the diseases under field conditions.

K 198 (Triloki Rajmash) : This variety was released for high hills (Cold and Dry temperate zone, Zone IV) of Himachal Pradesh. It matures in 98-100 days. Its average yield is 25-27 q/ha. It is resistant to bacterial blight, angular leaf spot and anthracnose except in Sangla valley, where it is moderately susceptible to anthracnose. It is tolerant to shattering also.

KRC 8 (Baspa) : It was recommended for Zone IV of the state. It matures in 110-120 days. Its average yield is 18-20 q/ha. It is highly resistant to anthracnose.

HPR 35 (Kanchan) : This variety was released for Zone II of the state. It is an early maturing variety, matures 7-10 days earlier than Him 1 and Jwala. It is suitable for both monoculture and intercropping with maize. Its average yield is 12-15 q/ha.

HPR 12 (Jwala) : This variety was recommended for cultivation in areas between 1100-1800 masl in district Kullu, Barot, Shimla and Chamba valley. It is an early maturing variety. It matures in 75-80 days and yields 12-15 q/ha. It is resistant to shattering.

Him 1 : It is recommended for mid and high hills. Its average yield is 9 q/ha. It is susceptible to the leaf spot diseases.

Kulthi

HPK 4 (Baizu) : It is suitable for all the kulthi grown areas in the state. It matures in 100-125 days. Its yield potential is 17-18 q/ha. But due to high rainfall and late maturity its average yield is 6-7 q/ha only. It is resistant to shattering and leaf spot diseases. It is highly tolerant to drought.

Cowpea

C 475 (Himachal Lobiya) : It is suitable for cultivation in Zone I of the state. Its plants are semi determinate, semi-dwarf with bushy growth habit. It matures in 80-85 days and due to its near synchronous maturity, it requires only one hand picking before final harvesting. Its average grain yield is 14-15 q/ha. It is resistant to *Cercospora* leaf spot, anthracnose, yellow mosaic virus and yellow golden mosaic virus.

Himachal Lobia 2 (C-519) : It is released for Zone I of Himachal Pradesh. It is a dual-purpose variety with average green pod yield of 45-50 q/ha and mean grain yield of 15-16 q/ha. It takes about 85-95 days to mature and 5-6 pickings can be taken before maturity. It is resistant to *Cercospora* leaf spot, anthracnose, web blight and cowpea yellow mosaic virus. It performes well when sown between 15-30 July with row to row spacing of 60 cm.

Summer Moong

DPM 8909 (Saketi 1) : It was released for Zone I of Himachal Pradesh. It matures in 85 days and due to its near synchronous maturity, requires only one hand picking before final harvesting. It yields 10-11 q/ha. It is moderately resistant to mosaic and tolerant to leaf spot diseases.

Pusa Baisakhi : It is suitable for both the summer and *kharif* cultivation under irrigated conditions. It matures in about 65-70 days. It gives an average yield of 6 q/ha. It is moderately resistant to yellow mosaic virus.

Shining Moong No.1 : It is an early maturing variety which is ready for harvest in 70-80 days. It has an average yield of 6 q/ha. It is moderately resistant to leaf spot diseases.

Varieties SML134, SML 357, SML 668 have been found suitable for summer/spring sowing under irrigated condition in Zone 1

SML 668 : It is released for the summer cultivation during the second fortnight of February to first week of April in Zone I (Sub mountainous low hills, Sub tropical zone of Himachal Pradesh) under irrigated conditions after ginger, potato and sugarcane harvesting or intercropping in sugarcane. It matures in 75-85 days and is harvested before onset of monsoon. It is synchronous in maturity and yields 11-12 q/ha. It is resistant to *Cercospora* leaf spot and Moong bean yellow bean mosaic virus.

Pigeonpea

Sarita (ICPL 85010) : It was released for Zone I and lower areas of Zone II of Himachal Pradesh. It is a determinate type variety and is suitable for the areas where wheat- pigeonpea cropping sequence is followed. It takes 150-155 days to mature. Its yield is about 15 q/ha. It is resistant to sterility mosaic virus.

Ricebean

There is no ricebean variety recommended for Himachal Pradesh. The farmers grow local land races of ricebean only. However, genotype **BRS 1 BRS 2, Naini, RBL 6** and **Chaukambha** have been found suitable for Himachal conditions. It takes 150-160 days to mature and yields 18-20 q/ha.

Gram

C 235 : It has a wider adaptability and is also suitable for rained agriculture. It matures in 171-175 days and on an average yields 8-10 q/ha. It is moderately susceptible to Ascochyta blight, wilt and root rot diseases.

HPG 17 : It is a bold seeded variety. It matures in 180-185 days and on an average yields 16-17 q/ha. It is resistant to Ascochyta blight, wilt and root diseases.

Himachal chana I (ICCX 810800) : This variety has been released for Zone 1. It matures in 180-185 days and on an average yields 11-12 q/ha. It is highly resistant to Ascochyta blight and moderately resistant to wilt, root rot and collar rot.

Himachal chana II (ICCV 90201) : It is a high yielding short duration variety released for Zone 1 of the state. It matures in 160-165 days and yields 13-14 q/ha. Being a short duration variety it escapes the attack of pod borer. It is resistant to Ascochyta blight and root rot.

GPF 2 : It is recommended for Sub montane low hills, Sub tropical zone of Himachal Pradesh (Zone I). Its average yield in experimental plots is 16-17 q/ha, whereas, it yields 8-10 q/ha on farmers fields. It is resistant to *Fusarium wilt,* root rot, BGM and Ascochyta blight and tolerant to cold stress and lodging.

Lentil

L 9-12 : It matures in 130-140 days. Its average yield is 13-14 q/ha. and is resistant to rust.

HPL 5 (Vipasha) : It is a small seeded variety with 100 grain weight of 2.5 g. It takes 140-150 days to mature and its average yield is 14-15 q/ha. It is highly resistant to rust and has field resistance to Ascochyta blight.

EC 1 (Markanday) : It is a bold seeded variety. It takes 160-165 days to mature and yields 10-15 q/ha. It is highly resistant to rust and tolerant to Ascochyta blight.

Peas

There is no recommended variety of peas in the state. Farmers generally grow local land races as mixed cultivation with gram or wheat. Kinnauri, a vegetable type variety, can also be sown for seed purpose. The varieties released from PAU, Ludhiana and HAU, Hissar i.e. 163, PG 3, Field Pea 48 and Aparna can also be cultivated in Zone I of the state.

Crop-wise Production Technologies

Himachal Pradesh Agriculture University has come up with some worthwhile production technologies for the cultivation of pulses in the state (HPKV 2008 a&b; Basandrai *et al.*, 2004; Kumar *et al.*, 2008; Rana *et al.*, 2003), which are summarized here.

Mash (Moong)

Land preparation : Well drained loam to clay loam soils with neutral pH are suitable for the mash cultivation. The field should be prepared by two or three cross harrowing followed by planking. It should be well leveled, free from weeds and left over stubbles.

Seed rate : Seed rate of 18-20 kg/ha is sufficient for *kharif* season mash. However, a seed rate of 25-30 kg/ha required for summer crop. In mixed cropping with maize and sesamum, 8-10 kg seed/ha is sufficient.

Sowing Management : Mash is generally grown during the *kharif* season, however, it can also be grown in summer/spring season in sub montane and low hills sub tropical zone (Zone I) under assured irrigation. *Kharif* crop is sown with the advent of the monsoon from the end of June to third week of July. *Kharif* crop requires lesser seed rate and wider spacing because of its profuse plant growth. A row to row spacing of 30 cm is appropriate for *kharif* crop. Summer crop is sown from second fortnight of March to first week of

April. During this season, vegetative growth is comparatively less so it is sown at row to row spacing of 20-25 cm. Seed treatment with bavistin/Thiram/ Captan at the rate of 3 g/kg of seed should be done to protect the crop from seed borne diseases.

Fertilizer application including micronutrients : Pulses in general require less amount of nitrogenous fertilizer. They can fix atmospheric nitrogen due to the presence of bacteria (*Rhizobium* spp.) in the nodules of the roots. For the crop sown for the first time in the field or after a long time, treatment of the seed with suitable *Rhizobium* culture is beneficial. Mash requires a starter dose of 15-20 Kg N and 45-60 Kg P_2O_5/ha at the time of sowing as basal placement. Thakur and Negi (1985) has also reported significant increase with 20 kg N and 60 kg P_2O_5/ha. Rhizobium inoculation also brought about significant increase in grain and straw yield of mash. The fertilizer applied to the main crop will be sufficient for mash grown as a mixed crop. Summer *urdbean* grown after potato/ginger requires no nitrogenous fertilizers as the residual nitrogen is sufficient to meet the requirement of mash.

Weed control and interculture : Weeds cause losses even upto 50-60 % in *kharif* crops. The initial growth of the crop is very slow and it suffers from severe weed competition upto 35 days of sowing. One hand weeding at 25-30 days followed by 40-45 days after sowing is sufficient. First weeding can be effectively done by wheel hand hoe. Where hand weeding is not possible due to continuous rains and scarcity of labour, fluchloralin 1 kg/ha as PPI or alachlor 0.75/ha is very effective. Fluchloralin should be sprayed 24 hours before sowing and should be mixed in the soil to avoid photodecomposition. Alachlor or pendimethalin should be sprayed within 48 hours of sowing.

Water management : *Kharif* crop generally does not require irrigation but drainage of excess water is essential. In summer crop, sufficient moisture should be ensured by pre-sown irrigation. First irrigation should be given at about 25-30 days after sowing and should be repeated at an interval of 7- 10 days. For synchronous maturity and higher grain yields, the last irrigation should be terminated about 50 days after sowing.

Sequential and intercropping systems : Mash is mainly grown as a sole crop or mixed crop with maize, sesamum etc. One or two rows of mash can be sown between two rows of maize. Maize is planted at a row distance of 60 cm, thus a lot of space remains unutilized which generally becomes infested with weeds. In this space, short duration mash varieties can be grown tocontrol weed competition. Defoliating the lower leaves of the maize plant had an added advantage in boosting legume component's productivity without adversely affecting that of maize (Rana *et al.,* 2009). The following crop sequences are economically beneficial to the farmers:

Maize/sesamum + Mash-Wheat; Maize-Mustard-summer mash; Maize-Potato-summer mash; ginger-summer mash; Mash-Wheat-summer mungbean; Maize+ Mash- Wheat- Green manuring cowpea.

Harvesting and threshing : The crop should be harvested at proper maturity to avoid losses due to shattering. The *kharif* crop is ready for harvesting when the plants turn half yellow and pods turn black or start drying in the first fortnight of October and the summer crop by last week of June. It may be harvested by using serrated sickles. The plants may be dried for 3-4 days before threshing. Threshing can be done manually or by tampling under the feet of bullocks or tyres of tractors. Grains should be dried under the sun and stored at less than 10% moisture in clean storage bins.

Rajmash/French Bean

Land preparation : It is grown in *kharif* in cold regions after monsoon rains or under irrigated conditions. It can be successfully grown in Kullu, Kangra, Mandi, Chamba, Shimla, Kinnaur and Lahaul Spiti and in areas at altitude of 1200-2000 m. The crop can be grown successfully in sandy loam to heavy clay soils. Although loam soils are the most suitable for its growth. One or two ploughings are sufficient for good tilth. Land should be leveled and well drained. Soil should be loose and fragile so that air can pass easily. There should not be any crust formation.

Seed rate : The optimum seed rate for small seeded varieties like Him 1 is 35-40 kg/ha, whereas, 50-55 kg/ha seed is required for bold seeded varieties like Jwala. In mixed cropping with maize, seed rate should be 25-30 kg/ha.

Sowing Management : Two crops of French bean can be grown in a year at the lower elevations. The first crop is sown by the last week of March to mid April and harvested in the last week of June or first week of July. It is followed by cultivation of *kharif* crop. The crop should be sown with the onset of the monsoon rains under rainfed conditions, whereas under irrigated conditions, the crop should be sown in the second fortnight of June. In dry temperate zone, it should be sown between 15 May to 15 June. The yield decreases in late sown crops due to frost in the month of September or October. The crop should be sown in rows 25-30 cm apart with plant to plant distance of 10-15 cm and at 2.5- 4.0 cm depth.

Fertilizer application including micronutrients : Rajmash is responsive to fertilizers (Rana *et al.*, 2003). Phosphorous plays a vital role in root development, nodulation, growth and yield of the crop. It requires about 20 kg N and 60 kg P_2O_5/ ha, which should be applied at the time of sowing. Rajmash also responds to 20 kg K_2O/ha in the dry temperate zone of the state. Before field preparation, 20 t/ha FYM should be added. Soil and foliar application of biozyme has also been found promising in the dry temperate region of HP (Rana *et al.,* 2006).

Weed control and inter-culture : The first 60 DAS or so are critical from weed removal point of view (Rana *et al.,* 2008). Weeding and hoeing should be done twice to control the weeds. If hand weeding or hoeing is not possible, pre-sowing application of fluchloralin at 1 kg/ha or pre emergence spray of alachlor at 1.50 kg/ha is highly effective. Integrated weed management with herbicides at low dose and hand weeding was more effective (Rana *et al.,* 1999). Combinations of alachlor and pendimethalin have also been found effective in rajmash (Rana, 2002).

Water management : During *kharif* season, irrigation is not required but 2-3 irrigations are required for summer or spring crop at flowering and pod filling stage. In the dry temperate regions, rajmash requires 4-5 irrigations.

Sequential and intercropping systems : In the dry temperate region, the sole crop of rajmash is rotated with barley and wheat in one year period and with maize or pea in two years rotation. In the lower elevations, rajmash - barley/wheat/pea/vegetable is common. It is grown mixed with maize and grain amaranth. Intercropping of rajmash with kalazira is remunerative (Rana, 2005).

Harvesting and threshing : Crop is ready for harvesting when the leaves turn yellow and pods start drying. It may be harvested at proper maturity by using serrated sickles. Plants may be dried under sun for 3-4 days and threshed manually by sticks or by trampling under the feet of bullocks or tyres of tractors. Grains should be sun dried and stored in air tight, clean storage bins.

Kulthi

Land preparation : Kulthi is generally grown on marginal lands during *kharif* season in region upto an altitude of 1000 m amsl. Well drained sandy/ clay soils are suitable for its cultivation. It can be successfully grown in slopy lands where water does not stagnate. Land is prepared by two to three ploughings followed by planking.

Seed rate : Recommended seed rate is 18-20 kg/ha and it is reduced to $1/3^{rd}$, when it is intercropped with maize.

Sowing management : This crop is sown with the onset of monsoon by the last week of June or beginning of July. The sowing should be done in lines 30 cm apart by '*kera*' method. However, farmers generally sow the crop by broadcasting.

Fertilizer application including micronutrients : Fertilizer should be applied @ 12-15 kg N/ha and 45 kg P_2O_5/ ha, at the time of sowing. Nitrogen is not required in crops sown mixed with maize.

Weed control : Two weedings and hoeings are necessary to control the weeds. Pre emergence spray of alachlor at 0.75 kg/ha or pendimethalin 0.9 kg/ha is highly effective. In general farmers do not pay any attention to interculture operations which leads to reduction in yield.

Sequential and intercropping system : In Himachal Pradesh, *kulthi* is generally sown mixed with maize.

Harvesting and threshing : The crop is harvested by the second week of November when the plants turn half yellow and pods start drying by pulling out the whole plant. The plants are dried and threshed by trampling under the feet of bullocks or by beating with wooden sticks.

Cowpea

Land preparation : Cowpea is a warm weather crop and can withstand drought and shade. It can be grown well upto 1500 m elevation and performs well on well-drained sandy loam or sandy soils with neutral pH. Soils should be ploughed deeply to ensure smooth penetration of tap roots. It is adversely affected by soil crusting. One deep ploughing followed by 2-3 harrowings and planking are sufficient to obtain good tilth. Soil should be well leveled to avoid waterlogging.

Seed rate : For *kharif*, the optimum seed rate is 15-20 kg/ha. For fodder purpose and green manuring, the seed rate should be increased to 30-35 kg/ ha. During summer, sowing is done with increased seed rate of about 30 kg/ ha.

Sowing management : Cowpea can be sown in summer/spring/*kharif* season. Grain purpose crop may be sown in March (summer) or from mid June to end of July (*kharif*). Indeterminate fodder or green manuring varieties can be grown in first week of June. The row to row spacing of 45 cm and plant to plant spacing of 8-10 cm may be maintained. The crop can be sown by '*kera*' method or drilled in furrows behind a plough. During summer, the row spacing of 30 cm and plant spacing of 7 cm is recommended.

Fertilizer application including micronutrients : The crop should be applied with 20 kg N, 40 kg P_2O_5 and 20 kg K_2O, at the time of sowing especially in organic matter deficient soils for proper development of the roots. Seed inoculation with cowpea specific *Rhizobium spp* would be beneficial when the crop is grown for the first time.

Weed control and interculture : One hand weeding is essential during the first month of the crop. Pre emergence application of alachlor 1.50/ha is highly effective during rainy season. Later on, the plants cover the soil and smoother the weeds.

Water management : It cannot tolerate cold or frost and is vulnerable to water logging. The most critical moisture requiring period for this crop is during flowering. The rainy season crop requires no irrigation whereas; five-six irrigations are required for summer crop.

Sequential and intercropping systems : In Himachal Pradesh, cowpea is generally grown mixed with maize. It is also intercropped with sesamum, soybean etc. One or two rows of cowpea can be grown in between two rows of maize. The following rotations are followed: cowpea - wheat; cowpea - barseem - maize + cowpea; cowpea + maize/sesamum-wheat/gram.

Harvesting and threshing : The variety Himachal Lobiya is almost synchronous in maturity. It requires one picking before final harvesting. Green tender pod can be harvested 50-60 days after sowing and may be consumed as vegetable. For grain purposes, the crop should be harvested after about 85-90 days. The crop should be sun dried and threshed and stored in seed bins at 12% moisture for short term storage and at 8 to 9% moisture for long-term storage.

Mungbean

Land preparation : It is grown both as a *kharif* and summer crop. It requires hot climate. It can be grown upto an elevation of 2000 m amsl. Mungbean can be grown on well drained sandy loam to loam soils. Waterlogged soils are not fit for this crop. A well prepared seed bed is required for proper germination and establishment of the crop. Two or three ploughings followed by planking are essential to make the seed bed free from clods and weeds.

Seed rate : The optimum seed rate is 30 kg/ha for *kharif* crop. It should be increased by 25% for summer crop.

Sowing management : Mungbean is generally sown with the onset of monsoon i.e. by the last week of June to first fortnight of July. The early sown crop suffers due to rains at maturity. Summer crop is sown from the second fortnight of March to first week of April and short duration varieties maturing in 60-70 days can be sown under irrigated conditions. The row to row spacing should be 22.5 and 30 cm for summer and *kharif* season mung, respectively, with plant to plant distance of 10 cm with the seed placed 4 to 5 cm deep. Crust formation should be avoided as it reduces germination. Sowing can be done with *kera* method or behind the plough or with seed drill.

Fertilizer application including micronutrients : On medium fertility soils 15-20 kg N and 40 kg P_2O_5/ha should be applied as basal dressing. Single superphosphate which contains sulphur should be preferred as the phosphorus.The treatment of seed with *rhizobium* increases yield up to 10-12% particularly in the fields where the crop is sown for the first time.

Weed control and interculture : Weeds may cause 30-40% losses in grain yield. Hoeing (25 and 40 days after sowing) should be done to control the weeds. The use of wheel hand hoe is effective. Pre sowing application of 0.75 kg/ha fluchloralin on a well prepared seed bed or pre emergence application of pendimethalin) (0.75 kg/ha using 600 l water/ha) and one hoeing 25 days after sowing is highly effective.

Water management : It has the capability to tolerate moisture stress. In summer season crop, 3-5 irrigations should be applied 20-25 days after sowing and for synchronous maturity and higher grain yields, irrigation should be terminated about 50 days after sowing. No irrigation is required for *kharif* crops but in case of drought conditions, irrigation is essential at flower initiation stage.

Sequential and intercropping systems : *Kharif* mung can be intercropped with maize or sesamum. The following crop rotations can be adopted for including new crop as catch crop in the cropping sequence i.e. Maize/Rice-Raya/Wheat -Summer Moong; Maize-Potato- Summer Moong; Kharif Moong-Wheat/Gram/Mustard and raya.

Harvesting and threshing : Mungbean is asynchronous in maturity. Therefore, the early maturing pods should be hand picked and the crop harvested with serrated sickle when 85% of the pods mature. Threshing may be done after sun drying the crop for three-four days.

Pigeonpea

Land preparation : It is mainly grown in tropical and subtropical climate. Loam to sandy loam soil is the best. The crop grows well in slopy lands in mid hills. It responds well to properly tilled and well drained seed bed. The land should be deep ploughed with soil turning plough followed by two to three cross harrowings. It must be properly leveled to ensure proper drainage.

Seed rate : Seed rate of 15 kg/ha is sufficient.

Sowing management : The crop should be sown in last week of May. Seed should be sown behind the plough or with seed drill with row to row spacing of 50 cm and plant to plant spacing of 15-20 cm.

Fertilizer management including micronutrients : It requires a starter dose of 15-20 kg N and 45 kg P_2O_5/ha as a basal dose.

Weed control and inter-culture : The growth of pigeonpea is quite slow during the first 45-60 days after sowing. On the other hand, the weed growth is very fast due to frequent rains, hence, timely control of weeds is necessary for proper growth of the crop. Two weedings and hoeings are essential for the crop. Weeds can also be controlled by using herbicides such as alachlor 1.0 kg/ha, pendimethalin 1.20 kg/ha and fluchloralin 1.0 kg/ha.

Water management : It can tolerate heavy rains in well drained soils and also tolerate moisture stress because of its deep root system. It is grown in the monsoon season, as a rainfed crop. Being a deep rooted crop, it can extract moisture from deeper layers of the soil. However, in an extended period of moisture stress, it responds well to irrigation.

Sequential and intercropping systems : The crop should be sown in paired rows 30 cm apart followed by two rows of sesame 20 cm apart in Zone I. One row of mash can also be successfully grown between two rows of pigeonpea (arhar). Arhar -wheat/gram/lentil cropping sequence can be followed.

Harvesting and threshing : Pigeonpea is an indeterminate crop. It should be harvested with serrated sickle when 2/3 to ¾ of pods turn brown. The harvested plants should be sun dried and threshed by beating the pods with sticks or trampled under bullocks or tractors. The grains should be sun dried to reduce the moisture content to 10% and stored in dry and clean storage bins.

Ricebean

Land preparation : Rice bean can be grown in varied climatic conditions upto an elevation of 1500 m altitude. Loam, sandy loam and slightly heavy soils with good drainage are suitable for its cultivation. The soil pH of 6.0 to 7.5 is favourable for root development, good nodulation and better nitrogen fixation. A well prepared field is required for better growth and development of this crop. A field with good tilth can be prepared by one or two ploughing followed by harrowing and planking. The field must be free from stubbles and weeds.

Seed rate : Seed size, spacing, method of sowing, germination and purity per cent of the seed governs the seed rate. For sole grain crop of rice bean 25 kg/ha seed is sufficient for the varieties with medium sized seed, however, in bold seeded varieties the seed rate can be doubled. In mixed or intercropping system, seed rate can be adjusted accordingly. A seed rate of 40-50 kg/ha is sufficient for fodder crop.

Sowing management : The crop is generally sown during the *kharif* season. The best time of sowing in mid hill areas is second fortnight of June. However, in sub-Montane and low hills sub-tropical zone, it can be sown upto first fortnight of July. Early sowing of crop before 15 June adversely affects the crop. In the early sown crop, initial drought and flower drop during peak rainy season causes a significant reduction in yield. The crop is generally sown by broadcasting, however, to get better yield, line sowing by seed drill or behind the plough is best. For pure grain crop, row to row spacing of 40-50 cm and plant to plant spacing of 8-10 cm is optimum. Under normal conditions, seed should be sown at a depth of 2-3 cm in heavy soil and 3-4 cm in light soil.

Fertilizer application including micronutrients : Crop should be fertilized with 20 kg N, 40 kg P_2O_5 and 20 kg K_2O/ha. The entire nutrient should be applied at the time of sowing and seed inoculation with *Rhizobium* has beneficial effects. In acidic soils (pH below 6) application of P_2O_5 at 90 kg/ha through ground Mussorrie rock phosphate is as effective as super phosphate. Lime application, depending on the soil reaction is essential in acid soils.

Weed Control : Weeds cause a drastic reduction in yield of the crop ranging between 15-50% depending upon weed density and their composition. Hoeing and weedings are required twice (25-30 and 40-45 DAS) in the early stages of growth. Pre-plant incorporation of fluchloralin 1.0 kg/ha in 600-700 litre of water is highly effective in controlling weeds. The weeds can also be effectively controlled by the pre emergence application of alachlor 1.50 kg/ha or pendimethalin 1.50 kg/ha .

Crop rotation and Intercropping : For better utilization of space, other resources and enhancing land productivity, rice bean can be sown as mixed/intercrop with maize or sorghum. Rice bean-Wheat; Rice bean-Potato-/summer moong or mash are some of the productive rice bean based cropping sequences.

Harvesting and threshing : The crop is harvested by cutting the stalk from the ground level with the sickle, when about 90% pods are matured. Severe beating or trampling may damage the seed coat resulting in reduced seed quality and viability. The threshed grains should be dried to 11% moisture. The grains should be stored in dried seed bins.

Chickpea

Land preparation : It requires cold climate for its growth and development and high temperature for maturity. Areas with moderate rainfall are suitable for chickpea cultivation, however, severe cold and frost can cause flower drop. Well drained loamy to sandy loam soils are suitable for chickpea cultivation. Saline, alkaline or waterlogged soils should be avoided. The crop should be sown in a well prepared seed bed for its proper root development and establishment.

Seed rate : The seed rate varies with the seed size. Seed rate of small seeded desi gram varieties like C 235, HC 1 and HC 2 is 40-45 kg/ha, whereas, it is 80 kg/ha for bold seeded *desi* (HPG17) and *'kabuli'* varieties.

Sowing management : Under irrigated conditions, crop can be sown from 25th October to 10th November. Chickpea mixed with wheat/barley should be sown with the main crop. The small seeded *'desi'* varieties are sown in rows 30 cm apart and 10-12.5 cm deep by *'kera'* method or with seed drill whereas bold seeded varieties should be sown at row to row spacing of 45 cm. Seed should be treated with bavistin + thiram (1:1)/captan at 3 g/kg to protect the crop from seed borne diseases.

Fertilizer application including micronutrients : Gram is responsive to fertilizers. On medium fertile soil, apply 30 kg N, 60 kg P_2O_5 and 30 kg K_2O at the time of sowing. As a mixed crop with wheat, no additional fertilizers are required. Pre sowing seed inoculation with recommended rhizobium culture helps to increase grain yieldly 10-12%.

Weed control and interculture : Two hoeings at 30 and 60 days after sowings are quite effective, alternatively pre emergence spray of Linuron at 0.9 kg/ha or pendimethaline/alachlor 0.75 kg/ha in 600 litre of water per ha is highly effective.

Water management : The crop should be sown after pre sowing irrigation. Thereafter, one or two irrigations are required depending upon soil and climatic conditions at two critical stages i.e. at flowering and pod development stage. No irrigation is required if there is plenty of soil moisture at the time of sowing followed by one or two winter showers. The crop sown after the rice crop in heavy soils does not require irrigation.

Sequential and intercropping systems : Chickpea is generally grown as a mixed crop with wheat, barley, rapeseed and mustard, taramira and linseed. Crop rotations like rice/maize-Chickpea; Mash/Moong/Cowpea - Chickpea; Sesamum-Chickpea can be followed.

Harvesting and threshing : The crop for green pod purpose is harvested by uprooting the whole plant. Crop should be harvested when the leaves become reddish brown and start shedding. Crop should be sun dried for few days and threshed by beating the plants with sticks or by trampling under the tyres of the tractor or feet of the bullocks.

Lentil

Land preparation : Lentil requires cold temperature during growth and warm temperature at the time of maturity. It is a hardy crop which can tolerate frost and can be grown up to an elevation of 3000 m amsl. Lentil can be grown in all types of soils but well drained loam soils are the best. Two to three ploughings are needed for well pulverized soil and the field should be free from weeds and clods.

Seed rate : Seed rate of 25-30 kg/ha is optimum, however, it should be increased to 35 kg/ha under late sown conditions.

Sowing management : The optimum time for sowing is from end of October to first fortnight of November. It can be sown up to last week of December as a contingent crop in case of late rainfall. Sow the crop in rows 22.5-25 cm apart by *'kera'* method and spacing reduced to 20 cm under late sown condition.

Fertilizer management : It requires 10 kg N and 40 kg P_2O_5/ha which may be applied at the time of sowing. Seed should be treated with the recommended rhizobium culture.

Weed control and inter-culture : Lentil is a short statured crop and its growth and productivity is affected by weeds. Weeding or hoeing once or twice is sufficient to control weeds. Pre emergence application of alachlor 1.0 kg/ha results in excellent weed control.

Water management : Lentil requires very limited irrigation. However, in the absence of winter rains, one irrigation may be given at pod formation.

Sequential and intercropping systems : Lentil is generally intercropped with mustard, barley and wheat. One row of mustard or barley sown after 4-6 rows of lentil is optimum for getting maximum productivity. The following crop rotations can be followed.

Rice/Maize-Lentil; Rice-Lentil-Green Manuring; Rice/Maize-Lentil + Mustard/Lentil + Wheat

Harvesting and threshing : The crop should be harvested preferably in morning hours to avoid shattering when the plants dry up and pods mature. Seeds should be cleaned and sun dried to 12% moisture.

Pea

Land preparation : Crop requires cool climate during its vegetative phase. However, severe cold and frost may adversely affect flowering and fruiting. Field pea can be grown on well pulverized, levelled and well drained sandy loam to clay loam soils. Two to three ploughings followed by planking are required.

Seed rate : 60-75 kg seed/ha.

Sowing management : The optimum sowing time is from end October to mid November. The crop should be sown in rows at 30 cm apart by *'pora'* or *'kera'* method.

Fertilizer application including micronutrients : On medium fertile soils, apply 30 kg N and 40 kg P_2O_5 at the time of sowing along the rows and inoculate seed with the recommended rhizobium culture results especially when it is being grown for the first time. Garden pea is more responsive to fertilizers (Rana *et al.,* 2003). Response of pea to seed soaking and coating with P, neem and sea weed extract was also significant under dry temperate conditions (Rana *et al.,* 2007).

Weed control : The crop should be kept free from weeds by giving two hoeings at 3 and 6 weeks after sowing. Pre emergence sprays of pendimethalin 0.75 kg/ha within two days of sowing in 600 litres of water is very effective. Herbicide combination (Rana *et al.,* 2004) or INM involving herbicides and hand weeding (Rana, 2002) is a better option than the herbicides or hand weeding alone.

Irrigation : The crop should be sown under good moisture conditions i.e. after pre sowing irrigation. Thereafter, one or two irrigations are required depending upon soil and climatic conditions. The crop sown in heavy soils after rice does not require irrigation.

Sequential and intercropping systems : In general, the field pea is grown mixed with mustard, barley, linseed and wheat etc. The following crop rotations can be followed: Maize/rice-Pea; Rice-Pea + Mustard/wheat; Rice-Pea -Okra. In Zone IV, pea - common buckwheat has emerged as an important cropping system in the recent past (Rana *et al.,* 2003; Rana *et al.,* 2006). Peas can be grown as a relay crop between standing row of *kalazira* in June (Rana *et al.,* 2003).

Harvesting and threshing : The crop is ready for harvesting from second to last week of April. Fully matured crop should be threshed after sufficient drying in the sun.

Plant Protection Measures of Pulses

Insect-pests

The important insect- pests of pulses and their management are as follows:

Pod borer (*Heliothis Armigera* (Hubner) : The larvae feed on leaves, flowers and pods as soon as they have developed. The young larvae feed on leaflets, scrapping the epidermal tissues and the older larvae feed on leaves, flower buds and pods and cause great loss in yield. The larvae also make a hole in the pod and feed on its seeds, which are left partially damaged.

The pest is reported to feed on a large number of hosts in the Himachal Pradesh. Amongst the pulses, it is a serious pest for chickpea and pigeonpea, and causes considerable damage in lentil, soybean, cowpea, greengram, black gram and pea.

The following package is recommended for its management.

- Proper field sanitation.
- Chickpea crop sown during middle of October usually records least damage of *H. armigera* as the crop matures before the peak period of pest incidence.
- Mixed/intercropping of chickpea with wheat, barley, mustard and linseed leads to reduced damage of *H. armigera* in chickpea.
- Sex pheromone traps should be used for monitoring of adult male population of *H. armigera.* A threshold level of 5-6 moths per trap is indicative of its peak activity period.
- Nuclear Polyhedrosis Virus (NPV) at 250 litre/ha alone.
- Spray the crop at 50% flowering with monocrotophos (625 ml Monocil 36 WSC) or carbaryl (1250 g Sevin 50 WP) or 1.250 l Thiodan/ Endocel/ Hildan 35 EC or fenvalerate (315 ml Fenval/ Sumicidin/ Agrofen 20 EC) or deltamethrin (625 ml Decis 2.8EC) in 625 l water per hectare. Repeat the spray at 7-15 days interval depending upon severity of the pest.
- Spray the crop with azadirachtine 0.03%. If there is another attack of insect, spray after 15 days.
- *Campoletis chloridae* is a major larval parasitoid of *H. armigera.*

Cut worm (*Agrotis Psilon* Huber and *A. Segetum*) : The caterpillars become active at night. They cut plants at base, just above the soil surface. They also cut branches of older plants. The cut plants or branches are dragged and partially buried in the soil for feeding by the larvae. Cutworms are polyphagous but among the pulses, gram and chickpea are main hosts. The following package is recommended for its management.

- Well rotted farmyard manure.
- Light trap for monitoring and mass-trapping adult moths.
- Foliar and soil surface application of 1250 ml chlorpyriphos (Durmet 20 EC) or 470 ml cypermethrin (Ripcord 10 EC) in 625 l water per hectare with the initiation of plant damage during evening hours.
- At field preparation, broadcast 2 l chlorpyriphos 20 EC mixed in 25 kg sand per hectare.

Termites (*Odontotermes Obesus* (Rambur)) and *Microtermes Obesi* (Holmgren) : Termites initiate damage soon after germination. The damaged plants wither and dry up completely and are easily pulled out. For its control, Broadcast 2 l chloropyriphos 20 EC mixed in 25 kg of sand/hectare after the last ploughing but before planking and seed treatment with chlor pyriphos 20 EC at 5 ml/kg seed is recommended.

Red hairy Caterpillar (*Amsacta moorei* Butler) : Young larvae are gregarious. They feed on leaves in clusters and skeletonize the plants completely and later on disperse in the field. They appear in epidemic form with the onset of the monsoons. The older larvae feed voraciously on all vegetation resulting in a disaster. Following package is recommended for its management.

- Collection and destruction of egg masses and the gregarious larvae in the first and second instar.
- Plough the field after harvesting to expose the pupae to let them desiccate and die.
- Use light trap for monitoring and mass-trapping adult moths.
- Give spot treatment by dusting 5% malathion dust at 35-40 kg/ha against gregarious caterpillars.
- Spray Thiodon/Endocel 35 EC at 1.250 l or dichlrovos (Nuvan 76 EC) at 625 ml in 625 l water/ha for the advanced staged larvae.

Bihar Hairy Caterpillar (*Spilosoma Obliqua* (Walker) : The caterpillars skeletonize the plants completely. They appear in epidemic form at the onset of monsoon. The advanced stage larvae nibble the leaves in solitary phase and eat away the entire leaf tissue, leaving only midribs. Hairy caterpillar is polyphagous. In addition to black gram, greengram, soyabean, cowpea and peas, it attacks a large number of other crops. The following package is recommended for its management.

- Collection and destruction of egg masses and the newly emerged caterpillars.
- Clean cultivation and destruction of alternate hosts of the pest.
- Spray 1.250 litre (Thiodon/Endocel 35 EC) or 1.250 litre quinalphos (Ekalux 25 EC) or 625 ml dichlrovos (Nuvan 76 EC) ml in 625 l water per hectare. The early stages can be controlled by dusting with malathion 5% at 35-40 kg/ha.

Blister Beetle (*Mylabris* spp.) : Adults voraciously feed on flowers and tender terminals, thus causing loss in pod formation in black gram, rajmash, pigeonpea etc. Follow the package as recommended below for its management.

- Collect and destroy adult beetles.
- Spray fenvalerate (315 ml Sumicidin/Fenval/Agrofen 20 EC) or cypermethrin (190 ml Cymbush 25 EC/470 ml Ripcord 10 EC) or deltamethrin (625 ml Decis 2.8 EC) or monocrotophos (625 ml Monocil 36 WSC) in 625 litres of water per hectare.

Pod Bug (*Clavigralla Gibbosa* Spinola) : The nymphs are gregarious. Both nymphs and adults suck sap from stem, leaves, flower-buds and pods and lower the vitality of the plant. Small, whitish and yellowish spots on the upper leaf surface confirm the damage. Pigeonpea (*arhar), lab-lab*, cowpea, chickpea, kidneybean and black gram are the main hosts.

Spray the crop with monocrotophos (625 ml Monocil/Milphos/ Monocrotophos 36 SL) or dimethoate (625 ml Rogor 30 EC) or methyl demeton (625 ml Metasystox 25 EC) or fenvalerate (315 ml Sumicidin/Fenval/Agrofen 20 EC) in 625 litres water per hectare before flowering or pod formation with the appearance of pests is recommended for its timely management.

Whitefly (*Bemisia Tabaci* Genn.) : Both nymphs and adults suck cell sap, remain at the underside of leaves, which results in lowering the vitality of plants. Damage by the insect is caused due to loss of cell sap and interference with normal photosynthesis due to the growth of the sooty mould on the honey dew excreted by the insect. The most serious damage is caused by transmitting the yellow mosaic virus. Black gram, greengram, French bean and cowpea are the main hosts.

Spray the crop with 1.250 litre Thiodon/Endocel 35 EC or 625 ml methyl demeton (Metasystox 25 EC) or 625 ml dimethoate (Rogor 30 EC) or acephate (415 g Starthane 75 SP) or imidacloprid (200 ml Confidor 200 SL) in 625 litre water/ha at the pre bloom and flowering stages is recommended for its timely management.

The pest is rapidly developing resistance to a large number of insecticides, therefore, insecticides belonging to different groups should be sprayed alternatively.

Bean Thrips (*Megalurothrips Distalis* (Karny) : Infested inflorescence becomes abnormal showing symptoms of flower drop. Infested flowers develop into abnormal pods with shrivelled and small sized grains in the mature pods. Feeding on the leaves leads to localized discolouration and may cause curling down of the leaves. In heavy infestation, infested leaves are shed resulting in devitalization of the plants affecting its growth. The plants attain bushy growth.

At flower initiation stage spray dimethoate (Rogor 30 EC) 625 ml or malathion 50 EC 625 ml or methyl demeton (Metasystox 25 EC) 625 ml in 625 litres of water per hectare is recommend for its management.

Black Aphid (*Aphis Craccivora* Koch.) : The nymphs and adults damage the plant by sucking the sap from the tender portion. Thus, there is heavy drain of vital plant sap and leaves, shoots and flowers deform. Due to secretion of honey dew, black moulds develop on infested plants which interfere in photosynthesis.

This aphid is recorded on a number of host plants including vegetables, ornamental plants and weeds. Among the pulses, it damages beans, pigeonpea, chickpea, lentil and peas. The recommended package for its timely management is given below.

- Spray the crop with 415 g acephate (Starthane 75 SP) or 625 ml methyl demeton (Metasystox 25 EC) or 625 ml dimethoate (Rogor 30EC) or 315 ml fenvalerate (Sumicidin 20 EC) or 190 ml cypermethrin (Cymbush 25 EC) in 625 litre water per hectare.
- Aphids are parasitised by *braconid hymenopterans* and predated upon by ladybird beetles, syrphids and lacewings.

Pod Fly (*Melanogromyza Obtusa* Malloch) : Maggots feed on the developing pods. The maggots eat away only a part of seed and the partially damaged seed becomes subject to bacterial and fungal infections. The development of pod is retarded. The attacked pods become twisted and deformed. The fully developed maggots make a mustard grain sized membranous demarcation on the inner surface of the pods, which is visible from outside. The adult fly emerges from this membraneous portion making a small hole of about 1 mm on the pod surface. Damaged seeds become unfit for consumption and sowing. Pigeonpea is the main host, but they also attack some other pulses.

Spray the crop with endosulphan (1.250 litre Thiodan/Endocel 35 EC or monocrotophos (625 ml Monocil/Milphos 36 SL) or carbaryl (1250 gm Sevin/ Hexavin 50 WP) in 625 litres of water at 50 per cent pod formation stage and repeat it after 10-15 days interval is recommended for its management.

Pod Borer (*Etiella Zinckenella* Treitschke) : Larvae of the pod borer damage the buds, flowers, pods and seeds in pods. The young larvae bores its way into the floral parts and once the larvae is inside the flowers or pods, it causes severe damage. A single larva devours many seeds in the pod and converts it into a mass of frass. These are held together by a loosely spun web. Lentil, pigeonpea, peas, cowpea and horse gram are the main hosts.

Spray 1250 ml Thodan/Hildan/Endocel 35 EC or 1.250 kg carbaryl (Sevin/Hexavin 50 WP) mixed in 625 litres of water/ha at the flower initiation stage followed by another spray after three weeks, depending upon the incidence of pest is recommended for its management.

Leaf Eating Weevil (*Strophosomoides Kumaoensis* Aslam) : The rajmash leaf weevil, *S. kumaoensis,* damages the crop right from the seedling to crop growth stages. The adults damage young and tender leaves and growing points immediately after germination. Under severe infestations, complete defoliation of seedlings occurs and resowing becomes mandatory (Sood and Rana, 2004). The following package is recommended for its timely management.

- Deep ploughing of fields after crop harvest should be done to expose the over wintering population.
- Sowing of crop should be done in the last week of May.
- Optimum dose of potash fertilizers (at least 20 kg/ha) reduces the incidence of weevil to some extent, whereas higher nitrogen increases the incidence of weevil.
- Soil treatment at sowing with fenvalerate 4% dust (20 kg/ha) reduces the weevil incidence.
- If the population of weevil touches the economic threshold of 5 weevils/plant, the crop should be sprayed with deltamethrin (625 ml Decis 2.8 EC) or fenvalerate (315 ml Sumicidin/Fenval 20 EC) or methyl demeton (625 ml Metasystox 25 EC) in 625 litre water per hectare 10-15 days after germination to manage the weevil population below the economic injury level of 6-8 weevils/plant.

Mite (*Tetranychus* spp.) : During dry weather, adults and nymphs which feed on underside of leaves, breed near the petiole in between the main veins. In case of heavy infestations, mites are found all over the leaves. The mites live gregariously and are protected by a silken web. Feeding by the mites results in characteristic blotches on leaves and crinkling of the leaves from below. The infested leaves show yellowish speckled appearance which later turns brown. Gradually the infested leaves dry up and a large number of webs are formed on the leaves. For its control, as soon as the attack is noticed, spray the crop with malathion. The pest can also be controlled by the spray of dicofol or wettable sulphur (3 g/l) or ethion.

Pea Leaf Minor (*Chromatomyia Horticola* (Goureau) : It is a polyphagous pest. Larvae make serpentine mines in the leaves of the host plants. The larvae feed below the epidermis and make a linear leaf mine on dorsal and ventral surface of leaf which results in loss of chlorophyll bearing tissues thereby, reducing photosynthetic activity. Adults feed on the cell sap with their sucking and sponging type of mouth parts. The maggots mine into the leaves, eating the green matter and pupate inside the mines. The conspicuous zigzag mines ruin the leaves. Flowering and fruiting capacity of the infested crop are reduced and the crop yield is adversely affected. Peas and lentil are the main hosts besides a large number of other crops. The following package is recommended for its timely management.

- Timely sown crop has lesser incidence of pea leaf minor.
- At 40 per cent leaf infestation, spray the crop with lambda-cyhalothrin (500 ml Karate 5EC) or cypermethrin (190 ml Cymbush 25EC) or dichlorvos (625 ml Nuvan 76 EC) or fenitrothion (625 ml Sumithion/Folithion/Accothion 50 EC) in 625 litre water per hectare. Repeat the spray at 15 days interval, if necessary.
- During latter stages of crop growth, the population of the pea leaf miner is kept under check naturally by a large number of larval and pupal parasites. These include Braconids (*Bacnusa* sp.) and eulophids (*Chrysonotomyia* sp.; *Diglyphus* sp.; *Pediobius* sp.).

Pea Pod Borers (*Etiella Zinckenella* T.) : Description and management of the pest is given under lentil pests. Besides this, larvae of other insects damage the pea pods by boring into them.

- Gram pod borer, *Helicoverpa armigera*: Discussed under chickpea pests.
- Pea blue butterfly: There are two types of blue butterflies.
    - *a.* *Lampides boeticus* : Larvae are yellowish green or light purple with light green color on ventral surface and small setae covering the entire body.
    - *b.* *Euchrysops cnejus* : Attacks beans, cowpea, peas and pigeonpea. Larvae feed on bud flowers, developing pods and bore into pods and also feed inside throughout larval stage contaminating all seeds. The tiny larvae enter the pods and entry holes are kept closed with faecal matter thus, leaving no sign of its attack. The pest can only be detected after opening the pods.

For its timely management spray the crop with carbaryl (1250 g Sevin 50 WP) or 1.250 litre Thiodan/Endocel/Hildan 35 EC or fenvalerate (315 ml Fenval/Sumicidin/Agrofen 20 EC) or deltamethrin (625 ml Decis 2.8EC) in 625 litre water per hectare. Repeat the spray at 15 days interval depending upon severity of the pest attack. Before consuming the pods, a waiting period of 7-10 days after application of insecticides must be observed.

Pea Stemfly (*Ophiomyia Phaseoli* (Tryon) : The infestation appears as soon as plants appear on the ground and the seedlings are severely damaged. On hatching the maggots mine the leaves and tunnel underneath the epidermis of the leaf until they reach one of the veins, which leads to midrib and finally to leaf stalk and stem. The first-two leaves become yellow and droop. The sides where larvae and pupae are present, become swollen and start rotting. The affected leaves fall down; seedlings remained stunted or die. In older plants, infestation leads to retarted plant growth. Beans, pea, cowpea, soybean and dolichos bean are preferred hosts.

For timely management follow the following schedule.

- Avoid early sowing of the crop. Sow it in second fortnight of October to escape damage of the pest.
- Apply 625 ml of monocrotophos (Monocil 36 WSC) or 625 ml methyl demeton (Metasystox 25 EC) in 625 litres water per hectare at 15 days interval during flowering stage.

Pea Aphid (*Acrythosiphon Pisum* Harris) : Colonies of young and old aphids are found on tender leaves, stems and pods. Both nymphs and adults suck the cell sap and the affected leaves often gets cupped or become irregularly distorted; shoots become stunted, malformed and give a sickly appearance. Leaves and pods are malformed with undersized grains. Attacked plants become stunted and yellow. The development of plant is also retarded resulting in loss of the crop. A wide range of leguminous plants are affected by this pest.

The pest can be controlled by spraying 1.25 kg of carbaryl (Sevin 50 WP) or 625 ml Dimethoate (Rogor 30 EC) when the attack starts and repeat it after 15 days interval.

Cowpea Aphid (*Aphis craccivora* Koch) : It is the major pest of cowpea causing damage to cowpea at all stages of crop development and has been dealt in detail under black gram and greengram.

Coreid Bug (*Clavigralla gibbosa* Spinola) : It causes widespread damage. Adults migrate from wild host plant and together with nymphs suck sap from developing pods and seeds. The nature of damage and management has been discussed under black gram and greengram.

Pod Borers : These are major insect pests of cowpea and cause extensive damage to flowers, pods and developing seed. There are three species of insects viz., *Helicoverpa armigera, Etiella zinckenella* and *Lampides boeticus* which devour the pods of cowpea. The nature of damage and management of *Helicoverpa armigera* has been discussed under gram, while *Etiella zinckenella* and *Lampides boeticus* has been dealt under peas.

Diseases

***Cercospora* leaf spot :** It is a widespread disease causing losses ranging from 25-37%. Two species C. *canescens* and C. *cruenta* are the most prevalent ones. The symptoms appear in about 45 days old plants. The symptoms of C.

canescens appear on upper leaf surface as irregular, light brown spots with a grey centre and red margin on upper leaf surface. These coalesce to cover the entire leaf area. Slightly sunken purple spots are produced on the pods which turn black. Infected pods produce shrivelled and discoloured grains.

The symptom due to *C. cruenta* appear as circular to irregular purplish brown spots with pale green centres on leaves. During humid conditions, spots increase and merge with each other to form large irregular lesions delimited by veins. The spots on midrib, veins and petioles are elongated, sunken, tan to brown in colour and depressed in the centre. They coalesce to girdle the stem and branches. The lesions on pods initiate as water soaked spots with bright red, yellow or orange margins and dark centres. Pink vicious sporulation can be seen in the centre.

To control the disease follow the following schedule.

- Spray the crop with mancozeb 75 WP (Dithane M 45/ Indofil M45)/ captan at 2.5 g/litre cabendazim 50WP (Bavistin 50 WP) or Kavach at 2 g/litre at fortnightly interval starting 45 days after sowing or with the appearance of the symptoms.
- Use disease free seeds produced in disease free area and destroy diseased crop debris.
- Sow the crop late in the season i.e. from 15 July to 20 July.
- Use field tolerant varieties Saketi (Moong), UG 218, Pant U 19 (Mash).

***Collototrichium* Leaf Spot :** The fungus attacks all the aerial parts. Seed borne infection produces dark brown, eye shaped lesions on hypocotyl which may enlarge and break the stems. Crescent shaped, brown spots appear on the under surface of the leaves involving midrib and veins. Later on, similar symptoms appear on the upper surface of the leaves. In severe cases, the lesions may coalesce resulting in the burnt appearance of plants. Water soaked spots with bright red, yellow or orange margins and dark centres appear on pods which enlarge under the favourable environmental conditions and become brown coloured. Sporulation of the fungus can be seen as pink viscous mass in the centre of the spots. Seeds from such infected pods are discoloured and reduced in size.

Control : Follow the control measures as in case of Cercospora leaf spot

Powdery Mildew : It starts as faintly dark areas on leaves and develop small, white specks which enlarge and coalesce to form complete coating of a white powdery mass on leaves, stems and pods. The powdery mass consisting of mycelium and conidia turn dirty white. The disease becomes more severe at flowering stage and persists until harvesting.

Control : Dust the crop with sulphur at 20 kg/ha/Spray with carbendazim 50 WP (Bavistin 50 WP) at 1 g/litre of water.

Mungbean Yellow Mosaic Virus : The symptoms appear as mild scattered yellow specks or spots. New trifoliate leaves show yellow patches alternating with green areas which turn yellow and white shaded and finally necrotic. The leaves show slight puckering. Affected plants remain stunted with poor flowering and seed setting.

To control the disease, follow the following schedule.

- Obtain seed from disease free crop grown in disease free areas eg. as high hills
- Intercropping with maize reduces the disease.
- Destroy diseased plants in the initial stages.
- Grow resistant varieties UG 218 and Pant U19 (mash), SML 668, SML 357 and HPBM 1(moong)
- Spray the crop with dimecron/metasystox/monocrotophos/ dimethoate at 2 ml/litre of water to control white fly.

Leaf Crinkle : Disease manifests itself at the second or third leaf stage (after 2-3 weeks of sowing) as light green colour of the leaf followed by typical crinkling. The leaves from second trifoliate leaf onwards gets enlarged, wrinkled, puckered and plant is markedly reduced in size. Petioles of the terminal leaflets are shortened giving the plant a bushy appearance. Big yellowish patches are formed on the foliage. Flowering is delayed or does not occur. Only few flowers develop into pods bearing a few small and deformed seeds. The plants remain green for a long time.

Follow the control measures as in case of mungbean yellow mosaic virus. Additionally, hot water seed treatment at 52 °C for 30 minutes followed by drying under shade is effective.

Anthracnose : All the above ground plant parts of Rajmash are attacked. The characteristic symptoms appear on pods as water soaked lesions which become brown and enlarge to form circular spots of varying sizes. Seed from infected pods show variation in colour. The spots are usually depressed with dark centres and bright red yellow or orange margins. In moist weather, a pink spore mass appears in centre of the lesion. Seed borne infection induces brown to black eye shaped lesions on the hypocotyl and cotyledons causing breakage of stem. In the older stems, 5-7 cm long sunken lesions appear. Leaf infection initiates as small lesions on petioles and lower leaf surface extending along the veins and becomes brick red to purple and blackened. Similar symptoms appear on upper leaf surface. Ton control the disease, follow the following schedule.

- Use healthy seeds obtained from disease free areas and treat seed with Bavistin + thiram 1:1/Benlate at 2.5 g/kg seed.
- Deep plough the crop debris and follow two years crop rotation.
- Spray the crop with mancozeb 75 WP Diathane M 45 or Indofil M 45 at 2.5 g/litre 35-40 days after seed germination and repeat it after 15 days followed by spray of disease resistant bavistin 50 WP at 1g/ litre of water, mancozeb 25 WP at 2.5 g/litre of water with the initiation of the disease.
- Cultivate varieties Kanhan, Baspa, etc.
- Well drained soil and proper spacing help in reduction of the disease.

Angular leaf spot : The disease occurs on all aerial plant parts. It starts at flowering stage as dark brown and angular round lesions with concentric rings surrounded by yellow halo on primary leaves. Several lesions coalesce resulting in general leaf chlorosis and premature defoliation. Oval to circular, reddish to dark brown lesions appear on the pods whereas, elongated and brown lesions appear on stems, branches and petioles enlarge and become brown in colour. Sometimes pods are deformed. Dark grey stromata and dark grey to black synnemata bearing conidia are produced in all types of lesions

Follow the control measures as in case of anthracnose except that spray the crop with Baycor 25 EC, Chlorthalonil, bavistin 50 WP at1 ml/litre of water mancozeb 75 WP/Zineb 75 WP/copperoxychloride 50 WP at 2.5g/ litre of water at fortnightly intervals starting from flowering or with the appearance of the symptoms and repeat it for 2-3 times

Floury leaf spot : White powdery growth resembling flour appear on lower surface of leaves at flowering stage. The diseased leaves are shed prematurely.

Control : Follow the control measure as in case of angular leaf spot.

Web blight : Rhizoctonia web blight and damping off occurs at the time of seedling emergence. Sunken reddish brown lesions appear near the soil line on hypocotyl and stem of the plant. It retards development of the plants. Under damp conditions, circular to irregular brown spots with distinct borders appear on leaves and under frequent rains, the lesions in contact with soil show pronounced leaf scald symptoms. The whole leaf turns grey, brown and dies. Brown to dirty grey fungal mycelium with sclerotia can be seen in the soil in vicinity of dead leaves.

To control follow the following schedule.

- Keep the field clean and follow multiple cropping rotations.
- Seed treatment with Bavistin 50 WP at 2.5 g/kg.

- Spray the crop with bavistin 50 WP at 1 ml/litre of water with the initiation of the symptoms and repeat it at 7-10 days interval in warm and humid weather conditions.
- Grow recommended resistant varieties.

Common Bacterial/Fuscous Blight : The symptoms of common and fuscous blight appear as translucent, water soaked lesions on the lower leaf surface. The lesions enlarge irregularly, coalesce and become surrounded by narrow chlorotic zone which becomes brown and dies. Small water soaked spots appear on pods which enlarge, turn reddish brown and become slightly sunken. Under humid conditions, a yellow exudate is produced which forms a yellow crust on drying. Heavy infections during podding and seed formation result in shrivelled seeds. Seed borne infection results in destruction of growing tips and primary leaves of the seedlings. Symptoms on older stems appear as water soaked spots which enlarge and become reddish in colour. It may girdle the stem at node resulting in its breakage.

To control follow the following schedule.

- Use of clean seed produced by following strict seed production standards.
- Seed treatment by dipping the seed in 0.1% streptocycline sulphate + hexacop solution (1g+ 2.5g/10 litres of water) for 4 hours
- Foliar spray with copper oxychloride 50 WP+ Zineb 75 WP/ mancozeb 75 WP/copper oxychloride 50 WP + streptocycline sulphate 30 g+ 1 g in 10 litre of water is effective and economically viable for seed production plots.
- Destruction and removal of plant debris and effective weed control.
- Crop rotation of 2-3 years and intercropping with maize.
- Avoid movement through crops when foliage is wet.
- Use of resistant varieties like K 198.

Common Mosaic Virus : Early symptoms appear as crinkle, chlorosis and stiffness of leaves. The leaves drop and petioles are much shortened. Typical symptoms comprise of distinct dark green sectors on lighter green background, accompanied by downward curling of leaf margins. Severe infection leads to distortion and puckering of leaves, stunted growth and distortion of flowers and pods. The plants remain pale.

To control fallow the following schedule.

- Timely sowing of crop with clean and healthy seeds and maintaining optimum plant density.

- Chemical control of insect vectors as has been given in mung yellow mosaic virus of mash.
- Grow recommended resistant varieties.

Phytophthora Blight : The symptoms resemble damping off. Young plants of pigeonpea and its tissues are more susceptible. Infected plants show water soaked lesions on leaves and brownish black, slightly sunken lesions on the stems and petioles. Infected leaves lose turgidity and become blighted and the seedlings suddenly die. Stem lesions expand rapidly and girdle the affected portion. The stem and branches break at the point of infection and many plants die. Plants which survive the disease produces large galls on stem especially on the edges of the lesions.

To control the diseases follow the following schedule.

- Cultivation of resistant varieties.
- Mulching, wider inter row spacing, crop rotation and intercropping with short leguminous crops e.g. moong and mash.
- Avoid sowing in low lying fields and provide good drainage
- Balanced use of fertilizers especially potassium.
- Seed dressing with Ridomil MZ at 3 g/kg followed by two sprays of Ridomil MZ at 2 g/litre of water starting 15 days after sowing at 10 days interval or 3 sprays of copperoxychloride at 3 g/litre of water.

Sterlity Mosaic : Sterility mosaic virus is the most important disease of pigeonpea in India causing losses upto 95%. The disease can be easily identified as patches of bushy, pale green plants without flowers and pods. The leaves of infected plants remain small with light green mosaics and mild mottling. The plants may be wholly or partially sterile. The entire plant becomes stunted with shortened internodes.

To control the disease follow the following schedule.

- Collect seed from healthy plants and follow crop rotation and avoid repeated cultivation in the same field to reduce inoculum levels and vector populations.
- Cultivate resistant varieties and grow sole crop with optimum plant density.
- Spraying the crop with metasystox/karathane at 1 ml/litre of water reduces the vector population and controls the disease.

***Ascochyta* Blight :** The initial symptoms appear near the tips of young shoots and subsequently spread to leaflets, petioles, pods, branches and stem in the form of brown spots. These dark spots become studded with minute black, dot like bodies (pycnidia). On pods and leaves, the lesions are circular with concentrically arranged pycnidia. The apical tips, twigs, branches and stem show girdling and branches above the girdled portion are killed and break off before drying. In severe infections, the whole plant may be killed and the whole field becomes blighted.

To control the disease follow the following schedule.

- Use disease free seed produced in disease free areas. Destroy the crop debris.
- Treat the seed with Indofil M 45/Bavistin/Captan /Bavistin + Thiram (1:1) at 3 g/kg of seed.
- Two to three foliar sprays of Captan, Captaf, Indofil M 45 at 2.5g/ litre of water or Bavistin 50 WP or Kavach at 1g/litre of water are highly effective.
- Grow resistant varieties HPG 17, ICCX 810800 (HC 1) and ICCV 90201 (HC 2)

Botrytis Grey Mould : The disease initiates as grey or dark brown lesions covered with erect hair like sporophores on stem, branches, leaves and pods. The lesions enlarge and girdle the stem. Tender branches break off at the point of infection and leaves turn yellow and dark brown before shedding. On the stem, grey mould is gradually replaced by grey or black mass containing black sporodochia. In case of heavy infection, flowers drop and pods are not formed and the infected pods bear small and shrivelled seeds.

To control adopt the following pacakge.

- Late sowing of the crop helps in reducing the disease.
- Dry seed treatment with a mixture of fungicides Bavistin + Thiram (1:1)/ Mancozeb 75 WP at 3 g/kg, Triademinol 25 EC at 2 g/kg can completely eliminate the seed borne infection.
- Foliar sprays of bavistin 50 WP/bayleton 25 EC at 1 g/litre or Indofil M 45 at 2.5 g/litre of water immediately after rains in February and March is effective in controlling the disease.

Wilt : The disease appears on 20 days old seedlings which collapse and lie flat on the ground. Older seedlings show drooping of the leaves and dull green colour in the initial stages. Plants turn yellow, straw coloured, wither and finally dry up. The roots become dark coloured resulting in complete rotting. Dark brown to black vascular discoloration is the most characteristic symptom. Some times, only one side branch is affected resulting in partial wilting.

To control the wilt, follow the following schedule.

- Destruction of crop debris, deep summer ploughing and late sowing help in decreasing the disease.
- Crop rotation with non leguminous crops.
- Use of healthy and disease free seeds produced in disease free areas and their treatment with Thiram + Bavistin (1:1)/Captan at 3 g/kg.
- Cultivation of wilt tolerant varieties like HPG 17, ICCX 810800 (HC 1) and ICCV 90201(HC 2).

Rust : Rust is a widespread foliar disease of lentil. The disease appears at the flowering stage with the formation of yellowish white, circular pycnidia and aecia singly or in small groups on leaflets and pods. Later on, oval to circular brown uredial pustules measuring up to 1 mm in diameter develop on either surface of leaflets, branches, stem and pods which coalesce to form larger pustules. The dark brown to black elongated telia are formed mainly on branches and stems late in the season. In the severe form of infection, plants dry up and die. Consequently, either there is no seeds formation or seeds remain small and shriveled.

To control the lentil rust adopt the following package.

- Foliar spray with Zineb (Dithane Z 78)/Mancozeb 75 WP at 2.5 g/ litre of water, Triademifon (bayleton 25 EC) + Propineb (anthracol 50 WP) at 1 g + 2.5 g/litre of water are effective in controlling the disease.
- Cultivate rust resistant cultivars *viz.* L 9-12, EC 1 , Vipasha and L 207.

Vascular wilt : The disease appears in patches both at the seedling and adult plant stage. Seedling wilt is characterized by sudden drooping, drying of leaves and seedling death. The roots appear healthy but show reduced proliferation. Symptoms at adult plant stage appear from flowering to late pod- filling stage. Top leaflets show sudden drooping, premature shedding, dull green foliage and wilting of the whole plant, or individual branches. The root system remains underdeveloped and appears light brown in colour, with reduction of lateral roots. No internal discoloration of the vascular system is visible. In adult plants, the wilt appears at flowering and podding resulting in shrivelled seeds.

To control the disease, follow the following schedule.

- Delayed sowing and addition of organic matter reduces disease incidence.
- A crop rotation of 2-3 years with paddy reduces the disease incidence.

- Seed treatment with Thiram + Carboxin /Bavistin at 3 g/kg reduces the disease.
- Biological control agents like *Trichoderma viride* Pers. ex Fr., *Trichoderma harzianum* Rifai and *T. koningii* Oudemans are antagonistic to *F. oxysporum* f. sp. *lentis*.
- Cultivation of resistant varieties.

***Rhizoctonia* Seedling Blight :** Water soaked lesions appear on the hypocotyl and epicotyl and reddish- brown lesions are formed above and below the cotyledonary node. Most characteristic and diagnostic symptom is the death of stem radicals on their emergence. Auxiliary stems are also affected and the plant never fully recovers. On older plants, reddish-brown, sunken lesions may appear on epicotyl, resulting in girdling and severe stunting of plants.

To control Follow crop rotation with cereals and clean tillage. Treat the seed with Carbendazim 50 WP (Bavistin 50 WP) at 3g/kg of seed.

White Rot : The disease appears when the canopy has completely covered the soil surface. Dense mycelial mat is formed at the soil surface and on the prostrate vines, pods and leaves. Affected areas become slimy and dark. Plants start decaying and drop down. Irregular sclerotia are formed within the mycelial mats. Free water or dew and nutrient source such as senescent flower petals or other decaying tissues lead to high infection and the disease become severe within two days. It affects the quality of pods.

To control follow the following schedule.

- Sclerotia should be deep ploughed after harvesting the infected crop and follow crop rotation with wheat or barley.
- Seed rate should be reduced and cultivars with open vine habit grown eg. semi leafless type and keep proper row to row and plant to plant spacing.
- Staking should be done in the normal and tall varieties.
- Treat the seed with Carbendazim 50 WP (bavistin 50 WP) at 2.5 g/kg and spray the crop with bavistin 50 WP at 1 g/litre of water at flowering stage. It is difficult for the chemicals to penetrate to the soil line under the dense canopy to arrest the spread of white rot.

Powdery Mildew : White, off coloured spots appear on the upper surface of the lowest and oldest leaves. The disease progresses until the entire plant is covered with white, powdery mycelial growth. Tissue beneath infected areas may turn purplish and black fruiting bodies are formed in the mature lesions. Severe infection results in early crop senescence reduced quality and decreased yields.

To control, follow the following schedule.

- Crop rotation and immediate ploughing and destruction of the infected crop debris.
- Grow resistant varieties like Palam Priya., Solan Nirog, Kinnauri etc.
- Spray the crop with flowable Elemental Sulphur at 3–4 kg/ha or Karathane/Calaxin at 5 ml/10 litres of water, Propiconazole (tilt 25 EC), Hexaconazole (contaf 5 EC), Tetraconazole (Folicur 250 EC), Bavistin 50 WP at 10 ml/litre of water with the initiation of the disease and repeat it at 10 days interval.

Ascochyta Blight : Three pathogens can cause seedling infection starting as blackening of collar region which extends upto 4-12 cm and results in its death. Leaf infection by *M. pinodes* results in appearance of small, purple to black lesions which increase in size, turn brown to black and become zonate. Affected leaves dry up and hang down while still clinging to the plant. Leaf blight usually starts on the lower leaves and spreads to upper ones. Basal portions of the stem show blackening and browning, the lesions enlarge and coalesce to girdle the stem giving the entire lower plant a blue-black appearance. Brown irregular spots with dark brown margins appear on immature pods which may coalesce and shrink. The seeds become deformed and remain small in size.

In contrast to *M. pinodes* and *P. medicaginis* var. pinodella, *A. pisi* seldom attacks the plant base and does not cause root rot. The leaf spots are large, pale brown, papery with grey or tan centre and are studded with dark, pin pointed pycnidia. The leaves are killed in severe infection. The spots on the pods are sunken, tan, circular and surrounded by dark margins with black pycnidia in the centre.

To control, follow the following schedule.

- Use clean and healthy seeds grown in dry areas and treat then with chemicals viz. Bavistin 50 WP at 2.5 g/kg of seed.
- Peas refuse should be disked and ploughed under immediately after harvest.
- Spray the crop at flowering stage with Bavistin 50 WP + Mancozeb 75 WP at 01 g + 2.5 g/litre at 10-15 days interval and repeat it 4-5 times.

Pea rust : The symptoms appear as yellow spots which gradually turn brown. The yellow spots with round or elongated clusters of aecia persist for long mainly on lower surface of the leaf. The light brown powdery uredo-pustules develop on both surfaces of the leaves and other green parts of the plant, i.e. stem, green pods and petioles. The dark brown or almost black teleuto-pustules occur in the same sorus as uredia.

To control follow the following schedule.

- Timely sowing of the crop is recommended.
- Destroy the diseased crop debris and weed hosts.
- Spray the crop with Propiconazoles (Tilt 25EC)/ Hexaconazoles (Contaf 5 EC)/ Tebuconazole (folicur 250 ECW at 10 ml/litre of water/Bitertinol (baycor 25 EC)/Difenconazole (score 25EC) at 5 ml/10 litre of water/Mancozeb 75 WP+Carbendazim 50 WP at 2.5 g +10 g in 10 litres of water followed by two-three more sprays at 10-15 days interval. The triazole fungicides are effective in controlling powdery mildew and Ascochyta blight also.

Cultivation of Pulses in Non-Conventional Seasons/Areas

- Presently, 56.1 thousand ha area is under current fallow whereas 16.5 thousand ha is under other fallows. In a total of 72.6 thousand ha of fallow areas, pulses should be encouraged depending upon their suitability.
- 35.7% of the total gross cropped area is under maize where cowpea, horsegram, pole rajmash, ricebean and mothbean can be successfully intercropped. Mash or pigeonpeacan be easily introduced on the side bunds of maize fields.
- 9.74% of the cropped area is under rice and mash can be raised on the side bunds. Mash can be dibbled on the bunds with little effort.
- 43.52% of the area is under wheat. With a proper crop geometry, gram, lentil and peas should find space in some wheat fields.
- In Zone I of Himachal Pradesh, summer moong bean and urd bean can be introduced in the rice-wheat cropping sequence in between the two crops.
- Buckwheat and grain amaranth are the main crops in the Zone IV of Himachal Pradesh. Rajmash can be cultivated as an intercrop with grain amaranth in additive series and in buckwheat, in replacement series. In the lower elevations, ricebean/urdbean can be introduced on the field bunds.
- 2.82% (128.4 thousan ha) and 33.3% (1515 thousand ha) of the total geographical area is culturable wastes and under permanent pastures and grasslands. With the aid of suitable production technologies, pigeonpea, horsegram and cowpea in particular, can find place in some of the culturable wastes as well as grasslands. Strategies suggested as above can only be implemented if farmers, policy makers, developmental agencies and above all governmental bodies make

serious attempts for successful implementation of the programme. This can be successful only if funds and manpower available in the prestigious schemes like MNREGA, RKVY etc are integrated with agricultural activities/development.

Future Thrust

- Revival of mixed or intercropping to reap complementarities of cereals and pulse crops.
- Pulse based diversification of rice-wheat and maize-wheat cropping system both in space and time dimensions at least every third year on 20% of the land.
- Collection of genetic diversity and breeding stress tolerance (moisture high or low, weeds, diseases, insect pests etc.) and short duration varieties.
- Niche based development and refinement of agrotechniques (sowing time, plant geometry, fertility management, water and weed management, IDM and IPM).
- Organic production package for different pulse crops.
- Quality seed under different pulse production situations.
- Strengthening scientist-farmer linkages through FLDs, on-farm trials, field days and other extension activities.
- Post harvest and value addition.
- Production- marketing chain.
- Small agricultural machinery and tools.

References

Basandrai D, AK Basandrai and HL Thakur (2004). Scientific pulse production in Himachal Pradesh, CSK Hill Agricultural Research and Extension Centre Dhaulakuan, Distt Sirmour, 72pp.

Gupta M. (2004). Pulse production in Himachal Pradesh: Trend, Potential and constraints. Scientific pulse production in Himachal Pradesh, CSK Hill Agricultural Research and Extension Centre Dhaulakuan, Distt Sirmour, (Eds Basandrai D, AK Basandrai and HL Thakur) p 68-72.

HPKV, (2008). Package of Practices for kharif crops, Directorate of Extension Education, CSKHPKV, Palampur.

HPKV, (2008). Package of Practices for rabi crops, Directorate of Extension Education, CSKHPKV, Palampur.

Important statistics of Himachal Pradesh (2011-12). Department of Economics and Statstics, Govt of Himachal Pradesh.

Kumar N, JC Bhandari, SK Thakur and R Chahota. (2008). Sucessful cultivation of ricebean in North-western Himalayan. Fodder Production and Grassland Management Centre, CSK Himachal Pradesh Krishi Vishvavidyalaya, Palampur.

Rana M C, Naveen Kumar, A Sharma and S S Rana. (2004). Management of complex weed flora in pea with herbicide mixtures under Lahaul valley conditions of Himachal Pradesh. Indian Journal of Weed Science 36 (1&2): 68-72.

Rana M.C., AD Bindra and S S Rana. (2006). Identification of suitable varieties of pea and buckwheat for double cropping under dry temperate conditions of northwestern Himalayas. Indian Journal of Agricultural Sciences 76 (4): 222-5.

Rana MC, S S Rana, G D Sharma and R S Rana (2006). Response of soil and foliar application of biozyme in French bean under dry temperate high hill conditions of Himachal Pradesh. Himachal Journal of Agricultural Research 3 (2): 15-18.

Rana R S, R Prasad and S S Rana. (2009). Effect of maize leaf defoliation on production and economics of maize + legume intercropping systems. Himachal Journal of Agricultural Research 35(2): 154-157.

Rana S S, K K Mondal and P Sood. (2003). Technology for Kalazira Production, Mountain Agriculture Research and Extension Centre, CSK Himachal Pradesh Krishi Vishvavidyalaya, Sangla, Kinnaur, HP.

Rana S S, P Sood, K K Mondal and K S Thakur. (2003). Rajmash evm matter, CSK Himachal Pradesh Krishi Vishvavidyalaya, Parvatiya Krishi Anusandhan evm Prasar Kendra, Sangla, Kinnaur pp 26.

Rana S S, P Sood, K K Mondal., and Y Singh. (2003). Influence of varieties and fertility levels on pea and succeeding common buckwheat in pea- buckwheat cropping sequence. Life Science Reporter 5 (1&2): 16-19.

Rana S S, Pankaj Sood, Pawan Pathania, Suresh Kumar, Sanjay Chadha and D R Chaudhary (2008). Crop-weed competition in French bean (Phaseolus vulgaris L.) under dry temperate conditions of North-western Himalaya. Himachal J. Agric. Res. 34 (2): 10-14.

Rana S S, R S Rana, B Jangpo and N N Angiras. (1999). Studies on integrated weed management in rajmash (Phaseolus vulgaris L.) under Sangla valley conditions of Himachal Pradesh. Indian Journal of Weed Science 31(3&4): 218-221.

Rana S S, Rajinder Pal, K K Mondal and P Sood (2005). Influence of seeding time of rajmash varieties intercropped with kalazira under dry temperate conditions. Indian Journal of Pulses Research 18(2): 219-221.

Rana S S. (2002). Evaluation of promising herbicide combinations for broad spectrum weed management in rajmash (Phaseolus vulgaris L.) under dry temperate conditions of Himachal Pradesh. Indian Journal of Weed Science 34 (3&4): 227-230

Rana S S. (2002). Integrated weed management in pea (Pisum sativum L.) under Sangla valley conditions of Himachal Pradesh. Indian Journal of Weed Science 34 (3&4): 204-207

Siddiq E A (2000). Yawning productivity gap. Survey of Indian Agriculture 2000. The Hindu, Chennai, India.

Sood P, S S Rana and K K Mondal. (2004). Evaluation and economics of some insecticides against black bean bug, Chauliops nigrescens Distant infesting rajmash under dry temperate conditions of Himachal Pradesh. Indian Journal of Pulses Research 17(1):101-102.

Sood Pankaj and S S Rana (2004). Insecticidal management of kidney bean leaf weevil, Strophosomoides kumaoensis Aslam, under dry-temperate conditions of Himachal Pradesh. Pest Management and Economic Zoology 12(2): 197-202

Statistical outlines of Himachal Pradesh. 1975-76 to 2007-08. Department of Economics and Statistics, Govt of Himachal Pradesh.

Thakur RC and SC Negi (1985). Effect of fertilizers and Rhizobium inoculation in balck gram. Indian J. Agron 30(4): 501-4.

❑❑❑

9

Islands

N. Ravisankar[1], B. Gangwar[2], T.P. Swarnam[3], S.K. Zamir Ahmed[4] and N. Bommayasamy[5]

[1]Principal Scientist (Agronomy), [2]Project Director, Project Directorate for Farming Systems Research, Modipuram, [3]Senior Scientist (Agronomy), [4]Senior Scientist (Agricultural Extension), [5]Subject Matter Specialist (Agronomy), Natural Resource Management Division, Central Agricultural Research Institute, Port Blair-744 101, A&N Islands

The geographical area of the union territory of Andaman and Nicobar Islands is 8, 24,900 ha with 86.9 % of it under reserve and protected forests. The humid tropical climate that prevails in the Andaman and Nicobar islands provides a unique opportunity to grow various crops. Owing to a number of factors, hardly 4.8% of the area is under cultivation. Maximizing the productivity and farm income mainly depends on effective management of natural resources like rain, soil, land as well as labour present in the Island ecosystem. On an average, the Islands receive 3074 mm of rainfall distributed over 8 months. The rainfall is intense from June to September and may have even up to 30 rainy days per month sometimes, however, from January to April, the rainfall is scanty or absent, particularly in March. The undulating terrain results in severe water crisis during the period which is aggravated by higher evapotranspiration. The Islands receive south-west monsoon from May to September and north-east monsoon during October to January. The topography of A&N Island is undulating with a stiff terrain exposed to soil erosion due to heavy rainfall. The crop growing season can be divided in to two major seasons *viz.,* wet and dry seasons. The wet season is usually from May to November receiving annual average rainfall of 2789.9 mm leading to water logging in the low lying areas. Even in the upland areas, the soil remains in saturated conditions throughout the season which forces farmers to grow only paddy crop in most of the area and sugarcane in limited areas. Dry season falls between

December – April which receives an annual average rainfall of 284.4 mm only in which pulses like blackgram and greengram are grown on the islands.

Soil Resources

The soil type of Andaman and Nicobar Islands ranges from sandy clay to sandy loam. These have developed under the dominant influence of vegetation and climate over diverse parent material. The uplands under forest cover are intensely leached, but runoff is very high, wherever forest cover has been removed completely. The valley floors comprise of depositional landforms and have been termed as low lands and have developed from the out wash of parent material from the surrounding hills. These soils are medium to heavy textured and moderately well drained and subjected to seasonal fluctuations in ground water. Most of the soils of these islands have medium to high organic matter status indicating that the organic carbon is >0.5% (Ganeshamurthy *et al.*, 2002). Parts of south and middle Andaman have low organic carbon status mainly because of severe erosion of the surface soil caused due to extensive deforestation and complete neglect of the deforested area. In general, soil fertility analysis indicates medium in available N, low in both available P and K.

Importance of Pulses in Islands

Indian diets are changing from grains to grams and greens. As stated earlier, Andaman and Nicobar Islands have heavy rainfall from May – December in which paddy is the main crop. Presence of light textured soil in islands leads to short time availability of moisture for rice fallow crops. Short duration pulses *viz.*, Black gram and Greengram can be grown successfully in the residual moisture. During the course of experimentation on pulses at Central Agricultural Research Institute, Port Blair, several technologies have been evolved which can lead to the increase in area, production and productivity of pulses. However, the productivity of the pulses remained almost static in the Islands (Table 1). The scope of increasing the productivity exists with the available technologies. The production can be increased by at least 1.5 times if proper post harvest practices and handling are adopted. The pulses produced from the Islands are not processed as such due to lack of processing facilities. The raw products are exported to either Kolkata or Chennai and then processed products are again imported to Islands. This causes two dimensional loss to the farmers as well as consumers due to cost of transportation of raw as well as processed products. Trend in production of pulses in the Islands during previous three years are given in Fig 1. Blackgram and Greengram are the two major pulses grown in the Islands. Projected demand of pulses in 2011 and 2021 are given in Table 2.

Table 1 : Area, production and productivity of pulses (Blackgram and Greengram) in Islands

Year	Area (ha)	Production (t)	Productivity (kg/ha)
2006-07	726	400	550
2007-08	2093	1369	650
2008-09	2119	1154	540

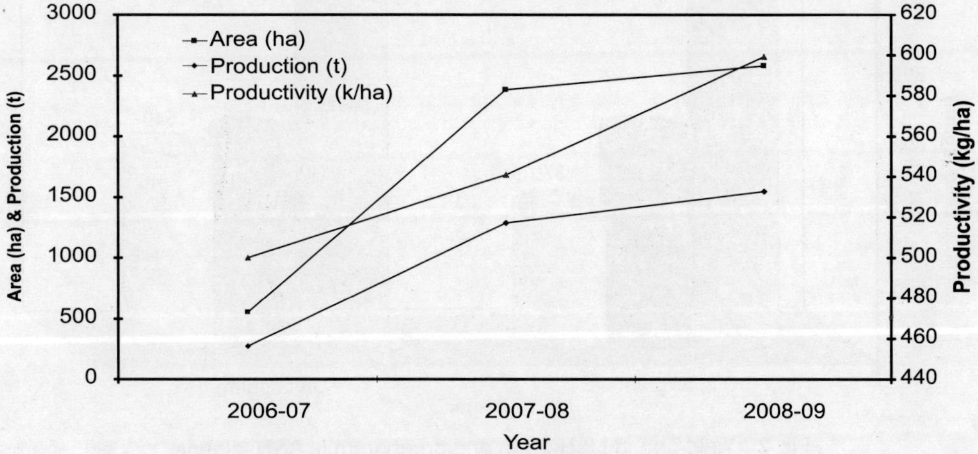

Fig 1 : Trend in pulses production in Island

Blackgram and greengram are highly suited for the Islands and at Central Agricultural Research Institute, Port Blair, several experiments have been conducted to develop package of practices for pulses under various cropping systems.

Table 2 : Projection of demand of cereals, pulses and oilseeds for Islands based on growth of population

Commodity	Present (2009-10)			Demand	
	Production (t)	Demand (t)	Deficit (t)	2011	2021
Cereals	13772	73295	59523	76596	95746
Pulses	280	4465	4185	4666	5832
Oilseeds	15	6150	6135	6427	8033

(Andaman and Nicobar Administration, 2005; 2006)

Yield gap in Blackgram and Greengram

Analysis of achievable and actual on-farm yield of blackgram and greengram indicates that there is a yield gap of 370 and 540 kg ha $^{-1}$ (Fig 2) in blackgram and greengram, respectively. It is expected that with the same area of cultivation (2119 ha), the production of pulses in the islands can be increased to 2118 t from the present level of 1154 t by bridging the yield gap through adoption of scientific package of practices.

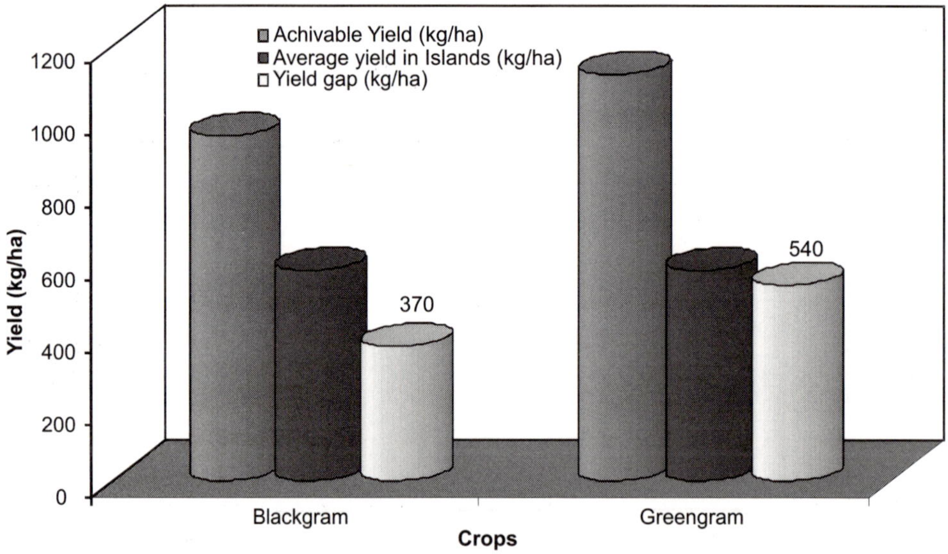

Fig 2 : Yield gap in blackgram and greengram in A&N Islands

Major Constraints

The major constraints in cultivation of pulses in Island conditions are

- Shorter length of growing period for pulses due to heavy rainfall for eight months (April to December)
- Extended period of rainfall in December does not allow soil to dry to take up the timely sowing
- Water stress during the peak and critical stages of the crop growth
- Unexpected and heavy downpour of unseasonal rainfall in dry season affects the growth and development of crops
- Timely availability of seeds of short duration varieties
- Unexpected early onset of rainfall in March or April when the crop is in maturity stage
- Maintenance of poor population due to broadcasting method of sowing
- Lack of proper intercultural operations like weeding, nutrient application etc
- Lack of post harvest processing facilities like Dal mills etc in the Islands
- Lack of procurement centres /policy in the Islands for pulses

Performance of Cropping Systems Involving Pulses

Performance of cropping systems involving pulses have been experimented from 1978 onwards at the Islands (Annual Report 1978 to 2006-07) and the gist of the same is presented. Six sequences of cropping *viz.*, Paddy (Jaya), Paddy (Ratna)-Pulse (Blackgram T-9), Paddy (Bala)-Rice bean (K_1), Paddy (CR-126-42-1)-Til (Vinayak), Paddy (Indira)–Mustard and Paddy (Jayanti)-Dhaincha were evaluated for yield potential. It is seen that crops like black gram, sesame, rice bean and dhaincha can be profitably grown after the harvest of first paddy crop under rain fed conditions. Among the varieties tried in first crop, the highest yield was recorded with CR-126-42-1 (32.3 q/ha) (Table 3). The yield of rice at different levels of nitrogen application after different preceding crops were estimated by using quadratic production function. The data revealed that there was significant increase in yield under different preceding crops over control. The maximum yield of rice was 44.4 q/ha followed by 42.5 and 42.0 q/ha when preceding crops were blackgram, maize and sesame, respectively. The optimum doses of nitrogen application for the rice were 87.4, 101.4 and 123.9 kg/ha when grown after black gram, maize and sesame respectively (Tables 4 & 5).

Table. 3 : Yield of different cropping sequences under rainfed conditions

Treatment	Yield (q/ha)	
	First crop	Second crop
Paddy (Jaya)- Monocrop	28.9	-
Paddy (Ratna)- Black gram	31.6	5.6
Paddy (Bala) - Rice bean (K_1)	31.6	450 (green fodder)
Paddy (CR-126-42-1)- Sesame (Vinayak)	32.3	3.5
Paddy (Jayanti) - Dhaincha	20.7	400 (green fodder)

Table 4 : Production functions of rice in relation to crop sequences and levels of nitrogen application

Name of previous crop	Production functions	R^2
Blackgram	Y =28.2584+3.3852X-0.1772X2	0.8936
Sesame	Y = 23.0708+3.0614X-0.1234X2	0.8008
Maize	Y = 26.9271+3.0903X-0.1506X2	0.8993

X = Rate of nitrogen (kg/ha), Y = Yield of crop (q/ha)

Table 5 : Total yield of rice (q/ha) under different preceding crops

N Levels (kg/ha)	Previous crops		
	Blackgram	Sesame	Maize
0	28.3	23.1	26.9
30	36.3	31.1	34.8
60	42.2	37.0	40.1
90	44.4	40.6	42.5
120	43.4	42.0	42.3

Significant increase in grain and straw yield of blackgram and cowpea were recorded where drainage was provided (Table 6). It is interesting to note that grain and straw yield of maize was less where black gram or cowpea were inter-cropped without providing drains. This is because the drainage and outflow of excess water from maize field was slowed down due to presence of legumes resulting into temporary water logging which had distinct harmful effect on maize. This was not true where drains were provided.

Table 6 : Effect of surface drainage on blackgram and cowpea

Treatment	Maize (q/ha)		Black gram (q/ha)		Cowpea (q/ha)	
	Grain	Stover	Grain	Stovr	Grain	Stover
Pure Maize without surface drains	10.3	53.0	-	-	-	-
Pure Maize with surface drains	22.0	67.5	-	-	-	-
Maize + Black-gram without surface drains	13.8	41.5	0.93	4.0	-	-
Maize + Black-gram with surface drains	31.9	81.5	4.28	10.0	-	-
Maize + Cowpea without surface drains	12.5	59.8	-	-	4.77	20.17
Maize + Cowpea with surface drains	21.2	77.8	-	-	5.16	21.00

Multiple cropping : Out of 56 farm families where emphasis was on field crops, 6 families were covered for multiple cropping. Rice was followed by Maize/Pulses/Oil seeds. The first short duration crop of rice (variety IET-4106) was transplanted in the last week of May and harvested in the last week of August. The second crop of rice (Variety-Bhavani) was transplanted in the same field in the middle of September and harvested in the last week of December. In the first crop, the average paddy yield was 42.7 q/ha while the second crop yielded 35.5 q/ha of paddy. Thus it was possible to obtain 78 q/ha from the same land under double cropping of rice. In upland areas, highest total grain yield of 98.7 q/ha was recorded in maize-rice-rotation followed by 89.3 q/ha in blackgram-rice-rice rotation. In lowland areas, after the harvest of rainy season long duration rice (CR-1009), maize, sorghum, greengram, sesame, and sorghum + cowpea (Forage) could be raised successfully in dry season with respective grain yield of 32.1, 15.8, 4.6, 6.0 and forage yield of 331.8 q/ha (Tables 7 & 8) (Gangwar *et al.*, 1991).

Table 7 : Highest, lowest and mean frequency distribution of yield and gross return over operating cost for various cropping sequence

Cropping sequence	No. of Demons -trations	Yield (q/ha)			Economics (Rs/ha)		
		Highest	Lowest	Mean	Gross Return Rs	Cost of produc -tion	Return over operating cost
Rice-rice-blackgram							
Rice	8	55.0	36.0	42.9	7224	3446	3778
Rice		48.8	30.0	37.3	6296	3390	2906
Blackgram		9.0	6.9	8.0	3168	1568	1600
	Total	112.8	72.9	88.2	16688	8404	8284
Rice-blackgram							
Rice	22	50.0	30.0	42.9	7182	3449	3733
Blackgram		9.0	6.3	8.2	3236	1563	1673
	Total	59.0	36.3	51.1	10418	5012	5406

Table 8 : Yield of pulses in various rotations

Crop rotation	Growing period	Yield (q/ha)	
		Grain	Straw
Rice-blackgram			
Rice	May-August	30.5	73.5
Blackgram	Dec / January- March / April	11.9	18.3
Rice-rice-blackgram			
Rice	May-August	33.9	69.2
Rice	September-December	29.8	67.0
Blackgram	January-April	13.2	22.4
Rice-rice-greengram			
Rice	May-August	37.9	77.1
Rice	September-December	38.2	79.3
Greengram	January-April	7.6	15.1

Another experiment involving 10 rotations using early, medium and long durations and C-14-8 as local check variety of rice with or without green leaf manuring before raising rice crop was repeated for the third consecutive year. In the experiment, instead of sorghum for fodder purpose and maize for grain purpose, were replaced with groundnut and maize for cobs respectively. All other treatments were same as during the previous years. On individual crop basis, maize for cobs purpose gave the highest net returns (Rs.16, 864/ha) closely followed by vegetable cowpea (Rs.15,600). Similarly the rotations involving these two crops gave distinctly higher returns (Table 9) mainly because of good market price. Though the cost of cultivation was comparatively higher in cases where three crops were raised during the year, the net returns were also good. However, the net returns can be increased considerably by raising vegetables in these rotations.

Table 9 : Yield of pulses q/ha in various rotations

Crop rotation	Rainy season		Dry season				Cost of cultivation	Net return
	1st crop (Rice) (May- August)		2nd Crop (Rice) (Sept. – Dec.)		3rd Crop (Jan.-April)		Rs/ha/yr	
	Grain	Straw	Grain	Straw	Grain	Straw		
Rice-fallow	43.3	80.2	-	-	-	-	2988	5111
Rice-Rice-Black gram	36.8	70.1	31.4	58.4	8.6	14.5	7656	9879
Rice-Rice-Cowpea (G)	43.0	81.3	36.0	61.6	18.7	34.6	7546	15680
Rice-Rice-Cowpea (veg)	45.8	87.2	36.6	62.2	30.0*	25.5	8206	16707
Rice-Rice-Cowpea (F)	42.0	78.4	32.1	56.4	-	261.7*	7106	9322
Rice-Rice-Greengram	36.8	85.6	34.9	62.3	8.3	16.1	7656	10690
CD at 5%	5.2	7.0	-	-	-	-	-	-

*Yield on fresh weight basis

Performance of Pulse Crops in Association with Sesame

Highest total grain was recorded when sesame was grown in association with blackgram in 2:2 ratio paired row system of planting (Table 10). On the other hand, it was found feasible to grow cowpea in alternate rows for forage purpose with minimum detrimental associative effect on sesame. Interestingly, considerable decrease in herbage yield of weeds was recorded when crops were grown in association. Even in case of main and associated crop of cowpea, where only one hand weeding was done at initial stages of crop growth, the herbage yield was less when compared with sesame alone. The weed infestation was found to be low when blackgram and greengram were grown with sesame in 1:1 ratio. Considering all the factors, raising of sesame with blackgram in 2:2 ratio for grain purpose and cowpea in 1:1 ratio for forage purpose appears to be better than other systems of intercropping (Gangwar and Singh, 1991).

Table 10 : Yield of pulses in association with sesame

Treatment	Yield (q/ha)		
	Sesame	Pulses (grain/forage)	Herbage yield
Sesame alone	2.24	-	5.80
Cowpea alone	-	96.53*	3.71
Blackgram alone	-	8.90	3.79
Greengram alone	-	3.42	3.70
Sesame + Cowpea (1:1)	2.04	91.15*	5.66
Sesame +Cowpea (2:1)	2.05	52.55*	5.70
Sesame + Cowpea (2:2)	0.98	59.36*	3.83
Sesame + Blackgram(1:1)	1.94	4.77	3.88
Sesame + Blackgram (2:1)	1.86	2.99	4.83
Sesame + Blackgram (2:2)	1.46	6.04	4.88
Sesame + Greengram (1:1)	1.76	2.18	3.33
Sesame + Greengram (2:1)	1.29	1.81	4.26
Sesame + Greengram (2:2)	1.14	1.85	3.63

*Fresh weight basis (harvested for forage).

Comparative performance of various cropping system involving pulses are presented in Fig. 3 which indicates that the net returns and B:C ratio were higher when blackgram was grown in residual soil moisture after either the traditional C14-8 variety of rice or improved high yielding long duration variety of rice namely Ranjit (Pramanik and Ravisankar, 2007).

Other General Observations on Cropping Systems Involving Pulses in Islands

- The agro-climatic conditions of the Islands are suitable for safely adopting rice based multiple cropping systems in valley areas.
- Rice–rice-pulse, particularly with cowpea for vegetable, was the most economic rotation giving more than Rs.10,000/- net return in a year. Rice-Rice-Blackgram with annual net return of about 8000/- was also good .
- Legumes in rotation with rice not only increased the productivity but reduced the nitrogen requirement (about 15 kg/ha) of following rice crop.

- Pulse and oil seed crops fit well in rice based cropping system and hence need encouragement.

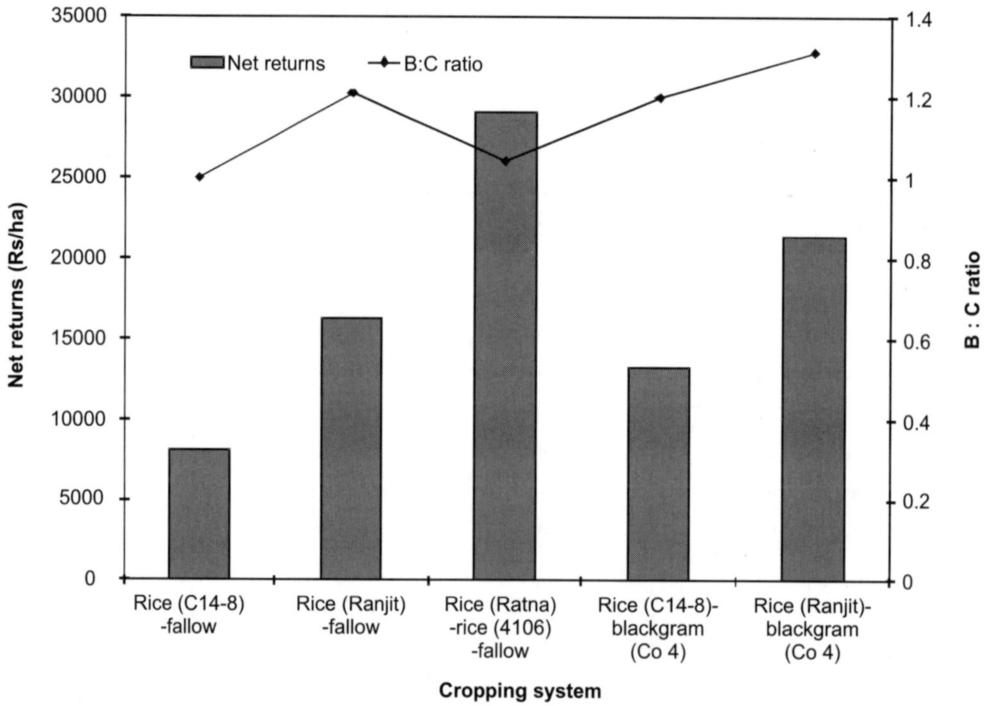

Fig 3 : Comparative performance of various cropping systems involving pulses

Package of Practices for Pulses

Blackgram

Varieties : There was a large variation in the yield of different black gram varieties (Table 11). Among the varieties evaluated, T9 registered significantly higher yield (0.95 t ha^{-1}) which is at par with TMV1 (0.94 t ha^{-1}). The other varieties namely, PDV1, Naveen and Co5, registered significantly lower yield compared to T9 and TMV1 under farmers field conditions. Significantly higher yield registered by T9 and TMV1 variety may be due to the tolerance of these varieties for drought in the later stages and short duration (< 80 days). Among the black gram varieties, T9 registered higher net returns of Rs. 19250.00 per ha, with B: C ratio of 4.24 followed by TMV1 registering the net return of Rs. 19150.00 with B:C ratio of 4.25. Higher net returns and B:C ratio of T9 and TMV1is attributed to higher grain yield recorded by these varieties. T9 variety of Black gram recorded higher out put energy of 13965 MJ followed by TMV 1. Energy efficiency was also higher with T9 in terms of conversion of input energy in to output energy. Specific energy is lower for T9 and TMV1 (0.78 and 0.79 MJ/kg) indicating that less energy is required by T9 and TMV1 varieties to produce one kg of grain. In another evaluation trial of four black gram varieties, viz. Pant U-30, Type –9 Pant U-19 and H-76-1 tried in R.B.D. with four replications under dry land conditions, the performance of Pant U-30 and Type –9 was observed to be better in visual estimation.

Table 11 : Yield and economics of black gram under participatory evaluation (pooled over 5 years i.e. 2001-02 to 2005-06)

Variety	Yield (t/ha)	Net Return (Rs./ha)	B :C Ratio
T9	0.95	19250	4.24
PDV1	0.87	17150	3.81
TMV1	0.94	19150	4.25
Naveen	0.85	16750	3.72
Co 5	0.77	14675	3.26
CD (P=0.05)	**0.02**	-	-

Seven cultivars of blackgram *viz* Co-3, Co-4, Co-5, KM-2, ADT-3, Type-9 and Pant-30, were sown after rice harvest in the second year of experiment in R.B.D and replicated four times. Among the six varieties compared, highest yield of 4.0 q/ha was recorded in Co-4 followed by Co-5 (Table 12). The lowest yield of 2.86 q/ha was recorded in variety KM-2. **The yield range of various varieties was low due to unusual cyclonic heavy rains in February, 1984.** In terms of productivity, highest grain of 5.05 kg grain/day/ha was recorded in variety Co-4 followed by Co-3 (4.1 kg/day /ha).

Table 12 : Yield of blackgram varieties

Variety	Yield (q/ha)	Days to maturity	Grain productivity (kg/day/ha)
Co-3	3.15	77	4.09
Co-4	4.04	80	5.05
Co-5	3.22	85	3.79
Km-2	2.86	76	3.76
ADT-3	2.98	76	3.92
Type-9	3.00	74	4.05

Planting density : Field experiments were conducted in rice fallows with greengram, black gram and cowpea using seed rate of 12, 20 and 28 kg/ha. Leaf water potential and net photosynthetic rate increased significantly up to a seed rate of 20 kg seed/ha in all the crops and ranged from 0.66 to 0.91 at 30 for blackgram (Table 13). Similarly the leaf water potential also varied from -6.2 to

-7.1 bars in blackgram at 30 days. In commensurate with these observations, the grain and straw yield of blackgram significantly increased up to a seed rate of 20 kg/ha and decreased further.

Table 13 : Effect of planting on yield and net photosynthetic rate of blackgram in residual soil moisture

Seed rates (kg/ha)	Relative leaf water content (-bars)		Net photosyn -thesis rate ($mg/m^2/sec$)		Yield (q/ha)	
					Grain	Straw
	30 DAS	50 DAS	30 DAS	50 DAS		
12	6.2	4.6	0.66	0.48	8.1	18.2
20	7.0	4.7	0.91	0.94	17.0	33.8
28	7.1	5.1	0.71	0.89	13.2	26.6
CD at 5%	-	-	-	-	3.3	3.6

Irrigation : The study on blackgram (T-9) comprised of seven irrigation treatments and four replications laid in a randomized block design. One and two irrigations of 2.5 and 5.0 cm depth were given at 25 and 45 days after sowing. The later irrigation corresponded with flowering and pod formation of the crop. Grain yield of blackgram, corresponding to different treatments, is given in Table 14. One light irrigation at 25 days after sowing or at flowering was at par with control. Significantly higher grain yields were obtained with one irrigation (5 cm depth) at flowering or 2 irrigations at 25 days and at flowering. Highest yield of 7.4 q/ha was obtained with one irrigation at flowering which is a critical stage in blackgram (Pramanik *et al.,* 2001).

Table 14 : Response of blackgram to depth and frequency of irrigation

Treatment	Grain yield (q/ha)
Control	5.8
One irrigation at 25 days with 2.5 cm depth	4.5
One irrigation at 45 days with 2.5 cm depth	5.2
One irrigation at 25 days with 5 cm depth	6.2
One irrigation at 45 days with 5 cm depth	7.4
Two irrigations at 25 & 45 days with 2.5 cm depth	7.2
Two irrigations at 25 & 45 days with 5 cm depth	6.6
C.D at 5%	1.8

Mulching : The experiment comprised of seven treatments in which black gram variety T-9 was evaluated. Three organic mulch materials equivalent to 10 q/ha were spread in between crop rows 10 and 30 days after sowing. The mulching treatment and grain yield of blackgram are given in Table 15. The data indicate that mulching at any stage invariably benefited the crop. However, early mulching had an edge over delayed application of mulch due to moisture conservation for longer duration. Control gave the lowest yield of 3.8 q/ha but yields of 6.2 to 6.6 q/ha were realized from mulching 10 days after sowing. The mulching materials did not differ significantly among themselves (Pramanik *et al.*, 2000).

Table 15 : Response of black gram to types of mulch and mulching time

Treatments	Grain yield q/ha
Control	3.8
Application of saw dust, 10 days after sowing	6.6
Application of forest leaves, 10 days after sowing	6.2
Application of paddy straw, 10 days after sowing	6.5
Application of saw dust, 30 days after sowing	6.8
Application of forest leaves, 30 days after sowing	5.8
Application of paddy straw, 30 days after sowing	4.7
C.D. at 5%	1.7

Nutrients : Application of 40 kg P_2O_5/ha along with 10 kg of N/ha was found to be better for obtaining higher yield of blackgram and also the yield of succeeding rice crop (Table 16).

Table 16 : Yield and yield attributes of black gram as influenced by NP levels

Treatments	No. of plants/m²	No.of branches /plant	No.of pods /plants	Yield (q/ha)		Grain yield of succeeding rice crop (q/ha)
				Grain	Straw	
N_0P_0	9.0	1.6	19.4	6.1	17.0	16.1
N_0P_{90}	9.0	1.8	22.4	7.6	19.2	18.6
N_0P_{49}	8.5	2.2	22.5	7.8	20.6	20.0
$N_{10}P_0$	8.3	2.1	19.6	7.2	23.1	18.9
$N_{10}P_{20}$	7.8	2.0	21.9	8.4	20.0	18.6
$N_{10}P_{40}$	8.7	1.7	20.8	8.5	21.9	20.3
$N_{20}P_0$	7.8	2.1	22.4	6.7	18.3	19.2
$N_{20}P_{20}$	8.7	2.3	22.5	7.9	20.3	19.8
$N_{20}P_{40}$	8.0	2.1	22.9	8.2	19.2	22.2
C.D.at 5%	0.4	0.41	NS	0.76	NS	-

Greengram

Varieties : The performance of short duration high yielding greengram variety PDM 54 was evaluated on 10 farmer's field of South Andaman villages. Plant growth and yield attributes of greengram variety PDM 54 was higher than the local variety. PDM 54 recorded 66.7% (7.5 q/ha) higher grain yield than local variety (4.5 q/ha) (Table 17). The net return and benefit cost ratio was also significantly more with higher yielding variety. However, participatory assessment showed that variety with 60-75 days duration were more preferred by the farmers for safe harvesting of crop before onset of early monsoon (1st week of May) in these islands (Gangwar and Ahmed, 1991).

Table 17 : Performance of high yielding greengram variety in paddy fallows

Performance Indicator	Local variety	PDM 54
Plant height (cm)	32.5	50.5
Length of pod (cm)	5.3	7.8
Seeds per pod (no.)	9.0	11.0
Yield (q/ha)	4.5	7.5
Gross return (Rs./ha)	7200	12000
Net return (Rs./ha)	3700	7800
Benefit cost ratio	2.6	4.1

Pooled analysis of 5 years participatory evaluation of greengram varieties revealed the significant influence of varieties on yield under residual moisture. Among the varieties evaluated, PDM 54 registered significantly higher yield of 0.86 t ha^{-1} which is statistically at par with T44 (0.85 t ha^{-1}). Co5 also registered a yield of 0.80 t ha^{-1} compared to other varieties. Branching habit of PDM54 and T44 coupled with short duration (< 75 days) and stress tolerance mechanism of these varieties contributed to increased yield. Greengram varieties PDM 54 and T44 recorded higher net returns and B: C ratio (Rs. 17175 and Rs. 16850/ha and 3.81 and 3.74, respectively). Higher yield (farmers field conditions) registered by these varieties led to higher net returns and B:C ratio (Table 18).

Table 18 : Yield and economics of greengram under participatory evaluation trials (pooled over 5 years i.e. 2001-02 to 2005-06)

Variety	Yield (t/ha)	Net Return (Rs./ha)	B :C Ratio
T 44	0.85	16850	3.74
P 105	0.75	14250	3.16
PDM 54	0.86	17175	3.81
Co 5	0.80	15525	3.45
Narendra Moong	0.82	16100	3.57
PDM 11	0.73	13850	3.07
CD (P=0.05)	**0.03**	-	-

Eleven varieties of greengram, namely, Co-3, Co-4, ADT-2,PS-16,S-8, ML-65,ML-70,P-103,P-104 and P-105, were evaluated in R.B.D and 3 replications for two consecutive years. Among the varieties tested, P-105, ML-70 and Co-3 performed better in the order of their merit, with an average yield of 6.58, 6.29, 6.07 q/ha respectively (Table 19). The grain yield of these varieties was significantly higher to compared with varieties with non-significant difference among themselves. Lowest yield of 3.1q/ha was recorded in variety ADT-2. In terms of productivity, highest grain yield of 8.66 kg/day was recorded in variety P-105 while it was lowest in variety ADT-2 (4.04 kg/day).

Table 19 : Yield of greengram varieties

Variety	Grain yield (q/ha)	Days to maturity	Grain productivity (kg/day/ha)
Co-3	6.07	86	7.06
S-8	3.95	85	4.65
PS-16	3.86	76	5.00
ADT-2	3.11	77	4.04
ML-65	4.50	75	6.00
ML-70	6.29	75	8.38
P-101	3.95	90	4.39
P-103	5.12	76	6.74
P-104	5.76	90	6.40
P-105	6.58	76	8.66
CD at 5%	1.03	-	-

Planting density : Field experiments were conducted in rice fallows with greengram using seed rate of 12, 20 and 28 kg/ha. Leaf water potential and net photosynthetic rate increased significantly up to a seed rate of 20 kg seed/ha in greengram and ranged from 0.36 to 0.70 at 30 days in greengram (Table 20). Similarly the leaf water potential was observed to be –4.1 bars at 30 days. Corresponding to these observations, the grain and straw yield of greengram significantly increased up to a seed rate of 20 kg/ha and decreased further.

Irrigation : In order to optimize the supplemental irrigation (SI) schedule for different crops for getting maximum profit during dry season, field trials were taken up with different supplemental irrigation treatments (No irrigation, one, two, three and four supplemental irrigations and farmer's practice) and crops (maize, greengram, Sesame, chilli and okra). The supplemental irrigation had a significant effect on growth of greengram (Table 21). With respect to greengram, providing three supplemental irrigations (I_3) resulted in higher yield (340 kg ha^{-1}) compared to all other treatments.

Table 20 : Effect of planting on yield and net photosynthetic rate of greengram grown on residual soil moisture.

Seeds rate (kg/ha)	Ralative leaf water content (-bars)		Net photosyn -thesis rate (mg/m²/sec)		Yield (q/ha)	
	30 DAS	50 DAS	30 DAS	50 DAS	Grain	Straw
12	4.1	3.6	0.36	0.74	8.9	26.8
20	4.4	5.0	0.70	0.93	11.2	42.5
28	5.6	5.2	0.51	0.86	10.1	33.8
CD at 5%	-	-	-	-	1.6	9.4

Table 21 : Yield of greengram with supplemental irrigation at critical growth stages

Supplemental irrigation stages	Yield (kg/ha)
Control	250
Flowering	276
Flowering and pod setting	295
Pre flowering (25 DAS), flowering and pod setting	340
Pre flowering (25 DAS), flowering, pod setting and early maturity	328
Pre flowering (25 DAS), flowering, pod setting, early maturity and maturity	304
CD at 5 %	25

Production technologies

Scientific packages developed and refined, can be adopted to realize higher yields and bridge the yield gap between on-station and on-farm conditions (Ravisankar *et al.*, 2008). The recommended package of practices for black gram and greengram in Island conditions are given in table 22.

Table 22 : Recommended package of practices for blackgram and greengram.

Cultural operation	Black gram	Greengram
Season	Dry season (Dec to April)	Dry season (Dec to April)
Sowing time	December or early January with in 10 to 15 days of cessation of rainfall	December or early January with in 10 to 15 days of cessation of rainfall
Varieties	Type 9	Pusa-105, Pusa-115 and ML 320
Soils	Well drained loamy and light soils	Well drained loamy and light soils

Contd.

Locations	South, Middle and North Andaman regions	South, Middle and North Andaman regions
Seed treatment	Treat the seeds with 5 g each of *Trichoderma viridie* and *Pseudomonas*/kg of seeds and or 3 g of Thiram or Captan /kg of seeds before sowing	Treat the seeds with 5 g each of *Trichoderma viridie* and *Pseudomonas*/kg of seeds and or 3 g of Thiram or Captan /kg of seeds before sowing
Seed rate	20 kg ha^{-1}	20 kg ha^{-1}
Spacing	30 X 10 cm	30 X 10 cm
Method of sowing	Line sowing	Line sowing
Intercropping	Sesame + blackgram in 2:2 ratio & Maize + blackgram in 1:1 ratio	Sesame + greengram in 1:1 ratio
Organic manures	12.5 t/ha of well decomposed FYM to be mixed with soil at least one week before sowing	12.5 t/ha of well decomposed FYM to be mixed with soil at least one week before sowing
Nutrient application	10:20:20 kg NPK ha^{-1} as basal application. Di Ammonium Phosphate (DAP) is preferred source for N and P.	10:20:20 kg NPK ha^{-1} as basal application. Di Ammonium Phosphate (DAP) is preferred source for N and P.
Weeding	One hoeing within 3 to 4 weeks of sowing	One hoeing within 3 to 4 weeks of sowing
Interculture operation	Small ridges and furrows after weeding to drain out excess water incase of heavy rain occurs during growth stage	Small ridges and furrows after weeding to drain out excess water incase of heavy rain occurs during growth stage
Mulching	Mulching with either paddy straw, saw dust and forest leaves at 10 DAS or 35 DAS to conserve soil moisture. Early mulching conserves moisture better.	Mulching with either paddy straw, saw dust and forest leaves at 10 DAS or 35 DAS to conserve soil moisture. Early mulching conserves moisture better.
Irrigation	One irrigation at flowering (45 DAS) with 5 cm depth should be give as it is the critical stage for the crop. If sufficient water is available, two irrigations at 25 and 45 DAS can be given with 2.5 cm depth of application.	Three irrigations at 25 DAS, 45 DAS (flowering) & 60 DAS (Pod setting) should be given as supplemental irrigation.
Foliar spray	2 % DAP (20 g/litre of water) at 55 & 65 DAS to improve the source-sink relationship	2 % DAP (20 g/litre of water) at 55 & 65 DAS to improve the source-sink relationship
Pest management	Hairy caterpillar and Jassids are major pests. Spray of 300 ml of Dichlorphos of 100 EC	Hairy caterpillar and Jassids are major pests. Spray of 300 ml of Dichlorphos of 100 EC in 400 lit of water ha^{-1} for

Contd.

	in 400 lit of water ha^{-1} for controlling aphids. Application of Carbofuran @ 10 kg ha^{-1} at the time of sowing is considered to be a preventive measure for checking major insects. If intensity of pest is more, 2 % Quinolphos can be applied.	controlling aphids. Application of Carbofuran @ 10 kg ha^{-1} at the time of sowing is considered to be a preventive measure for checking major insects. If intensity of pest is more, 2 % Quinolphos can be applied.
Disease management	Leaf spot and powdery mildew are major diseases reported in the Islands. Leaf spot can be controlled by 2 or 3 sprays of Zinc (2 %). The Powdery mildew can be controlled by 2-4 sprays of wettable sulphur @2-3 kg ha^{-1} in 100 litres of water.	Leaf spot and powdery mildew are major diseases reported in the Islands. Leaf spot can be controlled by 2 or 3 sprays of Zinc (2 %). The Powdery mildew can be controlled by 2-4 sprays of wettable sulphur @2-3 kg ha^{-1} in 100 litres of water.
Harvesting	The pods become black or brown on maturity in 75 to 85 days depending on variety. Staggered harvesting can be practiced by plucking the pods as soon as colour turns black or brown if no rainfall is expected during harvesting phase.	Plucking the pods can be done weekly or as soon as colour turns black or brown if no rainfall is expected during harvesting phase. After plucking of pods, plants should be ploughed in to the soil for enriching the soil fertility.
Yield	800 to 1200 kg ha^{-1} can be obtained with scientific package of practices	800 to 1200 kg ha^{-1} can be obtained with scientific package of practices

Contingency Measures

Delayed sowing due to excess moisture in soil

- Prefer early maturity varieties to avoid crop loss due to onset of rainfall during maturity
- Mulching with paddy straw

Excess rainfall during vegetative and flowering stage

- Excess standing water should be drained immediately after cessation of rainfall (draining of water will be easy if small ridges and furrows are made after first weeding)
- Application of starter dose of nitrogen @ 25 kg ha^{-1} should be done if yellowing symptoms are observed due to water logging.
- Spray of 2 % DAP can also be made to restore the crop from yellowing

Drought during vegetative and flowering phase

- Apply irrigation at critical stages (flowering) only depending upon water availability.
- Reduce the plant population to 66 %.
- Mulching with paddy straw.

Rainfall during maturity phase :

- Uproot the entire plants and heap on elevated places.

Conclusion

Pulses play a critical role in food production and enhancement of soil fertility besides contributing towards reducing malnutrition. However, due to the various factors, production of pulses has reached plateau on the Islands. The present level of technologies available will make a significant contribution towards enhancement in production. In the Islands, black gram and greengram are fitting well with rice based cropping system and residual moisture. The production of pulses can be doubled by simply bridging the yield gap through scientific package of practices. The islands are having a deficit of around 3000 t at the present level of population which is going to increase to 5000 t in the next 10 years. Hence, proper policy on pulses needs to be developed for the islands with concerted efforts on technology demonstration, upgradation and refinement. The technology in terms of varieties and package of practices are to be developed with modern tools and equipments which can reduce drudgery in farming. Promotion of local small scale processing units in the islands are very much needed to enhance the farmer's profitability and encourage them to grow pulses.

Future Thrust

- Research efforts should be made to introduce perennial pigeonpea in the islands to harness excess moisture in the rainy season and moisture stress during dry season.
- Growing of pulses during rainy season in the bunds of rice field to be worked out with proper management practices.
- Identification of extra short duration (<50 days) varieties of pulses are required to be under taken to take benefit of shorter effective growing period in dry season.
- Small scale *dal* mills are to be introduced in the islands for post harvest processing and value addition.

- Scientific technology delivery through researcher managed experiments in farmers field.

References

Andaman and Nicobar Administration statistics book. (2005; 2006). Directorate of Statistics, A&N Administration, Port Blair.

Annual Report (1978 to 2006-07). Central Agricultural Research Institute, Port Blair.

Gangwar, B. and R. Ahmed (1991). Performance of mungbean varieties under A&N Island conditions. Indian Journal of Pulses Research 4(1):105-106.

Ganeshamurthy A.N., R.Dinesh, N.Ravisankar, A.K.Nair and S.P.S.Ahlawat. (2002). Land resource atlas of Andaman and Nicobar islands, Central Agricultural Research Institute, Port Blair.

Gangwar, B. and N.T. Singh. (1991). Stratagy for self sufficiency of pulses and oil seeds in Bay Islands, Journal of Island, A&N Administration pp 48-50.

Gangwar, B., Mongia, A.D and A.K. Bandyopadhyay. (1991). Cropping systems and management practices for humid tropical conditions of Bay islands, Journal of Indian Society of Coastal Agricultural Research 9(1&2): 241-249.

Pramanik S.C. and N. Ravisankar. (2007). Productivity, profitability and sustainability of rice based cropping systems in Bay islands. J. Farming System Research & Development 13 (1): 93-98.

Pramanik, S.C., N.Ravisankar, A.K.Nair and S. Ghoshal Choudhuri. (2001). Rain water conservation technique for sustainable crop production during dry season in Andaman. In: Natn. Conference on resource conservation and watershed management, 23-25th May, 2001, CSWCR&TI, Dehradun, Uttaranchal.

Pramanik, S.C., R.P. Dubey, N. Ravisankar and A.K. Nair. (2000). Development of water resources and irrigation potential in A&N Islands. Yojana 44(6): 19-20, 32.

Ravisankar, N., S.K. Ambast and R.C. Srivastava. (2008). Crop diversification through broad bed and furrow system in coastal regions, Central Agricultural Research Institute, Port Blair.

❑❑❑

10

Jammu & Kashmir

Dileep Kachroo, Jyoti Kachroo, Mudasir Iqbal, N.P. Thakur and Ashok Gupta
Farming System Research Centre (AICRP-IFS), Sher-e-Kashmir University of Agricultural Sciences and Technology - Main Campus Chatha, Jammu. (J&K) 180009

Pulses are important crops in the State of Jammu and Kashmir as they are the only source of protein, an essential part of the vegetative diet. Besides that, pulses enrich the soil fertility by fixing atmospheric nitrogen in the root nodules and improve the soil structure (Thakur *et al.,* 2009). The tap root system explores the deeper soil strata and heavy leaf litter increases the soil organic matter. Pulses in J& K state are ideal crops for mixed and inter-cropping as well as vegetable and fodder crops. It has been estimated that by 2020 the pulse requirement for Jammu region of J&K alone will be 1.51 lakh tonnes (based on average per capita consumption of 68 grams for the projected population of 60.75 lakh) whereas the estimated requirement of J&K state as a whole works out to be 3.85 lakh tonnes (Digest of Statistics 2008-09).

As per the present scenario of Jammu region of J&K, the area under other crops (Rice, wheat, maize and oilseeds is maximum). However, during 2007-08 the area under the maize witnessed a decline trend due to replacement by the pulse crops from 2002-03 onwards (Table 1). The area under pulses during 1980-81 was maximum but in the later years, the area under the pulse crops has declined mainly because of more risk of diseases and lack of suitable genotypes (Digest of Statistics, 2008-09).

Table 1 : Area (thousand ha) under different crops in J & K

S. No	Year	Rice	Wheat	Maize	Pulses	Oilseeds
1	1980-81	265	202	276	49	53
2	1990-91	274	245	294	41	67
3	1995-96	273	243	303	33	68
4	1999-00	250	245	217	29	76
5	2000-01	244	280	330	27	74
6	2001-02	249	259	326	27	78
7	2002-03	236	248	329	29	66
8	2003-04	259	254	321	28	63
9	2004-05	250	252	323	31	64
10	2005-06	259	252	321	29	63
11	2006-07	252	266	323	29	64
12	2007-08	263	278	302	30	63
13	2008-09	258	279	316	30	65
14	2009-10	280	289	311	30	65
15	2010-11	261	291	308	29	64

Land Use Pattern in J&K

Net area sown in Jammu is 3.89 lakh ha whereas in Kashmir region, it is 3.39 lakh ha having a total net area of 7.48 lakh hectares (Table 2 and Fig 1). The cropping intensity also varies within the regions, it is 172% in Jammu and 123% lowest in Kashmir. The average land holding is more in Jammu (1.02 ha) as compared to Kashmir region (0.53 ha). However, the per cent irrigated area in Jammu is less (25.7%) than Kashmir (56.3 %) thereby indicating that on an average the J and K state is having about 70% of the area under rainfed ecosystems and this area can be utilized properly through intensification/diversification of pulse crops

Table 2 : Land Use Pattern in J&K (Lakh ha)

Particular	Jammu	Kashmir	J&K
Net sown area	3.89	3.39	7.48
Gross sown area	6.68	4.16	11.06
Cropping intensity (%)	172	123	148
Avg. land holdings(ha)	1.02	0.53	0.76
% irrigated area	25.7	56.3	41.5

Fig. 1 : Land Resource in J & K.

Trends in Area, Production and Productivity of Pulses

The area under the pulses in J&K during 1955–56 to 1980–81 (Table 3 and Fig. 3) was observed to have constantly increased but started declining from 1980-81. It decreased from 48.60 thousand ha in 1981 to 27.54 thousand ha in 2003-05, whereas from 2004-05 onwards the area under the pulse crop has shown a slightly positive trend. The data further revealed that the productivity (q/ha) increased from 6.07 to 6.49 q/ha during 1955-56 to 1990-91, but thereafter, decreased to 4.61 q/ha up to 2005-06. However, during 2006-07 and 2007-08, the pulse productivity recorded a slight improvement to the extent of 5-10 per cent.

Table 3 : Area, Production and Productivity of Pulses in J&K state

S. No.	Year	Area (000 ha)	Production (000 q)	Productivity (kg/ha)
1.	1955-56	44.00	267	607
2.	1960-61	52.00	335	644
3.	1965-66	44.00	231	525
4.	1968-69	55.00	361	656
5.	1974-75	52.20	318	609
6.	1980-81	48.60	337	693
7.	1985-86	44.76	285	637
8.	1990-91	41.32	268	649
9.	1995-96	32.57	152	467
10.	1999-00	29.27	145	495
11.	2000-01	27.45	128	460
12.	2001-02	26.75	125	467
13.	2002-03	28.96	142	490
14.	2003-04	27.54	132	479
15.	2004-05	30.90	152	492
16.	2005-06	29.27	135	461
17.	2006-07	29.06	141	485
18.	2007-08	30.15	153	507
19.	2008-09	30.00	139	463
20.	2009-10	30.30	172	568
21.	2010-11	28.91	169	584

Source : Digest of Statistics, Directorate of Statistics, Govt. of J&K

Growth Trends

The data for growth of area, production and productivity has been grouped for five years up to 1999-00 and for the last decade the annual growth has been worked out, (Table 4 and Fig 2). The area, production and productivity of pulses increased from 1955-56 to 1960-61 after which there was a decrease in all the parameters till the green revolution. A negative growth was observed in area after the green revolution till 2002-03. The growth rate in case of productivity showed a fluctuating trend with sometimes positive and sometimes negative rate. The growth was positive in 1980-81, 1990-91, and 1999-00, 2001-02, 2003-04, 2004-05, 2006-07 and 2007-08. It has been estimated that the per cent change in terms of area, production and productivity recorded during 2007-08 was 3.75, 8.51 and 4.51 per cent, respectively, which indicated that the demand of the pulses is increasing and needs to be included in a cropping system mode. The reason of low productivity is attributed to the fact the pulse production in Jammu is undertaken in rainfed areas, with poor soil fertility status, besides the constraint of farmers not getting high yielding varieties.

Table 4 : Growth of area, production and productivity of pulses in J&K state

S. No.	Period		Area		Production		Productivity	
	From	To	Change (000 ha)	Per cent change	Change (000 q)	Per cent change	Change (q/ha)	Per cent change
1.	1955-56	1960-61	(+)8.00	(+)18.18	(+)68.00	(+)25.47	(+)0.37	(+)6.10
2.	1960-61	1965-66	(-)8.00	(-)15.38	(-)104.00	(-)31.04	(-)1.19	(-)18.48
3.	1965-66	1968-69	(+)11.00	(+)25.00	(+)130.00	(+)56.28	(+)1.31	(+)24.95
4.	1968-69	1974-75	(-)2.80	(-)5.09	(-)43.00	(-)11.91	(-)0.47	(-)7.16
5.	1974-75	1980-81	(-)3.60	(-)6.90	(+)19.00	(+)5.97	(+)0.84	(+)13.79
6.	1980-81	1985-86	(-)3.84	(-)7.90	(-)52.00	(-)15.43	(-)0.56	(-)8.08
7.	1985-86	1990-91	(-)3.44	(-)7.69	(-)17.00	(-)5.96	(+)0.12	(+)1.88
8.	1990-91	1995-96	(-)8.75	(-)21.18	(-)116.00	(-)43.28	(-)1.82	(-)28.04
9.	1995-96	1999-00	(-)3.30	(-)10.13	(-)7.00	(-)4.61	(+)0.28	(+)6.00
10.	1999-00	2000-01	(-)1.82	(-)6.22	(-)17.00	(-)11.72	(-)0.35	(-)7.07
11.	2000-01	2001-02	(-)0.70	(-)2.55	(-)3.00	(-)2.34	(+)0.07	(+)1.52
12.	2001-02	2002-03	(+)2.21	(+)8.26	(+)17.00	(+)13.60	(+)0.23	(+)4.93
13.	2002-03	2003-04	(-)1.42	(-)4.90	(-)10.00	(-)7.04	(-)0.11	(-)2.24
14.	2003-04	2004-05	(+)3.36	(+)12.20	(+)20.00	(+)15.15	(+)0.13	(+)2.71
15.	2004-05	2005-06	(-)1.63	(-)5.28	(-)17.00	(-)11.18	(-)0.31	(-)6.30
16.	2005-06	2006-07	(-)0.21	(-)0.72	(+)6.00	(+)4.44	(+)0.24	(+)5.21
17.	2006-07	2007-08	(+)1.09	(+)3.75	(+)12.00	(+)8.51	(+)0.22	(+)4.54
18.	2007-08	2008-09	(-) 0.15	(-)0.50	(-)14.00	(-)9.15	(-)0.44	(-)8.68
19.	2008-09	2009-10	(+) 0.30	(+)1.00	(+)33.00	(+)23.94	(+)1.05	(+)22.68
20.	2009-10	2010-11	(-) 1.39	(-)4.59	(-)3.00	(-)1.74	(+).16	(+)2.82

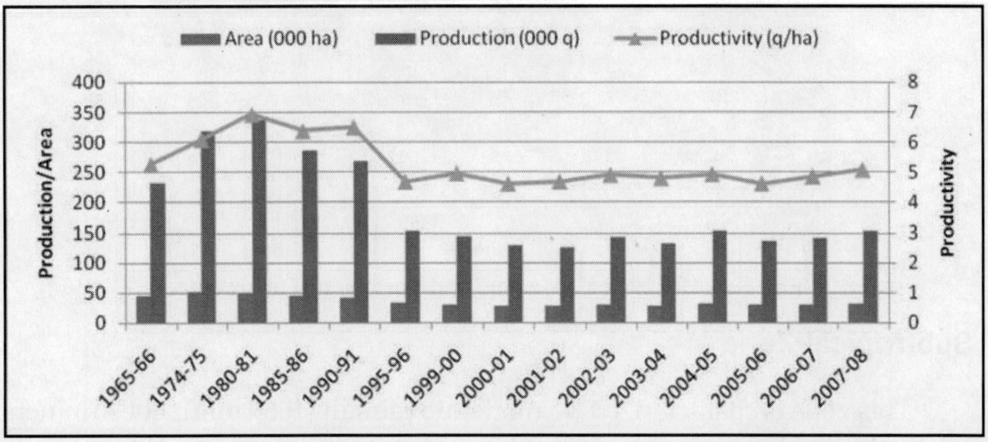

Fig. 2 : Area, Production and Productivity of Pulses in J&K state.

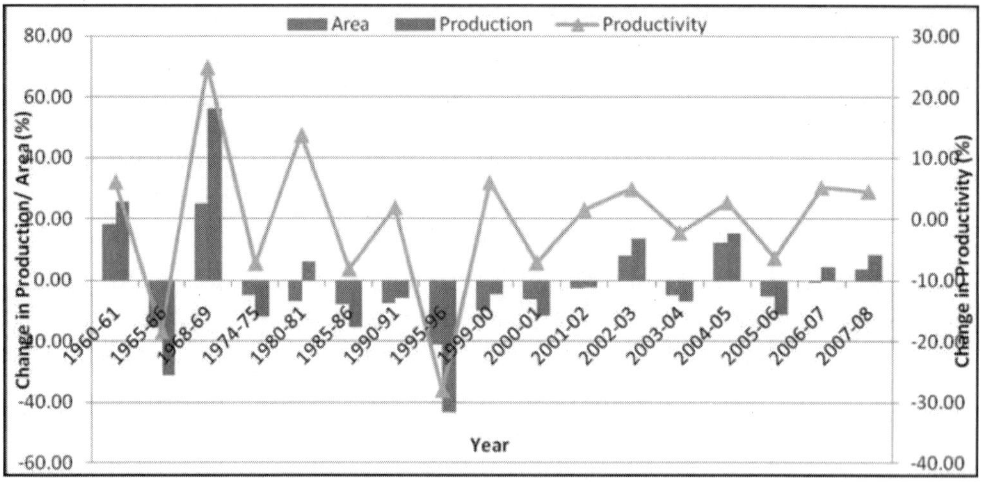

Fig. 3 : Per cent growth of area, production and productivity of pulses in J&K state.

Status of Pulse Growing Areas in Jammu and Kashmir

Jammu & Kashmir state comprises three provinces *viz*: Jammu, Kashmir, and Ladhak which comes under subtropical zone, intermediate zone, temperate zone and cold arid zone (Fig.4).

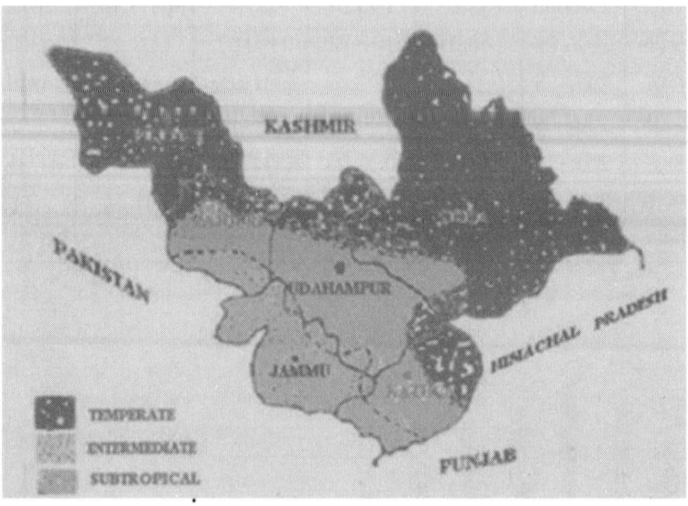

Fig. 4 : NARP zones of Jammu and Kashmir and study area.

Subtropical Zone

This zone is characterized by moderate rainfall (1069 mm), hot summers, relatively dry but pronounced winter and preponderance of alluvial soils, comprising of outer planes and hills of Jammu, Kathua and part of Udhampur district which an altitude range of 1000 feet to 4500 feet above mean sea level (AMSL). It has 10.61 lakh ha total geographical area with net sown area of 3.36 lakh ha. Rice-wheat is the prominent cropping system in irrigated areas, whereas maize/bajra-wheat including pulses in cropping systems in rainfed areas.

The maximum area of pulses in Jammu region is under lentil followed by *kulthi*, frenchbean and field pea, whereas, maximum production was of lentil followed by frenchbean, *kulthi* and field pea. (Tables 5, 6 and 7)

The maximum area under pulses in 2009 was in Jammu district followed by Kathua, Samba and Udhampur in Jammu division. The maximum area under pulses during 1980-81 in *rabi and kharif* was under Ravi Tawi command area, which was once a rain fed belt with no source of assured irrigation. With creation of canal water irrigation facility, the farmers of the area shifted to rice-wheat.

The area for *kharif* pulses of Jammu province comes under the rainfed dry land areas of district Kathua, Jammu, Rajouri and Udhampur, where Urdbean is the main crop. The shift from pulse to maize has been observed due to Yellow Mosaic Virus. Now with resistant varieties of Urdbean by SKUAST-J, the area under the pulse crop is gaining momentum.

The major pulses grown in Jammu region are lentil, Kulthi and French bean while in Kashmir region soyabean, French bean and peas are the major pulses grown.

Intermediate Zone

This zone is a transition zone between sub-tropical and temperate zone with high mean annual rainfall (1478 mm), relatively wet and cold winters and preponderance of spodosols, comprising of mountainous area above the outer hills including part of Kathua, Udhampur, Poonch, Rajouri and Doda districts of Jammu province with altitude ranging from 4500-1200 feet above mean sea level. This area experiences rugged terrain, steep slopes and lofty hills with no irrigation facility. It has a geographical area of 7.78 lakh ha with net sown area of 1.23 lakh ha out of which only 16 thousand hectare is irrigated.

The maximum area of 2583 ha with production (13740 qtls) of different pulses are in Udhampur district followed by Doda, Kishtwar, Rajouri and Reasi districts of intermediate zone. The data in table 5, showed that Urdbean occupied maximum area in all the six districts, while rajmash and kulthi were second and third, respectively area-wise. Although cultivated in four districts, cowpea is cultivated in 15 ha in Rajouri and 6 ha in Poonch district only. Similarly moong occupies 12 ha in Udhampur and 1 ha in Reasi, whereas

Table 5 : District wise area (ha) of various pulses in Jammu and Kashmir State during 2010

District	Cowpea (Rongi)	Soya bean	Rajmash (French bean)	Mash (Urdbean)	Moong (Green gram)	Lentil (Masoor)	Kulthi (Horse gram)	Field Pea	Other Pulses	Total pulses
Temperate zone (Kashmir region)										
Srinagar	-	8	31	-	-	-	-	48	4	91
Ganderbal	-	-	-	-	-	-	-	-	310	310
Budgam	-	224	635	7	122	171	-	458	-	1617
Anantnag	-	1098	152	41	123	-	-	16	17	1447
Kulgam	-	154	314	-	199	-	-	15	451	1133
Pulwama	-	164	39	5	100	-	-	56	-	364
Shopian	-	149	21	-	-	-	-	36	-	206
Baramulla	-	361	383	122	150	68	-	10	560	1654
Bandipora	-	622	74	-	149	762	-	182	3	1792
Kupwara	-	-	-	-	-	-	-	-	511	511
Cold Arid Zone										
Leh	-	-	-	-	-	1	18	187	100	306
Kargil	-	-	-	-	-	-	-	526	21	547
Total Kashmir Division	-	2780	1649	175	843	1002	18	1534	1977	9978
Sub tropical zone (Jammu region)										
Jammu	-	6	-	3605	-	120	981	55	928	5695
Samba	-	-	-	2706	34	26	7	-	95	2868
Kathua	-	2	1	3348	-	-	301	4	51	3707
Intermediate zone										
Udhampur*	-	-	75	2401	12	-	50	17	28	2583
Reasi	-	18	-	576	1	-	58	-	99	752
Rajouri	15	-	-	1217	-	-	-	-	-	1232
Poonch	6	-	43	7	-	-	-	-	-	56
Doda*	-	-	461	1120	-	-	374	42	-	1997
Kishtwar	-	-	349	594	-	-	2	101	338	1384
Ramban	-	-	-	-	-	-	-	-	-	0
Total Jammu Division	21	26	929	15574	47	146	1773	219	1539	20274
J&K State	21	2806	2578	15749	890	1148	1791	1753	3516	30252

*Part of Udhampur, Doda and Kashtiwar districts fall under the temperate zone in Jammu region.

Source: Directorate of Agriculture, Government of Jammu and Kashmir.

Table 6 : District wise production (q) of various pulses in Jammu and Kashmir State during 2010

District	Cowpea (Rongi)	Soyabean	Rajmash (French bean)	Mash (Urdbean)	Moong (Green gram)	Lentil (Masoor)	Kuthi (HorseGram)	Field Pea	Other Pulses	Total pulses
Temperate zone (Kashmir region)										
Srinagar	-	54	78	-	-	-	-	268	17	417
Ganderbal	-	-	-	-	-	-	-	-	1636	1636
Budgam	-	1792	3582	49	700	1371	-	5496	-	12990
Anantnag	-	5740	856	139	635	-	-	112	90	7572
Kulgam	-	805	1767	-	1026	-	-	105	2368	6071
Pulwama	-	1347	210	27	405	-	-	392	-	2381
Shopian	-	1224	114	-	-	-	-	252	-	1590
Baramulla	-	1812	2068	829	534	342	-	70	2962	8617
Bandipora	-	3123	401	-	531	3826	-	1093	16	8990
Kupwara	-	-	-	-	-	-	-	-	2686	2686
Cold Arid zone										
Leh	-	-	-	-	-	4	67	921	400	1392
Kargil	-	-	-	-	-	-	-	4088	95	4183
Total Kashmir Division	-	15897	9076	1044	3831	5543	67	12797	10270	58525
Sub Tropical zone (Jammu region)										
Jammu	-	33	-	15935	-	660	2649	1023	35231	55531
Samba	-	-	-	1196	170	142	19	-	507	2034
Kathua	-	15	6	10077	-	-	844	28	212	11182
Intermediate Zone										
Udhampur	-	-	375	12221	80	-	233	119	712	13740
Reasi	-	126	-	2933	7	-	270	-	528	3864
Rajouri	110	-	-	6095	-	-	-	-	-	6205
Poonch	45	-	215	35	-	-	-	-	-	295
Doda	-	-	5532	4705	-	-	1122	252	-	11611
Kishtwar	-	-	4188	2495	-	-	7	510	1758	8958
Ramban	-	-	-	-	-	-	-	-	-	0
Total Jammu Division	155	174	10316	55692	257	802	5144	1932	38948	113420
J&K State	155	16071	19392	56736	4088	6345	5211	14729	49218	171945

Source : Directorate of Agriculture, Government of Jammu and Kashmir.

Table 7 : District wise productivity (q ha^{-1}) of various pulses in Jammu and Kashmir State during 2010

District	Cowpea (Rongi)	Soyabean	Rajmash (French bean)	Mash (Urdbean)	Moong (Green gram)	Lentil (Masoor)	Kulthi (Horse Gram)	Field Pea	Other Pulses	Total pulses
Temperate zone (Kashmir region)										
Srinagar	-	6.75	2.52	-	-	-	-	5.58	4.25	4.58
Budgam	-	8.00	5.64	7.00	5.74	8.02	-	12.00	-	8.03
Anantnag	-	5.23	5.63	3.39	5.16	-	-	7.00	5.29	5.23
Kulgam	-	5.23	5.63	-	5.16	-	-	7.00	5.25	5.36
Pulwama	-	8.21	5.38	5.40	4.05	-	-	7.00	-	6.54
Shopian	-	8.21	5.43	-	-	-	-	7.00	-	7.72
Baramulla	-	5.02	5.40	6.80	3.56	5.03	-	7.00	5.29	5.21
Bandipora	-	5.02	5.42	-	3.56	5.02	-	6.01	5.33	5.02
Kupwara	-	-	-	-	-	-	-	-	5.26	5.26
Cold Arid zone										
Leh	-	-	-	-	-	4.00	3.72	4.93	4.00	4.55
Kargil	-	-	-	-	-	-	-	7.77	4.52	7.65
Total Kashmir Division	-	5.72	5.50	5.97	4.54	5.53	3.72	8.34	5.19	5.87
Sub Tropical zone (Jammu region)										
Jammu	-	5.50	-	4.42	-	5.50	2.70	18.60	37.96	9.75
Samba	-	-	-	0.44	5.00	5.46	2.71	-	5.34	0.71
Kathua	-	7.50	6.00	3.01	-	-	2.80	7.00	4.16	3.02
Intermediate Zone										
Udhampur	-	-	5.00	5.09	6.67	-	4.66	7.00	25.43	5.32
Reasi	-	7.00	-	5.09	7.00	-	4.66	-	5.33	5.14
Rajouri	7.33	-	-	5.01	-	-	-	-	-	5.04
Poonch	7.50	-	5.00	5.00	-	-	-	-	-	5.27
Doda	-	-	12.00	4.20	-	-	3.00	6.00	-	5.81
Kishtwar	-	-	12.00	4.20	-	-	3.50	5.05	5.20	6.47
Ramban	-	-	-	-	-	-	-	-	-	-
Total Jammu Division	7.38	6.69	11.10	3.58	5.47	5.49	2.90	8.82	25.31	5.59
J&K State	7.38	5.73	7.52	3.60	4.59	5.53	2.91	8.40	14.00	5.68

Source : Directorate of Agriculture, Government of Jammu and Kashmir.

soybean is cultivated only in Reasi district (Table 5). However, Ramban is the only district in intermediate zone where there is no area under pulses. The highest production of urdbean was recorded in Udhampur, Rajouri and Reasi districts with productivity of 5.09 q/ha, while Doda and Kishtiwar district recorded highest production of rajmash with high productivity of 12 q/ha (Table 7). Cowpea recorded productivity of 7.5 q/ha in Poonch, while productivity of soybean and moong was 7 q/ha each in Reasi districts. *Kulthi* recorded the lowest productivity of 4.66 q/ha in Udhampur as well as in Reasi but 3.5 and 3.0 q/ha in Kishtwar and Doda districts, respectively.

Temperate Zone

A temperate zone is a region on the Earth located between either the Tropic of Cancer and the Arctic Circle or the Tropic of Capricorn and the Antarctic Circle. The regions in the continental temperate zone generally do not receive much precipitation, receiving on an average less than 30 inches (75 cm) of rain a year, although they may experience thunderstorms. In this zone, it is very rare for the temperatures to drop below the freezing point (32° F) even in the dead of winter.

In the Pirpanjal and Higher Himalayas, the temperature during the winter season, drop to as low as -20^0 C. In these regions, the winters are prolonged for a long period (7 to 8 months) while the summers are of short duration. High altitudes in north-west Himalayas are enclosed within mountain ranges which give it distinctive character. The region has a diversified climate and moderate climate is found only in central longitudinal valleys at 1500 to 2100 meters above the mean sea level. It has been further observed that in Bhaderwah, the utilization of soil moisture is less and recharge is more as compared to the rest of the area. Bhaderwah dominates in terms of precipitation over potential evapo-transpiration throughout the year except for the month of June, August and October. The Himalayan highlands (Bhaderwah, Batote, Banihal) receive much larger amount of precipitation due to these western disturbances in the form of snow, whose melt water furnishes the indispensable irrigation water for the plains. The 15°C temperature is mainly confined to the mountainous tract like Batote, Bhaderwah and Banihal. Bhaderwah is also dominated by the impact of western disturbances as well as by topo-climatic setup.

In temperate zone, maximum area under pulses was recorded in Bandipora closely followed by Baramulla, Budgam and Anantnag, whereas it was lowest in Srinagar. Pulse crops like soybean, rajmash, moong and field pea exist in almost in all the districts of this zone. Soybean occupied maximum area of 1098 ha in Anantnag, rajmash (635 ha) and field pea (4.58 ha) in Budgam. Mash (122 ha) in Baramula, whereas moong (199 ha) in Kulgam, lentil (762 ha) in Bandipora (Table 5). The highest production (12990 q) of pulses was from Budgam district of Temperate zone followed by Bandipora (8990 q) and Baramulla (8617 q). However, the highest production was recorded for Soybean crop in Anantnag district. Followed by field Pea (5496 quintal) and rajmash (3585 q) in Budgam and lentil (3826 q) in Bandipora district with highest productivity of 12 q/ha in field pea in Budgam followed by soyabeen (8.21 q/ha) in Pulwama and Shopiana district. The lowest productivity of 2.52 q/ha was recorded under rajmash crop in Srinagar district.

Cold Arid Zone

This zone is confined to Leh and Kargil districts of Ladakh province, located geographically between 32^0 152 to 36^0 north latitude and 75° 152 to 80^0 152 east longitude covering approximately 82665 sq km area (SKUAST, Leh 1998). The altitude of the region ranges between 9000 feet to 15000 feet above mean sea level with mountain peaks as high as 26000 feet ASL. The precipitation in this region is low varying from 80-300 mm mostly in the form of snow with extremes of temperatures ranging from -40°C in winter to 40°C in summer, with oxygen concentration in air and atmospheric pressures low enough to effect the physiology of living organisms. Field pea, kulthi and lentil are the pulses mainly grown in this zone. The maximum area of 526 hectare and 187 hectare is under field pea in Kargil and Leh districts of Ladakh region (Table 5). While kulthi and lentil occupied 18 and 1 ha only in Leh district respectively, the highest production and productivity of 4088 quintal and 7.77 q/ha for Field pea, respectively, was in Kargil district and 921 quintal with productivity of 4.93 q/ha was recorded in Leh district. However, the productivity of lentil and *kulthi* was recorded as 4.0 and 3.72 q/ha in Leh (Table 6)

Improved varieties : The improved varieties of different pulse crops for J&K are as follows

Chickpea : C-235, Gaurav, Avrodhi, BG-256, PBG-1, GNG 469, BG 1053, RSG931

Lentil : L-9-12, PL-406, L-4147, DPL-15, DPL-62

Field pea : T-163, Rachna, HFP-4, HFP-8909, KPMR552, DDR27

Urdbean : Pant U-19, UG-218, Mash -338, Uttara

Mungbean : Ps-18, Pusa Baisakhi, PDM-54, ML-131, Ml-337, Pusa 105, ML818

Rajmash : C-235, K-468, Gaurav, PBG-1 and SCS-3:

Existing Cropping Systems

The predominant cropping systems involving pulse crops in different zones are highlighted here.

Subtropical Zone

Under Irrigated Conditions : Paddy - Wheat/Pea (Mid June-End Jul - Nov-Mid Dec), Maize-Wheat/Pea/Potato (Onset of Monsoon-Nov/Oct-Nov/Oct-Nov) and Cowpea-Wheat/Mustard (Onset of monsoon- Nov/Mid Oct-Nov).

Under Rainfed Conditions : Maize/Bajra-Chickpea (Onset of monsoon-Mid Oct-Mid Nov), Maize/Bajra-Potato/Pea (Onset of monsoon- Mid October-Mid Nov) and Moong/Mash-Wheat (Onset of monsoon- Nov-Mid Dec).

Intermediate Zone

Under Irrigated Conditions : Rajmash –Wheat/Barley (Apr-May - Oct-Dec) and Mash -Wheat/Barley (Last week of June - Oct-Dec).

Under Rainfed Conditions : Cowpea- Wheat and Mash- Wheat/Barley.

Temperate Zone

Rice-Peas, Maize + Cowpea – Mustard, Maize + Rajmash – Potato, Rajmash Potato, Mash – Mustard, Mash – Oats, Mash – Peas, Mash – Potato, Horse gram + Mustard and Soya bean- Mustard.

The temperate zone in particular is agriculturally dormant for a long spell of nearly four months. During this period only few crops are capable to grow which include rape seed, oat, berseem, wheat and pulses like peas and lentil etc. (Diagnostic survey report 2001-04).

Zone-wise Improved Cropping Systems Involving Pulses

Sub- Tropical Zone

Irrigated Area : Rice-Mustard-Greengram with B:C ratio of 2.45, Rice-Peas-Maize (cobs) with B:C ration 2.65, Rice-Potato-Rajmash with B:C ratio 2.64, Rice-Chickpea-Fodder with B:C ratio 2.56 (Thakur, *et al.* (2009), Rice-Potato-Blackgram with B:C ratio 1.41, Rice-Gobhi Sarson-Maize + Cowpea with B:C ratio 1.30 and Rice-Berseem-Moong/Mash (Annual report 2008-09 AICRP on CS).

Un-Irrigated Area : Maize + Cowpea (1:1)-Wheat + Mustard (4:1), Maize-Chickpea +Gobhi Sarson, Mash-Wheat, Mash-Mustard, Mash-Berseem,

Maize + Cowpea-Wheat, Moong-Wheat, Maize-Potato-Mash, Mash + Till-Wheat + Pea, Maize- Chickpea + Barley and Maize-Horse gram+ Wheat.

Intermediate Zone : Maize + Cowpea-Fodder, Mash- Wheat, Mash-Toria-Wheat and Rice-Potato-Mash {net return of Rs. 61783/ha} (Annual report of AICRP on CS 1998-99).

Intercropping of Cowpea in maize during *kharif* has been found as promising system under the dry land condition of Rajouri district which falls under intermediate zone of Jammu region (Diagnostic survey of rice and maize based cropping system. Report under NATP Project 2001-02).

Temperate Zone : Rice-Peas, Maize + Cowpea-Mustard, Maize + Rajmash-Potato, Rajmash-Potato, Mash Mustard, Mash-Oats, Mash-Peas, Mash-Potato, Horse gram + Mustard and Soya bean-Mustard.

Yield gaps : Generally, it has been observed that there is a lot of yield gap among the research experiment yields, demonstration yields and farmers' yields. The extent in percentage terms varies from 45-65 percent in almost all the pulse crops.

Production Technologies of Pulse Crops

Moongbean and Urdbean (Mash)

Moongbean and urdbean are the most important pulse crops of the rainy (*kharif*) season.

Climate : These crops thrive best under hot and humid climate of subtropical plains.

Soil : Theses pulse crops do well on all type of soils, but light sandy loam to heavy clay are ideal soils for mash/moong cultivation. Heavy soils are preferred for moong cultivation.

Preparation of land : Plough the land with *Tawi* plough followed by 1-2 ploughings with *desi* plough or disc harrow. Each ploughing should be followed by planking to get desired tilth.

Selection of Suitable Varieties

Urdbean Mash : Pant U-19 and Uttara are suitable for both *kharif* season and summer season.

Moong : PDM-54, ML-131, and ML-818 are most suited. However, varieties, *viz.*, PS-16, PS-7 and SML-668 can also be sown during summer season only.

Seed rate : Use 15 to 20 kg of seed per ha. Treat the seed with Thiram or Captan @ 3g/per kg of seed before sowing.

Inoculation : It is better to inoculate the seed with *Rhizobium* culture before sowing for getting higher yields.

Time & method of sowing : *Kharif* mash crop is sown with the onset of monsoon up to 15th July in sub-tropical area and end of June in intermediate area up to 750 m altitude. However up to 800 m altitude, it can be sown up to Ist week of July. Summer crop of mash is sown from 15th March to first week of April. However, summer moong can be sown up to 20th April. The crop should be sown in lines 30 cm apart by '*Kera*' method. The seed should be sown 4 to 6 cm deep.

Manures and fertilizer : It is better to get the soil tested and apply fertilizers according to the recommendations. In general, 16 kg N and 40 kg P_2O_5 through 90 Kg di-ammonium phosphate may be applied preferably drilled before sowing. The nutrient may also be supplied through SSP and urea.

Weed Management

i) **Cultural weed control practices** : Give one weeding and hoeing one month after sowing when the plants attain a height of 15-20 cm with hand blade hoe, *khurpa* or medium cultivator.

ii) **Chemical methods** : Use Fluchloralin @ 0.75 kg a.i/ha in 800-1000 L irrigation, but in case of failures of rains, the crop should be irrigated. In case of heavy rains, water should be drained off from the field; otherwise it will damage the crop.

In case of summer sowing, regular irrigation at an interval of 10 to 15 days may be given to crop. However summer crop requires 2 irrigations under heavy soil, 3 under medium and 4 under light soil conditions. Last irrigation should be given at pod formation stage.

Plant protection : Adopt the following measures to control the insect, pest and diseases.

Hairy caterpillar : Spray *Carbaryl* 50% WP @ 1.5 kg/ha in 750 L water. Collection and destruction of first and second larvae of hairy caterpillar present on underside of leaves.

White fly : These damage the crop by sucking the cell sap from leaves and pods. White fly also spreads mosaic virus from one plant to another. Spray the crop with Malathion 50 EC @ 1 ml/L of water.

Leaf spot diseases : Spray the crop with Zineb (0.2 %) or Mancozeb (0.25%).

Virus diseases : Remove the infected plants and burn them.

Harvesting : The crop may be harvested when the leaves fall off and most of the pods turn greyish black. Uprooting of the mature crop should be avoided. Threshing should be done by using conventional methods.

Cowpea

Cowpea is mostly grown in rainfed area during rainy *(kharif)* season as a mixed crop with maize or as a pure crop. It is an important pulse crop in the intermediate region of Jammu province. For getting higher yield from this crop, the following packages of practices are recommended.

Varieties : C-152, PS-42, CULTURE-1.

Climate : It requires warm climate and can be grown both during summer and *kharif* seasons.

Soil : It thrives best on light to heavy loam soils with good drainage especially during rainy season. However, for summer crop (March-June), heavy soils are better, as they retain water and consequently, the number of irrigations required is less.

Land Preparation : 2 to 3 ploughings followed by *"sohaga"* are sufficient to get good seed bed. The field should be free from root stubbles, grasses and clods.

Mixed cropping : For Zone III, cowpea can be sown mixed with maize while in Zone II with bajra. Sow one row of cowpea and two rows of maize /bajra.

Time of sowing : During summer, it should be sown between 15th to 30th March under irrigated conditions in the areas of Jammu plains. In fact, it can be sown after the harvest of *sarson* and barley as a third crop in sequence in paddy growing areas especially with Culture - 1 variety.

During *kharif* it should be sown with the onset of monsoon upto 10th July. However, its sowing can be extended upto 3rd week of July with Culture-1 variety in sub-tropical areas. In hills, its sowing time coincides with time of maize sowing.

Seed rate : Use 20-25 kg/ha of seed for pure crop and 10 kg/ha for mixed crop with maize.

Inoculation : Inoculate the seed with *Rhizobium* culture before sowing.

Fertilizer application :- As a pure crop, it requires 18 kg N and 46 kg P_2O_5/ha i.e 100 kg DAP, when grown mixed with maize use N, P, K doses as required for maize. For additional benefit use *Rhizobium* culture.

Irrigation : It should be grown as irrigated crop during summer season in Jammu plains. Under such conditions, it requires 4-6 irrigations under light soil and 2-3 under heavy soil conditions. During summer it should be sown in heavy soil where paddy is grown. One irrigation 15-20 days after germination and two during flowering are sufficient to get higher yield. During *kharif*, it is grown as rainfed crop only.

Harvesting : It requires two pickings and then final harvesting. At the time of harvesting, green stalks can be fed to the cattle as fodder after picking the mature pods during both the seasons.

Yield : The yield varies from variety to variety. The higher yield is obtained from C-152 variety, although its grains are of brown colour. It yields 15 to 20 q/ha during summer and 8-10 q/ha during rainy season. Other varieties yield less than this variety but grain quality is better, being milky white in colour.

Rajmash +maize (mixed cropping)

Maize and rajmash are grown in hilly areas of Jammu division. They can be grown as pure crops. Taking Rajmash as a pure crop is not economical because returns from a pure crop are meagre. Most of the Rajmash varieties are red seeded twining type which requires arrangements for support during the crop growth period. So when sown mixed it gets natural support from maize plants.

All the practices required for maize crop are needed for this crop. Additionally the following practices are to be followed for increasing the rajmash production from mixed crop.

Seed rate : Use 20 kg of maize seed and 1 ½ kg of rajmash seed per ha. Mixed cropping of maize and rajmash can be undertaken by using seed rate 8:1.

Sowing : The sowing should be done in proper moisture conditions at a depth of 5 cm. Rajmash seed should be sown within rows of maize plant.

Sowing time : Sowing time of rajmash coincides with the sowing time of maize.

Fertilizers : No additional fertilizers are required for rajmash crop. Fertilizers recommended for maize are sufficient to meet its requirement.

Harvesting : The pods from the earlier flushes of flowers mature earlier, which exhibit shattering tendency and should be hand picked. The pods from later flushes are to be harvested subsequently.

Plant Protection

Aphids damage the crops by sucking the cell sap from leaves and pods. Aphids also spread yellow mosaic virus from one plant to another.

To control the aphids, spray the crop with 750 L of water/ha in 1000 L of water. Anthracnose, Leaf spot and Rust diseases are most common in rajmash crop and can be controlled by spray treatment with Captan or Thiram @ 2.5 g per kg of seed or spray of Zineb (0.2%). If needed, the spray is repeated after 10 days interval.

Chickpea (Gram)

Gram is one of the most important pulse crop grown during *rabi* season. Production can be increased by adopting improved cultivation practices and it can be grown upto 2000 ft. altitude.

Soil : Gram thrives best on well drained sandy or light soils. It does not do well in salt affected or water logged soils.

Climatic requirement : It is primarly a crop of low rainfall regions but gives good returns under irrigated conditions also. Extreme climates like severe cold and frost are injurious to its growth. Excess rainfall, heavy soil moisture either at time of sowing or flowering stages is deleterious to crop growth. Early onset of summer and increased day length reduce the yield by hastening maturity.

Land preparation : Gram does not require fine tilth but the soil should be ploughed once or twice to make it loose and well aerated. Soil has a remarkable effect on the control of wilt and increases the grain yield considerably. First ploughing with soil turning plough followed by 2 to 3 ploughings with disc harrow or soil stirring plough will give the desired tilth to soil. Each operation should be followed by planking.

Varieties

C-235 : Highly stable variety; seeds are small and yellowish brown.

K-468 : Medium duration (155 days) variety; seeds are small and yellowish brown in colour.

Gaurav: Seeds are bold and yellowish brown in colour.

SCS-3 : Medium seeded brown colored seeds, semi dwarf moderately resistant to podborer.

Seed rate : For bold seeded varieties, 75-80 kg/ha and for small seeded varieties, use 60-65 kg/ha.

Inoculation of seed : The seed should be inoculated with *'Rhizobium* culture' before sowing.

Fertilizer requirement : Get the soil tested but in absence of soil test, 15 kg N and 40 kg P_2O_5 through 88 kg of DAP or urea and SSP can be applied.

Sowing time : Middle of October is the optimum sowing time. However, sowing can be done up to end of October if there is no option.

Method of sowing : Gram should be sown by *"pora"/"kera"* in rows 30 cm apart. The seed should be placed 5 to 7 cm deep. Avoid shallow sowing as it encourages wilt attack upon irrigation.

Hoeing and weeding : Weed control up to 4 to 5 weeks after sowing is very essential otherwise there will be reduction in yield.

Irrigation : If water is available, irrigation at branching and pod formation stage may be given when there is no winter rain.

Nipping : For gram, nipping is a very important farm operation. It will keep down excessive vegetative growth which otherwise reduces number of pods. Nipping regularizes the vegetative growth and ensures greater number of pods per plant. It should be done before branching.

Crop association in Gram : Gram can be sown mixed with barley in seed mixture ratio of 1:1 or inter-cropped with mustard with 4 rows of gram alternating with 2 rows of mustard.

Plant protection : To control the important insect-pests and diseases the following schedule in respect of specific pests may be adopted.

Insect-pests

Pod borer (*Helicoverpa armigera*): It is a polyphagus pest. The caterpillars damage the crop by feeding on leaves, flowers, buds and pods. It bore into developing seed.Use resistant/tolerant varieties. Ploughing after harvesting of crop for destruction of *Heliothis* pupae. Collect grown up larvae of *Heliothis* and destroy them. Conserve predatory wasps, parasites and spiders. Install bird perch stand to encourage predatory birds for picking larvae of *Helicoverpa.* Spray the crop with 0.1% carbaryl 50 % WP @ 1.5 kg in 750 L of water/ha at pod initiation stage. Repeat the spray after 15 days of spraying if need arises.

Pest monitoring : Economic threshold level is 1 larvae/5 plants or 5-6 months/trap for *Heliothis.* Visual observations at weekly interval at vegetative /flowering and poding stages. Use pheromone traps (5 traps/ha).

Cut worms *(Agrotis* spp.*):* It is also polyphagus pest and nocturnal in habit. The larvae cut the germinating plants at ground level or branches of grown up plants. To control Mix lindane 1.3 % Dust @ 20 kg or 2 L of

Chlorphyriphos 20 EC in 25 kg of sand/ha in rows of the plants at sowing time.

Diseases

1. **Blight (*Ascochyta rabie*) :** Dark brown spots studded with dot like bodies are produced on the stem, branches, leaf-lets and pods. The disease spreads rapidly due to favourable climatic conditions like rain and humid weather. In cash of severe infections, the whole plant may be blighted and killed. To control the disease, treat the seeds with Carbendazim 50 WP @ 0.2% or Thiram 75 WP @ 0.3% or Carbendazim + Thiram (1:1) @ 0.2% and use resistant varieties e.g PBG-1 and Gaurav.
2. **Grey mould (*Botrytis cinerea*) :** The fungus forms gray or dark brown lesions on leaflets, branches and pods. Lesions become covered with hairy sporophores of the fungus under high humidity and low temperature conditions. Growing tips and flowers are particularly susceptible to infection. The affected plant turns yellow and shedding of infected flowers takes place. To control, treat the seeds with Carbendazim 50 WP @ 0.2 % or Tgiram 75 WP @ 0.3% or Carbendazim + Thiram (1:1) @ 0.2%. and Spray Carbendazim @ 0.1%. Crop rotation with non host crops and Use resistant varieties e.g. PBG-1 and Gaurav is recommended.
3. **Dry root rot (*Rhizoctonia bataticola*) :** The tap root on pulling out is found to be dry and devoid of lateral and fixed roots. The dead root shows shedding of bark and is brittle towards tip. Very minute, dark brown sclerotia of fungus can be seen below the dark. To control follow crop rotation with non-host crops. Seed treatment with biocides like *Tricoderma viride* @ 5 g/kg of seed is recommended.
4. **Collar rot (*Sclerotium rolfsii*) :** Affected young seedlings turn yellow and collapse, but older seedlings may dry slowly. Seedlings, when pulled out, show rottening at collar region nd downwards. White mycelium coatings can be seen on the tap root of dried seedlings. Sometimes rape seedlike sclerotia (1 mm in diameter) can be observed to mycellial strands like around the collar. To minimise the effect drain out excess water from the field. Uproot the affected plants at initial stages. Use wilt resistant varieties. Seed treatment with Carboxin or carbendazim @ 0.2% or seed treatment with biocides *Trichoderma viride* @ 5 g/kg of seed is recommended..
5. **Wilt (*Fusarium oxysporum* f.sp. *ciceri*) :** The disease attacks both seedlings and grown up plants which dry up retaining normal green

color for sometime. Uprooted plants show almost intact root systems, but when split open show inner black discoloration in roots. Control measures are same as for the wilt.

6. **Stem rot (*Sclerotinia sclerotiorum*)** : Usually appears in thick crop canopy. The disease is characterized by appearance of drying of branches or whole plant. Such dried branches show white mycelia growth or formation of black sclerotia of fungus. To minimize the incidence avoid excessive irrigation. Seed treatment as in collar rot. Spray Carbendazim @ 0.1 % and repeat after 10-14 days if required.

7. Integrated disease management of Ascochyta blight, stem rot (*Sclerotinia* spp.), Dry root rot and wilt.

To minimize the incidence, deep ploughing during summer months shall minimize the inoculums. Crop rotation with host crop. Seed treatment with *Trichoderma viride* @ 5 g/kg of seed or seed treatment with Carbendazim + Thiram (1:2) @ 0.25 % followed by two spray of Mancozeb @ 0.1% (90 days and 120 days after sowing). Avoid dense crop canopy and ensure early sowing i.e in October is to be followed.

Harvesting : The crop should be harvested when the pods are mature and plants start drying up. None of legumes should be uprooted. It should be harvested 4-5 cm above ground.

Lentil (*Lens Esculantus*, Masser)

Among *rabi* pulses, lentil is another important drought tolerant crop that can withstand severe winter to a great extent. However, yield level is very low because due attention has not been given on management of this crop. The yield can be improved considerably by following the recommended practices.

Climatic requirement : It requires cold climate. It is very hardy and can tolerate frost and severe winter to a great extent as compared to gram.

Soil : Loam to clay loam is most suitable for lentil. Sandy soils are not good for this crop. Avoid growing of the crop on salt affected soil and/or with impeded drainage.

Varieties : Variety L – 9/12 and PL – 406 are recommended. The plants of these varieties grow erect and branch profusely. The grains possess good cooking quality, are palatable and delicious. After the harvest of *kharif* crops, substantial area remains fallow during *rabi* owing to shortage of water, low soil fertility and other similar reasons. A major portion of this *rabi* fallow area can be conveniently brought under lentil during *rabi* season with a view to meet shortage of pulses, increase cropping intensity and improve soil fertility.

Land preparation : Land should initially be ploughed with soil turning plough followed by 2 to 3 operations with *desi* plough/disc harrow. Planking should follow each operation.

Sowing time : The optimum time of sowing is from last week of October to second week of November but sowing can be extended up to first week of December. Delay in sowing may reduce the yield.

Seed Inoculation : Inoculate the seed with appropriate *Rhizobium* culture before sowing for higher yields.

Fertilizer requirement : For efficient use of fertilizers, get the soil tested from nearest soil testing laboratory and apply recommended dose of fertilizers. 15 kg N and 40 kg P_2O_5 through 88 kg of DAP or urea and SSP can be applied.

Seed rate : Usually 40 kg seed is required for sowing in one hectare area.

Method of sowing : The seed should be sown in lines 20 to 22 cm apart by *"Kera"* method. In paddy fields where excessive wetness does not permit good land preparation, sowing is often done by broadcast method using higher seed rate. However, line sowing is the best method.

Interculture : One or two hoeings and weeding between 20 and 40 days after sowing are beneficial for good yield.

Irrigation : This crop is generally grown under rainfed conditions but if water is available, irrigation at 45 to 50 days after sowing is beneficial as it increases the branching and final crop yield.

Mixed cropping : Lentil can be grown with barley in a seed mixture ratio of 4:1.

Plant Protection Measures

Lentil pod borer : It causes heavy damage to lentil crop. To control this pest spray the crop with 0.1% Carbaryl 50 WP @ 1.5 kg per hectare at flower initiation and repeat the spray after 15 days of first spray if necessary.

Wilt (*Fusarium oxysporum* f. sp. *Lentis*) : The tap roots get infected leading to stunted growth and drying up of leaves. The fungus is seed born and causes more damage towards flowering and pod formation stage. To minimize wilt seed treatment with Captan/Thiram or Carbendazim @ 0.2 % is recommended.

Rust (*Uromyces fabae*) : Under high humid conditions, yellow to brown rust pustules are formed on leaves, pods and stem. Dark brown *"telia"* are formed mostly on stem late in the season. In case of severe infection, plants dry up before seed setting. Pathogens are also perpetuated through seeds. To reduce the disease incidence, spray with Mancozeb @ 0.2 % at 10-12 days interval starting immediately on appearance of the disease is recommended. Use of resistant varieties is also desirable.

Collar rot (*Scleroitum rolfsii*) : The pathogen causes yellowish brown discoloration and rotting of tissue. The young seedlings show damping off symptom. The plants infected at the advanced stage turn pale gradually, dry and drop. White feathery growth of the fungus generally associated with dirty white to brown colored sclerotia can be seen. The seed treatment with Thiram, Carbendazim or Carboxin and Mancozeb @ 0.2% is recommended to control the disease.

Root rot (*Rhizoctonia solani* and *Rhizoctonia bataticola*) : The infected plants show dull green to pale yellow colour starts wilting and dry up. The use of resistant varieties and seed treatment with Thiram or Carbendazim @ 0.2 % is recommended to manage root rot.

Harvesting : The crop should be harvested when pods are mature or plants start drying up otherwise shattering of seeds will take place.

Field Pea

It is an important crop which usually serves both the purposes of vegetables and pulses. For higher production of peas, recommendations should be followed meticulously.

Soil : Field Pea thrive best on well-drained loamy soil.

Varieties :

T-163 : It is a field pea variety (Pulse type). The plant is trailing type and produces white flowers. The pods are generally born as twins, the seed is round, smooth and white with greenish tinge. For seed, the crop matures in 140-150 days. However, if desired the green pods can be collected after 120 days of sowing. This is the highest yielding variety.

PG-2 : Seeds are creamy, round to somewhat wrinkled, average yield 15-20 q/ha. The variety is pulse type with medium to late maturity.

Rachna : Variety is early maturing and resistant to powdery mildew with average yield of 12-15 q/ha.

Sowing time : Optimum time of sowing is from mid of October to mid November. Early sowing encourages wilt attack. Under dry land conditions, 3rd week of October is optimum time, however, sowing can be extended up to 1st week of December provided adequate moisture is available in the soil.

Seed rate : For pulse purpose (T-163 & PG-2), use 75 kg/ha.

Fertilizer : Apply the following fertilizer dose to the crop. 15 kg N and 40 kg P_2O_5 through 88 kg of DAP or urea and SSP can be applied.

Method of sowing : Peas must be sown in lines. For pulse varieties, the lines should be placed 30 cm apart and plant distance in rows should be maintained at 10 cm.

Seed Inoculation : Inoculation of pea seed with *Rhizobium* culture will increase the yield and quality of peas. In the absence of culture, there shall be poor nodule formation. On light sandy soil, even fertilizer N application can be increased to about 30 kg/ha for obtaining better harvest.

Interculture : Field should be kept free of weeds by giving two hand weddings and hoeing after four and eight weeks after sowing.

Irrigation : For field peas, if irrigation water is available, irrigation before flowering and at pod formation is very beneficial.

Mixed cropping : Field peas can be sown mixed with barley or wheat in 1:2 seed mixture ratio. In this way barley/wheat helps the pea in keeping the vines off the ground and peas help by fixing nitrogen which gives higher returns in comparison to pure crop of barley/wheat or peas. The mixed cropping is also a safe guard against total failure in dry farming areas where crop is grown on conserved moisture. Since wheat is often grown heavily irrigated, mixed sowing with wheat may not be adopted but mixed cropping with barley is the best bet.

Plant Protection

Insect Pest Management : To control the important insect-pests and diseases the following schedule in respect of specific pests may be adopted.

Insect-pests

1. **Pea maggot (*Melanagromyza phaseoli*):** It causes heavy damage to early sown pea crop. The maggot bores the under ground portion of stem which causes poor stand of the plants. To control mix Phorate 10 G @ 10 kg or Carbofuran 3 G @ 20 kg/ha after final ploughing before sowing.
2. **Pea leaf minor :** Larvae feed by making tunnel in leaves. Spray the crop with 0.05% Malathion 50 EC @ 1 L of water. This will control thrips and aphids also.
3. **Pea pod borer :** It is serious pest in temperate region. The larvae bore into the pod and eat developing seeds. Spray the crop with 0.1% carbaryl 50 WP @ 1.5 kg/ha in 750 L of water.
4. **Pea thrips :** Causes damage to young crop by sucking cell sap. Control measure same as in pea leaf minor.

5. **Pea aphid :** Sucks cell sap, leaves turn pale and dry. To control measure same as in pea leaf minor.

Disease

1. **Powdery mildew (*Erysiphe pisi*) :** White floury patches appear on both sides of leaves as well as on the tendrils, pods, stem etc. To control the disease, spray the crop with wettable sulpur @ 0.25% Benomyl/Carbendazim @ 0.1% Dinocap @ 0.05% and treat the seed with Carbendazim + Thiram (1:2) @ 0.2% followed by two sprays of Mancozeb @ 0.25% (75 days and 105 days after sowing).
2. **Wilt and Root rot (*Fusarium oxysporum* f.sp. *pisi* and *Rhizoctonia* spp.) :** Roots start rotting and plants show yellowing of lower most leaves followed by wilting. To control, treat the seed with Thiram or Captan @ 2.5 g/kg of seed or with *Trichoderma viride* @ 5 g/kg of seed.
3. ***Blight (Ascochyta* spp./*Alternaria* spp.) :** Ascochyta spp. leroduces lesions on leaves, stem and pods. The lesions are slightly sunken with concentric rings with pycnidia, whereas *Alternaria* spp. produces lesions with distinct concentric rings. Seed treatment with Carbendazim + Thiram (1:1) @ 2 g/kg of seed followed by two sprays of Mancozeb @ 0.25% (75 days and 105 days after sowing) is recommended for its control.
4. **Foot rot (*Phoma medicaginis* var *pinodella*) :** Lesions are produced with wider streaks of blue-black color on leaves also. To minimize the incidence, crop should be planted in well drained soil. Crop rotation with non host crops to be adopted Seed treatment with Captan or Thiram @ 2 g/kg of seed is recommended.
5. **Rust (*Uromyces fabae*) :** Rust colored lesions are produced on lower leaves, stems and tendrils. In advanced infection, diseased leaves fall off prematurely. To minimise incidence, field sanitation to destroy crop debris. Spray the crop with 0.05 Bayleton @ 0.05% Mancozeb @ 0.2% or calixin (0.1%) as the disease appears and repeat the same after 10 days if required.

Harvesting : If peas are grown for disposal as green pods, then vegetable type of peas should be grown. In case of pulse type, the crop should be harvested when the leaves turn yellowish in colour and start shedding. Care should be taken not to over seed or over trample the crop, which may damage the seed coat and reduce quality.

Future Strategies

- Evaluation of suitable cultivars for Jammu region and horizontal expansion through short duration pulses.
- Refining production technologies for new pulse varieties.
- Development of more productive and sustainable cropping systems involving pulses for various agro-ecological zones of Jammu region.
- Creating micro irrigation facility and establishment rainwater harvesting measures for recycling of water for irrigation purposes.
- Emphasis on Intergraded Pest Management (IPM) for efficient resource management.
- Exploitation of useful soil biota like VAM, PSB, *Rhizobium*, etc. and their efficient management.
- Identification of donor sources for development of multiple disease resistant varieties by pyramiding resistant genes for various diseases.
- On farm testing and transfer of technologies.

References

Thakur, N.P, Kachroo Dileep, Jai Kumar, Manpreet Kour and P. Kumar. (2009). Diversification of rice-wheat cropping system for enhancing resource use efficiency and profitability under sub-humid indo-gangetic plains of Jammu region Oryza 46 (2):108-112

AICRP (CS). (2008). Annual Progress Report. Cropping System Research Centre, SKUAST-Jammu, Chatha, edited by Dileep Kachroo: pp 86(03).

AICRP (IFS). (2009). Annual Progress Report. Farming System Research Centre, SKUAST-Jammu, Chatha, edited by Dileep Kachroo: pp-131(04).

Digest of Statistics J&K Government. (2009). Directorate of Economics and Statistics, Jammu

❑❑❑

11
Jharkhand

M.K. Singh, R. Thakur, R.R. Upasani and U.N. Verma
Department of Agronomy, Birsa Agricultural University, Ranchi, Jharkhand

Jharkhand, a newly created (15 November, 2000) state having total geographical area 79.72 lakh hectares, net sown area 18.1 lakh hectares with cropping intensity 114%, is rich in natural resources such as minerals, coal, soil, water, forest cover (29.3%), biodiversity but deficient (50%) in food grain production because of meagre irrigation facilities (10% of cultivated area) in absence of perennial rivers. This state is primarily rainfed and crop production largely depends on monsoon. The south-west monsoon contributes nearly 85% of rain water. Accordingly, the success or failure of crop production is closely linked with the onset, progress and withdrawal of monsoon. Delay in onset, intermittent dry spell (1-3 weeks), early withdrawal of monsoon are also usually observed. Failure of rain has adverse impact on the reproductive phase of rainy season crops and thus on crop yield as well as establishment of winter season crops. Sometimes high rainfall (140 mm/day) coupled with light soil texture and undulating topo sequence provides congenial environment for soil erosion with runoff. This also offers opportunity for water harvesting. Infact, soil and water are the two basic natural resources which must be conserved and utilized efficiently on long term basis.

The population density of state is 338/km^2 with an average land holding 0.67 ha and forest cover 0.09 ha/capita. The majority of farm community belongs to marginal (62.6%) with an average holding of 0.52 ha and small (17.5%) farmers with an average holding (1.52 ha). The status of semi medium and medium group is 14.3 and 4.5% with an average holding of 2.8 and 9.06 ha, respectively. The large farming group is only 1.1% with an average holding of 16.52 ha. The literacy rate of the state is only 54% and about 49% population lives below poverty line. Rural poverty is greater than urban property. The most important rural profession is agriculture and nearly 67% of the total labour force is dependent on it for their income and livehood security.

Jharkhand forms a part of agro-climatic zone VII of the country known as Eastern Plateau and Hill Region. Based on rainfall, temperature, terrain and soil characteristics, the state has been divided into three agro climatic sub zones i.e. IV Central and North Eastern Plateau, V-Western Plateau and VI - South Eastern Plateau.

On an average, the annual rainfall is 1289 mm that varies from 1208 to 1308 mm between different sub zones, out of which 80-85% is received during the four monsoon months (June-September). In the remaining 8 months, evapotranspiration is much higher than precipitation which requires supplemental irrigation for raising crops. Temperature starts to increase from March onwards and reaches its peak (43.6-44.6° C) in May and June. July and August are the active monsoon months during which the, mean temperature remains around 28-29° C. With the start in withdrawal of monsoon from mid September, a second peak of temperature is observed in October, with day temperatures touching up to 35- 40° C, however, nights are cooler. From November onwards, a gradual decrease in both day and night temperatures is observed. In winter season, low temperature occurs in January with mean temperature around 14°C and minimum temperature ranging from 4.8-6.6° C. As monsoon advances, the relative humidity increases which decreases with the withdrawal of monsoon. Relative humidity values are maximum (76.5-90.9%) during August and the lowest (25.5-59.9%) during April. Its values are usually high in morning and low during afternoon. Sunshine hours are greatly influenced by the presence or absence of clouds in the sky. Presence of clouds reduces the sunshine hour where as absence increases it. Sunshine hours vary from 3.5 5.5 hrs/day during *kharif* season, 7.4-8.7 hrs/day during *rabi* and 8.5-8.6 hrs/day during summer season. Highest sunshine hour (8.7 hrs/day) is recorded in February and the lowest (3.5 hrs/day) in July.

The farming situation in this state is based on the type of land. The land located on the upper slope is called *Tanr* and the land following it is called *Don* land. These two classes have been again sub divided into three sub classes *viz.* (a) *Tanr* I, *Tanr* II and *Tanr* III and (b) *Don* I, *Don* II and *Don* III. These classes and subclasses have been regrouped under three categories of land for technology generation. Upland consisting of *Tanr* I and *Tanr* II, medium land *Tanr* III and *Don* III and lowland consisting of *Don* II and *Don* I. This region, having undulating topo sequences and being primarily rainfed, poses several problems e.g. massive soil degradation (about 30 lakh hectares, constituting 40% of total geographical area), diverse agricultural practices with low productivity. The soils of upland and medium land are red and lateritic, highly permeable, course textured, shallow depth with low base saturation, cation exchange and water retention capacity, acidic in reaction (about 16 lakh hectares constituting 19% of total geographical area), low organic matter with low nitrogen status, deficient in phosphorous. Deficiency of potassium is on increase. The low land soil, (5-8% cropped area) have excessive wetness and stickiness after rice harvest *(Don* I) due to heavy texture associated with poor drainage, which restricts tillage operation and thus, the field remains fallow in winter season. In this situation, surface seeding technique provides an elbow space for growing wheat eliminating tillage. Crack formation on surface layers with depletion of moisture is the usual feature of lowland soil.

Present Scenario

The common pulses like pigeonpea, black gram, greengram, horse gram (kulthi) and cowpea in *kharif* season and chickpea, lentil, pea and lathyrus (khesari) in *rabi* season and greengram and cowpea in summer are grown in Jharkhand. About 60% of pulses area is covered during *kharif* season while 40% during *rabi* season. Pulses account for roughly 8.25% area of food grain crops and contribute about 2.53% of the total food grain production in the state. Actually pulse occupies 3.68 lakh hectares, produces 2.94 lakh tonnes with productivity 800 kg/ha during 2008-09. Non availability of suitable high yielding varieties (HYV) is the main constraint for low productivity of pulses in the state. Therefore, to meet the minimum requirement of the state, it has been proposed to increase the area under pulses and to raise productivity by adopting dry land farming technique, mixed and intercropping systems and use of quality seeds, inoculation of seed with *Rhizobium* and Phosphorus solubilising bacteria (PSB) as well as use of fertilizers.

Pulses Production Trends

Area and production of pulses has varied over the years (Table 1). It was 290.2 thousand hectares in 2004-05, and increased to 389.7 thousand hectares in 2007-08. Like-wise, the production also varied from 190.1 thousand tonnes in 2004-05 to maximum 297.8 thousand tonnes in 2007-08. The productivity increased from 655 kg/ha in 2004-05 to 800 kg/ha in 2008-09. In general, pulses area, production and productivity have increased over the years.

Major area is under *kharif* pulses which occupy nearly 59% of acreage and 52% of the total pulse production. The productivity is higher in *rabi* pulses. Amongst the different pulse crops, Pigeonpea has the maximwn area followed by Black gram, Horse gram and Greengram in *kharif* season and Chickpea, Field pea, Lentil and Lathyrus in *rabi* season. Field pea has the maximum average (2004-09) productivity of 1060 kg/ha followed by Lathyrus (994 kg/ha), Chickpea (904 kg/ha) and Lentil (639 kg/ha). Horse gram has the lowest productivity of 440 kg/ha. Pulses are grown in and around 65% of total area in Western plateau i.e. Ranchi, Khunti, Palamu, Gwnla, Lohardaga, Latehar, Simdega and Garhwa; 33% in Central and North eastern plateau i.e. Hazaribagh, Ramgarh, Chatra, Dhanbad, Bokaro, Giridih, Koderma, Jamtara, Deoghar, Godda, Dumka, Pakur and Sahebganj and rest 2% in South eastern

Table 1 : Trends of area, production and productivity of pulses crops in Jharkhand

(Area in hectares, Production in Metric tonnes and Productivity in kg/ha)

Year/ Head	*Kharif* Pulses					Total *Kharif*	*Rabi* Pulses				Total *Rabi*	Total Pulses
	Arhar	Urd	Mung	Kulthi	Others		Gram	Lentil	Pea	Others		
2004-05												
Area	88241	77795	8592	12014	22971	209613	33176	15469	19927	12010	80582	290195
Prod.	62461	41614	4398	5449	11078	125000	29260	9220	19800	6856	65136	190136
Yield	**708**	**535**	**512**	**454**	**482**	**596**	**882**	**596**	**994**	**571**	**808**	**655**
2005-06												
Area	85226	85931	7963	19085	19647	217852	30163	14118	16618	12270	73169	291021
Prod.	53922	45221	3901	7423	8464	118931	24513	7403	14967	6498	53381	172312
Yield	**633**	**526**	**490**	**389**	**431**	**546**	**813**	**524**	**901**	**530**	**730**	**592**
2006-07												
Area	102262	96678	19201	11614	19213	248968	76653	17847	20050	13318	127868	376836
Prod.	60243	54775	7930	5582	9958	138488	70752	10875	19786	8348	109761	248249
Yield	**589**	**567**	**413**	**481**	**518**	**556**	**923**	**609**	**987**	**627**	**858**	**659**
2007-08												
Area	116697	98504	14526	11614	19853	261194	74067	23677	21001	9776	128521	389715
Prod.	107586	52089	7175	5582	6604	179036	74515	15513	22358	6394	118780	297816
Yield	**922**	**529**	494	481	333	685	1006	655	1065	654	924	764
2008-09												
Area	104232	84114	15682	14145	0	218173	89535	19369	25393	15596	149893	368066
Prod.	90273	46013	9355	6155	0	151796	80309	15687	34368	12333	142697	294493
Yield	**866**	**547**	**597**	**435**	**0**	**696**	**897**	**810**	**1353**	**791**	**952**	**800**

Source: Directorate of Agril. Dept., Jharkhand

plateau i.e. East Singhbhwn, West Singhbhwn and Saraikela & Kharsanwa (Mishra, 2004). Potential districts of different pulses have been given as under:

Pigeonpea : Palamu, Garhwa, Latehar and Ranchi.

Black gram : Palamu, Garhwa, Gumla and Ranchi.

Horse gram : Palamu, Garhwa, Gumla, Dumka and Godda.

Chickpea : Palamu, Garhwa, Dumka and Giridih.

Lentil : Godda, Garhwa, Palamu and Pakur.

Field pea : Ranchi, Palamu and Lohardaga.

Greengram : Ranchi, Chatra and Garhwa.

Existing Cropping System

The cropping systems for different land situations are as follows :

Upland rainfed ecosystem

Sequential : Rice/Maize-Fallow; Sweet Potato/Ragi/Gundli-Fallow; Pigeonpea/Black gram/Greengram- Fallow; Little millet/Sesame/Sunflower/ Groundnut-Fallow and Fallow- Niger/Horse gram

Mixed cropping : Pigeonpea + Maize Maize + Black gram; Rice + Pigeonpea Ragi + Niger Maize + Cowpea/Castor

Intercropping : Crops grown as mixed cropping are also grown in intercropping system but limited to progressive farmers.

Medium Land

It is very potential land situation having limited to adequate irrigation facilities for sequential cropping.

Sequential : Rice/Maize-Chickpea/Lentil/Pea; Rice/Maize-Mustard/ Linseed/Barley Rice-Wheat; Rice/Maize-Potato; Maize + Black gram (Mixed)-Wheat; Rice/Maize-Wheat-Greengram (if summer irrigation); Rice-Pea/ Safflower/Linseed; Rice-Wheat + lentil/Chickpea/Mustard and Rice-Vegetables.

Low Land

Rice-Fallow; Rice-Rice (summer); Rice-*utera* Lathyrus/Linseed/Pea Chickpea/Lentil; Rice- Berseem and Rice-Summer vegetables

Yield Gap

There is a gap between potential and achievable yield (Farmers field demonstrations) which requires further modification in technology to bridge the gap (Table 2). Further, there is wide gap between achievable and average yield which needs to be bridged at the earliest by adoption of existing technology, thus increase pulse production by 40% in Jharkhand. Adequate availability of quality seed, inoculation with rhizobiun and PSB culture, use of lime, recommended fertilizer and application of boron in deficient soils can boost up the pulse production.

Table 2 : Yield gap analysis of pulses in Jharkhand

Crop	Yield (kg/ha)				
	Potential	Achievable	Average	Gap	% Gap
Pigeonpea	2200	1500	744	756	50
Black gram	1500	1000	541	459	46
Horse gram	1200	800	448	352	44
Greengram	1200	800	597	203	25
Chickpea	2000	1400	904	496	35
Lentil	1600	1200	639	561	47
Field pea	2200	1500	1060	440	29
Lathyrus	1800	1400	994	406	29

Pulses Based Sequential Cropping Systems

The replacement of a cereal crop through pulses appears difficult as this involves more risks and low productivity. As such, diversification of cereal based systems involving pulses appears to the the only alternative for improving the productivity of the system and soil health. The following systems are being suggested for different agro eco systems of Jharkhand.

Rainfed Upland Ecosystem : Black gram-Safflower/Linseed/Lentil/ Chickpea/Niger/Horse gram; Maize-Safflower/Linseed/Lentil/Chickpea and Soybean-Safflower/Linseed/Lentil/Chickpea.

Medium Land Ecosystem (Irrigated) : Maize-Early potato-Late wheat/ Late pea (green pods); Maize-Wheat-Greengram; Maize-Potato-Greengram; Maize-Early potato/Toria-Late wheat-Greengram Maize-Wheat + Lentil (4:2); Maize-Lentil + Mustard (5:1); Rice-Wheat-Greengram; Rice-Potato-Greengram; Rice-Early potato-Late peas; Rice Early potato/Toria-Wheat-Greengram; Rice-Wheat + Lentil (4:2) and Rice-Lentil + Mustard (5:1)

Low land (Don II) Eco System : Rice-Chickpea/ Lentil

Possibilities of Pigeonpea-Wheat System : Pigeonpea is also grown in sequential cropping because of the availability of short duration genotypes and flexibility in seeding times of wheat. Pigeonpea-wheat sequence is in practice under irrigated condition. Growing of black gram with pigeonpea in Pigeonpea-wheat sequence can further boost up the physical and monetary productivity. However, wheat crop following Pigeonpea (rainy season) is sown late in mid December, which affects productivity of wheat adversely, because the turn around time between harvesting of *kharif* sown Pigeonpea and optimum sowing time of wheat overlaps. This problem does not occur in April planted pigeonpea-wheat system as this crop vacates field well before the normal sowing time of wheat. Inclusion of greengram with pigeonpea further pushes up the physical and monetary productivity of the system (Table 3) in irrigated conditions.

Table 3 : Production potential of pigeonpea-wheat system

Location/cropping sequence	Production (Q/ha)		
	Pigeonpea	Wheat	WEY
Dholi (Bihar)			
April planted Pigeonpea-Wheat	17.5	28.7	81.2
June planted Pigeonpea-Wheat	13.5	20.0	60.5
Kanpur (U.P)			
April planted Pigeonpea-Wheat	26.0	67.2	145.2
June planted Pigeonpea-Wheat	17.4	53.4	105.6
Delhi			
April planted Pigeonpea-Wheat			
June planted Pigeonpea-Wheat	23.8	40.6	112.0

WEY= Wheat equivalent yield

Intercropping Systems : The potential of pigeonpea/maize based intercropping systems followed in Jharkhand has been given in Table 4.

Table 4 : Productivity of pigeonpea/maize based intercropping

Cropping System	Grain yield		Maize equivalent yield (q/ha)
	Main crop	Inter crop	
Maize + black gram (1:2)	21.21	4.00	31.16
Maize + soyabean (1:2)	14.50	8.00	32.37
Pigeonpea + black gram (1:2)	9.07	5.18	47.73
Pigeonpea + soyabean (1:2)	6.41	9.13	43.08
Pigeonpea + groundnut (1:2)	10.61	13.94	84.16
Pigeonpea + rice (1:3)	11.32	7.96	46.52
Pigeonpea + maize (1:1)	10.61	15.02	52.76

Source : Pande (2004), M.Sc. Thesis, Agronomy, BAU, Ranchi.

Cultivation of Pulses in Non-Conventional Seasons/Areas

Capitalizing on rice fallow: Jharkhand has approx 6 lakh ha of area under rice-fallow. The non availability of irrigation water, low water retention of soil as well as uncertainties of winter rains does not pennit double cropping. Rainfed *rabi* pulses in rice-fallow have great potential for improving food security and human nutrition. However, it is possible only with selection of appropriate varieties of rice and suitable succeeding pulse crops. Recently some short duration varieties of rice have been developed like Birsa Dhan 108 (70- 75 days), Birsa Gora 102 (90-95 days), Vandana (95-100 days), Anjali (95 -100 days), if these varieties are sown by mid June (pre monsoon! onset of monsoon) then field will be vacated by the end of August or 1st week of September and a crop of Horse gram can be grown without supplemental irrigation.

In Jharkhand, *Don* II land situation, *para* cropping of pulses followed in a very limited area otherwise it remains fallow due to unavailability of irrigation. In this situation, the top soil layer generally dries up at the time of harvest of rice and, thus, planting of post rainy season crop is not feasible. Under such conditions, relay cropping *(para)* of lentil or low toxin lathyrus genotype (Bio L 212) could convert these mono cropped areas in to double cropping and thus, increase pulse production as well as sustain productivity of the rice based system.

Crop Substitution in Uplands : In Jharkhand, direct seeded rice is the main crop being grown in uplands. Productivity of upland rice is low (10-12 q/ha) and there is need for growing other crops like Soybean, Groundnut, Maize, Pigeonpea, Black gram and Sweet potato. Reduction in acreage consequently production can be compensated by intensification of cultivation for which the following strategy may be followed:

- Intercropping of Pigeonpea + rice to increase land productivity and soil fertility.
- Need to promote cultivation of hybrid rice in medium and low land.
- Seed multiplication and distribution of upland rice variety Birsa Dhan 108, Birsa Gora 102, Vandana and Anjali.

Area Expansion : In irrigated situation, the most dominating cropping system is Cereal-Cereal, i.e. Rice-Wheat, Maize-Wheat. Continuous cultivation of cereals results in deteriorating soil health. In such a situation, if irrigation availability is assured after the harvest of wheat and before planting of rice, a catch crop of short duration pulse like greengram can be taken.

Strategies for Increasing Pulses Production

- Non availability of quality seeds in adequate quantity is one of the major constraints in pulse production. Hence, production of quality seed coupled with increasing seed replacement rate must be addressed properly.
- Transfer of improved pulse production technologies remains the most neglected component in the past. Consequently, the benefit of improved varieties and production technology has not been harnessed. This must be addressed properly for increasing productivity.
- Expansion of area under pulses is a major opportunity for increasing production of pulses in Jharkhand.
- Introduction and popularization of short duration varieties of chickpea and lentil in rice fallow.
- Popularization of black gram and greengram during Spring/ Summer as catch crop in intensive cropping systems.
- Popularization of pulses in intercropping systems in rainfed ecosystems.
- Refinement of integrated nutrient management including the components like soil amendment, rhizobium culture, PSB, organic manure, inorganic fertilizer (NPKS) and micronutrients (Boron) is the need of the day for higher productivity.
- Refinement of integrated pest management including synthetic pesticides, bio-pesticides, bio-agents, biological control, mechanical control and use of pheromone traps is also a need of the day.
- Use of farm machinery is an option for minimizing the cost of production in the light of increased wages of labour. Conservation tillage machinery and rotovator may be popularised for timely crop establishment to increase productivity of pulses.
- Pulses are grown in rainfed eco-systems in Jharkhand and thus, face severe moisture stress leading to low productivity. Quantum jump in productivity can be achieved by micro- irrigation systems especially in *rabi* pulses grown on residual moisture. Therefore, sprinkler irrigation can be popularized and used in the most efficient way with limited irrigation water.

Production Technologies

Pigeonpea

Land preparation : One deep summer ploughing of well drained upland soil reduces the incidence of weeds, diseases and pests. Two ploughings before planting produce a satisfactory seedbed. Application of Malathion 5% dust @ 25 kg/ha at the time of final ploughing can be effective for controlling termites. Apply lime @ 3q/ha in furrow at seeding for higher productivity. The rate should be doubled in case of Dolomite.

Sowing time : Early Mid June- mid July Variety:

Medium : ICPH2671, BirsaArhar-1. (190-200 days).

Late : Bahar, Basant, Laxmi , Narendra Arhar (240-250 days)

In undulating topo sequences of Jharkhand, farmers prefer short- duration vaieties for upland *(Tanr* II) where moisture recedes just after cessation of monsoon while they prefer long duration varieties for midland *(Tanr* III) which retains moisture even after cessation of monsoon.

Seed rate: HYV 20 kg/ha, Hybrid 10 kg/ha

Seed Treatment :

- Treat the seeds with 1 g Carbendazim or 2 g Bavistin 50 WP/kg seed against fungal diseases as well as with 3 ml Imidacloprid/kg seed against termites. This will control termite and foliage feeder aphids and thrips for 30 days.
- Inoculate seeds with suitable *Rhizobium* strain @ 20g/kg seed as well as with PSB (Phosphate solublizing bacteria) @ 25 g/kg seed. The sequence of treatment should be in order of Fungicide, Insecticide, Rhizobium and PSB. For inoculation, 100 g *'gur'* should be dissolved in 1 litre of water 'and boiled for about 20 minutes and cooled thereafter. Required culture should be mixed up with gur solution and finally seed should be mixed in, for coating on seed. The treated seed should be dried in shade and sown immediately.

Spacing : Row to row 75 cm for long and 60 cm for early duration crop. Similarly, plant to plant 25 cm for long and 20 cm for early duration crop. For Hybrid 90 cm row to row and 30 cm plant to plant.

Seeding depth : 4-5 cm below the surface soil

Fertilizer management : FYM or compost @ 5 tonnes /ha should be applied a fortnight before sowing. Inorganic fertilizer @ 20:40:20:20 kg NPKS/ ha (Nitrogen, Phosphorus, Potash and Sulphur) should be applied as basal in furrow.

Weed management : Apply fluchloraline 45 EC @ 0.75- 1 kg a.i. /ha in 600 litres of water just before sowing or alachlor 50 EC @ 1.5- 2 kg a.i./ha as pre-emergence (within 3 days of sowing) or hoeing by wheel hoe at 20-25 days after sowing or two hand weeding 15 and 30 days after germination.

Water management : Keep drainage system perfect to avoid water logging and irrigate the crop at pod formation, if needed.

Plant protection : Leaf spot can be controlled by spraying with dithane M-45 @ 2 g/litre water. Spray malathion @ 2 ml litre water at the time of 50% flowering followed by monocrotrophos @ 1 ml litre water 15 days after first spraying to protect the crop from pod borer. For wilt control, 1 g bavistin/ litre water should be sprayed at base of plants near root.

Harvesting : Harvest the crop when more than 50% pods mature.

Yield : Early duration crop : 8-10, Medium :12-15, late duration crop : 18-22 and Hybrid 30 q/ha.

In changing climatic scenario farmers are planting pigeonpea even up to end of July. Its cultivation can also be done successfully as post rainy season crop (September planting) as well as summer planted crop (April planting) in irrigated conditions. These crops give more yield then rainy season crop due to better plant stand, low incidence of diseases- pest and thus high input use efficiency.

Post Rainy Season Pigeonpea

Land preparation : Select well drained mid land having light to heavy textured soil. It is ploughed twice to obtain a satisfactory seed bed. Apply lime @ 3 quintal/ha in furrow at seeding for higher productivity.

Seeding time : First fortnight of September Variety : Basant, Bahar and Sharad.

Seed rate : 50 kg seed/ha

Seed treatment : Treat the seeds with Bavistin @ 2g/kg seed before sowing. Inoculate seeds with *rhizobium* and PSB.

Fertilizer management : 20 kg N, 50 Kg P_2O_5, 20 kg K_2O and 20 kg S/ha at the time of seeding.

Weed management : Apply 1 kg Pendimethalin as pre emergence (0-3 days of sowing) or handhoeing at 25 and 50 days after sowing.

Water management : Irrigate the crop at flowering and pod formation if required.

Plant protection : Leaf spot can be controlled by spraying with dithane M-45 @ 2 g/litre water. Spray malathion @ 2 ml/litre water at the time of

50% flowering followed by monocrotrophos @ 1 ml/ litre water 15 days after first spraying to protect the crop from pod borer. Add Tipol @ 2 ml//5 litre in spray solution. For wilt control 1 g Bavistin l/litre water should be sprayed at base of plants near roots.

Harvesting : Harvest the crop when more than 85% pods matured.

Yield : 20-22 q/ha

Black gram

Land preparation : Well drained light textured soils are suitable for the crop. One deep summer ploughing reduces the incidence of weeds, diseases and pests. Two ploughing before sowing produces satisfactory seed bed. Apply lime @ 3 q/ha in furrow at the time of sowing for higher productivity. The rate should be doubled in case of Dolomite.

Seeding time : Mid June to Mid July

Varieties : Pant U 19, Birsa Urd 1 Pant U-31, Shekhar-2, Azad Urd-l(KU92-1)

Seed rate : 30 kg/ha.

Seed Treatment :

- Treat the seeds with 1 g Carbendazim or 2 g Bavistin 50 WP/kg seed against fungal diseases as well as with 3 ml Imidacloprid/kg seed against termites. This will control termite and foliage feeder aphids and thrips for 30 days.
- Inoculate seeds with suitable *Rhizobium* strain @ 20 g/kg seed, as well as with PSB (Phosphate solublizing bacteria) @ 25 g/kg seed. The sequence of treatment should be in order of Fungicide, Insecticide, *Rhizobium* and PSB. For inoculation, 100 g 'gur' should be dissolved in 1 litre of water and boiled for about 20 minutes and cooled thereafter. Required culture should be mixed up with jaggery solution and finally seed should be mixed in, for coating on seed. The treated seed should be dried in shade and sown immediately.

Spacing : Row to row 30 cm and plant to plant 8-10 cm.

Depth of sowing : 4-5 cm below surface soil.

Fertilizer management : FYM or compost @ 5 tonnes /ha should be applied a fortnight before, sowing. Inorganic fertilizer @ 20:40:20:20 kg NPKS/ ha should be applied as basal in furrow.

Weed management : Spray Pendimethalin @ 1-1.25 kg in 600 litter of water after sowing (within 3 days of sowing) to control weeds. If herbicide is not used, two hand weeding is necessary 15 -30 days after germination.

Plant protection : Cercospora disease may be controlled by spraying dithane M-45 @ 2 g/litre of water. Bihar hairy caterpillar may be controlled by spraying Monocrotophos @ ml/ litre water, Tipol @ 2 ml/5 litre which acts as adhesive.

Harvesting : Harvest the crop after 80-85 days when pods have turned dark brown in colour.

Yield : 8- 10 q/ha.

Soybean

Land preparation : Well drained fertile sandy-loam soil is preferred for its cultivation. One deep summer ploughing is done to minimize the infestation of weeds, soil borne diseases and pests. After pre monsoon shower, 2 light ploughings with harrow should be done to pulverise the soil. Apply lime @ 3 q/ha in furrow at seeding for higher productivity. The rate should be doubled in case of dolomite.

Sowing time : Mid June- Mid July

Variety : JS 335, Birsa Soybean 1 and Birsa Safed Soybean 2

Seed rate : 75- 80 kg/ha

Seed treatment : Treat the seeds with 1 g Carbendazim or 2 g Bavistin 50 WP/kg seed against fungal diseases as well as with 3 ml Imidacloprid/kg of seed against termites. This will control termite and foliage feeder aphids and thrips for 30 days.

Inoculate seeds with suitable *Rhizobium* strain @ 20g/kg seed as well (as with PSB (Phosphate solublizing bacteria) @ 25 g/kg seed. The sequence of treatment should be in order of Fungicide, Insecticide, *Rhizobium* and PSB. For inoculation, 100 g *gur* should be dissolved in 1 litre of water and boiled for about 20 minutes and cooled thereafter. Required culture should be mixed up with *gur* solution and finally seed should be mixed in, for coating on seed. The treated seed should be dried in shade and sown immediately.

Spacing : Row to row- 30 cm; Plant to plant- 10 cm

Seeding depth : 3-4 cm below surface soil.

Fertilizer management : 40: 60: 40: 20 kg/ha N, P_2O_5, K_2O, S. Half N and full phosphorus, potash and sulphur should be applied as top dressing at 15 days after germination.

Weed management : Weeding should be done twice at 15 and 30 days after germination manually or apply F1uchloralin 1.25 Kg a.i /ha in 600 litres of water plus Tipol (adhesive) as pre emergence (0-3 DAS)

Plant protection : Spray malathion 35 EC @ 2 ml litre water to control Bihar hairy caterpillar. Monocrotophos @ 1 ml/litre water will be effective for controlling shoot fly and girdle beetle. Add Tipol @ 2 ml/5 litre spray solution for adhesiveness.

Harvesting : Crop should be harvested when leaves turn yellow and begin to fall.

Yield : 22-25 q/ha

Rice bean

Land preparation : Well drained fertile sandy loam soil is preferred for its cultivation. One deep summer ploughting is done to minimize the infestation of weeds, soil borne diseases and pests. After pre monsoon shower, 2 light ploughing with harrow should be done to pulverise the soil. Apply lime @ 3 q/ha in furrow at seeding for higher productivity. The rate should be doubled in case of dolomite.

Sowing time : Mid June - Mid July

Variety : RBL- 1, RBL-6

Seed rate : 20 kg/ha

Seed treatment : Treat the seeds with 1 g Carbendazim or 2 g Bavistin 50 WP/kg seed against fungal diseases as well as with 3 ml Imidacloprid/kg seed against termites. This will control termite and foliage feeder aphids and thrips for 30 days. Inoculate seeds with PSB (Phosphate solublizing bacteria) @ 25 gikg seed. The sequence of treatment should be in order of Fungicide, Insecticide and PSB. For inoculation, 100 g '*gur*' should be dissolved in I litre of water and boiled for about 20 minutes and cooled thereafter. Required culture should be mixed up with *gur* solution and finally seed should be mixed in, for coating on seed. The treated seed should be dried in shade and sown immediately.

Spacing : Row to row- 30; plant to plant 8-10 cm.

Seeding depth : 4-5 cm below surface soil.

Fertilizer management : 20: 40: 20: 20 kg N PO_5 K_2O/ha. All fertilizers should be applies as basal in furrow.

Weed management: weeding and hoeing twice at 15 and 40 days after germination are sufficient to suppress weed hazards.

Plant protection : Spray dithane M 45 @ 2 g/litre water to control powdery mildew. Monocrotophos @ 1 ml/litre water will be effective for controlling insects. Add Tipol @ 2 ml/5 litre spray solution for adhesiveness.

Harvesting : Crop should be harvested when pods turn brown and leaves start drying-falling

Yield: 15- 20 q/ha.

Lentil

Land preparation : Well drained, loamy soil is preferred for lentil cultivation. Plough the land 2-3 times to obtain good seed bed. Application of malathion 5% dust @ 25 kg/ha at the time of final ploughing can be effective to control termites. Apply lime 3 q/ha in furrow at sowing to harvest good crop. The rate should be doubled in case of Dolomite.

Seeding time : Timely: Mid October to mid November

Late: November end.

Varieties : Timely- PL 406, PL639, Late- PL 639 L 4076, KLS 218, K75.

Seed rate : Small seeded- 30 kg/ha, Large seeded (K 75)- 50 kg/ha, increase seed rate by 20% for late seeding.

Seed Treatment :

- Treat the seeds with 1 g Carbendazim or 2 g Bavistin 50 WP/kg seed against fungal diseases as well as with 3 ml Imidacloprid/kg seed against termites. This will control termite and foliage feeder aphids and thrips for 30 days.
- Inoculate seeds with suitable *Rhizobium* strain @ 20 g/kg seed, as well as with PSB (Phosphate solublizing bacteria) @ 25 g/kg seed. The sequence of treatment should be in order of Fungicide, Insecticide, Rhizobium and PSB. For inoculation, 100 g 'gur' should be dissolved in 1 litre of water and boiled for about 20 minutes and cooled thereafter. Required culture should be mixed up with gur solution and finally seed should be mixed in, for coating on seed. The treated seed should be dried in shade and sown immediately.

Spacing : Row spacing for timely sown crop 30 cm and late sown crop 25 cm.

Depth of seeding: 3-4 cm deep below the soil surface.

Fertilizer Management : Recommended fertilizer @ 20:40:20:20 kg N, P_2O_5, K_2O, S/ha should be applied as basal in furrow + Borax @ 10 kg/ha should be applied before sowing in soils with boron deficiency. Spray twice, 2% urea at pod formation stage and 10 days after pod formation for higher yield.

Water management : If required, irrigate at flower bud initiation and pod filling stage.

Weed management : One manual weeding at 25-30 days after seeding or spray Pendimethalin @ 1 kg a.i./ha in 600 litre water after sowing as pre emergence. (within 3 days).

Plant protection : Wilt-Use wilt resistant varieties i.e. PL 406 and PL 639. In standing crop spray Bavistin @ 1 g/ litre water at base of plants near roots.

Yield : Timely 12-15q/ha; Late 10-12q/ha.

Chickpea

Land preparation : Well drained heavy textured soil is preferred. Plough the land 2-3 times to obtain good seed bed free from clods, stubble of previous crop and weeds. Application of Malathion 5% dust @ 25 kg/ha at the time of final ploughing can be effective to control termites. Apply lime 3 q/ha in furrow at sowing for higher productivity. The rate should be doubled in case of Dolomite.

Seeding time (Timely sown) : Rainfed- 2nd - 3rd week of October.

Irrigated- by mid November

Late sown : End November to mid December

Varieties : Timely sown- BG256, Pant G 114, Radhey, KWR 108, KPG 59, BG 1003, KAK-2 (Kabuli), IIK 94-134 (Kabuli).

Late sown : Pant G 114, KPG 59, BG 372, BG 1053 (Kabuli).

Seed rate : Timely sown 70 kg/ha (Desi), 100 kg/ha (Kabuli). For late sown use 20% extra seed.

Seed Treatment :

- Treat the seeds with 1 g Carbendazim or 2 g Bavistin 50 WP/kg seed against fungal diseases as well as with 3 ml Imidacloprid/kg seed against termites. This will control termite and foliage feeder aphids and thrips for 30 days.
- Inoculate seeds with suitable *Rhizobium* strain @ 20 g/kg seed, as well as with PSB (Phosphate solublizing bacteria) @ 25 g/kg seed. The sequence of treatment should be in order of Fungicide, Insecticide, *Rhizobium* and PSB. For inoculation, 100 g *'gur'* should be dissolved in I litre of water and boiled for about 20 minutes and cooled thereafter. Required culture should be mixed up with gur solution and finally seed should be mixed in, for coating on seed. The treated seed should be dried in shade and sown immediately.

Spacing : Row to row: 30 cm, Plant to plant: 8-10 cm.

Depth of seeding: 5- 8 cm deep below the soil surface.

Fertilizer Management : Recommended fertilizer @ 20:40:20:20 kg N, P_2O_5, K_2O, S/ha should be applied as basal in furrow + borax @ 10 kg/ha should be applied before sowing in soils with boron deficiency. Urea should be sprayed twice @ 2% at pod formation stage and 10 days after.

Water Management : If required, irrigate at flower bud initiation and pod development stages.

Weed Management : Two weedings by hand hoe 20-25 and 45-50 days after seeding or pre emergence spray of Metolaclor @ 1.0 kg a.i/ha in 600 litre water after seeding. (within 3 days of sowing).

Plant protection : Control wilt by treating seed in standing crop, spray 1 g/ litre water Bavistin at base of the plants near roots, grow resistant varieties i.e. PG 114, avoid early or deep sowing (8-10 cm soil depth). For pod borer control spray Monocrotophos @ 1ml/litre of water at pod formation stage.

Yield : Timely- 12-15 q/ha; Late 10-12 q/ha.

Pea

Land preparation : Plough well drained sandy-loam soil 2-3 times after rainy season crop. Application of Malathion 5% dust @ 25% /ha at the time of final ploughing can be effective in controling termites. Apply lime 3 q/ha in furrow at sowing to harvest good crop. The rate should be doubled in case of Dolomite.

Sowing time : Vegetable Pea: 15 October to October end; Field Pea: October end to 15 November.

Varieties : Early maturing: Arkel, P-1542 (Pusa Pargati).

Medium maturing : Jawahar Matar 4, DDR-23(Pusa Prabhat), DDR-27, HUDP-15(Malviya).

Late maturing : Swarnrekha, Bounvilla, Rachna.

Seed rate : Vegetable Pea (60-65 days for pods): 80 to 100 kg/ha Field Pea 100 kg/ha.

Seed Treatment :

- Treat the seeds with 1 g Carbendazim or 2 g Bavistin 50 WP/kg seed against fungal diseases as well as with 3 ml Imidacloprid/kg seed against termites. This will control termite and foliage feeder aphids and thrips for 30 days.
- Inoculate seeds with suitable *Rhizobium* strain @ 20 g/kg seed, as well as with PSB (Phosphate solublizing bacteria) @ 25 g/kg seed.

The sequence of treatment should be in order of Fungicide, Insecticide, Rhizobium and PSB. For inoculation, 100 g '*gur*' should be dissolved in I litre of water and boiled for about 20 minutes and cooled thereafter. Required culture should be mixed up with '*gur*' solution and finally seed should be mixed in, for coating on seed. The treated seed should be dried in shade and sown immediately.

Spacing :

Early maturing: Row to row 20cm, Plant to plant 10 cm

Late maturing: Row to row 30cm, Plant to plant 10 cm

Fertilizer Management : Recommended fertilizer @ 20:60:20:20 kg/ha N, P_2O_2, K_2O, S/ha should be applied as basal in furrow + borax @ 10 kg/ ha should be applied in furrow as basal before sowing in soils with boron deficiency.

Water management : Irrigate crop at 45 days after sowing and at pod filling stage.

Weed management : Two hoeings and weeding after 3rd and 6th weeks of seeding or spray Metolaclor @ 1 kg a.i./ha in 600 litres of water as pre-emergence. (within 3 days).

Plant protection : Control powdery mildew by spraying Sulfex @ 4g / litre of water. Monocrotophos @ 1 mll litre is effective against sucking pests and borers.

Yield : Vegetable pea: 100-125 q green pods/ha, Field pea: 20-25 q grain/ha.

Horse gram

Land preparation : Upland having well drained soils are good for *kulthi* (Horse gram) cultivation. Two or three ploughings with country plough, each followed by planking are sufficient for land preparation. Application of Malathion 5% dust @ 25 kg/ha at the time of final ploughing can be effective in controling termites. Apply lime @ 3 q/ha in furrow at sowing to harvest good crop. The rate should be doubled in case Dolomite.

Sowing time : Mid August- mid September. Variety : Birsa kulthi- 1, Madhu.

Seed rate : 20 kg/ha.

Seed Treatment :

- Treat the seeds with 1 g Carbendazim or 2 g Bavistin 50 WP/kg seed against fungal diseases as well as with 3 ml Iimidacloprid/kg seed against termites. This will control termite and foliage feeder aphids and thrips for 30 days.

- Inoculate seeds with PSB (Phosphate solublizing bacteria) @ 25 g/kg seed. The sequence of treatment should be in order of Fungicide, Insecticide and PSB. For inoculation, 100 g '*gur*' should be dissolved in I litre of water and boiled for about 20 minutes and cooled thereafter. Required culture should be mixed up with *gur* solution and finally seed should be mixed in, for coating on seed. The treated seed should be dried in shade and sown immediately.

Spacing : Row to row- 30 cm; Plant to plant- 8-10 cm.

Depth of sowing : 4-5 cm below soil surface.

Fertilizer Management : 20:40:20:20 kg/ha N, P_2O_5, K_2O, S/ha at the time of seeding as basal application.

Weed management : Hoeing with dutch hoe at 20-25 days after seeding can take care of weeds.

Water management : Keep drainage system perfect to avoid water logging.

Plant protection : Hairy caterpillars feed on the leaf tissue and cause defoliation at a very faster rate. Monocrotophos @ 1ml/ litre water may be effective against this pest.

Harvesting : Harvest when 80-90% pods are ripe. Dry in sun for 2-3 days before threshing.

Yield: 10-12 q/ha.

Summer Moong

Land preparation : Plough the land having high water retention capacity 2-3 times after harvest of winter season crops. Application of Malathion 5% dust @ 25 kg /ha at the time of last ploughing can be effective to control termites. Apply lime 3 q/ha in furrow at sowing to harvest good crop.

Seeding time : Last week of February to mid April.

Varieties : Sunaina, K851, Pant M 2, PS 16, Pusa Vishal, Samrat (PDM 139), Meha-125 (IPM 99-125).

Seed rate: 30 kg seed/ha.

Seed Treatment :

- Treat the seeds with 1 g carbendazim or 2 g bavistin 50 WP/ kg seed against fungal diseases as well as with 3 ml imidacloprid / kg seed against termites. This will control termite and foliage feeder aphids and thrips for 30 days.
- Inoculate seeds with suitable *Rhizobium* strain @ 20 g/kg seed, as well as with PSB (Phosphate solublizing bacteria) @ 25 g/kg seed.

The sequence of treatment should be in order of Fungicide, Insecticide, *Rhizobium* and PSB. For inoculation, 100 g gur should be dissolved in I litre of water and boiled for about 20 minutes and cooled thereafter. Required culture should be mixed up with gur solution and finally seed should be mixed in, for coating on seed. The treated seed should be dried in shade and sown immediately.

Spacing : Row to row: 30 cm, **Plant to plant :** 8-10 cm.

Depth of sowing : 4-5 cm below surface soil.

Fertilizer Management : 20:60:20:20 kg NPKS/ha at the time of sowing.

Weed management : One weeding 20 days after sowing is sufficient.

Water management : Irrigate the crop at fortnightly interval starting from 15 days after sowing.

Plant protection : Monocrotophos @ 1ml/litre water is effective in controlling insect-pests.

Harvesting : Two pickings should be done after 55 and 65 days of sowing when pods turn dark brown.

Yield : 10-12 q/ha.

References

Mishra R K. (2004). Agricultural Statistics of Jharkhand at a Glance, 2004. Birsa Agricultural University, Ranchi, Jharkhand.

Mishra R K and Jayram V. (2006). Agricultural development of Jharkhand at a glance through figures.

Padmaja Pande. (2004). Production potential of pigeonpea! maize based intercropping system of Jharkhand. M.Sc. Thesis, Department of Agronomy, BAU, Ranchi

Pal S K., and Singh N N. (2010). Use and management of natural resources for sustainable agriculture in Development Concerns in the 21 st Century. Ed.

Sudesh Nangia, M M Jha, S Misra, H. Ramchandran and M Velayutham. Pub. Concept Pub. Compo Pvt. Ltd., New Delhi p 327 - 62

Raman K V. (2008). Report of the Commission on Agriculture Reforms, Research and Development, Jharkhand

Singh M K, Thakur R, Verma U N and Pal S K. (1995). Production in potential of crop sequences in plateau region of Bihar. Indian Journal of Agricultural Sciences 65 (4) : 6-9

Singh M K, Thakur R, Pal S K, Verma U N and Upasani R R (2000). Plant density and row arrangement of lentil + mustard intercropping for higher productivity under Bihar plateau. Indian Journal of Agronomy 45 (2) : 284 -87

Verma U N, Pal S K, Thakur R, Upasani R R and Singh M K. (2007). Cropping Strategies for Rainfed Agro- ecosystem of Jharkhand. BAU - Bulletin (Agron/ 2007).

❑❑❑

12

Karnataka

A.P. Vishwanath[1], M. Byre Gowda[2] and T.K. Prabhakara Setty[3]
[1]Professor of Agronomy AICRP-IFS; UAS, Banglore.
[2]Professor & PI, AICRP – Pigeonpea,
[3]Former Director of Research, UAS (B) & State Consultant NFSM Project, Department of Agriculture, Bangalore, Government of Karnataka.

In Karnataka, pulses are being cultivated in an area of about 20 lakh ha (about 17% of cultivated area) and the important pulse crops of the state are pigeonpea, chickpea, greengram and blackgram (Table 1). In addition, horsegram and field bean (Dalichos lab) need a special mention, as they occupy a unique position among pulse crops especially in Southern Karnataka. However, major pulses like pigeonpea chickpea, greengram and blackgram are confined to few districts in Northern Karnataka, although some area is distributed throughout the state. Compared to the average productivity of nation (560 kg/ ha), the Karnataka state productivity is low (505 kg/ha). In spite of the best efforts, the progress made in improving the overall production as well as productivity of these pulse crops is not significant as compared to the cereals.

One of the main reasons for the present low production levels is that these pulse crops are cultivated in the state under dry land situations characterized by low fertility level, non availability of sufficient soil moisture, deficiency of major and minor nutrients and inhospitable conditions for growth of beneficial soil micro organisms. In addition to the abiotic stresses, biotic stresses like diseases, insect pests and weeds are also other causes for low productivity.

Table 1 : Area and production of pulse crops in Karnataka (2000-01 to 2009-10)

S. No.	Crops	2000-01	2001-02	2002-03	2003-04	2004-05	2005-06	2006-07	2007-08	2008-09	2009-10
	Area (lakh hectares)										
1	Pigeonpea	5.83	4.82	5.14	5.32	5.62	6.00	5.96	6.81	5.97	6.04
2	Bengal gram	3.69	4.80	4.80	5.10	4.18	4.18	6.51	6.05	7.26	9.34
3	Horse gram	2.95	3.43	3.35	2.74	2.83	2.69	3.24	2.27	2.17	2.28
4	Black gram	1.46	1.63	1.65	1.41	1.33	1.11	1.45	1.63	1.12	1.18
5	Greengram	4.51	2.58	4.13	2.73	5.23	4.01	4.50	5.30	2.74	3.79
6	Cowpea	1.15	0.86	0.89	0.77	1.09	0.94	0.86	1.00	0.82	0.97
7	Fieldbean	0.88	0.81	0.65	0.69	0.80	0.87	0.64	0.79	0.79	0.75
	Total Pulses	**20.47**	**18.93**	**20.61**	**18.74**	**21.08**	**19.81**	**23.16**	**23.86**	**20.87**	**24.34**
	Production (lakh tonnes)										
1	Pigeonpea	2.64	1.47	2.41	2.00	2.91	4.08	2.63	4.51	3.15	2.82
2	Bengal gram	2.39	2.82	2.52	1.71	2.31	2.38	2.98	3.64	4.01	5.66
3	Horse gram	1.57	1.49	1.09	0.77	1.22	1.29	1.03	1.11	0.88	1.04
4	Black gram	0.56	0.60	0.23	0.39	0.16	0.25	0.24	0.70	0.29	0.14
5	Greengram	1.85	0.67	0.28	0.42	0.83	0.99	0.54	1.20	0.37	0.47
6	Cowpea	0.35	0.27	0.26	0.24	0.38	0.32	0.42	0.49	0.34	0.36
7	Fieldbean	0.20	0.19	0.15	0.16	0.19	0.21	0.55	0.68	0.68	0.51
	Total Pulses	**9.56**	**7.52**	**6.94**	**5.69**	**8.00**	**9.51**	**8.39**	**12.33**	**9.72**	**11.00**
	Productivity (t/ha)										
1	Pigeonpea	0.45	0.30	0.47	0.38	0.52	0.68	0.44	0.66	0.53	0.47
2	Bengal gram	0.65	0.59	0.53	0.34	0.55	0.57	0.46	0.60	0.55	0.61
3	Horse gram	0.53	0.43	0.33	0.28	0.43	0.48	0.32	0.49	0.41	0.46
4	Black gram	0.38	0.37	0.14	0.28	0.12	0.23	0.17	0.43	0.26	0.12
5	Greengram	0.41	0.26	0.07	0.15	0.16	0.25	0.12	0.23	0.14	0.12
6	Cowpea	0.30	0.31	0.29	0.31	0.35	0.34	0.49	0.49	0.41	0.37
7	Fieldbean	0.23	0.23	0.23	0.23	0.24	0.24	0.86	0.86	0.86	0.68
	Total Pulses	**0.47**	**0.40**	**0.34**	**0.30**	**0.38**	**0.48**	**0.36**	**0.52**	**0.47**	**0.45**

Major Constraints in Pulse Production

A variety of factors has hampered the expansion of area and productivity of pulses. The Agro-climatic conditions and vulnerability to moisture stress have been the key problem areas. Unlike rice and wheat, low productivity has led to low production and returns, retarding expansion of area. Support prices are often below open market prices and have not encouraged farmers. Low varietal turnover with lower yield gains combined with low resilience to moisture stress and pest infestation have contributed to variable production. At present more than 92 per cent of the area under pulses is confined to rainfed areas and in future too, the bulk of pulse production will continue to come from rainfed areas. Once irrigation is made available, the relatively low-productivity pulses will have to compete with other crops, particularly wheat and rice.

Agro-Ecological Constraints

Cultivation on marginal and sub-marginal lands : Pulses are mostly grown on marginal or sub marginal lands without much of inputs. As a consequence the yield potential is never achieved.

Mostly rainfed (uncertain and insufficient rainfall) : In addition to their cultivation on marginal or sub-marginal lands, pulses are generally grown under rainfed condition or as catch crops or relay crops. This greatly increases the exposure of the crops to various stresses particularly, terminal stress resulting in huge losses.

Biological Constraints

Crop domestication (survival under adverse situation) : The crop has been domesticated from marginal environment and therefore, possesses surviving ability through a mechanism of indeterminancy, perennial nature, greater production of polyphenols and lignin, all of which divert the primary products required for pod filling.

Inherent low productivity (genetic yield potential) : The genetic yield potential of the pulses is low due to higher concentration of protein which takes up greater energy equivalents compared to cereals that store energy in the form of carbohydrates.

Susceptibility to diseases and insect-pests (Field / storage) : Pulses are highly susceptible to pests and diseases and not much research has gone into these aspects of pulse production. Further, the marginal status of the farmer growing these crops makes him not to become a bottleneck in adoption of recommendations that could enhance production or the farmer just takes these crops as a bonus to the main crop. The crop, therefore, grows as an orphan crop susceptible to pests and diseases giving low yield returns. Pulses suffer heavy pest losses during storage also. The quality of seeds stored in the traditional storage structures also deteriorates very rapidly.

In legume crops, the major biotic and abiotic stresses (Table 2) which affect the crops at various growth stages are :

- Biotic stress likes Fusarium wilt, *Ascochyta blight*, root rots, YMV, rust, powdery mildew, stunt virus etc.

- Insect and pests like *Helicoverpa armigera*, bruchids, leaf minor etc.
- Abiotic stresses like drought, temperatures (low and high), water logging, salinity and toxicity etc.

Table 2 : The losses caused by biotic and abiotic stresses

Crop	Caused by	Losses (%)
Biotic stresses (Diseases and pests)		
Chickpea	Fusarium wilt, As. blight, BGM, Root rots, Stunt virus, Helicoverpa, Bruchid	20-35
Pigeonpea	Fusarium wilt, Phy. Blight, St. Mosaic,Helicoverpa, Bruchid	20-3025-40
Green Bean	MYMV, Circospora leaf blight,	15-20
Black gram	Powdery mildew, MYMV	20-35
Abiotic stress (Climatic extremes)		
Chickpea	Drought, Cold, Heat, etc	20-30
Pigeonpea	Drought, Cold, Heat, etc	20-30
Green Bean	Drought, Heat	15-20
Black gram	Drought, Heat	15-20

Among the pulses, pigeonpea is very sensitive to abrupt fluctuations of temperature either lower or higher extremes leading to massive flower drop. For cool season pulse crops, a daily maximum temperature above 25^0 C is considered as threshold level for heat stress, as it affects seed yield by reducing flowering, fertilization and seed formation.

Management Constraints

Lack of timely availability of quality seeds of improved varieties : Seed is the basic and most important input of any crop husbandry and good quality seeds varieties are lacking in pulses. With Seed Replacement Ratio (SRR) as low as 2-5%, there is severe non-availability of quality seed. Since hybrids are not available except in pigeonpea, private companies also do not take up much interest in seed production. The non-availability of high yielding improved variety seed in required quantities forces the farmer to rely on poor quality seed of old varieties, thus decreasing the yield potential of most of the pulse crops. The only way out seems through increasing seed replacement rate from 4% to 10% and production of quality seed from 40,000 mt to 100,000 mt.

Inadequate supply of quality insecticides and fungicides : On an average, 20-40% crop is lost annually due to damage caused by pod borers in pigeonpea and chickpea (Table 3). Pod fly also causes 10-15% loss especially in North India. Wilt and root rots cause heavy loss to pigeonpea and chickpea crops. Effective IPM modules are available for management of targeted pest and diseases.

Table 3 : The extent of losses/damages caused by important insects and diseases

Sl.No	Pulse crops	Name of the Insect/pest/diseases	Extent of damage (%)
	Insects pests		
1.	Pigeonpea	*Helicoverpa*	45-50
		Podfly	5-15
		Field infestation of bruchids	10-25
2.	Chickpea	Pod borer	35-40
3.	Greengram	*Maruca*	5-15
		Stem fly	5-40
		Helicoverpa	10-20
		Psydia, Apion	5-15
4.	Blackgram	*Helicoverpa*	10-20
		Stem fly	30-40
		Psydia, Apion	5-10
		Sucking pests	5-10
5.	Horsegram	Jassids & White fly (transmission)	10-15
6.	Fieldbean (Avare)	*Helicoverpa*	20-40
		Adisura	5-20
		Field infestation of bruchids	10-25
7	Cowpea	*Aphids,*	5-10
		Podborers (Maruca & Helicoverpa)	12-15
		Stemfly	2-5
		Bruchids	10-25
	Important diseases		
1.	Pigeonpea	Fusarium wilt	23
		Sterility mosaic	10
		Phythophthora blight	
2.	Chickpea	Fusarium wilt	50
		Root rot	25-30
3.	Greengram	MYMV	4-5
		Cercospora leaf spot	4-15
		Powdery mildew	20-40
4.	Blackgram	Cercospora leaf spot	5-10
		Powdery mildew	20-30
5.	Horsegram	HYMV	5-10
		Powdery mildew	30-40
6.	Fieldbean (Avare)	Yellow mosaic	5-10
		Anthracnose	5-10
7	Cowpea	Rust *(kharif)*	30-40
		Bacterial blight	5-10
		Yellow mosaic	5-10
		Powdery mildew	5-10

Limited post harvest processing units : Further, there are no small processing units to convert pulse grains into *Dal* and other by-products. This compels the growers to dispose of their produce immediately after harvest at low price. Low cost *dal* mill and metal storage bins are available but they need to be popularized.

Poor transfer of technology system : Transfer of improved pulse production technologies has generally remained the most neglected component in the past, and consequently the benefit of improved varieties and production technology could not be harnessed. This calls for aggressive transfer of production technologies through frontline demonstrations and Block demonstrations involving SDA, KVKs, NGOs, SAUs and private sectors for dissemination of knowledge on improved technologies.

Non-availability of proper farm implements : Poor drainage/water stagnation during rainy season causes heavy losses to pigeonpea on account of low plant stand and increased incidence of *Phytophthora* particularly in the states of U.P., Bihar, W.B., Chhattisgarh, M.P. and Jharkhand. Ridge planting has been found very effective in ensuring optimal plant stand and consequently higher yield. A simple ridger already available can effectively be used for this purpose.

Technical Constraints

Cultivars : Poor genetic potential, susceptibility and pests.

Technology : Poor adoption of integrated crop production and management technology. In recent years, wide spread deficiency of sulphur and zinc has been noticed in pulse growing regions, which constrains productivity of pulses. In the major pulse growing areas, 44 districts have shown 0-60% sulphur deficiency and 82 districts with 50-60% zinc deficiency. Very encouraging response to application of S and Zn has been recorded with cost benefit of 10-21%. As mentioned earlier, the lack of quality seed of high yielding cultivars and seed replacement ratio below 5% are other factors associated with low productivity.

Socio-economic : Poor linkage between breeders-extension workers-workers-farmers, poor market access and information system, lack of quality storage facility and an absence of reliable weather forecasts and advisories.

Production Technologies

The season wise pulse crops and cropping systems recommended for different zones of Karnataka are detailed in table 4 while the crop wise production technologies are discussed below.

Table 4 : Season-wise pulse crops and cropping systems recommended for different agro climatic zones of Karnataka

Zone no. name and districts (Taluks)	Soil Type	Rainfall range (mm)	*Kharif* (May-Sept)	*Rabi* (Sept-Dec)	Summer (Jan-April)	Total geographical area (ha)
1. North Eastern Transition Zone						
Bidar (Aland, Bhalki, Basavakalyan, Bidar, Chincholi, Humnabad, Aurad)	Shallow to medium black clay soils in major areas. Red lateritic soils in remaining areas.	830-919	Black gram, Green gram, Tur	Bengal gram	--	871036
2. North Eastern Dry Zone						
Gulbarga, Raichur (Afzalpur, Chitapur, Gulbarga, Jewargi, Sedam, Shahapur, Shorapur, Yadgir, Deodurg, Manvi, Raichur)	Deep to very deep black clay soils in major areas. Shallow to medium black soils in minor pockets.	633-807	Tur, Green gram, Black gram,	Bengal gram	--	1762604
3. Northern Dry Zone						
Koppal, Bagalkote, Bijapur, Bellary, Bellary, Gokak (Gangavathi, Koppal, Kushtagi, Lingasugur, Sindhanur, Yelburga, Badami, Bagalkote, Bagewadi, Bilgi, Bijapur, Hungund, Indi, Jamkhandi, Mudhol, Muddebihal, Sindhagi, Bellary, Hagaribommanahalli, Harapanahalli, Hadagali, Hospet, Kudligi, Sandur, Siruguppa, Ron, Navalgund, Naragund, Gadag, Mundargi, Ramdurga, Gokak, Raibag, Soundatti, Athani)	Black clay medium and deep in major areas, sand loams in remaining areas.	465-786	Green gram, Horse gram, Cowpea, Tur	Bengal gram	--	4783642
4. Central Dry Zone						
Chitradurga, Davanagere, Tumakur, Hassan, Chikkamagalore (Challakere, Chitradurga, Davanagere, Harihara, Hiriyur, Hosadurga, Holalkere, Jagalur, Molkalmuru, Arasikere, Kadur, Madhugiri, Pavagada, Koratagere, C.N.Halli, Sira, Tiptur)	Red Sandy loams in major areas, shallow to deep black soil in remaining areas.	456-717	Tur, Field bean, Horsegram,	Horse gram	--	1943830
5. Eastern Dry Zone						
Tumakur,Bangalore Urban, Bangalore rural, Ramanagar, Kolar, Chikkabalapur (Gubbi, Tumkur, Anekal, Bangalore South, Bangalore North, Channapatna, Devanahalli, Doddabalapur, Hosakote, Kankapura, Magadi, Nelmangala, Ramanagar, Bagepalli, Bangarpet, Chikkabalapur, Chintamani, Gudibanda, Gowribidanur, Kolar, Malur, Mulbagal, Sidalaghatta, Srinivasapura)	Red loamy soils in major areas, clay lateritic soils in remaining areas.	679-889	Avare, Horse gram	Horse gram	--	1808217

Contd.

6. Southern Dry Zone						
Mysore, Mandya, Chamarajanagar, Tumakur, Hassan (K.R.Nagar, T.Narasipur, Mysore, Kollegal, Nanjangud, Turuvekere, Kunigal, Nagamangala, Srirangapatna, Malavalli, Maddur, Mandya, Pandavapura, K.R.Pet, Channarayapatna, Hassan, Chamarajanagar, Yelandur, Gundlupet)	Red sandy loams in major areas and in remaining areas, pockets of black soils.	671-889	Tur, Horse gram, Cow pea, Avare	Horse gram, Cow pea.	--	1739430
7. Southern Transition Zone						
Mysore, Hassan , Chikkamagalur, Davanagere, Shimoga (H.D.Kote, Hunsur, Periyapatna, H.N.Pura, Alur, Arkalgud, Belur, Tarikere, Bhadravathi, Shimoga, Honnali, Shikaripura, Channagiri)	Red sandy loams in major areas and in remaining areas, red loamy soils.	612-1054	Tur, Horse gram, Cowpea	Horse gram, Bengal gram,	--	1218029
8. Northern Transition Zone						
Belgaum, Haveri, Dharwad (Hukkeri, Chikodi, Bailhongal, Belgaum, Haveri, Shiggaon, Shirahatti, Kundagol, Savanur, Hubli, Dharwad, Byadgi, Hirekerur, Raneebennu)	Shallow to medium black clay soils and red sandy loamy soils in equal proportion.	618-1303	Tur, Green gram, Horse gram, Avare, Cowpea	Bengal gram, Horse gram	--	1194941
9. Hilly Zone						
Chikkamagalur, Shimoga, Kodagu, Hassan, Uttara Kannada, Dharawad, Haveri, Belgaum (Sirsi, Siddapura, Yellapura, Supa, Haliyal, Mundgod, Khanapur, Soraba, Hosanagar, Sagar, Thirthahalli, Koppa, Sringeri, Mudigere, Narasimharajapur, Chickmagalur, Kalaghatagi, Hangal, Sakleshpur, Virajpet, Somwarpet, Madikere)	Red clay loamy soils in major areas.	904-3695	--	Paddy, Black gram, Green gram,	Cow pea	2560727
10. Coastal Zone						
Dakshina kannada, Udupi, Uttara kannada (Karwar, Kumta, Honnavar, Bhatkal, Ankola, Bantwal, Udupi, Belthangadi, Karkala, Kundapura, Mangalore, Puttur, Sulya)	Red lateritic and coastal alluvial.	3011-694	--	Black gram, Horse gram	--	1167380
TOTAL						**19049836**

Pigeonpea (*Tur*)

Pigeonpea is an important pulse crop of Karnataka. It occupies an area of 6.80 lakh ha with production of 4.79 lakh tonnes and an average productivity of 704 kg/ha which is less than national average (824 kg/ha). In Karnataka, Gulbarga district alone contributes to more than 60 per cent of the state's production and hence, is called as the pulse bowl of Karnataka. Majority of the area is rainfed.

Crops and cropping systems for pigeonpea : Pigeonpea is grown as sole crop or intercropping with *ragi*, ground nut and jowar. Replacement of uneconomic crops like cotton, millets with pigeonpea and also increasing area through intercropping of pigeonpea with soybean, cotton, sorghum millet and groundnut is one of the key strategic research and extension agenda. Under late sown situations and in small areas, transplanting of 25-30 days old seedlings of pigeonpea will give higher yields and seed treatment of Neem powder at 50 g/kg of seed will control *kosha jantu roga* (nematode) problem (Technologies released from UAS Dharwad).

Zones : Southern dry zone, Hilly zone, Central dry zone, Southern transition zone, Eastern dry zone

Suitable varieties : The most suitable varieties for different NARP zones of Karnataka are : TTB-7 (150-200 days), HYD-3C (160-210 days), BRG-1 (170-190 days), BRG-2 (150-170 days) and ICP-7035 (160-170 days).

Sowing time : Mainly sown in May-July and a harvest of 12.5 to 15 q/ha as sole crop and 2.5 to 5 q/ha as inter crop is generally obtained.

Seed treatment and fertilizer application : Treat the seed with *Rhizobium* 500g/ha. Apply 7.5 t/ha FYM for obtaining good yields.

Weed management and inter culture *:* One hand weeding, 2-3 inter cultivation and pre emergence application of alachlor/metalachlor/ pendimethallin @1kg/ha.

Water management : Ensure 2 irrigations at flowering and pod filling stage.

Plant protection measures : To control pod borer, apply- chlorpyriphos @ 2 ml/ltr or indoxicarb 14.5 sc 1 ml/ltr. To control diseases like sterility mosaic, apply dicofal 20 Fc 2.5 ltr/ha and for Fusarium wilt, apply carbendaizin *kosha jantu roga*-50 gm neem powder/kg seed.

Strategies for Improvement

- Crop is highly sensitive to water logging. Therefore, broad beds and furrows or ridge and furrow system are useful to prevent water logging. Crusting should be avoided as it prevents seedling emergence.

- Good quality seed from certified sources or that of previous season should be used.
- Bold seeds should always be used for sowing as their germination and vigour is better.
- A uniform seeding depth should be maintained using bullock-drawn or mechanical seed drill.
- Shallow sown seeds are exposed to pest damage and water deficits.
- Always choose varieties suitable to the location and having resistance to wilt and mosaic.
- Seed treatment with thiram (3 g/ kg seed); inoculation with *rhizobuim* @ 2g/kg seed is essential.
- Rhizobium treatment should be at least half an hour after thiram treatment and just before sowing.
- Pigeonpea is nodulated by rhizobia belonging to the "cowpea-miscellany"
- Estimates of fixed N in pigeonpea genotypes of different maturity ranged from 6 to 69 kg N ha^{-1}
- A long-duration pigeonpea grown in northern India could fix up to 200 kg N ha^{-1} over a period of 40 weeks.

Transplanted Pigeonpea

This practice involves rising the seeding in polythene bags and transplanting. This is an innovative technique practiced to mitigate late onset of monsoon. The most optimum date of sowing for pigeonpea is in the month of June. On many occasions, the optimum date of sowing will be delayed as rain will be delayed beyond second fortnight of July.

This technique involves raising of pigeonpea seedlings in a polythene bag. The seeds are sown in the polythene bag during first fortnight of June and watered for their better establishment. The seedlings will be ready for transplanting after 30 to 40 days in the main field which usually coincides with the onset of rainfall in the month of July. Wider spacing of 5′x 3′ is found to be optimum. This technique is economically feasible and large number of farmers adopting this technique in Gulbarga and Bidar districts and the expected yield is about 20-25 quintals per hectare. The technique is siuted for those farmers who are having protective irrigation facilities.

Integrated Nutrient Management

- Basal application of micro-nutrient of 5 kg borax (0.5 kg B ha^{-1}) and 50 kg zinc sulphate (10 kg Zn ha^{-1}) depending on soil test values helps to augment yields.
- Treat the seeds with rhizobium @ 500 grams per hectare.
- Treat the seeds with Phosphate Solubalizing Bacteria (PSB) @ 500grams per hectare.
- Apply 500 kg of well decomposed powdered FYM along with fertilizer while sowing.
- Apply dimmmonium phosphate (DAP) @ 100 kg per hectare.
- Spray 2% DAP and growth regulator to avoid flower dropping (can be mixed with plant protection sprays)

Weed Management

- Low initial crop growth makes weed control essential. In the absence of weed control, there can be upto 90% crop loss.
- Fluchloralin (1 Kg ai/ha) or pendimethalin (0.7 kg ai/ha) are efficient per-emergence weedicides.
- Field should be weed free for the first 30 days of crop growth. Two hands weeding at 25 and 45 days after sowing will take care of the weeds.
- Follow frequent inter- cultivation for proper control of weeds, proper aeration and better infiltration of rainwater.

Management of Fusarium wilt

- Seed and soil borne fungus which affects the medium and late types more.
- The symptoms include patches of dead plants at flowering or podding in the field. When the stem is cut open there is a purple band extending upwards from the base of the main stem and blackening of the xylem when the main stem or primary branches are split open.
- The control measures include
    - Selecting disease free fields, avoid ratooning, crop rotation every 3 years.
    - Solarize the field in summer to reduce inoculums.
    - Grow varieties resistant to Fusarium wilt.

- Seed treatment with 4 g Trichoderma viride formulation + 3 g thiram per kg seed and application of 2 kg T. viride formulation mixed with 125 kg FYM per ha.

Integrated Pest Management practices

- Solarize the field in summer to help reduce inoculums.
- Prior to sowing, soil application of trichoderma harzianum @ 10 g 1 kg of FYM for controlling the pigeonpea wilt. .
- Sowing of pigeonpea crop in ridges for suppression of *Phytophthora* disease.
- Installation of pheromone traps @ 10/ha in the month of September.
- Erection of bird perches @ 25/ha for facilitating predation of Helicoverpa larvae.
- Two applications of neem seed kernel extract @ 5% in September and October.
- Spray of NPV @ 250ml (2 * 109 POB/m)/ha in September and October.

Value Addition

- Green pods can be harvested, shelled, canned and frozen and sold during off-season.
- Increasing super market presence helps in direct disposal
- Farmer co-operatives in this regard can be formed.
- **Vegetable:** Fresh green pods – Shelling – Cleaning.
- **Green pods for canning:** Shelling - Cleaning – Blanching – Cooling – Filling cans.
- **Freezing :** Shelling – Cleaning – Washing – Cooling – Packing in polythene bags – Storage leaning – Washing – Cooling – Packing in polythene bags – Storage.

The Strategies

- Identification of short duration cultivars that are able to escape drought due to early maturity.
- The latest technology available should be transferred to Farmer's field through front line demostration and on farm trials along with training.
- UAS, Raichur, KSDA in collaboration with ISRO is establishing a forecasting centre to announce the outbreak in advance to take the

necessary precautionary measures. Plant health clinic can also be set up in collaboration.

- KSSC, NSC and Seed Unit of UAS Raichur to take up seed production to provide good quality seeds of high yielding varieties.
- Yield maximization in pigeonpea through ICM in farmers' fields.
- Conduct large scale demonstration using innovative equipments for seeding, spraying and harvesting operations.
- Identification of suitable methods for control of storage pests and value addition through various innovative approaches.

Chickpea

Chickpea is also another important *rabi* pulse crop of Karnataka. It occupies an area of 6.50 lakh ha and production of 3.10 lakh tonnes with an average productivity of 473 kg/ha which is less than national average (759 kg/ha). Major chickpea growing districts are Dharwad, Bijapur, Bagalkot, Belgaum, Gadag and Haveri. It is mainly grown on medium to deep black soils under rain fed conditions. The crop matures in 90-110 days and records high yields, when one or two supplemental irrigations are provided. It has a yield level of 8 – 10 q/ha under rain fed conditions, and recording around 20-25 q/ha under irrigated conditions.

Zones : Southern dry zone, Coastal dry zone, North eastern dry zone, Northern dry zone, and Northern transition zone

Suitable varieties : The most suitable varieties for different NARP zones of Karnataka are: Annegere-1 (95-100 days), JG-11(95-100 days), KAK-2 (80-85 days) and VISHAL (100-110 days).

Sowing time : Mainly sown in post monsoon and a harvest of 8-10 q/ha as rainfed and 15-20 q/ha as irrigated.

Seed treatment and fertilizer application : Seed treatment with *rhizobium* is essential when the crop is taken for the first time. However, in all legume crops, *rhizobium* application as seed treatment is advantageous for good harvest. Treat the seeds with PSB 1250 g/ha. Apply 7.5 t/ha FYM. Also apply 25-50-50 kg NPK/ha under irrigated while 13-25-25 kg NPK/ha in rainfed conditions as basal.

At the time of last ploughing 100 kg DAP as a basal dose should be applied. Application of sulphur @ 30 kg/ha or Gypsum is also recommended. Zinc sulphate @ 20-25 kg/ha should be applied in zinc deficient soils.

Weed management : For effective control of weeds a suitable pre-emergence weedicide like Stomp should be applied after one or two days of planting at the rate of 1.5 litres/ha in 600 litres of water. First manual weeding at 20-45 days after planting is also recommended to ensure a healthy crop.

Diseases and pest management : For control of seed and soil borne diseases, seed should be treated before planting by Thirm or Bavistin or Captan at the rate of 2 g per kg of seed. Only the resistant variety should be planted where soil borne diseases are regularly occurring.

The pod borer is a major problem in chickpea therefore, increased monitoring of pod borer at vegetative and flowering stages is essential. The installation of *Helicoverpa armigera* pheromone traps at the time of vegetative stage for intensive monitoring is important. The application of insecticide like Thiodan at the rate of 2 liter/ha in 800 liter of water is recommended whenever damage is visible. The second spray of insecticide should be carried out 20 days after the first spray. The insecticide spray should be carried out in the morning when sky is clear. Application of appropriate botanical pesticides such as plant derived products, hormones, microbial agents and pathogenic fungi can also be applied whenever possible.

Harvesting : Harvesting should be carried out after physiological maturity when crop turns completely yellow, to avoid shattering.

Threshing : Multi-crop thresher at low speed is suitable for threshing.

Storage : Proper storage conditions are essential to protect seeds from bruchid damages. For this purpose a concrete made separate room should be used. Proper fumigation by fumigants is recommended at different intervals.

Production of quality seeds by farmers : Generally, farmers depend for quality seeds on seed agencies and the timely availability is not ensured. Therefore, farmers should try to produce their own quality seeds. The quality seed production can be taken up by farmers if the following steps are followed :

- Each variety should be planted separately with the required isolation.
- Harvesting should be carried out carefully at proper stage of maturity.
- Threshing should be done carefully and the produce should be kept separately.
- After threshing, the seed should be dried properly and moisture level should not be more than 8-10%
- The grading should be carried out manually and small seeds may be removed during grading operation.
- Seed should be bagged properly in the water proof or plastic bags and each bag should be properly labelled.
- The storage of seed bags should be in a separate room where fumigation can be done and humidity maintained.

Greengram (Moong)

The importance of greengram as the source of vegetable protein and its role in sustainable agriculture under Indian situation is well known. In Karnataka, greengram is cultivated in an area of about 5.30 lakh hectares. The area is mostly confined to northern part of Karnataka although they are distributed throughout the state. The productivity of the state is low (206 Kg/ha) compared to the average productivity of the nation (413 Kg/ha). One of the main reasons for low yield of greengram is that the crop is cultivated in dry land situations characterized by low fertility. Also vagaries of monsoon, deficiency of major and micro nutrients and inhospitable conditions for growth of beneficial microorganisms leads to low productivity.

Land preparation : Land should be leveled and should be brought to fine tilth before the sowing operation is carried out.

Suitable varieties : Improved varieties play a vital role in improvement of productivity and production. The farmers should use good quality seeds of new improved varieties. The most suitable varieties for different NARP zones of Karnataka are: PS-16 (65-70 days), PUSA BAISAKI (65-70 days) and PDO-84-178 (65-70 days).

Sowing time : The variety PS-16, is mainly sown in April-May and Jan-Feb; Pusa Baisaki sown in Jan-Feb; and PDO-84-178 sown in April-May and Jan-March. Harvest of 5-7.5 q/ha is obtained in rainfed and 10-12.5 q/ha in irrigated conditions.

Seed treatment and fertilizer application : The seed is treated with Rhizobium 500 g/ha and PSB 1250 g/ha. Apply 7.5 t/ha FYM. Also apply 25-50-50 kg NPK/ha under irrigated while 13-25-25 kg NPK/ha in rainfed conditions as basal. At the time of last ploughing or at the time of sowing, 100 kg DAP has to be given as a basal dose. As a source of sulphur, 100 kg gypsum has to be applied to provide 20 kg sulphur per hectare.

Weed management and inter culture : Weeds are now becoming a major problem because of scarcity of labour. For better crop production, the crop should be free from weeds throughout the growing period. For control of weeds, a pre-emergence weedicide Pendimethalin @ 1.5 lit/ha dissolved in 700 litres of water may be sprayed on the day of sowing or the after day. One hand weeding, 2-3 inter cultivations and application of pendimethaline @ 1kg/ha. are generally required for a good crop stard.

Water management : Two irrigations one at flowering and the other at pod filling are recommended.

Plant protection measures : Before sowing the seeds should be treated with 2 g of Bavistin per kg of seeds. If a disease like powdery mildew is noticed at crop growth stage, spraying of Bavistin or Hexaconazole @ of 1 g/liter of water is essential. For control of pod borer in greengram, chemicals like Chlorpyripnos may be used @ 2 ml/liter of water. For control of caterpillars, apply Chlorpyripnos @2ml/l of water and for control of Aphids/whiteflies, apply dimethoate-1.7 ml/l. Similarly, for control of disease like leaf spot and rust, apply Bavistin @500g/ha.

Harvesting : In crop like greengram, care should be taken to harvest at proper time of maturity otherwise shattering of pods is a major problem. This may be done using labourers or multi crop thresher.

- Harvesting should be done after crop has matured and most of the pods have turned black/brown.
- Harvesting should be done in the early hours to minimize shattering losses.
- Over maturity may result in shattering losses.

Storage of produce : The seeds or the commercial produce should be stored in dry place with minimum moisture content in the seeds. Otherwise, stored grain pests attack within a short period. In case of seeds, they may be treated with castor oil to protect from the bruchids.

Importance of quality seeds : Good, bold and pure seeds play important role in productivity. Farmers can purchase seeds of greengram varieties and take up sowing. Further they can retain the seeds of that variety for next season. Therefore, they can also attempt to take up seed production of the chosen variety. For effective seed production, the following important steps are necessary :

- Once the variety is planted, proper crop production measures are necessary.
- At harvesting, the plants are selected which are uniform in maturity and the seed stored for next season. This will help save lot of money and time.
- The seeds should be hand cleaned and stored in a proper place to avoid attack of stored grain pests.
- Seeds should be bagged properly in water proof or plastic bags.

IPM Module

- Visual observation at weekly intervals at vegetative, flowering and poding stages for Heliothis.
- Use pheromone traps (5 traps/ha) for Heliothis monitoring.

- Deep ploughing of the field, early sowing, growing resistant / tolerant varieties, timely irrigations and soil application of Phorate or Carbofuran granules @ 1.0 kg a.i. per ha.
- Timely sowing, proper seed rate, irrigation and fertilizer management to have desirable crop stand.
- Intercultural and hand weeding to keep to crop weed free initially for 4-6 weeks.
- Look for the plants infested with $1^{st}/2^{nd}$ in star larvae of hairy caterpillar, collect and destroy them.
- Light traps are very effective against hairy caterpillar.

Black Gram (Urad)

The importance of blackgram as the source of vegetable protein and its role in sustainable agriculture under Indian situation is well known. Though tremendous progress has been made in achieving quantum jumps in cereal crops, which ushered in an era of green revolution, the grains made in improvement of productivity of pulses in general and black gram in particular are very meagre.

In Karnataka, black gram is cultivated in an area of about 1.60 lakh hectares. The area is mostly confined to northern part of Karnataka although they are distributed throughout the state. The productivity of the state is low (395 Kg/ha) compared to the national average (869 kg/ha). One of the main reasons for low yield of black gram is that the crop is cultivated in dry land situations characterized by low fertility. Vagaries of monsoon, deficiency of nutrients and inhospitable conditions for growth of beneficial microorganisms also affects the crop yield.

Land preparation : Land should be brought to fine tilth before sowing operation is carried out.

Suitable varieties : Improved varieties play a vital role in improvement of productivity and production. The farmers should use the seeds of new improved varieties with good quality. The most suitable varieties for different NARP zones of Karnataka are : Karagav-3 (85-90), T-9 (75-80 days) and Rashmi (lbg-625) (70-75 days).

Sowing time : Karagav-3 variety is mainly sown in April-May and Jan-March; T-9 sown in April-May and Jan-March; and Rashmi (lbg-625) sown in May-Aug and Jan-March. Harvest of 5-7.5 q/ha as rainfed and 10-12.5 q/ha as irrigated.

Seed treatment and fertilizer application : *Rhizobium,* is known to fix atmospheric nitrogen in pulses, and PSB play an important role in making phosphorus available to plants. Therefore, treat the seed with rhizobium 500 g/ha and PSB 1250 g/ha. Apply 7.5 t/ha FYM and 25-50-50 NPK kg/ha for obtaining good yield. Recommended doses of nutrients have to be given before sowing is taken up. At the time of last ploughing or at the time of sowing 100 kg DAP has to be given as basal dose. As a source of sulphur 100 kg gypsum has to be applied to supply 20 kg sulphur per hectare.

Weed management and inter culture : Controlling weeds has now become a major problem because of the scarcity of labour. For better crop production, the crop should be kept weed free throughout the growing period. For effective control of weeds, a pre-emergence weedicide, pendimethalin @ 1.5 lit/ha dissolved in 700 liters of water, may be sprayed on the day of sowing or the next day. One hand weeding 2-3 inter cultivations, are pendimethalin @ 1kg/ ha are needed for optimum production.

Water management : One irrigation at pod formation.

Plant protection measures : For agromyzid fly-chlorpyriphos @ 2ml/ltr; thrips-methyl parathion @1ml/ltr. For control of disease like leafspot-Mancozeb @ 2g/ltr. Before sowing the seeds should be treated with 2g of Bavistin per kg of seeds. If diseases like powdery mildew is noticed at crop growth stage, spraying of Bavistin or Hexaconazole @ of 1g/litre of water is essential. For control of pod borer in blackgram, chemicals like Chlorpyripnos may be used @ 2ml/litre of water.

Harvesting : In a crop like blackgram, care should be taken to harvest at the proper time of maturity. This may be done using labourers or multi crop thresher. Harvesting should be done after crop is matured and most of the pods turn black/brown. Harvesting should be done in the early hours to minimize shattering losses. Over maturity may result in shattering losses.

Storage of produce : The seeds or the commercial produce should be stored in dry place with minimum moisture content in the seeds. Otherwise stored grain pests attack within a short period. In case of seeds, they may be treated with castor oil to protect from the bruchids.

Importance of quality seeds : Good, bold and pure seeds play an important role in productivity. Farmers can purchase seeds of black gram varieties and take up sowing. Further they can retain the seeds of that variety for next season.

Therefore, they can also attempt to take up seed production of the chosen variety. For effective seed production, The following important steps are necessary, i.e.,

- Once the variety is planted, proper crop production measures are necessary.

- At harvesting, they can select the plants which are uniform in maturity and bulk as seed for next season. This will help to save lot of money and time.
- The seeds should be cleaned and stored in a proper place to avoid attack of stored grain pests.
- Seeds should be bagged properly in water proof or plastic bags.

Horsegram

In Karnataka, horse gram occupies an area of 3.24 lakh ha with a production of 1.03 lakh tones and productivity of 335 kg/ha. it is grown as a main crop after *ragi*.

Major Production Constraints

- Powdery mildew & YMV diseases.
- Photosensitivity.

Zones : Southern dry zone, eastern dry zone and hilly zone.

Suitable varieties : The most suitable varieties for different NARP zones of Karnataka are: KBH-1(BGM-10) (90-100 days) and PHG-9 (90-100 days).

Fertilizer : Apply 7.5 t/ha FYM and 25-38-25 NPK kg/ha for obtaining good yield.

Weed management and inter culture : 1 hand weeding of 2 inter cultivation.

Water management : Rainfed.

Plant protection measures : Caterpillar and pod borer: Chlorpyripnos 2ml/litre; Powdery mildew diseases: Calixin @ 0.5 ml/litre.

Field bean (Avare)

Is cultivated in an area of 0.64 lakh hectares with a production of 0.55 lakh tonnes and productivity of 197 kg/ha. The field bean is grown as sole crop or it can be grown as a intercrop with crops like *Ragi* and Groundnut.

Zones : Southern dry zone, eastern dry zone and hilly zone.

Suitable varieties : The most suitable varieties for different NARP zones of Karnataka are: HA-3 (90-100 days) and HA-4 (95-105 days).

Seed treatment fertilizer and application : Treat the seed with *Rhizobium* 500 g/ha and PSB 1250g/ha. Apply 7.5 t/ha FYM also apply 25-50-50 NPK kg/ha under irrigated while 13-25-25 kg NPK/ha in rainfed conditions.

Sowing time : Both the varieties are more suitable for summer and farmers can harvest 10-12.5 q/ha.

Weed management and inter culture : 1 hand weeding, and 2-3 inter cultivation pendimethalin @ 1kg/ha for optimium production.

Water management : Two irrigations, one at flowering and the other at pod filling.

Plant protection measures : For Catterpillars- Chlorpyripnos 2ml/l; Aphids, whitefly- dimethoate-1.7 ml/l; Diseases: for leafspot, rust, YMV- Bavistin 500g/ha.

Strategies for Pulse Production in Karnataka

- Farmers' participatory varietal selection of improved varieties with all the recommended packages (Seed treatment, integrated nutrient management, integrated pest and disease management).
- Formal and informal seed distribution through large scale seed production and supply of quality seeds of all the new and high yielding varieties of the important pulses.
- Yield maximization through integrated nutrient management, integrated pest management and integrated disease management.
- Large scale demonstrations in the farmers field with integrated pest and disease management.
    - Using suitable chemicals.
    - Use of botanical pesticides like neem and other plant extracts.
    - Use of micro organisms in combating diseases like *Trichoderma viridae, Pseudomonas* etc.,
- Large scale production of bio-agents like *Rhizobium*, Trichoderma, PSB, NPV etc.
- Horizontal expansion of area under different cropping systems by developing suitable genotypes, intercropping, relay cropping, catch crops paddy follows, Agri-Horti system, expanding area under protective irrigation & cultivation during summer.
- Popularization of mini *dal* mills for *dal* making.

References

Bridging yield gaps in Major Agricultural crops : Strategies for Karnataka, (2010) K.S.D.A., Bangalore, Government of Karnataka.

Karnataka at a Glance (2009-2010). Directorate of Economics and Statistics, Bangalore, Government of Karnataka.

❑❑❑

13

Kerala

Kuruvilla Varughese, Jacob John, Rani, B. and Thomas Mathew
Cropping Systems Research Centre, Kerala Agriculture University, Karamana P.O., Thiruvananthapuram 695 002, Kerala, E-mail : csrc@kau.in

The total area under pulses in Kerala during 2008-09 was 3943 ha with a production of 2982 tonnes. The average yield per hectare is 756 kg. Cowpea, greengram, black gram and red gram are the important pulse crops traditionally grown in Kerala. Nearly 75% of area under pulses cultivation is occupied by cowpea. The remaining 25 per cent of area under this group of crops in the State is shared mainly by greengram, black gram and red gram. Among these, red gram is mainly confined to the Palakkad district and the crop season extends over *kharif* and *rabi* together. The other pulses are cultivated in all the districts of the State. Bengal gram is totally imported and there is little scope for its development in Kerala.

Due to various socio economic reasons, the scope for bringing additional area exclusively for pulses cultivation is limited and no area is specifically earmarked for cultivation of pulses in Kerala. They are grown as intercrop in coconut, tapioca, banana sugarcane etc. and also as a rotation crop in rice fallows during summer. Cultivation of pulses in rice fallows, which includes vast stretches of lowlands and terraced uplands that are left fallow after the harvest of paddy crop is the only source for increasing the area and production of pulses. Cowpea, greengram and black gram are important pulse crop traditionally grown in such situations. In addition to these, photo insensitive horse gram varieties also deserve special attention, as it is one of the hardiest crops among pulses and may perform well in tracts unsuitable for other pulses on account of extreme water scarcity. Season wise distribution of pulses cultivation is 36 % in *kharif*, 24% in *rabi* and 40% in summer i.e in rice fallows (Sreenivasan *et al.*, 2002).

In terms of consumption, the important pulses are red gram, greengram, bengal gram, black gram, cowpea and horse gram. Red gram and bengal gram top the list. Red gram constitutes almost one third of the total pulses consumed but the area devoted to this crop is less than 8 per cent of the total area under pulses. The average consumption of greengram is almost double that of black gram; but the area under cultivation is less than black gram. Bengal gram constitutes almost 15 per cent of the total pulses consumed in the State. The per capita availability of pulses in Kerala is less than 3 g day^{-1} only. More than 90 per cent of the pulses required for internal consumption is met through import from the neighbouring States. Hence, there is an urgent need to increase the pulse production in the State. The increase has to come mainly through increase in productivity and to some extent by utilizing the summer fallows.

Table 1 : Trend of pulses production

Year	Area (ha)	Production (MT)	Productivity (kg/ha)
2002-03	5764	4615	801
2003-04	5604	4272	762
2004-05	8428	8390	995
2005-06	10562	7940	752
2006-07	6870	5211	759
2007-08	4355	3294	756
2008-09	3943	2982	756

Kerala agricultural economy is undergoing structural transformation from the mid seventies by switching over a large proportion of its traditional crop area which was devoted to subsistence crops like rice and tapioca to more remunerative crops like coconut and rubber.

Existing Cropping Systems in Kerala

Rice

Rice is the dominant cropping system in valley lands. Some field crops including pulses and vegetables are grown in rice fallows while some in homesteads. The following pulses are in general find the place in existing rice-based systems.

Cowpea

- *Pure crop of cowpea in homesteads during kharif season* : In *kharif* season, cowpea can be grown as a pure crop in the homesteads. The varietal requirement during this period is a vegetable or dual purpose type with least tendency for trailing because a grain type variety during the heavy rainfall season will create problem in drying and processing of pods for grains. During this season almost all varieties show a tendency for trailing or twining, necessitating provision of supports.

- ***Intercropping of cowpea during kharif*** : Cowpea can be grown as intercrop with coconut, banana, tapioca, etc. Intercropping cowpea in semi-dry sown rice is another good practice that helps in controlling early weed growth. It also adds organic matter and improve nitrogen status of the soil.
- ***Fringe cropping*** : Fringe cropping of cowpea along the plastered rice field bunds during *rabi* and summer seasons is another popular practice that helps to generate additional income to the farmers with least input and efforts. Short duration varieties without much branching/trailing tendency are suitable for these conditions.
- ***Cowpea as a fodder/green manure crop*** : Both the shaded areas of coconut plantations as well as the open rice fallows offer scope for such a cropping in Kerala where both fodder and organic manures are becoming scarce. Cowpea under shaded situation can produce more than 40-50 tonnes of green matter per hectare, which can be conveniently used as green manure or as a protein- enriched fodder.
- ***Summer cultivation of cowpea*** : Summer rice fallows offer a large potential for expansion of area under pulses in the State. Grain type cowpea varieties can be extensively grown under this cropping system.

Greengram : Short duration varieties of greengram fit well in many intensive crop rotations. It can be grown in all three seasons. It is mainly grown as a crop in rice fallows after the harvest of *rabi* crop. It can also be grown as a mixed crop with tapioca, colocasia, yam, banana or as an intercrop in coconut garden.

Black gram : Black gram may be grown as pure crop in rice fallows after the harvest of the first or second crop of paddy. It can also be grown as pure or mixed crop during *kharif* season.

Red gram : In Kerala, red gram is cultivated along the bunds in rice fields as a fringe crop and also as a mixed crop with groundnut in Palakkad and Kozhikode Districts.

Zone wise Efficient Cropping Systems Involving Pulses in Kerala

Southern Zone : Rice-Rice-Cowpea, Rice-Rice-Greengram, Rice-Rice-Black gram, Rice-Red gram and Cowpea as intercrop in coconut (homesteads).

Central Zone : Rice-Red gram, Rice-Rice-Cowpea, Rice-Rice-Black gram, Rice-Rice-Greengram, Rice-Horse gram, Banana+Cowpea+Amaranthus, Banana+Amorphophallus+Cowpea, different pulses as intercrops in coconut (homesteads) and Cowpea as fringe crop in rice fields

Northern Zone : Rice-Rice-Cowpea, Rice-Rice-Black gram, Rice-Horse gram and Cowpea as intercrop in coconut and arecanut (homesteads).

High Range Zone: Rice-Rice-Cowpea, Rice-Cowpea and Arecanut + Cowpea.

Yield Gaps

Crop	Grain Yield realized normally (q/ha)	Best yields (q/ha) under very good management	Yield gap (%)
Cowpea	13.5	19.5	31
Greengram	6	11	45
Black gram	7.5	15	50
Red gram	17.5	22.5	22

The best yields in all pulse crops were obtained under irrigated conditions. Except in case of red gram, the highest yield was attained in rice fallows during summer season for all other pulse crops. Controlling weeds and rice seedlings in the early vegetative growth phase of the crop coupled with nutrient and water management practices help in obtaining highest yield. The exorbitant cost involved in manual weeding is the most important limiting factor in attaining the highest yield. Weedicides like Tok-E-25 may help in reducing the cost involved in weeding.

Improved Varieties of Pulse Crops

Cowpea

Vegetable type : (a) Bushy: Bhagyalakshmy, Pusa Barsathi, Pusa Komal. (b) Semitrailing: Kairali, Varun, Anaswara, Kanakamony (PTB-1), Arka Garima. (c) Trailing type: Sharika, Malika, KMV-1, Lola, Vyjayanthi, Manjeri Local, Vyalathur Local, Kurutholapayar.

Grain type : C-152, S-488, Pusa Phalguni, P-118, Pusa Do Fasli, Krishnamony (PTB-2), V-240, Amba (V-16), GC-827, CO-3 and Pournami (in summer rice fallows).

- *Dual purpose type:* Kanakamony (PTB 1) and New Era
- *As companion crop with tapioca:* V-26
- *Floor crop*: Gujarat V-118, Cowpea-2

Greengram : Madiera, Pusa Baisakhi, NP-24, Co-2, Pusa-8973 (Pusa-8973 is suited to the summer rice fallows of Onattukara; tolerant to pod borer; duration 66 days).

Black gram : T-9, Shyama, Sumanjana, CO-2, S-1, TAU-2. TMV-1 and KM-2

Red gram : SA 1

Production Technology

Cowpea (*Vigna Unguiculata*)

Land preparation : Plough the land thoroughly 2-3 times and remove weeds and stubbles. Make channels of 30 cm breadth and 15 cm depth at 2 m apart to drain off excess rainwater.

Seed Rate

For vegetable type Bushy:	20-25 kg/ha
Trailing:	4-5 kg/ha
For grain and dual purpose type	
Broadcasting:	60-65 kg/ha (45 kg for *var* Krishnamony)
Dibbling:	50-60 kg/ha (40 kg for *var* Krishnamony)

Sowing management : For grain type and dual-purpose type, if dibbling is adopted, spacing of 25 cm between rows and 15 cm between plants is recommended with two seeds per hole. If broadcasting is adopted, the seeds can be broadcast over the field and channels drawn after sowing. For bush vegetable type, spacing of 30 cm between rows and 15 cm between plants is suitable. For semi-trailing varieties, provide a spacing of 45 x 30 cm. Trailing varieties can be sown in pits (@ 3 plants/pit) at 2 x 2 m spacing for trailing on pandal or in channels at 1.5 m x 45 cm spacing for trailing on trellis (Kerala Agricultural University, 2007).

Fertilizer Application

FYM	20 t/ha
Lime	250 kg/ha or dolomite 400 kg/ha
N	20 kg/ha
P_2O_5	30 kg/ha
K_2O	10 kg/ha

Lime may be applied at the time of the first ploughing. Half the quantity of N, whole of phosphorus and potash may be applied at the time of final ploughing. The remaining N may be applied 15-20 days after sowing. The economic optimum dose of phosphorus was 30.0 and 36.8 kg P_2O_5 for the production of grain and green pods, respectively (Santhakumari *et al.*, 1994).

[For vegetable cowpea grown as an intercrop in the reclaimed alluvial soils of Kuttanad, N, P_2O_5, and K_2O at the rate of 10, 20 and 10 kg/ha are recommended. For vegetable cowpea, fertilizers can be applied in several split doses at fortnightly intervals]

Season

- Cowpea can be grown during any season.
- As a rainfed crop, sowing is done in the month of June. The most suitable time is after the first week of June.
- During the second crop season (*rabi*), i.e., September to December, cowpea can be grown as a fringe crop along the rice field bunds. Sowing can be done on either side of bunds on the day of transplanting the paddy crop.
- During summer, cowpea can be grown as a pure crop in rice fallows after the harvest of paddy.

After cultivation : Hoeing at the time of application of the second dose of N will give adequate aeration to the soil and help the root system to spread easily. For grain and dual-purpose varieties, decapitation is found to be advantageous as the crop shows trailing tendency. For vegetable types, provide trellis or *pandal* for trailing.

Water management : Giving two irrigations is highly beneficial; i.e., at 15 days after sowing and at the time of flowering. Irrigation at the flowering stage induces better flowering and pod set. IW/CPE ratio 0.075 is a better schedule for both grain and green pods (Santhakumari *et al.*, 1994).

Plant Protection

Diseases

- ***Anthracnose*** : Treat seeds with Carbendazim 50% WP (2 g/kg of seed) or with bio-control agents like *Trichoderma viride* (4 g/kg of seed) or *Pseudomonas fluorescens*. If the plants are severely affected, spray Carbendazim 50% WP (0.1 %)
- ***Collar rot and web blight*** : Mulch the field with dry leaves. Spray carbendazim 50% WP spray (0.1 %) at the time of appearance of the the disease.

- ***Dry root rot and stem rot*** : Treat seeds with Carbendazim 50% WP (2g/kg of seed) or bio- control agents *viz., Trichodenna viride* (4 g/kg of seed) or *Pseudomonas fluorescens* (10 kg/kg of seed). Collect and destroy infected plants and adopt crop rotation with non-leguminous crops. Apply organic amendments in soil.
- ***Powdery mildew*** : Spray the affected crop with Carbendazim 50% WP (0.1%) or wettable sulphur 80 WP (0.3%)
- ***Leaf spot*** : Remove the infected plants and debris. Spray the crop with copper oxychloride 50 WP(O.2 %) or mancozeb 75 % WP (0.2%).
- ***Rust*** : Spray the affected crop with wettable sulphur 80 WP (0.3 to 0.5%).
- ***Leaf blight*** : Grow resistant genotypes and adopt crop rotation with non-leguminous crop for two to three years.
- ***Bacterial leaf spot*** : Sow disease free seeds and resistant cultivars
- ***Mosaic*** : Remove the infected plants. Spray periodically with systemic insecticide like Dimethoate 30 EC (0.03%) or methyl demeton 25 EC (0.05%) or tobacco decoction to control the vector.

Insect Pests

- ***Pea aphid*** : Apply tobacco decoction (@ 1.5%) or anyone of the following insecticides *viz.,* Malathion 50 EC @ 0.05% or Dimethoate 30 EC @ 0.03% or Metasystox 25 EC @ 0.05% as soon as the pest appears. The fungus *Fusarium pallidoroseum* can be used for controlling black pea aphid. Bran based fungus can be applied at the rate of 3 kg per 400 m^2 immediately after infestation is observed. Only one application is necessary.
- ***Spotted and other Pod borers*** : Spray Carbaryl 50 WP (0.2%) or Fenthion (0.05%) at the time of 50 % flowering to protect the crop from pod borer and repeat the spray 15 days after if the infestation persists. Apply the insecticides after harvesting mature pods and pick the pods only 10 days after application of insecticides.
- ***Tobacco caterpillar*** : Collect the prominently visible egg-masses as well as young larvae feeding gregariously and destroy mechanically. In severe cases, spraying with dichlorvos 100 EC (0.05%) or dusting with carbaryl 5% D @ 2 kg a.i./ha.
- ***Cow-bugs*** : Generally no control measures are adopted against these bugs. If the infestation is severe spray 0.05% Dichlorvos or Malathion.

- ***Pod bugs*** : Dust with Carbaryl 5% D (@ 2 kg a.i./ha) or spray 0.05% Malathion.
- ***Lab-lab bug*** : Management is same as as pod bugs
- ***Hawk moth*** : Collect and mechanically destroy the larva in early stage of attack. It will check the pest population. To check the population build-up, plough the infested fields immediately after the harvest to expose and kill the pupae in the soil.
- ***Green jassid*** : Spray thevetia leaf extract (4%) with soap water or Dimethoate 30EC (0.05%) to check/the population build up of the pest.
- ***Bihar Hairy caterpillar*** : Destroy young larvae by pulling out the infested plants and burying them in the ground. Destroy the grown up caterpillars by crushing them under feet or by picking and putting them into water containing kerosene. If the population is high, spray Quinalphis 25 EC (0.05%) or Dichlorvos 100 (0.05%) or Carbaryl 50 WP (0.2%).
- ***Green semilooper*** : Spray Fenitrothion 5.0 EC (0.05%) or Dichlorvos 100EC (0.05%) or dust with carbaryl 5%0 @ 2 kg a.i./ha to check the pest population.

Non-insect Pests

- ***Root-knot nematode and reniform nematode*** : Apply neem and eupatorium leaves @ 15 t/ha, two weeks before sowing.

Cropping systems : Cowpea can be grown as a floor crop in coconut gardens and as an intercrop in tapioca during May-Sept. It can also be grown as a pure crop in single-crop and double-crop rice fallows during *rabi* and summer seasons. Cowpea can be grown in homestead garden throughout the year and in *kole* lands of Thrissur district during summer when rice crop cannot be raised due to water scarcity. Cropping systems involving tapioca with cowpea, greengram and black gram gave comparable yield of pure crop and were found suitable for the early part of crop growth (Varughese *et al.*, 1986b).

Greengram (*Vigna Radiata*)

Land preparation : One deep ploughing followed by two or three harrowings and plankings are considered optimum for greengram cultivation. Field should be well levelled and completely free from weeds. For summer season crop, give a pre-irrigation immediately after harvest of *rabi* crop. When the field comes in condition, prepare it by giving one to two ploughings followed by planking to level the field and minimize the loss of moisture by evaporation from the soil surface. For summer crops, after ploughing, a pre-sowing irrigation should be given it ensure good germination. Channels of 30 cm width and 15 cm deep are made 2 m apart to drain excess rain water during *kharif* and provide irrigation during summer seasons.

Seed rate : A seed rate of 20-25 kg/ha is recommended for pure crop but in mixed/inter cropping only 5-6 kg seed is needed. Greengram is generally broadcast. Sowing of greengram in rows has been found better as it ensures proper spacing for canopy. Sowing can be done behind the local plough with the help of a seed drill.

Sowing management : In *kharif* season, sowing is done on the onset of monsoon in the second fortnight of June to first fortnight of July. For *rabi* crop, the best time of sowing is October-November. Sowing in summer should be done in the last week of February to middle of March. Timely sowing of summer greengram is most important. Delay in sowing affects the yield adversely. The late sown crop may be impacted adversely by the early monsoon rain at harvest time, which may create problems for harvesting and threshing, and ultimately the grain quality besides yield reductions.

A spacing of 25-30 cm between rows and 15 cm between plants should be followed for sowing depending upon the time of sowing and varieties.

Fertilizer application : Greengram is generally grown on the basic fertility of soil. However, if available, 20 tonnes of compost or farm yard manure should be applied as basal. Apply lime 250 kg/ha or dolomite 400 kg/ha at the time of field preparation. General fertilizer recommendation for greengram is NPK @ 20 :30:30 kg/ha. Response of greengram to potassium is negligible. Application of 100 kg diammonium phosphate (18-46-0) per ha is comparatively cheaper on nutrient basis and can supply required quantity of N & P to the crop. Lime may be applied at the time of first ploughing and organic manure as basal. Half the quantity of N, whole of P and K are applied at the time of last ploughing. The remaining ½ N can be applied as foliar spray as 2 per cent urea solution in two equal doses on the l5th day and 30th day after sowing. Fertilizer placement should be done by drilling the fertilizer at the time of sowing using a seed cum fertilizer drill or just before sowing in such a way that the fertilizer is placed about 5-7 cm below the seed. *Rhizobium*, phosphorus solubilizing bacteria and VAM are effective in increasing the yield of greengram.

Weed control & interculture : Greengram being a dwarf statured crop, suffers to a great extent if the weeds are not controlled in time. One hand weeding after 20-25 days of sowing followed by another after about 20 days will keep the crop weed free. Weeds can also be controlled with the use of herbicides. Fluchloralin or Pendimethalin @ 1.0 kg a.i/ha applied in combination with one hand weeding controls weeds effectively and increases the yield of greengram.

Water management : Rainy season greengram does not require irrigation unless there is a prolonged drought. There should be adequate drainage in the field because this crop is sensitive to water logging. *Rabi* and summer greengram may be grown only in areas where adequate irrigations facilities are available. For *rabi* and summer crop, 5 to 6 irrigation may be given. The field should be irrigated when soil is seen deficient in moisture. Irrigation at IW/CPE at 0.6-0.7 is found to be optimum in greengram. No irrigation should be given when the crop is at full bloom stage. It was also noticed that moisture stress at early stage of crop was highly detrimental to the crop growth and yield (Varughese and Iruthayaraj, 1993).

Plant Protection

Diseases

The diseases *viz.*, anthracnose, dry root rot, powdery mildew, leaf spot and rust are similar to that of cowpea and treated accordingly. The rest of the diseases are mentioned below.

- ***Yellow mosaic*** (i). Remove the diseased plant (ii). Periodic spraying with systemic insecticide like dimethoate 30 EC (0.03%) or Neem seed kernel extract (0.3%) or Neem oil 3% will reduce the vector population.
- ***Phyllody*** : Remove the infected plants and spray with dimethoate 30 EC (0.03%) to control the vector.

Insect Pests

- ***Stem fly*** : Apply 10 percent lindane dust in soil before sowing. Timely sowing also minimizes the attack. Treat seeds with carbofuran @ 2 kg/ha or phorate 50 % SD @ 1 kg per 50 kg seed
- ***Spotted pod borer*** : Management as under cowpea
- ***Gram pod borer*** : Management as under red gram
- ***Leaf miner*** : Spray Fenitrothion 50 EC (@ 0.05%) or Quinalphos 25% EC (@ 0.025%) or neem oil emulsion (2%)
- ***Green semilooper*** : Management as under cowpea
- ***White fly*** : Use yellow sticky traps to attract the adults. Spray 0.5% soap solution on the under surface of the leaf. If the infestation is severe, spray either Malathion 50 EC (0.05%) or Dimethoate 30 EC (0.03%)

Cropping systems : Short duration varieties of greengram fit well in many intensive crop rotations. In Kerala, the crop is raised in all the three seasons. It is mainly grown as a crop in rice fallows after the harvest of second crop of paddy. It can also be grown as a mixed crop with tapioca, colocasia, yam, banana or as an intercrop in coconut garden.

Black gram (*Vigna Mungo*)

Land preparation : Plough the land 2-3 times thoroughly and remove the weeds and stubbles. The field should be well levelled. For summer crops, after ploughing a pre sowing irrigation should be given to ensure good germination.

Seed rate : For pure crop, 20-25 kg of seed/ha are recommended depending upon the season and variety. Only 5--6 kg of seed/ha is required for mixed cropping.

Sowing management : During *kharif* season, it can be grown as a pure or mixed crop. The *kharif* crop is generally sown late to synchronize the harvest with short day periods of late September or early October. During summer, the sowing could be done from the 3rd week of February to 1st week of April. Sowing done subsequently reduces the yield due to rain at the time of maturity. Black gram is generally sown by broadcasting. Sowing in rows in furrows behind the local plough or using a seed drill is however a better practice than broadcasting. A spacing of 25-30 cm between rows and 15 cm between plants should be given for sowing depending upon the time of sowing and varieties.

Fertilizer application : Apply lime 250 kg/ha or dolomite 400 kg/ha at the time of field preparation. If available 20 tonnes of compost or farm yard manure should be applied as basal. Black gram needs only a small quantity of N for the early growth period. Soils, which are poor in organic matter, should get about 15-20 kg N/ha as a starter dose. However. phosphatic and potassium fertilizers should be applied as per soil test values. In case soil test values are not available, one can apply 30-60 kg P_2O_5 and 30-40 kg K_2O per hectare.

Lime may be applied at the time of first ploughing and FYM as basal. Half the quantity of N, full P and K is applied at the time of last ploughing. The remaining half of N can be applied as foliar spray of 2% urea solution in two equal doses at 15 and 30 days after sowing. Fertilizer placement by drilling it at the time of sowing is also practised. *Rhizobium* inoculation should be done if black gram is being sown for the first time the field or after a long duration. Black gram responds well to inoculation of VAM and phosphate solubilizing bacteria.

Weed control & inter culture : Weeds pose a serious problem in rainy season. Losses even up to 50-60 per cent have been recorded due to weeds. One or two hand weedings up to 40-45 days of crop growth may be sufficient to control the weeds. Weeds can be controlled by the use of chemicals too. Pre plant/pre emergent spray of basalin 1 kg a. i. /ha is effective in controlling the weeds (Kerala Agricultural Universty, 2007).

Water management : Irrigation is not needed during rainy seasons but good drainage is essential as the crop is sensitive to water logging. Black gram grown during summer, requires 4 to 5 irrigations. Number and frequency of irrigation depends upon the soil type and weather prevailing during the growth period. Generally, the crop should get irrigation at an interval of 10-15 days or at 0.6 IW/CPE (Varughese *et al.,* 1986a). From flowering to pod development stage there must be sufficient moisture in the field.

Plant Protection : Diseases of greengram affect black gram also. Hence, the control measures are as that of greengram.

Cropping systems : Black gram may be grown as pure crop in rice fallows after the harvest of the first or second crop of paddy. It can also be grown as pure or mixed crop during *kharif* season.

Redgram (*Cajanus Cajan*)

Land preparation : Being a deep-rooted crop, red gram responds well to proper tilth. A deep ploughing with soil turning plough followed by two to three harrowings and proper levelling is essential to ensure uniform irrigation and proper drainage. It requires a deep and well pulverized field which is free from weeds and clods.

Seed rate : A seed rate of 15-20 kg/ha is sufficient for pure crop. In mixed cropping, seed rate is adjusted accordingly to the proportion of red gram and companion crops to be grown.

Sowing management : Red gram is traditionally a *kharif* crop. The duration of the crop extends over *kharif* and *rabi.* The seeds are sown in the first crop season and harvested at the end of second crop season. Extra early duration varieties can be cultivated during summer season in irrigated areas. In Kerala, sowing can be done in two seasons. As a mixed crop in groundnut or tapioca, sow the seeds in June-July. Red gram can be sown in plastered field bunds of *rabi* rice as a fringe crop. Red gram can also be sown in paddy fields after the harvest of *rabi* crop as a fringe crop. Long duration varieties of red gram are tall, spreading and occupy the fields for about 250-270 days. These varieties are planted at a spacing of 90-120 cm x 30 cm under rain fed condition. Early maturing varieties are planted at a spacing of 50-75 x 15-20 cm. For pure cropping, it is either broadcast or dibbled. Seed should be sown behind the plough or with the help of a seed drill.

Fertilizer application : One of the important reasons for poor yield of red gram is neglect in use of manures and fertilizers. However, this crop is a heavy feeder and hence, care should be taken to ensure that the crop does not suffer from lack of nutrients. Apply lime 500 kg/ha and FYM @ 3000 kg/ha. A starter dose of nitrogen @ 30-40 kg per hectare gives yield response in most of the soils depending upon the initial nitrogen status of the soil. Application of phosphorus and potassium should be done on the basis of soil test values. In case phosphorus is applied on poor soils, it would result in a conspicuous increase in yield of red gram. About 80-100 kg phosphate per hectare would be needed on soils poor in available phosphorus. Similarly on potassium deficit soils, about 40-60 kg/ha potash may be applied for good yield. Red gram shows susceptibility to zinc deficiency. Zinc sulphate @ 20 kg/ha should be applied at the time of sowing in zinc deficient soils. Zn deficiency in the standing crop can be rectified by spraying 5 kg zinc sulphate and 2.5 kg lime dissolved in 800-1000 litres of water per hectare. Bio-fertilizers like *Rhizobium* and phosphorus bacteria are also found to be beneficial in increasing the yield of red gram.

Weed control and interculture : The growth of red gram is quite slow during first 45-60 days after sowing. During this period if weeds are not controlled, they cover the red gram plants. The first 60 days are very critical in red gram. Two mechanical weedings, one at 25-30 days and another at 45-60 days after sowing gave excellent weed control. Pre-emergence application of lasso @ 110 kg a.i./ha or basalin @ 1 kg a.i./ha has been found quite effective.

Water management : Being a deep-rooted crop, red gram can tolerate drought to a great extent. In *kharif* crop, one or two pre monsoon irrigations may be given, if required. After the onset of monsoon, there is no need of irrigation. However, in case of prolonged drought during the flowering/pod setting stages one or two irrigations may be beneficial. Proper drainage is equally important. Ridge planting is effective in areas where subsurface drainage is poor. This helps in providing enough aeration for the roots during the period of excess rainfall. Water should not be allowed to stand anywhere in the field during rainy season.

Diseases

- ***Wilt*** : Rotate with non- leguminous crops and apply heavy dose of organic manures. Treat seeds with Carbendazim 50% WP (2 g/kg of seed).
- ***Powdery mildew*** : Spray the crop with Carbendazim 50 % WP (0.1%)
- ***Leaf spot*** : Spray the affected plant with Mancozeb 75% WP or copper oxychloride 50 W(0.2%)
- ***Blight*** : Grow plant resistant varieties. Provide good drainage to avoid the stagnation of water.
- ***Sterility mosaic*** : Remove and destroy diseased and self-sown plants to minimize the incidence of the disease. Spray the crop with wettable sulphur 80 WP (0.1 %) to control the vector.

Insect Pests

- ***Gram pod borer*** : Attract and kill the nocturnal adults using light trap or pheromone lure. Spray NPV @ 1.5 x 1012 POB/ ha. Spray Carbaryl 50 WP @ 0.2% at the time of 50% flowering and repeat the spray after 15 days or Deltamethrin 2.8 EC @ 400 ml /ha or fenvel 0.4 %@ 20 kg/ ha.
- ***Plume moth*** : Management is same to that of gram pod borer.
- ***Pod fly*** : Attract and kill the fly using fish meal trap or spray carbaryl 50 WP @ 0.2% or dimethoate 30EC @ 0.05% in the early stages of the attack.

Cultivation in Non-conventional Areas

The availability of suitable area for pulse cultivation is the limiting factor in expanding pulse production in Kerala. Due to climatic constraints, Bengal gram is not suited to the State. In the irrigated command areas, rice fallows can effectively be utilized for intensification of pulse cultivation. Lack of varieties suited to the shaded conditions of crops like coconut, tapioca etc is also a problem in extending pulse cultivation. Hence, varieties suited to shaded situations may be developed.

Future Thrust

- Development of varieties for shaded conditions in the uplands.
- Varieties maturing with shorter duration and which facilitate to completing harvest with one or two pickings.
- Agronomic management for different shade tolerance.
- Identification of suitable herbicides to control the weeds in the early vegetative phase of the crop.

References

Kerala Agricultural University (2007). Package of Practices Recommendations. 'Crops' 13th edition. Kerala Agricultural University, Thrissur pp. 334

Santhakumari, G., Pillai, G.R., Varughese, K and Jose Mathew (1994). Response of Cowpea to Irrigation and Phosphorus. Journal of Tropical Agriculture 32: 30-32

Sreenivasn, E., Anitha, S., Purushothaman, S.M. and Mani, C. (2002). Pulse Crops of Kerala. Kerala Agricultural University, Thrissur pp.67

Varughese, K., Jose, M. and Pillai, G.R (1986a). Response of black gram to different levels of irrigation in summer rice fallows. Agric. Res. J. Kerala 24 (2): 175-178

Varughese, K., Jose, M. and Pillai, G.R (1986b). Response of cassava to irrigation under pure and mixed stands. Agric. Res. J. Kerala 24 (2): 179-184.

Varughese, K. and Iruthayaraj, M.R. (1993). Effect of water stress at different phenophases of greengram-Co-3. Agric. Science Digest 13 (2): 87-89.

❑❑❑

14

Madhya Pradesh

V.K. Shukla, S.K.Vishwakarma, Lokesh Dubey and H.S. Yadav*
AICRP on Integrated Farming Systems, J.N.Krishi Vishwa Vidyalaya
Jabalpur - 482 002 (MP)
*R.S.V.K.V.V., Gwalior (M.P.)

Madhya Pradesh is a major pulse producing state of the country and more than half a dozen of pulses are grown on 48.99 lakh ha area with an annual production of 41.73 lakh tonnes and average productivity of 852 kg/ha (Table 1). The pulses production share of Madhya Pradesh is 28.6% to national pulses production. Amongst the pulses, chickpea, pigeonpea, moong, urad, lentil and field pea are the important crops of the state. These crops contribute about 3296 thousand tones to national pulse basket from an area of about 4318 thousand hectares (Table 2). Among the pulse crops grown in the state, chickpea accounts 76% in production from 60% area. Second important pulse is lentil (13%) followed by uradbean (12%), pigeonpea (8%), field pea (5%) and mungbean (2%) with percent share of 8, 5, 7, 3 and 1 in production, respectively. The area and consequent production of pulse crops in the state showed, in general, an increasing trend since 1981 till date (Table 2 and Fig.1). Linear increase in yield of these crops from 531 kg/ha to 764 kg/ha in this period is also witnessed as the impact of technological advancement in these crops in the recent past. Natural calamities mostly drought in few to most of the districts is the main reason of low productivity in triennium ended in 2004.

Pulse Based Sequential Cropping

Under traditional cropping system practiced in Madhya Pradesh, there is yield stability at lower level but adaptable to both favorable and adverse agro-climatic pulse enhancement conditions / prevailing over time. These cropping systems are adopted by the farmers considering the market prices and family requirement. The agro-climatic zone wise existing cropping systems followed in Madhya Pradesh by the farmers are as under:

Table 1 : Area, production and productivity of pulse crops in Madhya Pradesh (2011-12)

Pulse Crop	Area(000 ha)	Production (000 tonnes)	Productivity (kg/ha)
Chickpea	3043.7	3290.3	1082
Pigeonpea	543.9	334.2	622
Black gram	551.4	148.8	269
Lentil	620.5	230.0	371
Greengram	84.6	21.3	235
Field pea	263.9	94.2	357

Source: Commissioner land records, M.P. Gwalior

Table 2 : Trend in area, production and yield of pulse crops in Madhya Pradesh

Years	Area(000ha)	Production(000t)	Yield(Kg/ha)
1981-83	3800	2027	531
1984-86	3860	2177	564
1987-89	3734	2250	603
1990-92	3680	2414	654
1993-95	3861	2749	710
1996-98	4026	2842	707
1999-01	4000	3027	751
2002-04	4301	2961	686
2005-07	4318	3296	764

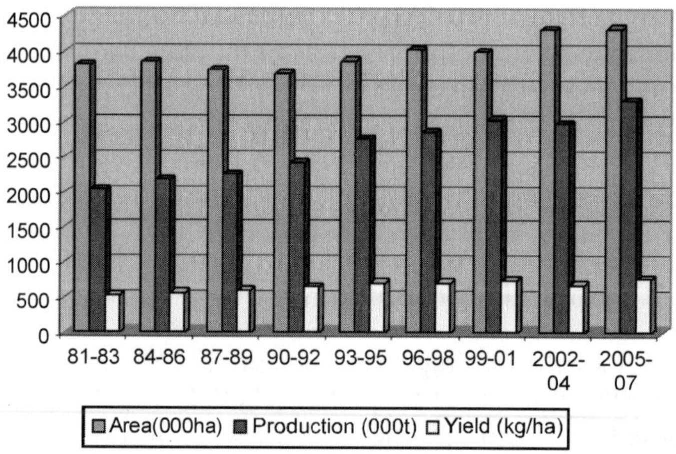

Fig.1 : Trend of area, production and yield of pulse crops in Madhya Pradesh

I. Chattisgarh plains (Balaghat):

Predominant cropping systems: Rice-wheat

Low land: Rice-rice, rice-wheat

Upland: Moong/urad-rice, rice-fallow

Diversified cropping systems: Rice-chickpea, rice-wheat + chickpea (4:2) and pigeonpea-wheat

II. Northern hills of Chattisgarh (Mandla, Dindori, Parts of Shahdol, Umaria and Sidhi):

Predominant systems: Rice-chickpea, rice-fallow and small millets-fallow

Diversified cropping systems: Pigeonpea-til and soybean-wheat + chickpea (4:2)

III. Kymore plateau and Satpura hills (Jabalpur, Seoni, Katni, Panna, Satna, Rewa and Parts of Sidhi, Umaria and Shahdol):

Predominant systems: Rice-wheat, rice-fallow, soybean-wheat, moong/urad-wheat/ chickpea, maize/ sorghum-wheat.

Diversified cropping systems: Rice-wheat + chickpea (4:2), soybean + moong (4:2)-chickpea, pigeonpea-moong /urad and small millets-chickpea

IV. Vindhyan plateau (Sagar, Damoh, Raisen, Bhopal, Sehore and Vidisha):

Predominant systems: Moong/urad-wheat, soybean-wheat and soybean-chickpea

Diversified cropping systems: Soybean-chickpea, soybean-wheat + chickpea (4:2) and sorghum-chickpea + mustard (6:1).

V. Central Narmada Valley (Narsinghpur, Hosangabad and Harda):

Predominant systems: Rice-wheat/chickpea, soybean-wheat and soybean-sugarcane

Diversified cropping systems: Pigeonpea-wheat, soybean-chickpea + mustard (6:1) and moong/ urad-wheat.

VI. Gird Morena, Shivpuri, Bhind, Gwalior, Guna and major part of Sheopur:

Predominant systems: Pearl millet-mustard, soybean-wheat and sorghum-chickpea

Diversified cropping systems: Soybean-mustard + chickpea (6:2) and Pigeonpea-chickpea.

VII. Bundelkhand (Tikamgarh, Chharturpur, Datiya and parts of Shivpur):

Predominant systems: Soybean-wheat, sorghum-chickpea and small millet-wheat

Diversified cropping systems: Soybean-chickpea, sorghum-chickpea + mustard (4:1) and sorghum + moong / urad (4:l)-wheat + chickpea (6:2).

VIII. Satpura plateau (Chhindwara and Betul):

Predominant systems: Soybean-wheat, maize-chickpea and vegetables-lentil

Diversified cropping systems: Moong/ urad-wheat and arhar + maize (4:2)-chickpea/lentil

IX. Malwa plateau (Ujjain, Mandswar, Neemach, Ratlam, Shajapura,

Indore, Rajgarh, Devas and major parts of Dhar):

Predominant systems: Soybean-wheat, maize-wheat and maize/ sorghum-wheat

Diversified cropping systems: Cotton-chickpea/ lentil and maize-wheat+chickpea (4:2)

X. Nimar valley (Khargone, Khandwa and Badwani):

Predominant systems: Cotton-lentil, sorghum-chickpea and groundnut-wheat

Diversified cropping systems: Cotton-lentil/ chickpea and cotton-wheat+ chickpea (4:2)

XI. Jhabua hills (Jhabua and parts of Dhar):

Predominant systems: Soybean-wheat and maize/sorghum-wheat/ chickpea

Diversified cropping systems: Maize/Jowar-chickpea

Yield Gaps

The yield gap of different pulse crops was computed based on yield of a crops in efficient cropping system/ research stations and average yield of Madhya Pradesh as depicted in Table 3.

Yield gap of 68, 68, 59, 50, 67 and 75 was observed in Moong, Urad, Pigeon-pea, Chick-pea, Lentil and Field-pea, respectively. This indicated that there is relatively more scope of yield improvement in chickpea followed by pigeon-pea.

Table 3 : Yield gap of important pulses crops grown in Madhya Pradesh

Crops	Actual yield(q/ha)	Yield potential(q/ha)	Yield gap (%)
Moong	3.25	10-11	68
Urad	3.53	10-12	68
Pigeonpea	7.39	16-18	59
Chickpea	9.26	18-19	50
Lentil	5.10	15-20	67
Field pea	4.88	10-22	75

Major Constraints in Pulse Production

1. **Unfavorable weather conditions:** High rainfall (1200-1600 mm) especially for *kharif* moong, urad and pigeonpea, Continuous heavy rains at flowering (August) during *kharif*, Short span of winter and sudden rise of temperature in the month of February, and susceptibility to environmental fluctuations disfavour *rabi* crops
2. **Abnormal soil conditions:** Unbounded black soils (Vertisol) have poor drainage. Field preparation after paddy harvest in Vertisol is difficult. Initial soil moisture during *rabi* sowing is poor. Cultivation on marginal and wastelands.
3. **Seed constraints:** Inadequate availability of good quality seed of improved varieties. Faulty system of indenting and seed production. High price. Untimely supply
4. **Varietal constraints:** Lack of high yielding, disease resistant varieties: Lack of short duration, synchronous, photo-thermo insensitive varieties of summer moong and urad. Narrow adaptability of cultivars. Poor response to inputs specially high doses of fertilizer and irrigation.
5. **Indeterminate growth habit:** Non-availability of genetic resistance particularly against yellow mosaic virus in moong & urad and pod borers in moong and urad; thrips in grass pea and lentil.
6. **Highly prone to insect, pests and diseases:** Inadequate plant protection measures adopted by farmers.
7. **Agronomical constraints:** Heavy weed population especially in *kharif* season, Lack of post emergence herbicides, Lack of irrigation, Insufficient / No use of organic and inorganic fertilizers, Supplementation of N and P through DAP induces the sulphur deficiency, Non-availability of granular single super phosphate, Water logging and soil moisture stress cause seedling mortality, most of the pulses are grown as *utera* in Balaghat, Dindori and Mandla district of MP.

8. **Constraints in use of culture:** Less awareness about Rhizobium and phosphorus solubilizing bacteria (PSB) culture, Most of the strains of bio-fertilizers are ineffective due to improper storage facility, head of unawareness of application techniques.

Crop-wise Reasons for Low Productivity

Chickpea

The productivity of chickpea in the state is higher than national and world average productivity of chickpea. However, wide extension gaps also exist in the productivity of this crop. The yield level of this crop can be nearly doubled with existing technological status by mitigating the following yield reducers:

- Occurrence of moisture stress during germination and moisture stress along with high temperature at grain filling stage of crop growth in rainfed farming.
- Increasing severity of soil born diseases like *Fusarium* wilt, dry root rot, collar rot, foot rot and pest like *Helicoverpa* because of predominance of soybean – chickpea cropping sequence.
- Low seed replacement rate.
- Cultivation of low yielding, late maturing, extra bold seeded and wilt susceptible local varieties of kabuli chickpea.
- Low adaptability of varieties under late sowing in rice fallows and light soils of Nimar valley, Jhabua hills and Satpura plateau.
- Nutrient stresses in marginal and sub marginal lands due to low and imbalance use of fertilizers.
- Delayed sowing due to erratic rains and sequential cropping system.
- Inadequate use of available irrigation resources.
- Indiscriminate use of pesticides with inadequate doses and methods.
- Slow awareness to developed technologies.
- Price fluctuations.

Pigeonpea

- Cultivation of low yielding, disease and pest susceptible short duration varieties in medium textured soils.
- Low yield potential of recommended varieties in light to medium textured hilly soils.
- Very low seed replacement rate.

- Low plant population due to occurrence of *Phytophthora* blight and water logging in medium textured soils.
- Nutrient stresses in marginal and sub marginal lands.
- High weed infestation.
- Indiscriminant use of pesticides with inadequate dose and methods.
- Slow awareness to developed technologies.

Lentil

- Low yielding, bold seeded and wilt susceptible local land races are mostly in cultivation.
- Very low seed replacement of local varieties with newly recommended varieties.
- Cultivation in marginal lands with minimum use of inputs.
- Occurrence of wilt complex in lentil based cropping system.
- No or very low and imbalance use of fertilizers.
- Limited use of bio-fertilizers.
- Slow awareness to developed technologies.

Field pea

- Low yield potential of adopted varieties due to terminal drought, powdery mildew and wilt.
- Low seed replacement rate.
- Low response to inputs like fertilizers and irrigation.
- Slow awareness to developed technologies.

Mungbean and Urdbean

- Cultivation of old varieties that are susceptible to YMV, powdery mildew and leaf spot diseases.
- Non- synchronous maturity of mungbean varieties with pre-harvest sprouting and pod shattering in the varieties under cultivation.
- Existence of traditional package.
- Discourage cultivation due to fast spread of soybean in the state.
- Very low seed replacement rate.
- Low remunerative crop as compared to soybean.

Strategies to Improve Productivity

Chickpea

- Fixing the demand based targets of chickpea production for the state that can be 3.42, 4.34 and 5.57 million tones during the years 2010-11, 2015-16 and 2020-21, respectively.
- Production enhancement by harvesting the optimum potential of existing technologies.
- Solving the researchable needs by creating new centers for location specific research/ verification of technologies.
- Bridging the yield gaps between inter district realized yield by location specific technological verification.
- Enhancement in seed replacement rate and assured supply of foundation and certified seed of newly released varieties.
- Efficient utilization of water resources for life saving irrigation.
- Timely supply of critical inputs even in remote areas.
- Preventing post harvest losses.
- Aggressive transfer of generated technologies.

Pigeonpea

- Development/ Identification of varieties suitable for cultivation in light textured hilly soils where occurrence of moisture tress is frequent feature.
- Enhancement in seed replacement rate of newly recommended medium to late maturing varieties.
- Popularization of CMS based pigeonpea hybrids.
- Adoption of IDM for *Phytophthora* blight management techniques.
- Adoption of IWM techniques for weed management and IPM for pest complex.
- Increasing awareness to developed technologies.

Lentil

- Popularization of early maturing, large seeded, high yielding varieties having resistance to *Fusarium* wilt by assured supply of certified seeds and enhancing the seed replacement rate.

- Increasing awareness towards balanced use of fertilizers/ IPNM techniques.
- Timely sowing in changing cropping scenario.
- Solving the local needs by technological verifications in different agro-climatic zones.

Field pea

- Popularization of newly recommended high yielding, early maturing and disease resistant varieties.
- Enhancement in seed replacement rate by assured supply of certified seed.
- Increasing awareness towards balanced use of fertilizers/ IPNM techniques.
- Solving the local needs by technological verifications in different agro-climatic zones

Mungbean and Uradbean

- Popularization of recently recommended varieties like TJM 3, JM 721, PDM 11, etc of mungbean and RBU 38, JU 8-6, PDU 1, etc of uradbean having better resistance against disease complex.
- Enhancement in seed replacement rate by assured supply of certified seed.
- Increasing awareness to developed technologies.
- Standardization of production techniques for spring and summer cultivation.
- Ideotype for inter cropping situations.

Weed Management in Pulses

Emergence of weeds in pulses begins simultaneously with the crops leading to severe competition between weeds and the crop right from the very early stage. Keeping the field free of weeds throughout the cropping season undoubtedly gives higher crop yield but it may not be economical. Removing weeds at any time during growing season may not be beneficial. Stage of weed removal is as important as removal per se. Therefore, it is necessary to identify the critical period of crop-weed competition to render the weed control practices more effective and economical. Information is available on the active duration when the presence of weeds make their deleterious effect on crops. Critical period varies in different pulse crops (Table 4). Weeds emerge beyond competitive period do not affect the productivity but these weeds should be removed before settings of seeds to avoid further weed seeds spread.

Based on the research work done at different AICRP-WC Centre's and elsewhere in the country, some herbicides have been found selective, effective and economical in different pulse based cropping systems (Table 5).

Table 4 : Critical period of crop-weed competition and losses due to weeds in different pulse crops

Crops	Critical period (DAS)*	Reduction in seed yield (%)
Pegion pea	15-60	20-40
Moong	15-30	25-50
Urad	15-30	30-50
Cowpea	15-15	15-30
Chickpea	30-60	15-25
Peas	30-45	20-30
Lentil	30-60	20-30
Frenchbean	30-60	15-30

Table 5 : Promising herbicides for different pulse crops

Herbicide	Dose (kg ai/ ha)	Crops	Weed Controlled
Pre-plant incorporation			
Flucholalin	0.75-1.0	All pulses	BLW* and grasses
Trifluralin	0.75-1.0	All pulses	BLW and grasses
Pre-emergence			
Alachlor	1.5-2.0	Rainy season pulses	Grasses
Linuron	1.0-1.5	Chickpea peas, lentil	BLW and grasses
Metolachlor	1.0	Pigeonpea, moong, urad	Grasses
Metribuzin	0.50-0.75	Peas	BLW and grasses
Oxyfluorfen	0.10-0.20	Urad, moong,	BLW and grasses
Pendimethalin	1.0	All pulses	BLW and grasses
Post-emergence			
Clodinafop-Propagyl	0.060	Chickpea, peas, lentil	Grasses
Quizalofop ethyl	0.050	Chickpea, peas, lentil	Grasses
Imazethapyr	0.10	Urad moong and pigeonpea	BLW and grasses
Mtribuzin	0.30-0.50	Peas	BLW and grasses

BLW= Breed learned weed

Improved Varieties

Improved varieties of pulse crops with their characteristics are given in Table 6. Only the varieties which are high yielding and best suited to the cropping systems in the area needs to be selected for cultivation.

Table 6 : Improved recommended varieties of pulses for M.P.

S. No.	Variety/ Hybrid	Maturity duration (Days)	Special features	Yield (q/ha)
Pigeonpea				
1	Gwalior 3	240	Light maroon color seed, 100 seed wt/ g, tall, branched, suitable for inter and mixed cropping	15-20
2	Khargoan - 2	155	Light maroon color seed, 100 seed wt/ g, non synchronous growth habit, early maturing	10-12
3	JA- 3	210	Maroon color seed, 100 seed wt/ 9-10g, early maturing, suitable for inter and mixed cropping	15-17
4	JA- 4	175	Red strip on flower, 100 seed wt/ 9-10 gm, pod drop resistant, 2-3 seed/ pod	18-20
5	JKM - 7	180	Maroon color seed, 100 seed wt/ 9-10g, wilt resistant	18-20
6	No 148	165	Semi-spreading, tolerant to wilt	9-10
7	ICPL 87 (Pragati)	116	Determinate, spreading, brown seeded	11-19
8	Jagriti (ICPL 151)	110	Determinate, semi-spreading, medium bold seeded	18-20
9	Asha (ICPL 87119)	170	Indeterminate, spreading, bold seeded, wilt and SMD resistant	16-18
10	Jawahar (JKM 7)	175	Tolerant to wilt and phytophthora blight	18-20
11	Maruthi (ICPL 8863)	160	Wilt resistant indeterminate, semi-spreading	12-14
13	Laxmi (ICPL 85063)	200	Semi-spreading Suitable for Rabi planting also	18-20
Urad (black gram)				
1.	Gwalior 2	100	Black seed, 100 seed wt 4 g, YVM resistant	8-10
2.	Khargoan- 3	90	Black seed, 100 seed wt 4 g, YVM resistant	7-9

Contd.

3.	JU- 2	87	Black shinning seed 100 seed wt 4 g, YVM resistant, suitable for unirrigated condition	13-15
4.	JU- 3	70	Black seed 100 seed wt 3-7 g, YVM resistant, suitable for Madhya Pradesh	13-15
5.	Pant.U.-30	70	Yellow mosaic resistant	10-12
6.	Barkha	70	Yellow mosaic resistant	12-15
7.	T.P.U.-4	75	Yellow mosaic susceptible	10-15
8.	K.U.96-3	75	Yellow mosaic resistant	12-15
Moong				
1.	J - 45	80	Green shinning seed, 100 seed wt. 3 g	6-8
2.	Khargoan - 1	65	Ovary shaped, green, shining seed, brinjal colour stem, suitable for summer season	6-7
3.	JM - 721	80	Powdery mildew resistant, suitable for kharif and summer seasons	12-13
4.	K.-851	75	Yellow mosaic, and sercospora leaf spot resistant	10-12
Chickpea				
1.	JG B2 (62)	120	Double podded, 100 seed wt. 16 g, suitable for reinfed conditions	12-15
2.	JG-5	140	Pinkish, shining seed, 100 seed wt. 22 g, white flower	14-16
3.	JG-315	125	100 seed wt. 15 g, resistant to fusarium wilt, suitable for late sown conditions	12-15
4.	JG-74	120	100 seed wt. 15 g, resistant to fusarium wilt, suitable for late sown conditions	15-18
5.	JG-218	120	100 seed wt. 15 g, resistant to fusarium wilt, suitable for late sown conditions	15-18
6.	JG-322	115	Medium sized seed, resistant to vascular wilt	15-18
7.	JG-1	125	Round, pinkist seed, suitable for bry seed	13-15
8.	JG-130	115	Branched light coloured leaves, 100 seed wt. 24.5 g resistant to fusarium wilt	18-20
9.	JG-11	115	100 seed wt 22.5 g resistant to fusarium wilt, suitable for irrigated and unirrigated condition	16-18

Contd.

Lentil				
1.	JL-3	115	Wilt resistant, bolder seed	12-15
2.	K-75	125	Wilt & rust resistant	12-16
4.	IPL-81	120	Wilt & rust resistant	12-15
5.	JL-1	120	Blue flowers, bold size seed, drought resistant, suitable for unirrigated conditions	10-12
6.	JL-3	120	Branched, 100 seed wt. 32 g resistant to wilt and drought	10-13
Field pea				
1	Arkel	80	For vegetable pea	70 q green pod
2	JM-1	125	Green wrinkled seed, 100 seed wt 18 g	100 q green pod
3	JM-2	125	Green, wrinkled seed wt 18 g, sickle shaped seed	100 q green pod
4	JM-3	90	Green, wrinkled seed 100 seed wt 18 g, 4.5 seeds/pod sickle shaped seed	60 q green pod
5	JM-4	90	Early podding, costly, 100 seed wt.	100 q green pod
6	JM-5	115	Yellow wrinkled seed, 5-6 deeds/ pod, 100 seed wt. 18 g, powdery mildew resistant	110 q green pod

Cropwise Production Technologies

Pigeonpea (*Cajanus cajan*)

India stands first in production of pigeonpea in the world. After chickpea, pigeonpea is the second most important pulse crop in India. It is a rich source of protein and supplies a major share of the protein requirement of the vegetarian population. The husk of pods and leaves obtained during threshing constitute a valuable cattle feed. It is a legume crop and, consequently, possesses valuable properties as restorer of nitrogen to the soil. In M.P., pigeonpea is cultivated in 0.36 lakh ha in Madhya Pradesh with a production of 0.21 lakh tonnes.

Climatic requirement: Warm and humid climate is beneficial for better growth of pigeonpea. Dry and sunny days are essential for flowering and pod formation. This condition prevails in some districts of Madhya Pradesh and, therefore, it is a potential crop for the state under upland heavy soil conditions.

Land preparation: A weed-free and firm seedbed ensures better plant stand and early vigour to the crop. This is achieved by opening the soil through soil-turning plough or disc harrow, followed by the cross harrowing or ploughing with desi plough on or before the onset of monsoon.

Seed rate: The seed rate of good quality of pigeonpea is 15 to 20 kg per/ ha. The row to row and plant to plant spacing for early pigeonpea and pre *rabi* crop are 45 cm and 15 cm, respectively where as for late pigeonpea, these are 60 x 20 cm, respectively.

Sowing management: The optimum time of sowing of pigeonpea is between 25th June to 15th July. In rainfed areas, sowing is done with the onset of the monsoon in *kharif*. However, sowing from the first week of June to the second week of July are considered best. Generally, delay in sowings beyond 15 July not only reduces the plant height and the total dry matter but also decreases the number of pods per plant.

Fertilizer application including micro-nutrients: Fertilizer should be applied based on soil testing and the general practices of soil test based fertilizer application are commonly followed. The recommended fertilizer dose is 20 kg N and P_2O_5 60 kg P/ha. Conveniently, 100 kg / ha diammonium phosphate is applied. Use of PSB culture improves the available phosphorus status in the soil. In addition, application of 20 kg sulphur/ha, 15 kg $ZnSO_4$/ha, 10 kg borax/ ha,

Weed Control and inter-culture operation: During rainy season, weeds pose a serious problem. As the early growth of pigeonpea is very slow, first 60 days are very critical in the life cycle of the plant. Two mechanical/hand weedings, one at 25 days and another at 45 days after sowing, give excellent weed control. The pre-emergence application of alachlor @ 3 liters per hectare kills the germinating seedlings of weeds and keeps the field free from the weeds for the first 50 days. Pendimethalin or metolachlor @ 1-1.5 kg ha^{-1}dissolved in 500-700 liters water may be sprayed before sowing or within 72 hrs of sowing and/or germination. It should be well incorporated in the soil before sowing.

Water management: Being a deep-rooted crop, it can tolerate drought. In crop planted in June, one or two pre-monsoon irrigations should be given as per requirement. After the start of monsoon, there is no need of irrigation but in case of prolonged drought during the reproductive period of growth, one or two irrigations may be needed. A pre-requisite for the success of pigeonpea is proper drainage. Ridge planting is effective in areas where sub-surface drainage is poor. This provides enough aeration for the roots during the period of excess rainfall. During rainy season, water should not stand anywhere in the field.

Plant Protection Measures

Insect pests

Pod-borer (*Heliothis armigera*): It causes considerable damage to pigeonpea, especially the early-maturing varieties. Spraying insecticides like Malathion (0.05%) provides good protection.

Pod-fly (*Melanagromyza ebtusa*): It is a very serious pest of pigeonpea all over. It causes extensive damage to the crop as consumption.

Hairy-caterpillar: The hairy-caterpillars are voracious feeders and defoliate the plants causing heavy losses to the crop. Dusting of chlorpyrifos 20 EC 40% @ 15 -can control the pest incidence.

Diseases

Vascular-wilt: Wilt caused by the fungus *fusarium udum* Bulter var. *cajani* Padwick is characterized by gradual wilting of plants. Addition of manganese (100-200) to the infested soil prevents infection. Sanitary measures like rouging and burning of plants reduce disease incidence. As the fungus is soil-borne, rotation of crop for 4 to 5 years effectively reduces pathogen population and the extent of damage.

Stem-blight: Blight caused by the fungus, *Phytophthora drechsleri* var. *cajani,* produces irregular to circular grey spots surrounded by brown lesions on the stem. Poor drainage and stagnation of rainwater in the fields during the monsoon need to be avoided. Application of potash fertilizer K_2SO_4 @ 25-50 kg/ha reduces the disease incidence.

Powdery-mildew: Sulphur dusting @ 30 kg/ha checks the spread of diseases.

Sterility-mosaic: Mites can be controlled by spraying with 0.1% Metasystox. Removal of old and volunteer plants before sowing of crop would help in restricting the population of mites.

Intercropping system: Planting of castor and tall sorghum or maize varieties on borders conserves natural enemies, *viz.,* predators of birds, etc. To facilitate spraying, short-statured crops, e.g. cowpea, moong, urad, fodder soybean, etc. may be grown in 1 m wide strip land after every 8-10 rows of pigeonpea. Soybean, urad, moong, cowpea and groundnut are very common intercrops. Being deep rooted, the crop is very well suited for mixed cropping and intercropping with the shallow rooted ones. Intercropping with urad (Type-9), moong (Pusa baisakhi) and cowpea (Pusa do phasali) gives about 4-7 quintals per hectare additional yield without affecting yield of pigeonpea.

Chickpea (*Cicer arietinum*)

Chickpea in the most important pulses crop of Madhya Pradesh and occupies 30.86 lakh hectares with a production of 33.04 lakh tonnes. The average productivity of chickpea is 1072 kg/ha. It is mainly grown in all the district of Madhya Pradesh but especially in Indore, Ratlam, Ujjain, Vidisha, Sajapur, Gunah, Bhopal, Raisen, Narsingpur, Sagar and Jabalpur districts of MP

Land preparation: Gram is highly sensitive to soil aeration. A rough seedbed is required for gram. In case, it is taken after a *kharif* fallow, it would be desirable to go for cross cultivation and cross discing with one leveller for the field preparation as the same would help in larger conservation of rainwater in the soil profile. Very fine and compact seedbed is not good for chickpea. It requires a loose and well-aerated seedbed.

Seed rate: The crop is sown in rows, about 30 cm apart. Seeds are placed 7-10 cm deep in soil; the rate depends on seed size from 60 to 80 kg/ha.

Sowing management: Rice crop dominates in Madhya Pradesh state during *kharif* seasons and immediately after harvesting of rice crop, chickpea is sown. The second fortnight of October to second fortnight of November is the best time for gram sowing. Any delay beyond this period results in conspicuous reduction in yield. The crop may be sown by seed-drill or local plough at a row spacing of 30 cm. The seed should be placed 5-7 centimetre deep because the shallow sown crop is more liable to be damaged by wilt. Seed should be treated with 0.25 per cent Thiram before sowing followed by *Rhizobium* culture.

Fertilizer application including micro-nutrients: On soils with low to medium phosphorus status, crop responds to phosphatic fertilizer resulting in higher yields. About 20 kg N, 60 kg P_2O_5 and 20 kg K_2O /ha are the general doses of fertilizers. Normally, 100 kg/ha diammonium phosphate is conveniently placed below the seed. Phosphate solubilizing bacteria (PSB) have been found effective in increasing the efficiency of applied phosphorus.

Weed control and inter culture: One hand-weeding 30-40 days after sowing or application of the herbicide pendimethalin @ 1 kg a.i./ha immediately after sowing successfully controls of the weeds.

Water management: Two irrigations are generally required for completing the life cycle i.e. first at 45 days after sowing (before flowering)/ and second 60 days after sowing (early podding stage). If the winter rains occur, one irrigation is sufficient.

Plant protection measures: Pod-borer is the major insect. It requires monitoring at weekly intervals at flowering and podding stages. Sex-pheromone traps (5 traps/ha) should be put in January-March. Economic threshold limit may be 1 larva/5 to 10 plants.

Mechanical practices: Collect and destroy the grown-up larvae.

Biological control: Spray nuclear polyhedrosis virus (NPV) @ 250 larval equivalents (LE)/ha on noticing the first-or second-instar larvae or eggs of *Helicoverpa armigera.* Apply 3 sprays at weekly intervals.

Chemical control: Selective pesticides, *viz.,* Endosulfan (0.07%), Monocrotophos (0.04%), and extract of neem-seed kernel (5%) or other neem-based formulations, *viz.,* Nimbecidine, Repelin, Achook, Neemguard, etc., should be used as soon as economic threshold limit is crossed.

Wilt and root rot: Seed should be treated with fungicides, *viz.,* Carbendazim+Thiram (1 g+2 g/kg) at sowing. Seed-dressing with bioagent *Gliocladium* virens @ 0.5 g/kg or Trichoderma viride @ 4 g/kg seed+fungicide Carboxin (Vitavex) @ 1 g/kg seed should be followed.

Ascochyta blight: Seed treatment with Hexacap (3 g/kg), Carbendazim (2 g/kg seed) or Calixin-M (Tridemorph 2 g/kg seed) is effective. Foliar sprays fo Calixin-M (Thridemorph 0.1%) or Hexacap (0.02%) or Carbendazim (0.1%) may be adopted.

Intercropping system: Deep summer ploughing, clean cultivation, timely sowing (up to mid of November), and intercropping with linseed, mustard, wheat, coriander etc. minimizes damage by pod-borers and nematodes and provide good yield of main and intercrops.

Lentil (*Lens culinaris*)

Lentil has great promise as a pulse crop in the state as it gives higher price in market. It may serve as a good alternative to Lathyrus under *utera* cultivation. At present, lentil occupies 541 thousand ha with production of 284 thousand tonnes and productivity of 525 kg/ha with positive growth rates. Generally, it is a temperate season crop, but it can be successfully grown in all the districts where is chickpea is successfully grown. The major production constraints are infestation of local varieties with wilt, poor crop stand and terminal drought.

Soil requirement and land preparation: Medium to heavy type of soils are suitable for lentil cultivation. The soil should be made friable and weed-free so that seeding could be done at a uniform depth. On heavy soils, one deep ploughing followed by two to three cross harrowings should be given.

Seed rate: The optimum seed rate for normal sown crop is 30-40 kg per hectare. Seed rate should be increased to 50-60 kg per hectare in case of late sowing.

Sowing management: Timely planting is the key factor in the full realization of the yield potential of improved varieties of lentil. Middle of October is the most suitable time for sowing lentil. Delay in planting causes reduction in yield but the magnitude of reduction is large after 15th November. The reduction in yield could be minimized up to a certain extent by relatively closer spacing and use of higher seed rate. Sowing should be done in rows 30 centimeter apart. This could be done either by using a ferti-seed drill or by seeding behind desi (country) plough.

Fertilizer application including micro-nutrients: Generally lentil is grown without fertilizers and manures. Soils deficient in nitrogen and phosphorus, give significant responses to application of starter nitrogen (20kg /ha) and moderate level of phosphorus (60 kg /ha).

Weed control & inter culture: Two manual weedings, one at 25-30 days and another 45-50 days after sowing, should be done. Weedicide like Basalin and Tribunil can be used safely for weed control. Basalin 0.75 kg a.i. per hectare in 800-1000 litres of water as pre-planting spray may be used as an effective herbicide.

Water management: First irrigation should be given at 45 days of planting and second, if needed, at pod filling stage. More irrigations may affect the crop performance adversely.

Plant protection measures: The major disease of lentil are rust in Madhya Pradesh plains and wilt in Northern hills, Use of resistant varieties is helpful in controlling the disease. Pests are not a major problem in lentil. In some years aphid population rises due to rise in temperature during January or February, and causes heavy damage. Spray of Monocrotophos (0.04%) is effective for aphid control.

Intercropping system: In intercropping it is grown with barley and mustard. It is also grown as an intercrop in autumn-planted sugarcane. In north-eastern plains, it is also grown as utera crop after rice. The seeds are broadcast in standing crop of rice just before harvest.

Moong (*Vigna radiata*)

In Madhya Pradesh, moong occupies 0.84 lakh ha and contributes to 0.28 lakh tonnes in pulse production with productivity of 325 kg/ha. The grains (whole or split) are used as *dal* or make into flour. The straw and husk are used as fodder for cattle. The germinated grains are also used as sprouts. Unlike other pulses, it does not produce heaviness or flatulence.

Land preparation: Field is prepared by giving two or three cross harrowinges followed by planking. The field should be well leveled and free from weeds. For summer season crop, give a pre-irrigation immediately after harvesting of *rabi* crop. When the field comes in condition, prepare it by giving two to three ploughings by local plough or harrowings. Each ploughing/

harrowing should be followed by planking to make the field leveled and to minimize the loss of moisture by evaporation form the soil surface.

Seed rate: During *kharif* season, 12-15 kg seed per hectare should be sown in rows 45 centimeter apart while during *rabi* and summer seasons, 20 kg seed per hectare should be sown in rows 30 centimeter apart. Before sowing the seed should be treated with Thiram or Captan at the rate of 2.5 g per kg of seed. It is also desirable to treat the seed with suitable Rhizobium culture as well, if the crop is being taken in the field for the first time or after a long period.

Sowing management: The sowing time is the first week of March for the spring moong and last week of March to first fortnight of April for the summer crop. In *kharif*, seed is sown with the onset of monsoon, preferably first week of June to mid-July. In the event of the late onset of monsoon, it can be sown by July end. In *rabi*, seed can be sown from September to December or January, depending on the time of rice harvest and availability of irrigation. During in *kharif* season, sowing is done with the onset of monsoon in the second fortnight of July. In *rabi* it should be sown in October-November. In summer the moong is grown in summer season too under irrigated conditions. It is sown after the harvest of sugarcane, wheat, potato etc. Sowing is done in the middle of March to middle of April.

Fertilizer application including micro-nutrients: Apply 20 kg nitrogen and 50 kg P_2O_5 per hectare at the time of sowing. If available, drill 100 kg of diammonium phosphate to meet the requirement of both nitrogen and phosphorus. The fertilizer should be applied by drilling either at the time of sowing or just before sowing in such a way that they are placed about 5-7 centimeter below the seed.

Weed control & inter culture: The control of weeds during 30-40 days after sowing is very beneficial. The young crop is given 1 hand weeding or hoeing. Use of weedicide pendimethalin or alachlor @ 1.0-1.5 kg a.i./ha immediately after sowing, is also useful.

Water management: For rainy season crop, irrigation is not needed but drainage is very important. There should be adequate drainage in the field because this crop is sensitive to water logging conditions. For *rabi* and summer crops, five to six irrigations may be given. The field should be irrigated when soil is seen deficient in moisture. During summer, because of high temperature and low relative humidity more irrigations are needed as compared to *kharif* crop. The first irrigation should be given about 20-25 days after sowing. The subsequent irrigations should be given at an interval of 12-15 days. No irrigation should be given when the crop is in full bloom stage.

Intercropping Systems: In *kharif*, intercropping with maize, sesame, pigeonpea and cotton is common.

Urad (*Vigna moongo*)

Urad differs from other pulses in its peculiarity of attaining a some what mucilaginous pasty character, when ground up with water giving additional body to the mass. It is also fried to serve as a savoury dish. Urad dal is also used in the preparation of '*halwa*' and '*imarti*'. It is used as a nutritive fodder especially for milch cattle. It is also used as a green manuring crop. Urad plant possesses deep root system which binds soil particles and thus, prevents soil erosion.

Land preparation: Field is prepared by one ploughing with mouldboard plough and followed by two to three harrowing followed by planking. The field should be well levelled and free from weeds. For summer crop, after harvest of *rabi* crops, give two to three harrowing followed by planking. To ensure good germination, a pre-sowing irrigation should be given.

Seed rate: During *kharif* season 12-15 kg seed is sufficient for sowing one hectare. The crop should be sown in furrows opened at a distance of 45 centimeter. About 20-25 kg seed is required for sowing of one-hectare area. Sowing should be done in furrows opened at a distance of 20-25 centimeter. Seed drill could be used for this purpose. Before sowing, seed should be treated with Thiram at the rate of 2.5 g per kg of seed. Seed should also be inoculated with suitable Rhizobium cultute, if urad is being taken for the first time in the field or after a long duration.

Sowing management: In *kharif* season, sowing is done with the onset of monsoon in later part of June or early part of July. In summer the sowing could be done form the third week of February to first week of April. In sowings done thereafter, yields are low due to rains at the time of maturity.

Urad is planted under flat bed system where the plant population is reduced very often due to accumulation of water in the field. By adopting raised bed system, water stagnation can be avoided in field. Plant population under such conditions can be maintained by adopting raised bed system and manful adopting seed rate.

Fertilizer application including micro-nutrients: Being a leguminous crop, urad needs a small quantity of nitrogen for early growth period on those soils which are poor in organic matter. Such soils should get about 15-20 kg nitrogen per hectare as a starter dose. However, phosphatic and potassic fertilizers should be applied as per soil test values. In case soil test facilities are not available, one can apply 50-60 kg P_2O_5/ha and 30-40 kg K_2O /ha. The fertilizers should be applied by drilling at the time of sowing in such a way that they are placed about 5-7 centimeter below the seed.

Weed control & inter culture: One or two hand weeding should be done up to 40 days of sowing depending upon the weed intensity. Weeds can be controlled by the use of chemicals too. Use fluchloralin @ 1 kg a.i. per hectare in 800-1000 liters of water as pre-planting treatment.

Water management: For rainy season crop, irrigation is not needed but good drainage is essential. Irrigation facilities should be available for raising urad during summer season. The crop should get irrigation at an interval of 10-15 days. From flowering to pod development stage, there is need of sufficient moisture in the field.

Plant protection measures: The major diseases of greengram and urad are yellow mosaic virus, Cercospora leaf spot, powdery mildew and Macrophomina blight. The integrated pest management strategy consists of cultural, mechanical, biological and chemical control measures.

Cultural practices: Deep ploughing, early sowing and timely irrigations are good for crop health. In disease prone areas, resistant varieties should be used.

Mechanical practices: Hairy caterpillars have a definite egg-laying pattern in masses. The first stage of larvae remains restricted to leaves of plants where eggs were laid. Collect and destroy them.

Biological control: Natural enemies and birds reduce the pest population.

Chemical control: For control of thrips, stem fly, jassids and whitefly apply Phorate or Carbofuran granules in soil before sowing @1 kg a.i./ha. Seed treatment with phorate or Disulfotone @ 2% on weight basis is also effective. Spraying with Monocrotophos (0.05%) or Quinalphos (0.05%) controls the leaf defoliators and hairy caterpillars. Dusting with 5% Malathion @ 35-40 kg/ha is also effective against hairy caterpillars.

Intercropping Systems: In *kharif*, intercropping with maize, sesame, pigeonpea and cotton is common.

Field pea and table pea (*Pisum sativum*)

It is a high-yielding pulse crop. Unripe pods are used as green vegetable and dry seeds are used for *dal* purpose and *chat* after boiling. Other parts are used for feeding cattle. It is cultivated on 2.45 lakh ha area with a production of 1.15 lakh tonnes. The productivity of field pea is 469 kg/ha in Madhya Pradesh. Field pea is getting popularity in the state as it offers high remuneration it fruits early in the season and can be sold at a premium price in local market.

Land preparation: After harvest of *kharif* crops, the land should be ploughed deep with soil-turning plough and then prepared with 2-3 harrowing followed by levelling. In rice fields, efforts must be made to break clods and attain friable structure. Optimum moisture must be ensured for proper germination.

Seed rate: Second fortnight of October is the optimum time of sowing of field peas. Sowing after October results in drastic yield reduction. Seeds are sown at the rate of 60-80 kg per hectare in rows 30 cm apart. Before sowing, seed should be treated with Thiram at the rate of 0.25 per cent. In case of early maturing dwarf varieties, crop should be sown in rows 20 cm apart and @ 100-125 kg seed per hectare.

Fertilizer application including micro-nutrients: The recommended dose of fertilizers is 20 kg N and 60 kg P_2O_5/ha. Use of PSB culture improves the status of available phosphorus in the soil. In addition, application of 20 kg sulphur/ha, improves the yield considerably.

Weed control & inter culture: One hand-weeding 25-30 days after sowing or application of Pendimethalin @1.0-1.5 kg a.i./ha soon after sowing helps in controlling the weeds.

Water management: In the absence of winter rains, irrigate the crop at 40-45 days after sowing and at pod formation stages.

Plant protection measures: Field-pea is attacked by a variety of insect pests, like leaf-minor, semi-looper, aphids, trips, pod-borer and leaf-eating caterpillar; and diseases, like powdery-mildew, wilt, foot-rot, downy-mildew, etc.

Insect pests

- **Leaf-miner [*Phytomyza horticola* Gour. (-atricornis)],**
- **Semi-looper (*Plusia orichalcea* Fb.).** Spraying with 10% Chlorpyrifos as soon as the symptoms appear controls the leaf-miner. In case of severe infestation, 0.25% Methy1 Parathion spray is recommended.
- **Aphids (*Aphis craccivora* Koch., A. gossypii GI.).** Spraying the crop with Nicotine Sulphate (40%) at 1:600 dilutions in water by volume or Malathion 0.03% reduces the incidence. About 650-850 liters of spray solution and 25-40 kg of dust are required per hectare.
- **Thrips [*Caliothrips indicus* (Bagnall)].** For the control of thrips, the crop may be sprayed with 1,250 g DDT (50 WP) in 650 liters of water per hectare and the spraying may be repeated after 2 weeks, if necessary.
- **Pod-borer (Etiella zinckenella Treit).** Dusting BHC 10% or DDT 10% @ 20-25 kg/ha or spraying with a mixture of 0.05% Diazinon and 0.25% DDT twice during pod-formation stage, minimizes the incidence of pod-borer.
- **Leaf-eating caterpillar [*Spodoptera exigua* (Hubner), *Diachrysia orichalcea* (fabricius)].** Dusting of BHC @ 10% or DDT 10% @ 20-25 kg/ha is very effective in controlling the pest.

Diseases

- **Powdery-mildew (*Erysiphe polygoni* DC.):** Seed treatment with 0.5% Benomy1 gives protection to the seedlings up to 60-65 days after emergence. Four sprays of Sulfex (Wet table sulphur) @ 3 kg/ha in 1,000 liters of water at intervals of 7 days starting immediately at the appearance of disease, controls powdery mildew effectively.
- **Wilt and foot rot:** Treatment of seeds with organo-mercurial compounds, like Ceresan, Captan or Bavistin @ 2.5 to 3.0 g/ha of seed, is recommended.
- **Downy mildew:** The disease is favoured by a temperature of 15^0C and heavy dew, and can be controlled by spraying with copper fungicide as a preventive measure.

Intercropping System: Field pea is more adapted than chickpea crop for late sown conditions as well as high input conditions. With these stipulations, it could replace chickpea in crop sequences and intercrops especially as pea-rice, pea-cotton and pea-summer Moong.

Strategies and Future Thrust

- Madhya Pradesh has diversified climate with the pulses production can be doubled excellent policy support from the State Government.
- Growing of rainfed variety of rice and upland maize, sorghum and soybean, the sowing of pulses and intercropping of pulses with upland crops are more remunerative. Therefore, to be encouraged.
- Training the farmers for using improved production technology including use of recycled byproducts of farms which reduce the dependency on costly inputs.
- Pulses are sensitive to heat and cold. Developments of varieties which are tolerant to abiotic factors are needed on priority basis.
- Selection/development of short duration varieties suitable for double cropping systems needs special attention.
- Pulses are rich in protein so easily targeted by pests and, provide congenial conditions for weeds as they are slow growing. Therefore, timely management is very important.

Conclusions

In short, to increase area and production of pulse crops we need crop specific and region specific approaches, which should be adopted in the overall framework of systems approach. The major thrust areas to be addressed are as follows (i) Replacement of cereal crops in the prevailing rice-wheat and rice-chickpea cropping systems with high yielding varieties of pulses; (ii) Inclusion of short duration varieties of pulses as catch crop; and (iii) Development of multiple disease and pest resistant varieties.

References

Annual Report of AICRP on IFSJNKVV, Jabalpur.

Ali, M. and Lai, S. (1989). Priority inputs in pulse production. Fertilizer News, 34 (11):17-21.

Ahlawat I.P.S., Singh, A. and Mohta, N.K. (1982). Studies on weed control in solid and mixed stands of pigeonpea and soybean: Indian Journal of Agronomy 27:191-194.

Kumar, N. and Singh, K.K. (2010). Weed management in pulses: why, when and how. Indian Farming July 2010, pp 9-12.

Madhya Pradesh Agricultue Statistics (2009-10). Government of Madhya Pradesh.

Panda, S.C. (2005). Agronomy: Agrobios (India) Jodhpur.

Panda, S.C. (2004). Cropping and Farming Systems: Agrobios (India), Jodhpur.

Singh, S.S., Gupta, P and Gupta, A.K. (1996). Hand book of Agriculture Sciences, published by Kalyani Publisher, Ludhiana.

Tiwari, A.N. and Rathi, K.S. (1997). Integrated weed management for sustainable production of pigeonpea based cropping system In Proc. Brighton Crop Protection Conference, Brighton, England, Weeds 1997-3c:pp.167-172.

❑❑❑

15

Maharashtra

A.G. Wani, W.N. Narkhede, B.V. Saoji and L.G. Panwar
Agricultural Universities Maharashtra, AICRP on IFS, MPKV Rahuri, MAU, Parbhani, PDKV, Akola and BSKKV, Dapoli, Maharashtra

Maharashtra state is located in the north centre of Peninsular India. The topography of the state is the outcome of its geological structure. The state area, barring the extreme eastern Vidarbha region, parts of Kolhapur and Sindhudurg, is practically co-terminous with the Deccan Trap. The state enjoys a tropical monsoon climate; the hot scorching summer from March onwards yields to the rainy monsoon in early June. The rich green cover of the monsoon season persists during the mild winter that follows through an unpleasant October transition, but turns into a dusty, barren brown as the summer sets in again. The seasonal rains from the western sea-clouds are very heavy and the rainfall is over 400 cm on the Sahyadrian crests. The Konkan on the windward side is also endowed with heavy rainfall, declining northwards. East of the Sahyadri, the rainfall diminishes to a meagre 70 cm in the western plateau districts, with Solapur-Ahmednagar lying in the heart of the dry zone. The rains increase slightly, later in the season, eastwards in the Marathwada and Vidarbha regions.

The highly pulsatory character of the monsoon, with its short spells of rainy weather and long dry breaks, floods, as well as droughts add much to the discomfort of the rural economy. Based on the rainfall distribution, the state has been divided into nine zones. The predominant cropping systems under rainfed and irrigated situations are given in Table 1.

The soils of Maharashtra are residual, derived from the underlying basalts. In the semi-dry plateau, the *regur* (black-cotton soil) is clayey, rich in iron, but poor in nitrogen and organic matter; it is moisture-retentive. Kali soils redeposited along the river valleys are deeper and heavier, better suited for *rabi* crops. Farther away, with a better mixture of lime, the morand soils form the ideal *kharif* zone. The higher plateau areas have *pather* soils, which contain more gravel. In the rainy Konkan and the Sahyadri Range, the same basalts give rise to the brick-red laterites productive under a forest-cover, but readily stripped into a sterile *varkas* when the vegetation is removed. By and large, soils of Maharashtra are shallow and somewhat poor.

Table 1 : Salient features of agro-climatic zones of Maharashtra State

S. No.	Zone	Districts	Net sown area (000'ha)	Rain fall (mm)	Crops grown	Major soil type
Zone 1	South Konkan coastal zone	Ratnagiri, Sindhudurg	3856	> 2500	Paddy and millets	Laterites and Lateritic (Alfisols)
Zone 2	North Konkan coastal zone	Thane, Raigarh	4331	> 2000	Paddy, Hill millets (*Ragi, wari*), Sapota, Mango, Forest	Non lateritic loams (Inceptisols, Alfisols)
Zone 3	Western Ghat zone	Sangli, Satara	11695	> 3500	Paddy, Hill millets	Mostly forest land (Entisols)
Zone 4	Sub montane zone	Pune, Kolhapur	14018	1750 to 2500	*Kharif*-Paddy, groundnut sorghum *Rabi*-Sorghum, wheat, gram, fruits-grapes	Mainly red to brown soils (Entisols)
Zone 5	Western Maharashtra plain zone	Nasik, Dhule	16213	700 to 1750	*Kharif*- groundnut, Jowar, Bajra, Pulses *Rabi*- Jowar, safflower & Sugarcane	Predominantly grayish black soils of varying textures (Entisols Inceptisols)
Zone 6	Western Maharashtra Scarcity zone	Ahmednagar, Beed, Solapur	29615	700 towards west and east	*Kharif*- Sorghum, groundnut pulses, sunflower *Rabi* – Jowar Safflower, gram	Medium black calcareous soils (Inceptisols/ Vertisols/ Entisols)
Zone 7	Central Maharashtra plateau zone	Jalgaon, Buldhana, Aurangabad, Jalna, Akola, Amravati, Osmanabad, Latur, Parbhani	62872	700 to 900	*Kharif* – Sorghum, Cotton Groundnut, Safflower, Pulses	Medium to deep black calcareous soils (Inceptisols/ Vertisols)
Zone 8	Central Vidarbha zone	Nanded Yotmal, Wardha, Nagpur	24746	900 to 1250	*Kharif*-Sorghum Cotton, pulses oilseeds, *Rabi* – Wheat, gram	Brown to dark calcareous soils (Inceptisols/ Vertisols/ Entisols)
Zone 9	Eastern Vidarbha zone	Chandrapur, Bhandara, Gadchiroli	9970	> 1250	*Kharif* – Paddy *Rabi*- Sogrhum, Linseed, gram	Yellowish brown to red soils (Alfisols/ Inceptisols)

The Maharashtra state has four agricultural universities *viz.*, Mahatma Phule Krishi Vidyapeeth, Rahuri; Dr. Punjabrao Deshmukh Krishi Vidyapeeth, Akola; Marathwada Agricultural University, Parbhani; and Dr. Balasaheb Sawant Konkan Krishi Vidyapeeth, Dapoli. These universities represent nine agro-climatic zones.

Present Scenario of Pulses

Agriculture continues to be the most predominant sector of the state economy, as 71% of the population is engaged in agriculture and allied activities. The predominant crops of Maharashtra are paddy, maize, sorghum, pearlmillet, soybean, groundnut, sunflower, sugarcane, cotton, vegetables, fruits crops and pulses. Among the pulses crops, blackgram, greengram, horse gram and pigeonpea, french bean and chickpea, are grown on most of the areas in Maharashtra. Cowpea, lablab, lentil, lythyrus are the other pulses grown to a small extent. There is an increasing trend in area under pulses and enhansed pulses production and productivity trend has been noticed over the years (2000-2008) in Maharashtra (Fig. 1)

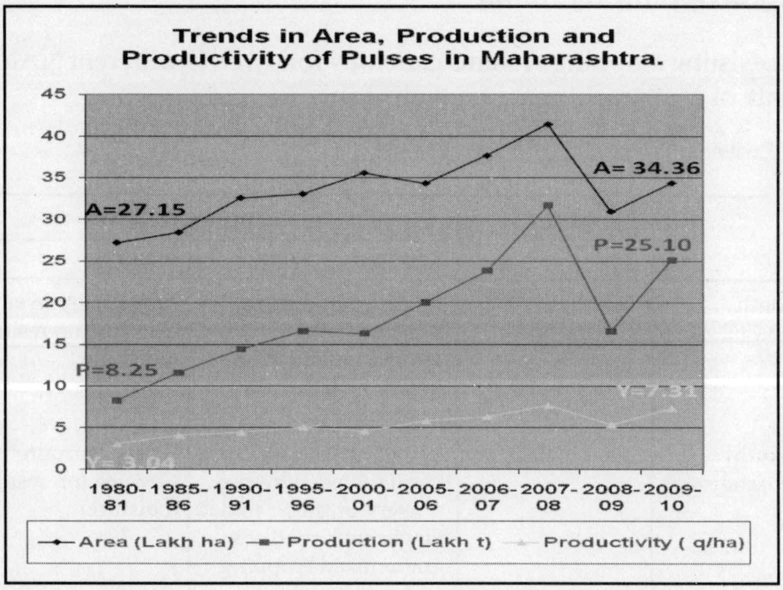

Fig. 1 : Trends in Area, Production and Productivity of Pulses in Maharashtra

In Maharashtra, pulses are grown on 30.92 lakh ha with production and productivity of 16.61 lakh tonnes and 537 kg ha^{-1}, respectively. Out of which, *kharif* pulses occupy 60% area, while *rabi* pulses 40% with production and productivity of *kharif* and *rabi* pulses being 50.50%, 49.51% and 85.54%, 122.90% respectively (Table-2). Though area under *kharif* pulses is more than *rabi*, the productivity of *rabi* pulses is higher (122.90%). It may be due to availability of protective irrigation for *rabi* pulses cultivation.

Table 2 : Area, production and productivity of pulses in Maharashtra (2008-09)

Pulses	Area (lakh ha)	Production (lakh tonnes)	Productivity (kg ha^{-1})
Major *Kharif* Pulses			
Pigeonpea	10.09	6.05	600
Mungbean	4.27	1.06	248
Urdbean 3.23	0.99	308	
Other kharif pulses	0.88	0.28	317
Total 18.46 (59.73)	8.38 (50.49)	454 (84.54)	
Major *Rabi* Pulses			
Chickpea 11.43	7.73	677	
Other *Rabi*	1.02	0.48	475
Total 12.45 (40.27)	8.22 (49.51)	660 (122.90)	
Total *Kharif* & Rabi	30.92	16.61	537

Existing Cropping Systems

The existing cropping systems generally followed in different NARD zones in the state of Maharashtra are given in Table 3.

Table 3 : Existing pulse based cropping systems

Zones	Districts	Cropping systems	
		Rainfed	**Irrigated**
Zone 1: South Konkan coastal zone	Ratnagiri, Sindhudurg	Horse gram during late *kharif* -rice nursery /finger millet Finger -millet + black gram mixed cropping	Rice-L. *purpureus* Rice-Cowpea (on residual moisture)
Zone 2: North Konkan coastal zone	Thane, Raigarh	Red gram during late *kharif* followed by rice nursery or finger millet. Finger millet + Black gram mixed cropping	Rice-L. purpureus Rice-Cowpea (on residual moisture)
Zone 3: Western Ghat zone	Sangli, Satara	Pearl millet-Chickpea Black gram-Rabi Sorghum Pearl millet + Pigeonpea (2:1) Sunflower + Pigeon pea (1:2)	Pigeonpea-Wheat
Zone 4: Sub montane zone	Pune, Kolhapur	Pearl millet - Chickpea Pigeonpea - Pigeon peaSoybean + Pigeonpea (4:1)	Pigeonpea-Wheat Black gram-Rabi Sorghum

Contd.

Zone 5: Western Maharashtra plain zone	Nasik, Dhule, Nandurbar	Black gram - Rabi Sorghum Greengram - Rabi SorghumPearl millet – Chickpea Cotton + Black gram (1:2)Cotton + Greengram (1:2)	Pearl millet-Chickpea Pigeonpea-Summer vegetables
Zone 6: Western Maharashtra Scarcity zone	Ahmednagar, Beed, Solapur	Pearl millet – Chickpea Black gram - Rabi SorghumPearl millet + Pigeonpea (2:1) Soybean + Pigeonpea (4:1)	Soybean-Chickpea Green gram-Rabi SorghumBlack gram-Rabi sorghum
Zone 7: Central Maharashtra plateau zone	Jalgaon, Buldhana, Aurangabad, Jalna, Akola, Amravati, Osmanabad, Latur, Parbhani	**Cropping sequence** Greengram-Rabi sorghum Greengram -Sunflower Black gram -Chickpea Black gram -Rabi sorghum Maize-chickpea Pearl millet-chickpeaSole pigenonpea (90 x 20 cm) Soybean + Pigeon pea(4:2 ratio at 45 cm.) Pigeonpea + Greengram (2:4) Pigeonpea + Cotton (2:8) Pigeonpea + Maize (2:4) Pigeon pea + Groundnut (2:4) Black gram + Pigeonpea (4:2) Black gram + Cotton (1:1) Greengram + Cotton (1:1)Chickpea + sunflower (4:2) Chickpea + Sunflower (4:2) Chickpea + Linseed (4:2)	Greengram-Wheat Green gram-Chickpea Black gram-Sunflower Sunflower-Chickpea
Zone 8: Central Vidarbha zone	Nanded Yeotmal, Wardha, Nagpur	Pigeonpea + Sorghum (3:3) Pigeonpea + Greengram/Black gram/Soybean (1:5) Pigeonpea + Groundnut (2:4) Greengram/ Black gram- Chickpea / sunflower	Pigeonpea- Chickpea + mustardGreengram/Black gram/ Soybean-Wheat-Groundnut
Zone 9: Eastern Vidarbha zone	Chandrapur, Bhandara, Gadchiroli	Chickpea + PeaBlack gram-Chickpea	Chickpea + mustard Soybean-Chickpea/ Wheat-Sesamum

Yield Gaps

Yield gap in pulses continues to be one of the major constrains in pulse production in Maharashtra. The Department of Agriculture and Main Research Centres of Agricultural Universities studied the yield gap in various pulse crops. A wide gap exists both at progressive and average farmers' level compared to the experimental station yield (Table 4).

Table 4 : Yield gap pulses in FLD and farmers practice in Maharashtra

Crop	Highest productivity (q/ha)	Productionin FLD (q/ha)	Average production (q/ha)	Difference (q/ha)
Greengram	12.0	6.24	2.26	3.98
Blackgram	12.0	4.50	2.47	2.03
Pigeonpea	25.0	11.17	5.89	5.28
Chickpea	27.0	15.27	9.27	6.00

The Constraints

The reasons for low productivity of pulses in Maharashtra are as follows.

- Major area under rainfed cultivation.
- Lack of adoption of production technology.
- Nil or very little fertilizer application to these crops.
- No attention to weed management.
- Lack of plant protection measures to control pest and diseases.
- Mostly farmers cultivate these crops for their consumption and surplus produce sold in the market.
- Most of the farmers using local varieties of pulses.
- Seed treatment is not followed on a large scale.
- Non availability of high yielding varieties/hybrids and quality seed material.
- Number of storage grain pests cause considerable losses.
- Poor storability and lack of storage facility.

Efficient Zones for Pulses

Among the nine agro-climatic zones in Maharashtra, Central Maharashtra zone was categorized as the most efficient zone with little year to year variation followed by Central and Western Vidharbha zone. Western Maharashtra is grouped as the next efficient zone with little variation in area, production and productivity and coastal Maharashtra (South and North Konkan) is the last. Among the various pulse crops, pigeonpea and chickpea are the major pulses of Maharashtra.

Improved Varieties of Pulses

The improved varieties of different pulses crops for Maharashtra state are given in Table 5

Table 5 : Crop varieties of pulses released for cultivation in Maharashtra

Crops/Varieties	Characteristics
Pigeonpea	
BDN-708 (Amol)	Seed coat red, suitable for light type of soil. Duration of 155-165 days, productivity of 14-17 q/ha, resistant to wilt and sterility mosaic viruses.
BSMR-736	Seed coat red, flower colour yellow. It is resistant to wilt and sterility mosaic disease, duration 175-180 days. Productivity 14-16 q/ha under rainfed, 18-23 q/ha under irrigated, protein content 17-18 %.
BSMR-853 (Vaishali)	Seed coat white in colour, flower colour external red and inner colour yellow. Duration 175-180 days, productivity under rainfed 17-18 q/ha, under irrigated 18-20 q/ha, protein 18 %.
ICPL-87	Extra early maturity, determinate growth habit, inflorescence bears pods in clusters, red coloured medium seeds, suitable for multiple cropping, released for Maharashtra. Yield varies from 10-12 q/ha.
Vipula	Resistant to *Fusarium* wilt, moderately resistant to sterility mosaic disease. Released for Maharashtra, indeterminate growth habit, red colour medium seeds. Yield varies from 12-14 q/ha.
Konkon Tur -1	Semi-spreading, early maturity with dark green leaves and 4-5 cm pod length, suitable for cultivation on rice bunds. Yield varies from 10-12 q/ha.
AKT- 8811	Short duration (135-150 days) with yield of 14-16 q/ha.
Greengram	
Kopergaon	It matures within 65 to 70 days. Resistant to blight and yellow vein mosaic diseases. Grains are of bigger size. The index 3 to 3.2 g and productivity ranges from of 8 to 10 q/ha.
BM-4	This variety was released by MAU, Parbhani from Agril. Research Station, Badnapur during the year 1992. It matures in 60 to 65 days. Seed index 2.5 to 3 g. The average productivity ranges from 10 to 12 q/ha. Contains 17.85 % protein.
BPMR-145	This variety was released from Agril. Research Station, Badnapur during the year 2000. It is of 65 to 70 days duration, productivity 12 to 14 q/ha. Resistant to powdery mildew disease.
Pusa Vaishakhi	Suitable for Summer season as short duration crop.
PKV Green gold	Resistant to powdery mildew, medium bold seed.
AKM - 8802	Early maturity, resistant to powdery mildew, medium bold seed.

Contd.

Blackgram	
BDU-1	Bold seeded, tolerant to mildew.
TAU-1	Seeds are medium sized, tolerant to mildew.
TPU-4	Black bold seeds, erect mid tall, compact plant, tolerant to powdery mildew and yellow mosaic virus. Released for Maharashtra, Madhya Pradesh and Gujarat.
PKVU - 15	High yielding, resistant to powdery mildew, uniform maturity.
Rajmashbean	
Varun	High yield, bold seeds, (39-40 g/100 seeds) attractive seed coat colour i.e. purple streaks on seed, early maturity, suitable for *kharif* season. Released for Rajmah growing areas i.e. Satara, Pune, Sangli and Kolhapur districts.
Cowpea	
Konkan Sadabahar	Dwarf, determinant type, umbel growth habit, early maturity.
Konkan Safed	Short plant, dark green foliage, white flowers, green pods and brown grain colour, resistant to yellow mosaic virus.
Horsegram	
SEENA (K-42)	Dull white seeds. Released for Western Maharashtra.
MAAN (D-40-1)	Dark brown seeds. Released for Western Maharashtra.
Dapoli kulthi - 1	Semi spreading, semi-determinant growth, smooth dark green foliage with light brownish seeds.
Chickpea	
BDN - 9 -3	Early maturity, sustains water stress conditions, resistant to wilt.
BDNG - 797	Best for dryland cultivation, Bold seeded.
Phule G-12 (Vikas)	Attractive yellowish brown medium size seeds, suitable for rainfed and irrigated conditions. Released for Maharashtra.
Vijay	High yield potential, wilt resistant, drought tolerant, suitable for rainfed, irrigated and late sown conditions. Released for Maharashtra, M.P., Gujarat.
Phule G - 5 (Vishwas)	Bold seeds, responsive to irrigation and fertilizer. Released for MS, MP, Gujarat, W.Bengal, East UP, Orissa and Bihar.
Vishal	Attractive yellowish brown bold seeds, wilt resistant, suitable for rainfed, irrigated conditions. Released for W.Maharashtra.
Virat	Extra bold seeded kabuli variety, wilt resistant, fetches higher market price. Released for Maharashtra.
Digvijay	Yellowish brown bold seeds, wilt resistant, suitable for rainfed, irrigated and late sown conditions. Released for Maharashtra.
Jaki 9218	Bold seeded suitable for rainfed and irrigated conditions under zero till (no till) cultivation, resistant to wilt disease.
Lablab purpureus	
Konkan wal 1	Dwarf plant, dark green smooth leaves, bold size seed with brown seed coat.
Konkan wal 2	Short plant, dark green foliage and white flower colour. Green pod with greenish brown grain colour. The pods are produced on lower half portion of the shoot.

Efficient Cropping Systems Involving Pulses

Field experiments were conducted at different centers in different zones of Maharashtra to diversify traditional cropping systems by inclusion of pulses as compared to the existing sequence based on production and economic efficiency. Based on studies conducted at MPKV, Rahuri during 2001-2006, soybean-chickpea and pearl millet + pigeonpea (2:2) intercropping systems are recommended for replacement of the existing Pearl millet-wheat cropping system for medium black soils of Western Maharashtra under irrigated conditions (Table 6).

Table 6 : Pooled mean gross, net monetary returns and B:C ratio as influenced by different cropping systems and irrigation levels (2001-02 to 2005-06)

Cropping Systems	Gross monetary returns (Rs/ha)	Cost of cultivation (Rs/ha)	Net monetary returns (Rs/ha)	B:C ratio
Pearl millet-Wheat	46875	21247	25628	2.21
Sorghum-Wheat	47755	23786	23969	2.01
Pearl millet- *Rabi* Sorghum	39379	19774	19605	1.99
Pearl millet-Chickpea	49605	18345	31260	2.70
Sorghum-Chickpea	47592	20921	26671	2.27
Pearl millet+ Pigeonpea (2:2)	49679	14737	34942	3.37
Groundnut + Pigeonpea (2:2)	40506	17195	23311	2.35
Soybean – Chickpea	63016	20056	42960	3.14
S.E.±	1065	--	1065	0.05
C.D. at 5 %	3012	--	3012	0.15
General Mean	48051	20823	28543	2.50

Considering sustainability, profitability, feasibility, production efficiency and soil fertility (Table 7), the following four sequences of five crops including pulses in two years are recommended for general adoption in Marathwada region.

- Blackgram-onion-maize (F)/Chilli-summer groundnut.
- Bajra-cabbage-groundnut/Cotton-summer sunflower.
- Sorghum-Wheat-greengram/Cotton-summer groundnut.
- Sorghum-Chickpea-cowpea (F)/Chilli summer groundnut.

Table 7 : Agronomic productivity, economics, land use efficiency and production efficiency of various crop sequences.

Treatment	Season	Crop and variety	Yield (q/ha)	GMR (Rs. th./ha	NMR (Rs. th./ha)	C:B ratio
T1	K	Cotton (NHH-44)	18.84	12.3	7.8	
	S	G. nut (SB-XI)	19.46	11.3	8.2	
	K	Sorghum (CSH-9)	33.23	7.2	4.7	
	R	Wheat (HD-2189)	29.32	8.9	6.2	
	S	Cluster bean (PNB)	64.51	10.1	7.1	
		Total		**49.8**	**34.0**	**2.15**
T2	K	Bajra (BK-560)	15.64	3.1	1.5	
	R	Wheat (HD2189)	28.64	8.6	6.1	
	K	G.Nut (PM)K-4-11	15.33	10.8	7.9	
	R	Sorghum (SPV-86)	22.08	7.4	5.3	
	S	Cluster bean (PNB)	58.74	9.0	6.3	
		Total		**38.9**	**27.1**	**2.30**
T3	K	Sorghum (CHS-9)	29.74	5.6	3.3	
	R	Potato (KCN)	133.83	17.1	0.4	
	S	Cowpea (HFC142)	218.98	5.1	3.7	
	K	Cotton (NHH44)	23.65	17.5	13.1	
	S	G.nut (SB-XI)	20.73	14.1	11.1	
		Total		**59.4**	**31.6**	**1.14**
T4	K	Paddy (PBN-1)	25.40	5.2	2.9	
	R	Wheat (HD-2189)	27.14	8.1	5.5	
	S	G.gram (S-8)	9.91	4.3	3.0	
	K	Turmeric (Rajapuri)	140.6	24.5	13.9	
	S	Cowpea (HFC142)	164.49	4.1	2.8	
		Total		**46.2**	**28.1**	**1.55**
T5	K	Bajara (Bk-560)	14.70	2.7	1.4	
	R	Cabbage (GA)	184.13	21.8	16.5	
	S	G.nut (SB 11)	20.20	11.9	8.8	
	K	Cotton (H-4)	18.44	14.4	9.9	
	S	Sunflower (Mordern)	8.53	5.2	3.5	
		Total		**56.0**	**40.1**	**2.52**
T6	K	Blackgram (T-9)	8.78	3.7	2.3	
	R	Onion (N123-7-11)	255.19	15.1	12.9	
	S	Maize (A.Tall)	272.58	8.7	6.9	

Contd.

	K	Chilli (Jwala)	89.65	11.16	7.0	
	S	G.nut (SB-11)	21.01	14.8	11.7	
		Total		**53.9**	**40.8**	**3.11**
T7	K	Sorghum (CSH-5)	27.04	5.0	2.8	
	R	Ratoon (CSH-5)	22.38	4.3	2.1	
	S	Cluster bean (PNB)	45.13	6.8	4.4	
	K	Turmeric (Takarpetha)	138.77	23.9	13.6	
	S	G.gram (S-8)	6.25	3.0	1.8	
		Total		**43.0**	**24.7**	**1.35**
T8	K	Sorghum (CSH-9)	28.98	5.3	3.1	
	R	Wheat (HD-2189)	26.54	7.9	5.2	
	S	G.gram (S-8)	9.60	4.2	2.9	
	K	Cotton (NHH-44)	23.93	17.8	13.3	
	S	G.nut (SB 11)	21.87	15.3	12.0	
		Total		**50.5**	**36.5**	**2.60**
T9	K	Cotton (SRT-1)	12.38	7.9	3.9	
	S	G.nut (SB-11)	19.65	11.8	8.7	
	K	Pigeonpea (ICPL-87)	15.17	9.1	7.5	
	R	Wheat (2189)	28.44	8.8	6.0	
	S	G.graim (S-8)	5.72	2.9	1.6	
		Total		**40.5**	**27.7**	**2.1**
T10	K	Sorghum (CSH-9)	29.90	5.6	3.3	
	R	Chickpea (BDN-93)	18.12	7.8	6.4	
	S	C.pea (HFC-142)	224.91	5.2	3.8	
	K	Chilly (Jwala)	102.96	13.8	9.5	
	S	G.nut	21.71	15.5	12.4	
		Total		**47.9**	**35.4**	**2.83**
SE				1.8	1.8	
C.D. 5%				5.4	5.4	

Based on studies conducted at Akola, the chilli (green)-groundnut system was identified to be the most profitable system for the Vidarbha zone (Table 8). To identify need based cropping systems under irrigation, the experiments were conducted during 1992-93 to 1994-95. Amongst several cropping sequences chilli (green)-groundnut gave highest net returns followed by cotton-summer groundnut and hence, recommended under irrigated conditions.

Table 8 : Productivity, Net monetary returns and Benefit /Cost (BC) ratio for various cropping sequences (1992-93 to 1994-95)

Treatment/Sequence		Grain Productivity (q/ha)	Net returns (Rs/ha)	B/C ratio
T1	K – Chilli (Green)	86.75	56895	
	R - Continued	-	-	
	S - Groundnut	23.04	23718	
	Total	**109.79**	**80613**	**3.88**
T2	K - Cotton	20.87	31296	
	R - Continued	-	-	
	S - Groundnut	23.81	24830	
	Total	**44.68**	**56126**	**2.99**
T3	K - Pigeonpea	9.44	10548	
	R - Chickpea	17.46	16818	
	S - Groundnut	19.04	18106	
	Total	**45.84**	**45472**	**2.12**
T4	K - Soybean	19.07	11443	
	R - Wheat	37.36	15107	
	S – Clusterbean (V)	53.19	13621	
	Total	**109.62**	**40171**	**1.76**
T5	K - Sorghum	36.35	16834	
	R - Wheat	37.10	14896	
	S – Cowpea (V)	24.21	5790	
	Total	**97.66**	**37520**	**1.73**
T6	K - Sorghum	35.74	16560	
	R - Chickpea	15.44	13864	
	S - Groundnut	18.71	17858	
	Total	**69.89**	**48282**	**2.13**
	CD 5%		6314	

Source : Thosar, (1998)

Studies conducted at Dapoli under North and South Konkan Costal Zone conditions (Table 9) for four years, indicated that Rice-water melon-cowpea system gave highest gross returns of Rs. 77,284/- per ha followed by Rice-Dolichos bean-fallow system (Rs.59,418/- per ha), Rice-mustard-cowpea (Rs. 59021/- per ha), Rice-groundnut-fallow (Rs. 54,410/- per ha), Rice- mothbean-cowpea (Rs. 44826/- per ha). The minimum gross returns were realized in case of Rice-sunflower-fallow system (Rs. 32715/- per ha).

Table 9 : Economics of treatments based on four years pooled data

Crop Sequence	Gain yield (kg ha^{-1})	Straw yield (kg ha^{-1})	Grain Value (Rs. ha^{-1})	Straw value (Rs. ha^{-1})	Gross returns (Rs. ha^{-1})
Rice	5432	7703	24389	2580	26969
Mustard	1119	1782	15511	202	15713
Cowpea	1024	1617	16040	299	16339
Total	**7575**	**11102**	**55940**	**3081**	**59021**
Rice	5332	7244	23172	2363	25535
Sunflower	732	2156	6937	243	7180
Total	**6064**	**9400**	**30109**	**2606**	**32715**
Rice	5247	7189	22795	2410	25205
Mothbean	420	840	5173	157	5330
Cowpea	897	1429	14029	262	14291
Total	**6564**	**9458**	**41997**	**2829**	**44826**
Rice	5068	6925	21989	2307	24296
Dolichos bean	6375	5570	34132	990	35122
Total	**11443**	**12495**	**56121**	**3297**	**59418**
Rice	5332	6973	23132	2319	25454
Watermelon	18523	3165	37973	254	38227
Cowpea	847	1427	13340	266	13706
Total	**24702**	**11565**	**74445**	**2839**	**77284**
Rice	5150	7117	22320	2366	24686
Groundnut	2923	3422	27986	1738	29724
Total	**8073**	**10539**	**50306**	**4104**	**54410**

Production Technologies

Maharashtra Agricultural Universities and the State Department of Agriculture periodically update and publish the Crop Production Guide for the benefit of researchers, extension functionaries and farmers. The important technologies are briefly discussed here:

Soil : Pigeonpea, mungbean, urdbean, rajmahbean and chickpea grow well in well drained medium to deep soils having 45 to 60 cm depth. However, horsegram and mothbean grow better on medium to light soils. Well drained soils are essential, as pulses cannot withstand water logged conditions. Lack of aeration results in poor nodulation and low rhizobial activity.

Climate : For *kharif* pulses like pigeonpea, mungbean, urdbean, horsegram and mothbean, the favourable daily mean, temperatures range from 30 to 35^0 C. During flowering and podding, crop requires dry weather with bright sunshine for minimum diseases and pests. Chickpea requires well drained medium to deep soils. It requires fairly cool dry weather and adequate soil moisture. The period with minimum temperature in the range of 10-15^0 C, maximum temperature in a range of 25 to 30^0 C during early phase favours the satisfactory and fast growth with better branching, flowering and podding. It grows well in areas receiving 750 to 1000 mm rainfall, while horsegram and moth bean are grown mostly in drought prone areas where rainfall is about 500 to 600 mm.

Preparatory tillage : As the tap root system of pulse crops penetrates deep into the soil, it is necessary to have deep ploughing and one to two harrowings for better pulverization. This facilitates good germination and healthy crop growth.

Sowing time : It is of crucial importance particularly in a state like Maharashtra because of uneven and untimely rainfall pattern. The optimum sowing period for *kharif* pulses is second fortnight of June to first week of July while rainfed chickpea crop should be sown at the end of September or beginning of October to utilize the available residual soil moisture. The optimum period for sowing irrigated chickpea is 20th October to 10th November.

Seed rate and spacing : For obtaining desired yield levels, adequate plant population is essential. For this, it is necessary to use optimum seed rate. The seed rate depends upon size of seeds of different pulse crops. The optimum seed rate of different pulses is given in Table 10.

Table 10 : Agronomic practices for different pulse crops.

Crop	Maturity Group	Duration (days)	Seed rate (kg/ha)	Spacing (cm)	Recommended Varieties
Pigeonpea	Extra early Maturity	120-130	18-20	45x10	ICPL-87, ICPL-151, TAT-10
	Early maturity	130-145	20	45x20	T-Vishakha-1, AKT-8811
	Medium maturity	145-165	15	60x20	BDN -1, BDN -2
	Late maturity	Above 165 days	12-15	60x20	BSMR- 736, BSMR-853, Asha (ICPL-7119)
Greengram	Early	60-65	12-15	30x10	Pulse-M-2, Vaibhav
Blackgram	Early	65-70 &80-85	12-15	30x10	TPU-4, T-9, TAU-1, TAU-2, BDU-1
Horsegram	Normal	110-120	12-15	30x10	Seena, Maan
Mothbean	Medium	125-130	12-15	30x10	MBS-27
Rajmahbean	Early	65-70	120	30x15	HPR-35, Varun
Chickpea	Medium	110-120	60-65	30x10	BDN-9-3, Phule G-12
			65-70	30x10	Vijay
			100	30x10	Vishwas, Vishal, Virat, KAK -2

Seed treatment : For protecting the germinating seed and seedlings from the soil borne pathogens, the seed should invariably be treated with fungicides before inoculation by biofertilzier. For this 2 g Thirum + 2 g Carbendenzim or 5 g Trichoderma per kg of seed should be used. The seed should be inoculated with *Rhizobium* and PSB strains @ 250 gm peat culture/10 kg seed, to ensure better nodulation and rhizobial activity.

Manures and fertilizers : Application of 5 tonnes of well decomposed farm yard manure/compost per ha at the time of preparatory tillage is necessary for pulse crops. The optimum fertilizer dose for chickpea and pigeonpea is 25 kg N and 50 kg P_2O_5/ ha, for greengram and black gram 20 kg N + 40 kg P_2O_5/ ha, for horsegram and mothbean 12.5 kg N + 25 kg P_2O_5/ha, while for rajmahbean 30 kg N and 80 kg P_2O_5/ha should be applied at the time of sowing and remaining 30 kg N/ha may be applied 20 days after sowing. In addition to these fertilizers, 20 kg/ha elemental S is recommended for pulses. The application of 75 % RDN (45 kg N/ha) through inorganic fertilizer + 25 % RDN (15 kg N/ha) through FYM to *kharif* pearlmillet and 100 % RDF (100:50:50 NPK) to *Rabi* onion is recommended for pearlmillet-onion cropping sequence for maintaining the soil health (Table 11). Foliar spray of 2 per cent DAPS or urea at 50 per cent flowering is recommended for most of the pulses.

Table 11 : Pooled mean economics (Rs/ha) of cropping systems (2001-02 to 2005-06)

Treatments	Gross returns (Rs/ha)	Cost of cultivation (Rs/ha)	Net returns (Rs/ha)	B:C ratio
Cropping systems				
Pearlmillet-Chickpea	55167	20486	34680	2.74
Pearl millet -Onion	108925	26969	82138	4.21
Soybean -Chickpea	68199	23233	44127	2.98
S.Em ±	1486	–	1489	0.06
C.D. at 5 %	4464	–	4464	0.18
Fertilizers to *Kharif* Crops				
100 % RDF	77273	24283	53687	3.33
75 % RDF + 25% N through FYM	79859	24362	56207	3.39
60% RDF + 25 % N through FYM + Biofertilizer	75160	23664	52202	3.26
S.Em ±	1486	–	1290	0.06
C.D. at 5 %	4464	–	2631	0.18
Fertilizer to *Rabi* crops				
100 % RDF	79399	24447	55758	3.36
75 % RDF	75461	23947	52318	3.27
S.E.m. ±	979	–	978	0.01
C.D. at 5 %	140	–	140	NS

For obtaining maximum productivity, net returns and B:C ratio in North Konkan Coastal Zone, "Rice-Sweet corn" cropping system should be fertilized with recommended fertilizer dose (RFD) using inorganic nutrient sources while "Rice-Groundnut" system should be fertilized with RFD using either inorganic nutrient sources or using 50 % of RFD through inorganics and 50 per cent recommended dose of N (RDN) through organic sources (Table 12).

Table 12 : Effect of nutrient management systems on Rice equivalent yield (REY), returns and B:C ratio of different rice based cropping systems based on three years pooled data

Treatments	REY (q ha^{-1})	Gross returns (Rs. ha^{-1})	Net returns (Rs. ha^{-1})	B:C ratio
Nutrient Management Treatments				
100 % Organic	77.18	77135	-39504	0.66
100 % inorganic	103.45	97307	14726	1.14
Integrated	100.34	94390	-6053	0.91
SE (m)±	0.49	449	547	0.01
CD at 5%	1.59	1464	1784	0.02
Cropping Systems				
Rice-groundnut	103.99	99046	5160	1.07
Rice-sweet corn	153.83	143937	25056	1.26
Rice-mustard	44.07	44567	-26632	0.66
Rice-Dolichosbean	72.75	70893	-44692	0.62
SE (m)±	1.98	1836	1409	0.01
CD at 5%	6.10	5658	4340	0.04

Considering the sustainability, profitability and fertility of soil, the adoption of cereal-green manuring (sorghum-wheat-sunhemp) sequence and soybean-wheat with 100% recommended fertilizer dose for both the crops was more remunerative and beneficial, hence recommended for cultivators of Marathwada region (Table-13) representing central Maharashtra plateau zone.

Table13 : Long term studies on the relative efficiency of Legume-Cereal or Cereal-Legume rotations over Cereal-Cereal rotations.

Trmt No.	Crop sequences			Fertility levels of NPK	
	Kharif	*Rabi*	Summer	*Kharif*	*Rabi*
T1	Sorghum	Wheat	Fallow	100% Recommended	100% Recommended
T2	Sorghum	Chickpea	Fallow	100%	100%
T3	Sorghum	Chickpea	Fallow	75%	100%
T4	Soybean	Wheat	Fallow	100%	100%
T5	Soybean	Wheat	Green Manuring	100%	75%
T6	Sorghum	Wheat	Green Manuring	100%	100%
T7	Sorghum	Wheat	Green Manuring	75%	75%
T8	Sorghum	Wheat	Green Manuring	50%	50%

Weed management / Inter culture : Use of Fluchloralin as pre-plant incorporation or Pendiamethilin as pre-emergence application @ 1 kg a.i./ha to be applied for successful weed control.

Pulses suffer severely due to competition by weeds during first 25-30 days. For a desirable performance, it is important to keep the field weed free during this period. For this, it is recommended to carry out first hoeing during 20-25 days after sowing and second hoeing 10-15 days after first hoeing for providing better aeration in root zone. After hoeing, hand weeding is also necessary to remove remaining weeds between plants

Water management : The *kharif* pulses *viz.* pigeonpea, mungbean and urdbean, are generally raised under rainfed conditions. Irrigations at flowering and pod filling stages are critical as the requirement of soil moisture to the crop is very high during these stages. Irrigations are needed for obtaining good yield.

Chickpea, a prime crop of *rabi* season, is mostly grown as rainfed crop in many parts of the State. In deep black soils, two irrigations are sufficient to obtain good yield from chickpea crop. First irrigation should be given 30-35 days after sowing and second after 60-65 days as per the soil condition. For medium type of soil, it is necessary to apply three irrigations i.e. first at 20-25 days, second at 45-50 days and third at 70-75 days after sowing of crop. Excess and untimely irrigation should be avoided. However, it should be desirable to maintain adequate soil moisture level during crop growth period, as well as appropriate intervals of irrigation to avoid cracks in the soil.

Sprinkler irrigation system is quite effective and a safe system for chickpea crop. Chickpea is highly water efficient crop but is sensitive to soil moisture status. The frequent and heavy irrigation by surface flow leads to a great reduction in yield due to root rot, while sprinkler irrigation is highly favourable for cultivating chickpea. It can fulfill the demand of limited water of the crop that to, at the critical stages of growth. It does not allow over irrigation. This also leads to adequate aeration of the soil and prevents the root rot, as well as excessive growth of the crop. Sprinkler irrigation system provides exact quantity of water required when needed by the crop which favours growth and ultimately results in profuse branching, flowers and pods, leading to very good yield.

Integrated Pest and Disease Management

To increase and maintain stability in production of pulses, it is necessary to manage the diseases and pests more effectively as they take a heavy toll of the production every year. Several insect pests are known to feed on pulse crops causing enormous losses in quality and quantity of produce. Among them pod borer (*Helicoverpa armigera*) is a key pest in chickpea and pigeonpea crop.

It feeds on tender buds, leaves, flowers and pods of chickpea crop. For effective control of these pests, it is necessary to carry out integrated pest management programme timely.

IPM Practices

- Crop rotation with least preferred host like pearl millet, sorghum or groundnut helps to minimize the pest and disease incidence.
- Deep ploughing to expose and destroy the soil inhibiting pest and pupae of *Helicoverpa* to sunlight, predators and parasites.
- Clean cultivation and removal of plant debris to check termite infestation.
- Use of improved package of practices such as timely sowing, seed treatment, recommended fertilizer dose, interculturing etc.
- Inter/mixed cropping with wheat, linseed, mustard or coriander helps in reducing pod borer damage.
- Installation of bird perches @ 50 to 60 per ha help to attract the predatory birds.
- Use of sex pheromone traps 20-25 per ha reduces the pest population significantly.
- Spraying of neem seed kernel extract (NSKE) @ 5 per cent i.e. 25 kg/ha at the time of flower initiation. Followed by the second spray of Heliokill 500 ml/ha should be applied 10 to 15 days after first spray and, if required, another spray of chemical insecticide i.e. Monocrotophos 36 WSC @ 550 ml/ha or Spark 1250 ml/ha may be applied to the crop.

In pigeonpea-wilt and sterility mosaic virus; in mung and urdbean-powdery mildew and YMV; and in chickpea-wilt/root rot, collar rot; are the important diseases. For control of powdery mildew disease of mungbean and urdbean crop, it is recommended to spray the crop with monocrotophos 550 ml + Carbendenzim 250 g at the time of occurrence of disease.

Integrated Control of Diseases

- Cultivation of wilt/root rot resistant varieties.
- Seed treatment with protectant fungicides i.e. 2 g Thiram + 2 g Carbendenzim or 5 g Trichoderma per kg of seed.
- Following suitable crop rotations with non host cereal and non leguminous crops at least every 3-4 years.

- Deep ploughing of land in summer to expose the soil-borne pathogens to high temperature.
- Judicial application of irrigation, excess irrigation should be avoided, only well drained soils should be preferred.
- Infected plants should be promptly collected and destroyed.

Harvesting

- Picking the matured pods, drying and processing.
- Uprooting or cutting the whole plants, heaping, drying and processing.

Future Strategies

To increase production and productivity of pulses in Maharashtra, the following strategies are recommended :

- Development of short duration varieties for multiple cropping systems.
- Awareness among the farmers regarding use of pressurized irrigations systems *viz.*, drip and sprinkler irrigation systems.
- Improvement in yield and production through efficient transfer of technology.
- Intensifying inter varietal and interspecific hybridization Programme for biotic and abiotic stresses tolerance.
- Bringing more pulse area under irrigation.
- Improved seed size, CMS based hybrid pigeonpea.
- Ensuring timely availability of critical inputs.
- Policy matter related to minimum support price.
- Development of mechanization in cultivation of pulses.
- Development of storage facilities at village levels.
- Organization of FLD's and fields days for pulses.
- Undertake seed production Programmes on large scale through village seed production campaigns.
- Encourage the farmers to popularize mini dal mills.
- Develop technology of moisture conservation for long duration pulses in rainfed areas.

References

Annual Report of AICRP on Cropping System MAU, Parbhani. 1989, 2000.

Annual Report of AICRP on C.S. MPKV, Rahuri 2005-06.

Anonymous (1997) 'Research Highlights 1972-97' Konkan Krish Vidyapeeth, Dapoli.

Annual Report, (1996) AICRP on Cropping System Research, RARS, Karjat, DBSKKV, Dapoli (1995-96) pp. 3-4.

Annual Report (2009) Network Project on organic farming, RARS, Karjat, DBSKKV, Dapoli.

Jambhale, N.D., Thaware, B.L., Pethe, U.B. (2004). Crop Varieties. Developed by DBS KKV, Dapoli. pp.26-30.

Jadhav *et al.,* (1998). Research Achievements and Technology transferred bulletin on Cropping System Research. pp. 16-17.

Lakhadive *et al.* (2000). Research Achievements for Dry land Agriculture (1971-1996) pp. 17-180.

Wanjari. K.B. (2007). Pulse crop. Improved Production Technology. Dr. PDKV, Akola.

❑❑❑

16

North-East States

A.S. Panwar, S. Gupta and S.V. Ngachan
ICAR Research Complex for NEH Region, Umiam - 793 103, Meghalaya.

Pulses occupy an area of 0.18 m ha with total production of 0.12 mt in NE Region. This region is one of the potential areas where pulse production can be increased horizontally by utilizing a part of 16.72 lac ha area under *jhum* cultivation (Bujarbaruah, 2004). Considering that pulses could be grown mixed with other crops and their feasibility for relay cropping; the area and production of pulses could be increased vertically. Intercropping system not only covers the risk but also give a bonus yield of pulse crop in addition to normal return of the main crop. Maximum area and production of 0.11 m ha and 0.60 mt which is 60.90 % and 51.28% of the total area and production of the region, respectively is contributed by Assam only. After Assam, Nagaland and Tripura are the major pulse growing states having 30 and 8 thousand hectares of area with 28 and 5 thousand tonnes of production, respectively. In Meghalaya, pulses are grown on 5 thousand ha area and 4 thousand tonnes of pulses were harvested. Other states of this region grow pulses in about 6-7 thousand hectares area and produce 3-6 thousand tonnes (Table 1). The productivity of pulses in all the states is 1000 kg/ha except Assam and Manipur where its productivity is only 500 kg/ha. The average pulse productivity of all the NE States (661 kg/ha) is higher than that of whole India (637 kg/ha).

The major pulses grown in the region are greengram (*Vigna radiata*), Black gram (*V. mungo*), pigeonpea (*Cajanus cajan*), rice bean (*Vigna umbellata*) and cowpea (*V. unguiculata*) in *kharif* and french bean (*Phaseolus vulgaris*), chickpea (*Cicer arietinum*), lentil (*L. culinaris*) and field pea (*Pisum sativum*) in *rabi*. In hills, various other beans such as faba bean (*V. faba*), adzuki bean (*Vigna angularis*), moth bean (*V. aconitifolia*) and broad bean (*Dolichos lablab*) are also used as pulse. In Meghalaya, pigeonpea is major pulse in *kharif* season and chickpea is major pulse in *rabi* season. In *jhum* area, rice bean is a predominant pulse crop in *kharif* season. Beans in *kharif* and pea in *rabi* are also grown in all the districts of Meghalaya except Ri-Bhoi district but are mostly consumed as vegetable.

Table 1 : Area, Production and productivity of pulses in NE India

States	Area (000 ha)	Production (000 tonnes)	Productivity (kg/ha)
Arunachal Pradesh	6.0	6.0	1000
Assam	111.0	60.0	540
Manipur	6.0	3.0	500
Meghalaya	5.0	4.0	800
Mizoram	5.0	5.0	1000
Nagaland	30.0	28.0	933
Sikkim	6.0	6.0	1000
Tripura	8.0	5.0	625
Total	177.0	117.0	661

Existing Cropping Systems

Pulses occupy a very small place in the farming practices in the region as well as state of Meghalaya. Small area is under different pulse crops at various places and no systematic cropping system is being followed.

Although pulses are grown on marginal areas with zero input, however, these crops can be successfully grown in sequential cropping after upland rice and maize. Pulse crops sown during summer season generally encounter rains at maturity resulting into *in-situ* seed germination and ultimately poor quality yield is obtained. This problem can be overcome by sowing these crops during last week July to Mid August. The promising cropping systems identified for agro-climatic conditions of Meghalaya are presented in Table 2.

Table 2 : Promising cropping systems in Meghalaya

Maize based cropping system	Rice based cropping system	Vegetables based cropping system
Maize-Blackgram	Rice-Pea	Okra-Rajmash
Maize-Greengram	Rice-Lentil	Okra-Cowpea
Maize-Rice bean	Groundnut + Pigeonpea	Tomato-Frenchbean-Frenchbean
Maize-Rajmash		Capsicum-Rajmash
Maize-Cowpea		
Maize + Pigeonpea		

Yield Gaps

Pulse crops are generally grown in marginal area with no management. These crops are not given importance in NE Region because pulse proteins are not given due place as compared to animal based proteins. Field demonstrations conducted at different farmers field indicated a yield gap of 110 and 78 % in blackgram and pea, respectively with improved varieties but with existing local varieties the gap was 91.5% because the crops are grown mainly under stress situation, in marginal lands and are susceptible to a wide range of pest and diseases. Secondly, pulse such as moong bean, urd bean etc grown during summer and main *kharif* season encountered daily rains causing *in-situ* seed germination leading to reduction in yield and quality of produce.

Table 3 : Estimated yields and yield gaps (kg/ha) in pulses in NEH region of India.

Crop	Expt. Station		On- farm		Yield gaps	
	Improved variety	Local	Improved variety	Local	Improved variety	Local
Blackgram	1565	1216	746	635	819 (110)	581 (91.5)
Pea	1824	1315	1021	725	803 (78.6)	590 (81.4)

Figures in parenthesis shows the percentage.

Various management factors responsible for yield difference are soil acidity, weed control, use of improved variety seeds, line sowing method, use of plant protection chemicals and use of fertilizers. The soil acidity is the major cause of concern and all pulses are sensitive to soil pH. Weeds cause drastic yield reduction in pulse crops. A weed free stage upto 45 days was observed critical for pulse crops. A relative contribution of weed control method (20 and 40 days after sowing) was estimated to be 39 percent to increase pulse productivity in Meghalaya followed by use of improved variety seeds (25 percent), line sowing method (20 per cent), use of plant protection chemicals (12 per cent) and fertilizer application (4 percent), under rain-fed upland condition.

Improved Varieties of Pulse crops

Greengram

Greengram is grown as *kharif* pulse in NE Region and pod decay in *in-situ* seed germination in the pods due to rains is the most severe constraint in cultivation of this pulse. Sowing time of greengram is critical for successful harvest. It should be sown in first week of August so that at the time of pod maturity the rains are almost over. Local collections of greengram are semi bushy to bushy types exhibiting much variation in leaf size, inflorescence nature and pod type. Considerable variation prevails in grain colour also (green, yellow, brown, black and mottled), the common being green. These crops have been found suitable for low as well as medium altitude. Greengram genotypes, namely, MH 309, PDM 11, PDM 54, ML 5, PDM 14, ML 326, PDM 62, have been found suitable in low and mid altitudes of Meghalaya. The yield potential of these genotype is 9-10 g/ha at low altitude and 6-9 g/ha at medium altitude (up to 1300 m) in a duration of 70-80 days.

Blackgram

Like mungbean, its sowing should also be completed in first week of August. This crop is better adapted to this region than mungbean. It is one of the most extensively grown pulses in NE Region especially in Assam. Local collections are semi spreading types. Much variability in days to maturity occurs usually in medium to late maturity group. The indigenous types are observed to be photoperiod sensitive. One of this forms '*Pahelo Dal*' characterized by its green seed coat, is very common in this region. The genotypes of blackgram namely, T-9, UG 203, B3-8-8, Pant U 19, UG 301, PDU 2, LU 4, JU 77-41, have been found suitable in low and mid altitudes of Meghalaya. The yield potential of these genotypes is 9-10 q/ha at low altitude and 6-9 q/ha at middle altitude (up to 1300m). The blackgram matures in 80-90 days.

Cowpea

Cowpea is rarely grown as a pulse crop in NE region. It is mostly grown for vegetable purpose. The native genetic resources possess much wider variability for growing in the plains, hills and even in stress environments. Good variation is exhibited in growth habit, photoperiodic and photothermic sensitivity, seed size, pod shape and colour. Varying selection pressures had resulted in enriching the variability of grain, vegetable and fodder types of cowpea in NE region. Lobia is grown both for vegetables as well as for grain purposes. Promising genotypes identified for state of Meghalaya are Amba, V 385, C 152, V 16, V 154, S 488, V 271 and V 385. These genotypes are capable of yielding 8-12 q/ha in 80- 90 days. Lobia can also provide fodder in the lean season.

Pigeonpea

Pigeonpea is the most widely grown pulse in the region. Pod borer and pod weevil are major threat to its cultivation in the region. A wide array of diversity in *Atylosia s.*, a wild relative of pigeonpea occurs in India with maximum concentration in the North Eastern region. Out of 22 species of *Atylosia*, 4 species viz. *A barbata*, *A villosa*, *A volubilis* and *A scarabaeoides*, are often found in NE region. In Meghalaya, perennial types are commonly grown especially in Garo hills. There are variations, in seed size and colour in local perennial germplasm. Pigeonpea is found growing up to 1000 m altitude. The promising pigeonpea genotypes for low altitude (less than 100 m) are : Tripura local, T 21 and Pant A-3 and for medium altitude are : UPAS 120, H 76-44, AL-1 and H 76-11 (Table 4). Tripura local is a late maturing genotype while others are of early maturity group (140-150 days). These genotypes can be grown at lowest altitude as well as on dry terraces and hill slopes. These can be sown with the onset of monsoon in April/May. These genotypes are also suitable for intercropping with maize, soybean, groundnut etc.

Table 4 : Suitable varieties of pigeonpea for Northeastern hill region

Variety	Duration (days)	Yield (q/ha)
UPAS	120-150	11-15
WB 20	160-170	18-21
Bahar	240-250	20-25
Pusa 9	210-248	22-26
Sarita (ICPL 85010)	130-140	10-12

French bean / Rajmash

The *Phaseolus* group contains a number of important pulses. It is now generally agreed that the *Phaseolus* originated in the New World and that the Old World *Phaseolus* is designated as *Vigna. P vulgaris* commonly known as common bean, hariotbean, navybean, frenchbean, kidneybean etc, is generally grown as vegetable in Meghalaya. Much variation occurs in pod length, grain size and colour in this crop in both dwarf and pole types. In higher altitudes (2100 -2700m) *P. coccineous* (runner or scarlet bean) is grown. The varieties identified for green pod as well as seed purpose are given in the table 5.

Table 5 : Promising varieties of frenchbean /rajmash

Variety	Expected yield (q/ha)		Duration (days)
	Green pod	Seed	
Bush type			
VL 63	-	15-20	110-115
PDR 14	-	20-25	120-125
Amber	-	20-25	120-125
Arka Komal	90-100	8-10	70-75
Contender	80-90	8-10	70-75
Pole Type			
Meghalaya Local Selection	90-100	12-15	85-90
Naga Local	90-100	10-15	90-95
Manipur	85-90	10-12	95-100

Rice bean

Vavilov (1951) put rice bean in the Indian Centre of Origin including Assam and Myanmar. Rice bean assumes considerable importance as an underutilized and under exploited legume and has recently gained attention as a supplementary food crop with potential economic importance for the future. Rice bean possesses high nutritional quality and grain yield. It has resistance to yellow mosaic virus, good storage quality, rich genetic diversity and grain yield and is a hardy crop with wider acceptability. Rice bean has adapted to humid conditions of North East through its highly photosensitive nature. It is a short day plant (Sarma *et al.*, 1995) and irrespective of sowing, it flowers when day length is reduced in September. This ensures pod formation and maturation in the post rainy season.

In addition to its use as a pulse crop, it is also used as a forage crop. The seed protein of rice bean varies from 14-25%, the amino acid composition specially methionine and trytophane is considerably high in rice bean compared to other *Vigna* species. It contains high quality of vitamins, thiamine, niacin and riboflavin. Singh *et al.* (1980) reported that calcium and iron content are also appreciably high. Phytin phosphorus content which generally inhabits the phosphorus availability and lowest protein digestibility in most of the Asian pulses is low in rice bean. Genetic resources of rice bean have largely been collected from NE region. Enormous variability occurs in this less known pulse crop. The indigenous collections of this crop are mainly viny types but some semi bushy types have also been collected from Meghalaya. Wide variability in pod length, growth habit, seed size and colour and maturity occurs in materials available. Materials resistant to pod boring weevil (*Avion clavipes*), the major insect pest and rust (*Uromyces appendiculatus*), the major disease of rice bean, are also found in indigenous collections from North Eastern region. A number of suitable genotypes of rice bean have been indentified for Meghalaya. Among these RCRB 1-6 (19-23 Q/ha), RBL 1(15-18 Q/ha), MNPL 1 (13 -16 Q/ha), PRR 1 (10-12 Q/ha), RBL 6 (12 -16 Q/ha) and MNPL 2 (9-10 Q/ ha) are promising ones. These genotypes mature in 120 – 150 days.

Field Pea

Pea is grown in all the states of NE region mainly used as vegetable and only a part of it is used as dal. Pea is the most successful *rabi* crop in the state. Lowland rice fallow with ample water offers tremendous potential to increase its area in the state. In *rabi* crops, there is a general problem of maturity because of low temperature and rains during March/ April but pea has the capability to mature under these conditions also. Open grazing during winter is the major obstacle in raising pea. This crop suffers heavy losses due to rust and powdery mildew and there is a need to breed resistant varieties suitable for this region.

Varieties DMR 7, DDR 50, TRCP 8, TRCP 9, Arkel, Rachana and Ambika, have been found suitable for both mid and low altitudes. TRCP 8 and TRCP 9 are late maturing genotypes (130-140 days) and produce 18-20 q/ha while other varieties mature in 100 -125 days and produce 15-17 q/ha.

Lentil

Lentil is the most favorite pulse, from the consumption point of view among the local people of the region but, presently, it is not grown on a larger scale. Trials conducted in ICAR Research Complex Barapani have shown the promise of this crop for the region. Like pea, this crop also has the potential to mature in low temperature and cloudy conditions. Varieties L 4147, PL 639, PL 4, VL 1 and VL 125, have been found suitable for mid altitude. These varieties mature in 120-130 days and produce 8-10 q/ha. Varieties Pant L 406 and Sapana, have been found suitable for low altitude. These varieties mature in 135 -140 days and produce 14-15 q/ha.

Chickpea

Chickpea is grown only in 4 thousand ha in NE region. But in Meghalaya, it is the second most important pulse crop and is grown in all the three districts of Garo hills. Chickpea is also a favorite pulse among the local people of the state. Chickpea exudates malic and other organic acids through their roots, which dissolve insoluble P compounds in the soil which are normally not available to other plants (Bolland *et al.*, 1999). This crop exhibits excellent vegetative growth, flowering and pod setting but unlike pea and lentil, it fails to mature when temperature is less and rains are there at the time of maturity. Its seed has poor shelf life and high humid conditions prevailing throughout the year in the state markedly reduce the seed germination.

Production Technology

Chickpea

It is also an important *rabi* pulse crop grown in 4 thousand ha area with production level 3 thousand tonnes production. It is grown only in the state of Meghalaya, Assam and Tripura. Assam is the state having very low productivity of 500 kg/ha, which is below the national productivity (728 kg/ha) of gram. The major cropping systems involving chickpea are : maize-gram, rice-gram, toria + gram - fallow and cabbage + gram - fallow.

Land Preparation : Gram does not require as a good preparation of soils as wheat. One or two ploughings are sufficient. The land should be a bit cloddy as it will help the aeration of the soil in root zone. Well drained loamy and sandy loam soils are suitable for gram cultivation.

Seed rate : Seed rate of 40 to 45 kg seed per hectare is required for crop of bold seeded varieties and 80 kg/ha for other varieties. Seed rate in intercropping systems can be adjusted keeping in view the nature of crops in the system.

Sowing management : The optimum time of sowing of gram is the middle of October but for hills areas it should be from the last week of September to first week of October. It helps to control the wilt disease. Early sown crop suffers from wilt due to high temperature at the sowing time. Such sowing is also associated with excessive vegetative growth which results in poor yield. When gram is mixed with mustard, wheat and vegetables, the time of sowing will depend upon the main crop. The crop should be sown in lines 30 cm apart. Seed should be placed at a depth of 5 to 10 cm because the shallow crop is more liable to be damaged by wilt.

Manures and fertilizers : Gram requires 30: 60: 40 kg NPK for optimum growth and yield. The field should be applied with 5 tonnes of FYM 15-20 days before sowing. The entire amount of NPK should be applied at the time of sowing. In case of mixed crop with wheat, additional fertilizers should not be added.

Water management : Pulse crops do not need any irrigation if there is adequate soil moisture at sowing time followed by one or two winter showers. Legumes need not to be irrigated in their early vegetative stages as it may harm by interfering with the nodulation and oxygen requirement of roots. However, if irrigation is available, then two irrigations-one at flowering and the second at pod development stage, should be given.

Weed control and inter culture : Weeds compete with gram for space, nutrients and water. A weed free condition is a must up to 45 - 60 days after sowing for which two weedings are helpful.

Plant Protection

Insect-pests

Pod borer : Initially caterpillars feed on the top foliage and later bore into the pods and feed on the developing grains. For controlling the insect, spray at 50% flowering 0.875 litre Monocrotophos (Monocil 36 EC/SL) or Carbaryl 50 WP (Sevin) or 1.250 litre Thiodan/Hindon/Endocel in 625 litre water per ha. Spray can be repeated at 15 days interval, if the pest still persists.

Cutworm : The dirty grey caterpillars remain hidden in soil and cause damage to seedlings immediately after germination by cutting the stem of young plant at ground level. They damage of plants is much greater then what they actually consume. Mix 2 litres Chlorpyriphos 20 EC in 25 kg sand and apply in field before sowing.

Diseases

Blight : The disease appears in the form of dark brown spots studded with minute, black, dot-like bodies on the stem, branches, leaves and pods. On pod and leaf lesions, the pycnidia are concentrically arranged. Whole of the plant may be blighted or killed in case of severe infection. For controlling this insect, use disease free seed of resistant variety and treat seed with Dithane M-45 or Vitavax @ 2.5 g/kg seed. If the insects attack is observed in the field, spray crop with Indofil M-45/ Dithane M-45 (0.25%) on the appearance of the first symptom of a disease and repeat the spray at 15 days interval.

Wilt : This is a serious disease in gram causing partial or complete loss of yield depending upon the severity of the infestation. The affected plants first turn yellow wither and finally dry up. The roots become dark coloured resulting in complete rotting. The field should be deeply ploughed with adjusted sowing time.

Lentil

Lentil or *masur* is the most potential important *rabi* pulse and grown in an area of 500 hectares with total production of 1500 quintals. The average yield per hectare is 300 kg which is quite low. It is mostly grown, a limited extent, by a few cultivators for their own use. The major Sequential Cropping Systems involving lentil are Maize - lentil, Rice - lentil and Toria + lentil-fallow system.

Preparatory tillage : Two to three ploughings are needed for pulverizing the soil. After ploughing, planking should be done for breaking clods and the field levelled.

Seed rate : 25 to 30 kg seed per hectare is optimum but under late sown conditions, the seed rate should be increased by 50%.

Sowing management : End of September to first fortnight of November is the best time for sowing of lentil. However, it can be sown up to first week of December in foot hills and plain areas. The crop should be sown in lines 25-30 cm apart in furrows behind the plough or seed drill.

Manure and fertilizers : Lentil is generally sown after maize as upland crop without any fertilization leading to low yield. For getting good harvest, application of 20 kg N/ha and 60 kg P_2O_5 in the form of SSP and 30 kg potassium are essential. Application of well decomposed FYM @ 5 t/ha at the time of field preparation helps in maintaining the soil fertility and balancing the nutrition of the crop. Full dose of N,P and K should be placed in furrows below the seed at the time of planting. The region is not experiencing the micro-nutrient deficiency which has been confined so far to a few pockets in Assam.

Weed control and inter culture : Lentil is sensitive to weeds infestation and hence, weed free condition at least up to 45-60 days was observed to increase its yield. To keep the weeds under control, one or two weedings and hoeing are helpful.

Water management : Lentil is also grown without irrigation. However, one irrigation at the time of pod formation increases the yield.

Plant Protection Measures

Insect-pests

Pod borer : Initially caterpillars feed on the top foliage and later bore into the pods and feed on the developing grains. The insect can be controlled by spraying of 0.875 litre Monocrotophos (Monocil 36 EC/SL) or Carbaryl 50 WP (Sevin) or 1.250 litre Thiodan/Hindon/Endocel in 625 litre water per ha at 50% flowering stages of crop growth. The spray can be repeated at 15 days interval, if the pest still persists. Caution : Don't pluck green pods upto 15 days after spray for consumption.

Cutworm : The dirty grey caterpillars remain hidden in soil and cause heavy damage to seedlings immediately after germination by cutting the stem of young plant at ground level. They damage the plants much more than what they actually consume. To control Mix 2 litres Chlorpyriphos 20 EC in 25 kg sand and apply in field before sowing.

Diseases

Blight : The disease appears in the form of dark brown spots studded with minute, black, dot-like bodies on the stem, branches, leaves and pods. On pod and leaf lesions, the pycnidia are concentrically arranged. Whole of the plant may be blighted or killed in case of severe infection.

Control measures : Use disease free seeds of resistant variety. The seed should be treated with Dithane M-45 or Vitavax @ 2.5 g/kg seed (0.25%) with first symptoms of disease and repeat the spray at 15 days interval. All the crop residues and weeds biomass should be replaced from the field.

Wilt : The affected plants first turn yellow, wither and finally dry up. The roots become dark coloured resulting in complete rotting. For controlling the disease, the field should be ploughed deep and sowing time need to be changed.

Pigeonpea

The major cropping systems involving pigeonpea are : Pigeonpea-Mustard, Pigeonpea-Blackgram/Greengram, Pigeonpea-Potato, Rice + Pigeonpea-fallow,

Maize + Pigeonpea- fallow, Finger millets + Pigeonpea-fallow, Pigeonpea + Greengram/Blackgram-fallow, Pigeonpea + Groundnut-fallow and Pigeonpea + soybean – fallow.

Land preparation : Pigeonpea is grown in mixed cropping system as intercrop or strip crop. So land preparation for the main crop is sufficient. Generally, deep ploughing and 1 or 2 harrowings are sufficient.

Seed rate : Quality seed is the key to get good harvest. Seed should be replaced by quality seed for every 3-4 years. Seed rate for short duration varieties is 15-20 kg/ha, while for long duration varieties the seed rate can be 8-10 kg/ ha.

Sowing management

Sowing time	:	Summer season	:	April
		Kharif season	:	May-June
Spacing	:	Short duration varieties-		40 - 60 x 10 - 15 cm
		Long duration varieties-		60 -120 x 15 -20 cm

Manures and fertilizers : For getting good harvest, application of 15-20 kg N/ha and 40-50 kg P_2O_5 ? in the form of SSP is essential. Application of well decomposed FYM @ 5 t/ha at the time of field preparation helps in maintaining the soil fertility and balancing the nutrition of the crop. Full dose of N and P should be placed in seed furrows below the seed at planting. The state has not experienced any micro –nutrients deficiency so far.

Weed control and inter culture : Suppression of weeds through cultural practices is safe and involves low inputs. When weeds are managed by hoeing and hand weeding till the crop starts branching, subsequently the close canopy suppress the weeds. For chemical control measures, pre-emergence application of Pemdimethlin @ 1.0 kg a.i. /ha in 500 litres of water 1-2 days after sowing was found effective.

Water management : The state of Meghalaya receives rains from April to October, by that time the root system of pigeonpea is so developed it can absorb water from deeper layers of soil and hence, does not requires irrigation. Pigeonpea is a drought tolerant crop. However, branching, flowering and pod filling stages are critical stages of crop growth and moisture stress at these stages can cause yield loss.

Plant protection measures : Major diseases and pests and their control measures are as follows:

Insect Pests

Pod borer (*Helicoverpa Armigera*)

Pod borer is a major pest of pigeonpea. Larvae of the insects feed on almost all the parts of the plant. They mainly bore into pods after pod formation and damage grains.

Management

- Dusting of 10% Malathion or Carbaryl @ 25 - 30 kg/ha
- Spraying of Monocrotophos (0.04%) or Dimethoate (0.03%) at flowering and pod formation stage.

Blister beetle (*Mylabris Pustulatus and M. Phalerata*)

The adult beetles feed heavily on the flowers and cause damage to flowers, which results in lesser number of pod formation per plant and ultimately lesser yield.

Management

- Hand picking and killing of the.beetle is most effective.
- Dusting of Malathion (5%) or Carbaryl (10%) @ 25 - 30 kg/ha at flowering stage.

Diseases

Wilt : This disease is caused by fungus *(Fusarium udum)*. Plant starts yellowing accompanied with drying of leaves. Subsequently, the entire plant dries up. Black streaks can be seen at the roots and base of the plants, particularly under the bark. The wilt can be managed by growing resistant varieties and following long crop rotations and mixed cropping.

Sterility mosaic virus : Stunted growth, pale green, bushy leaves, mild mottle or ring spots on the leaves. No reproductive part is formed. The virus is transmitted by *Aceria cajani*, a mite. The disease can be managed by spraying of Metasystox (0.1) as soon as symptoms appear and growing of resistant varieties.

Field Pea

The major Cropping systems involving Fieldpea are : Maize-Pea, Rice-Pea, Groundnut-Pea, Soybean-Pea. In upland conditions, Pea + Toria and Buckwheat + Pea while under lowland conditions, Pea + Mustard, Pea + Cauliflower/cabbage and Pea + Methi are practiced.

Land preparation : One deep ploughing followed by 2-3 harrowings and planking should be done to get the land well pulverized and leveled. In lowland rice fallows, raised beds of 20 – 25 cm high, 4-5 m in length and 1.2 to 1.5 m wide should be prepared and soil should be well pulverized.

Seed rate : Late maturing tall varieties - 60-80 kg/ha

Early maturity dwarf varieties - 100-125 kg/ha

Sowing Management

Sowing time : First fortnight of October is the optimum time for sowing of pea. Delay in sowing beyond October results in drastic reduction in yield.

Spacing : Tall varieties - 30 cm x 10 cm

Dwarf varieties - 20 cm x 10 cm

Manures and fertilizers : FYM @ 10 t/ha and lime @ 2 t/ha should be incorporated in the soil at the time of land preparation. A starter of 20-30 kg N/ha along with phosphorus @ 60 kg/ha and Potash @ 40 kg/ha should be applied in seed furrows below the seed. The amount of lime can be minimized up to 500 kg/ha by its application in furrows at the time of sowing of crops.

Staking : Tall varieties need support for their proper growth and development. Bamboo sticks of 1.5-2.0 m length can be used for staking the crop between two rows at tendril formation stage.

Weed management : Field should be kept free from weeds during initial period of growth later on the crop itself suppress the weeds. One hand weeding at 25-30 days after sowing should be done to manage the weeds. Weeds can also be controlled by pre-emergence application of Pendimethlin @ 1.0 kg a.i. / ha in 500 litres of water 1-2 days after sowing.

Water Management : Pea crop needs assured irrigation for higher yield. Special care should be taken to provide atleast one irrigation at flowering stage under upland conditions. In lowland rice fallows, generally irrigation is not required rather proper drainage is a must to avoid water logging.

Plant Protection Measures

Insect pests : Leaf miner and pea stem fly are the major pests of pea in this region. For their control, spray of 250 ml Phosphamidon or 1 litres of Metasystox in 500 litres of water per hectare should be done.

Diseases : *Rust and powdery mildew* are major disease in North Eastern Hill Region. However, sometimes incidents of wilt and root also cause severe losses.

Rust : Rust is caused by a fungus, *Uromyces fabae*. Firstly yellow spots appear on lower leaves and later these spots spread on stem and in severe conditions on pods also. 2-3 sprays of Dithane M-45 @ 2 g/liter water at 10 days interval from disease appearance stage should be done to control the disease.

Powdery mildew : Powdery mildew is caused by a fungus, *Erysiphe polygoni*. The symptoms first appear on leaf and then other parts of the plants. Leaves show white powdery growth on both the surfaces. Late planting should be avoided to reduce the disease incidence. Two to three sprays of wettable Sulphur like Sulfex or Elosal @ 3 g/liter of water at 10 days interval may control the disease.

Wilt and root rot : Wilt and root rot are caused by the fungi *Fusarium oxyporum* and *Rhizoctonia solani*, respectively. Plants show yellowing of lower leaves followed by wilting. For control of these diseases, seed should be treated with Captan 2 g/kg seed before sowing.

Greengram

The major cropping systems involving Greengram are : Maize-Black gram and Okra-black gram, Pigeonpea + Greengram - fallow and Maize + Greengram-fallow .

Land preparation : Greengram requires fine seed beds for which 2-3 ploughings are necessary to make the land weed-free and getting good tilth for better germination and healthy plants.

Seed rate : 15-20 kg/ha is optimum to provide adequate number of plants in the field.

Sowing management : First fortnight of August is the optimum time for sowing of greengram. Further delay in sowing beyond the middle of August results in reduction of yield due to moisture stress and drop in temperature at grain filling stage. Treat the seeds by Bavistin @ 3 g/kg seed for 24 hours before sowing. Seed should be sown in furrows at a row spacing of 30-40 cm keeping 10 cm distance from plant to plant.

Weed control and inter culture : Suppression of weeds through cultural practices is safe and involves low inputs. When weeds are controlled by hoeing and hand weeding for the first half of the season, close canopy suppresses the weeds. Weed can also be controlled by application of Pendimethlin @ 1.0 kg a.i./ha in 500 liters of water 1-2 days after sowing followed by one hand weeding 30 days after sowing.

Manures and fertilizers : For obtaining high yield of greengram, FYM (5t/ha) and N, P& K 20: 60:40 kg/ha (43 kg Urea, 375 kg Single Super Phosphate and 67 kg Muriate of potash) is required. FYM and N, P and K should be placed in seed furrows below the seed at planting.

Water Management : The crop is grown during late *kharif* season and at this stage there is no moisture stress in the field. Hence no irrigation is required.

However, during heavy rainfall, water should not be allowed to stagnate in the field and adequate drainage provision should be made.

Plant Protection Measures

Insect-pests

Hairy caterpillar : They first feed on leaves in clusters and later on disperse in the field. Often, the plant is completely skeletonized. Appear in epidemic form at the onset of monsoon.

Control measures : The insect can be controlled by giving spot treatment by dusting 5% Malathion at 30-40 kg/ha. Spray 1250 ml Thiodan/Endocel 35 EC or 625 ml Dichlorvos (Nuvan 100) in 625 L water/ha for advance staged larvae. Collect and destroy the young gregarious caterpillars.

Blister beetle : Adults feed voraciously on flowers and tender terminals. To control this insect, spray 1250 ml Thiodan/Endocel 35 EC or 625 ml Methyl parathion (Metacid 50 EC) in 625 L water/ha at the pre-bloom and flowering stages.

Bean bug : Both nymphs and adults suck sap from foliage and lower the vitality of the plant. Small whitish or yellowish spots on the upper leaf surface confirm the damage by the pest. Spray 625 ml Monocil/Milphos 36 SL (Moncrotophos) in 625 L water/ha.

Diseases

Cercospora leaf spot : Brown or black, circular to angular spots are formed on leaves. During humid weather, the spots become so numerous that they merge with each other to form large irregular spots. At this stage the leaves wither, die and fall. The crop should be sprayed with Indofil M-45 (0.25%) or Bavistin (0.05%) at fortnightly intervals starting after 45 days of sowing or when the first symptoms of the disease appear. 2-3 sprays give better and economical control.

Ascochyta leaf spot : Dark coloured, circular lesions with concentric rings are present on leaves. In severe cases, shot holes surrounded by remanants of concentric rings are formed. The crop should be sprayed with Indofil M-45 (0.25%) or Bavistin (0.05%) at fortnightly intervals starting after 45 days of sowing or when the first symptoms of the disease appear. 2-3 sprays give better and economical control.

Collectotrichum leaf spot : Crescent-shaped, dark brown lesions are formed on leaves. In severe cases, the lesions may coalesce resulting in burnt appearance. The crop should be sprayed with Indofil M-45 (0.25%) or Bavistin (0.05%) at fortnightly interval starting after 45 days of sowing or when the first symptoms of the disease appear. 2-3 sprays give better and economical control.

Powdery mildew : A white powdery growth is seen on the leaves which later on spread and cover the stem and other plant parts. The disease is more severe when the plants are at the flowering stage and persists until harvest. Dusting Sulphur @ 20 kg/ha found to check the disease. If the crop has been sprayed with Bavistin, there is no need to sulphur dusting.

Black gram

The major cropping systems involving Black gram are : Maize - black gram and Okra- Black gram, pigeonpea + blackgram - fallow and maize + blackgram - fallow.

Land preparation : Black gram does not require fine seed bed but for good and uniform germination 2-3 ploughings are necessary. Ploughing makes the land weeds free and incorporates the previous crops residues into the soil.

Seed rate : To obtain desired number of plants in the field, a seed rate of 15-20 kg/ha is sufficient.

Sowing management : First fortnight of August is the optimum time for sowing of black gram. Further delay in sowing beyond the middle of August results in yield reduction due to moisture stress and drop in temperature at grain filling stage. Treat the seeds in Bavistin @3g/kg seed for 24 hours before sowing. Seed should be sown in furrows at a row spacing of 30-40 cm keeping 10 cm distance from plant to plant.

Weed control and inter culture : Suppression of weeds through cultural practices is safe and involves low inputs. When weeds are controlled by hoeing and hand weeding for the first half of the season, close canopy suppresses the weeds. Weed can also be controlled by application of Pendimethlin @ 1.0 kg a.i./ha in 500 liters of water 1-2 days after sowing followed by one hand weeding 30 days after sowing.

Manures and fertilizers : For obtaining high yield of greengram, FYM (5t/ha) and N, P_2O_5, & K_2O: 60:40 kg/ha (43 kg Urea, 375 kg Single Super Phosphate and 67 kg Muriate of potash) is required. FYM and N, P and K should be placed in seed furrows below the seed at planting.

Water management : The crop is grown during late *kharif* season and at this stage there is no moisture stress in the field. Hence no irrigation is required. However, during heavy rainfall, water should not be allowed to stagnate in the field and suitable drainage mechanism should be made.

Plant protection measures : As in case of Greengram

Cowpea

It is a leguminous fodder as well as grain/pods crop suitable for the areas below 1200 m altitude. It can be grown as a pure crop or as a mixture with maize, teosinate or napier-bajra hybrid. It is quite rich in protein (18%) and supplements the nutritive value of non-leguminous fodder. Being a leguminous crop, it enriches the soil too.

Land preparation : It can be grown successfully on any type of soil from sandy loam to heavy clays. However, loam soils are the most suitable ones for its growth. Land should be leveled and well drained. One cross ploughing is sufficient for successful cultivation of cowpea.

Seed rate : The optimum seed rate is 25-30 kg /ha for fodder as well as grain/pod purpose.

Sowing management : The ideal time of sowing is the beginning of May but can be extended till mid July. Sowing should be done in rows 30 cm apart with 5-7 cm deep placement.

Manures and fertilizers : The crop is sensitive to aluminum toxicity, hence, soil pH should not be less than 5.4. Apply manure @ 5 t/ha before sowing and 15:60:30 kg NPK at the time of sowing is optimum for bumper harvest of crop.

Weed control and inter culture : The crop requires weed free field up to 35-40 days stages, thereafter, its foliage controls weed by smoothening them. One weeding at 20-25 days stage is sufficient to check weeds in the field.

Water Managements : The crop is grown during *kharif* season, hence, does not require irrigation due to sufficient moisture as a result of frequent rains during *kharif* season.

Rice Bean

The major Sequential cropping systems involving Rice bean : are Maize-Rice bean, Ricebean-Carrot-Tomato and Maize + Ricebean-fallow.

Land preparation : Fine seedbed preparation is necessary for ricebean. Generally, 2-3 ploughings are done to make the land weed free and getting good tilth for better germination and healthy plants.

Seed rate :	Bush type	:	50-60 kg/ha
	Pole type	:	25- 30 kg /ha

Treat seeds with Bavistin @ 3 g /kg seed for 24 hours before sowing.

Sowing management : When the crop is grown for green pod (vegetable) purpose, sowing should be done in February to March (lowland) and August to middle of September (upland) while for seed purpose the sowing done during August- middle of September (upland) is found optimum.

Sowing method : Generally sowing is done with flat bed sowing, in which seed is sown at a row spacing of 30-40 cm keeping 15-20 cm distance from plant to plant. *'Bun'* method of cultivation on slopy land is quite prevalent in NE region. In *Bun* sowing, seed should be sown on bun (raised bed) having 60-80 cm width and 20-30 cm height at a row spacing of 30-40 cm keeping 15-20 cm distance from plant to plant.

Staking to pole type : Plant should be supported by bamboo sticks (1.5-2.0 m height) 15-20 days after sowing. Staking increases pod yield as well as seed quality.

Manures and fertilizers : For obtaining high yields of French bean, FYM (10t/ha) and N, P & K (60:60:40 kg/ha) (130 kg Urea, 375 kg Single Super Phosphate and 67 kg Muriate of Potash) are required. Half of the nitrogen and full dose of P and K should be placed in seed furrows below the seed at planting and remaining half amount of nitrogen should be top dressed at flowering.

Weed control and Inter culture : Suppression of weeds through cultural practices is safe and involves low inputs. When weeds are controlled by hoeing and hand weeding for the first half of the season, close canopy suppresses the weeds. Pre-emergence application of Pendimethlin @ 1.0 kg a.i./ha must be applied 1-2 days after sowing. Herbicide should be applied after mixing with 500 litres water/ha.

Water management : Being a legume, rice bean does not require frequent irrigations. However, during flowering and pod development stages, water stress may reduce yields.

Plant Protection Measures

Insect Pests

Mites : Mites caused drastic yield reduction. It can be managed by spraying of Malathion 0.05% (100 ml Malathion 50 EC/100 litres of water) at the interval of 10 days.

Blister beetle : Adults feed voraciously on flowers and tender terminals. To control use, Carbaryl 0.1% (200 g Sevin 50 WP per litre of water) at the flowering stage.

Diseases and their Management : Rust, powdery mildew, rhizoctonia bean blight and *Cercospora* leaf spot are some of the common diseases that occur in the North Eastern Region.

For the control of the diseases, the seeds should be treated with Mancozeb or Carbendazim @ 2 g/kg seed 24 hours before sowing. Spraying of Dithane M 45 @ 2 g/liter water controls rust. Dusting of Malathion 5% @ 25 kg/ha gives effective control of pests.

Future Strategy

Pulses play a pivotal role in human diet. In North Eastern Region, the animal based protein is the preference of the local people. However, a number of major as well as minor pulses are preferred in daily diet justifying the importance of pulses in the region. But the productivity of pulses in the region is very low which is discouraging their cultivation. The following strategy deserves due attention for promoting pulses production in NEH region.

- Increase production per unit area and time. Usually spring (before *kharif* rice sowing) and autumn (after *kharif* rice harvesting) seasons in this region go as fallows. Short duration pulses like mung, urd and rajmash offer an opportunity for these seasons.
- Incorporation of pulses in rice and maize based systems of this region would increase the soil fertility in a sustainable way.
- Seed is the basic input for a bumper crop harvest and must be supplied to farmers. Quality seed production and distribution is a major constraint which is coming in the way of improving production of pulses in this region.
- A sound and co-coordinated seed production programme is urgently required in the region. Production of breeder and foundation seed has to be taken up by institutions like ICAR, AAU, NSC, etc. For the production of certified seed, active participation of farmers is required. Progressive farmers are required to be identified and trained in various seed production activities.
- Integrated nutrient management is necessary for increasing pulse production in this region. Pulses are capable of meeting out their nitrogen requirement through symbiotic nitrogen fixation by Rhizobium. These crops requires initial starter dose of nitrogen. Seed inoculation with biofertilizer (*Rhizobium* and PSB) - low cost inputs- can increase pulse productivity by 1 – 12 %.
- Frequent rains encourage the growth of a number of weeds, which compete with main crop and reduce their yield. Integrated weed management techniques need to be transfered to the farmers of the region.
- Lack of quality culture in adequate quantity is another of the major constraints in popularization of biofertilizers. Their production at regional level is required for easy availability to the farmers.
- Prevalent favorable conditions for the growth of microbes and insects pests result in significant yield losses due to diseases and insect-pests. Breeding of varieties resistant against diseases and insects is the need for boosting the pulse yield.

- Characterization of pathotypes, screening of individual genotypes with individual pathotypes, identification of resistant genotypes and using these genotypes for breeding are required for development of resistant varieties. There is need to supplement the resistant breeding work with molecular markers for effective incorporation of multiple genes for resistance (gene-pyramiding).
- Utilization of *rabi* season for growing crops may boost pulse production in this region. Lowland rice fallow with ample moisture and bright sunshine offers great potential for the production of *rabi* pulses like pea, lentil and chickpea.
- Post harvest losses especially during storages, discourage pulses in the region. Most of the post harvest losses are because of lack of proper storage structures. This causes rapid deterioration in seed quality. High moisture conditions throughout the year result in heavy disease and insect incidence during storage. Lack of small processing units compels the farmers to sell their produce as such without any value addition.

References

Anonymous (1997) Annual Report of ICAR Research Complex for NEH Region, Barapani.

Bolland, M.D.A., Siddique, K.H.M., Loss, S.P. and Baker, M.J. (1999). Comparing responses of grain Legumes, wheat and canola to applications of superphosphate. Nutrient Cycling in Agroecosystems, 53,157-175

Bujarbaruah K.M. (2004) Organic Farming: Opportunities and Challenges in North Eastern Region of India. In: Nature 2004, Souvenir of International Conference on Organic Food 14-17 Feb at ICAR Research Complex for NEH Region, Barapani pp 13-23

Gupta S., Bujarbaruah K.M., Dhiman K.R., Pathak K.A., Das S.P., Das, Alpana, Kumar Arvind and Vijaypal (2006) Pulse Production Technology for North East India. Technical Bulletin No 23, ICAR Research Complex for NEH Region, Barapani (Umiam), Meghalaya.

Kannaiyan, S. (2000) Perspective of increasing Pulse Productivity in Tamil Nadu. In pulses Production Strategies In Tamil Nadu, Centre for Plant Breeding and Genetics, Tamil Nadu Agricultural University, Coimbatore.

N I (1951) Origin, variation, immunity and breeding of cultivated plants. Chron. Bot 13: 1-364.

Raychaudhury, M and Ngachan, S.V. (2001). Selection and application of biofertilizers in North East Region. In Souvenir, IInd NER Conference on biofertilizers held at AAU, Jorhat w.e.f. 22-24 Jan, 2001 pp 59-62.

Sarkar, B.B. Bhattarcharya, N.R and Roy, S (1989) Rhizobium culture helps increase the yield of blackgram. Indian Farming, 38: 13-14.

Sarma B.K., Singh, M., Gupta H.S., Singh G. and Srivastava L.S. (1995). Studies in rice bean germplasm. Research Bulletins # 34. ICAR Research Complex for North East Hill Region, Barapani (Umiam), India pp 70.

Singh S.P, Mishra B.K., Chandel K.P.S and Pant K. C. (1980) Major food constituents of rice bean (Vigna umbellate). J. Food Sci. Techno. 17: 238-240.

Singhal, V (1999) Indian Economic Data Research Centre, New Delhi. pp. 105-110 Vavilov.

❑❑❑

17

Orissa

L. M. Garnayak, B.S. Rath, B.K. Mahapatra, A.K. Mohanty and M. Mohanty
AICRP on Integrated Farming Systems, Orissa University of Agriculture and Technology, Bhubaneswar 751 003

Pulses are the second most important group of crops next to cereals in Orissa occupying about 22% of total cropped area as compared to 54% for cereals. The state grows ten pulse crops, namely, greengram, blackgram, pigeonpea, horsegram, lentil, gram, cowpea, Rajmash, *Lathyrus* and ricebean out of which the first four crops cover 88% of total pulses area and 85% of total pulses production.

During 2008-09, pulse crops in Orissa were grown in about 20 lakh ha having production of 9.94 lakh tonnes and average productivity of 497 kg/ ha (Table 1). Although pulses are grown in all the districts, the major pulse growing districts are Kalahandi, Ganjam, Cuttack, Bolangir, Angul, Nawapara, Keonjhar, Rayagada, Kendrapara, Mayurbhanja and Bargarh, which together account for 59% of the pulses area and 63% of total production. Ganjam has the maximum area (10.8%) while Kalahandi has maximum production (1.29 lakh tonnes) and Rayagada the highest productivity (678 kg/ha). Per hectare yield is >600 kg/ha only in 3 districts (Rayagada, Kalahandi and Bhadrak), between 500-600 kg/ha in 7 districts (Gajapati, Mayurbhanja, Keonjhar, Cuttack, Ganjam, Boudh and Sundargarh), <400 kg/ha in one district i.e., Nayagarh (385 kg/ha) and between 400-500 kg/ha in the rest of 19 districts. Area under *kharif* (wet) and *rabi* (dry) pulses is in the ratio of 37:63. The corresponding ratio with respect to production is 39:61. Average productivity of *kharif* pulses is 527 kg/ha as compared to 479 kg/ha for *rabi* pulses (Orissa Agricultural Statistics, 2008-09).

Table 1 : Area, production and productivity of pulse crops in Orissa (2008-09)

Crop	Area (lakh ha)	% of total	Production (lakh tonnes)	% of total	Yield(kg/ha)
Greengram	7.56	37.8	3.41	34.3	409
Blackgram	6.07	30.4	2.76	27.8	425
Pigeonpea	1.38	6.9	1.19	12.0	825
Horsegram	2.54	12.7	1.09	10.9	362
Cowpea	0.90	4.5	0.62	6.2	670
Gram	0.38	1.9	0.25	2.5	659
Peas	0.26	1.3	0.19	1.9	714
Lentil	0.13	0.7	0.07	0.7	520
Other pulses	0.78	3.9	0.35	3.5	454
Total	20.00	100	9.94	100	497

Trend of Pulse Production

A perusal of the Pulse related statistics since 1951-52 in Orissa reveals that acreage under pulse crops in the state has increased 3.56 times and production by 3.30 times, but the yield rate has declined by 6% from 527 kg/ha in 1951-55 to 497 kg/ha during 2008-09 (Table 2).

Table 2 : Trend in area, production and productivity of pulse crops in Orissa (Quinquennial average)

Period	Area lakh ha	Yield kg/ha	Production (lakh tonnes)
1951-55	4.39	527	2.31
1956-60	4.84	528	2.56
1961-65	7.51	490	3.68
1966-70	8.45	472	3.99
1971-75	9.22	488	4.50
1976-80	13.38	444	5.95
1981-85	17.82	536	9.55
1986-90	19.93	540	10.77
1991-95	21.08	534	11.27
1996-00	17.00	440	7.47
2001-05	15.48	376	5.82
2008-09	20.00	497	9.94
% increase/decrease	+ 356	- 6	+ 330

Further, pulse area in the state has remained almost constant a for the last 25 years and pulse production is less than what it was 25 years back. Therefore, the per capita availability of pulses in Orissa, which had increased progressively from 32 g/ day in 1950-51 to 82 g/ day during 1990-91, has declined again to 32 g/ day during 2007-08 as against the requirement of 52 g/ day (Table 3). In rural areas of Orissa, expenditure on pulses and their products is Rs.13.60/ head/ month as against Rs.22.70 at national level (Economic Survey, Orissa, 2009-10). Estimated pulse requirement in Orissa is 22.3, 27.2, and 49.4 lakh tonnes for the years 2010, 2020 and 2050, respectively, (Lenka, 2000) as against the present production level of 9.94 lakh tonnes.

Table 3 : Per capita availability of pulses in Orissa

Year	Availability (g/day)
1951-52	32
1960-61	28
1970-71	47
1980-81	74
1990-91	82
2000-01	30
2007-08	32

Existing Cropping Systems

Cropping systems having a pulse component are known as pulse based cropping systems. Pulses fit well in various multiple, relay and mixed/ intercropping systems due to their wide range of duration, diverse growth habits, as soil fertility restorers, are deep rooted, drought resistance, rich sources of quality protein, nutritious fodder and weed smothering capacity. The pulse based cropping systems for rainfed uplands, double cropped systems for rainfed and triple cropped systems for irrigated conditions are detailed in Tables 4, 5 and 6.

Paira cropping: Pulses like greengram, blackgram, pea, gram, lentil and *Lathyrus* are grown as *Paira /Uttera/Amuhan* crop in medium and low land rice fields to take advantage of residual soil moisture.

Catch crops: Cultivation of greengram, blackgram, fieldpea and gram as catch crop in between two rice crops in rice-rice system in irrigated land will give additional pulse yield, improve soil health and reduce incidence of pests and diseases due to monocropping of rice.

Multi storey/Multi-tier cropping: Pulses like greengram, blackgram, horsegram and rice bean are grown in between rows of agroforestry systems and newly established orchards.

Table 4 : Pulse based Intercropping systems in rainfed uplands

Intercropping system (Row ratio)	Set specification (cm)	Inter-row distance (cm)	Population of component crops (%)
Pigeonpea + Groundnut (2:6)	30-210-30	30	50+75
Pigeonpea + Greengram/ Blackgram (2:3)	30-120-30	30	80+60
Pigeonpea + Sesame/Niger (2:4)	30-150-30	30	66.7+66.7
Pigeonpea + Rice (2:5) 30-90-30	15	100+62.5	
Pigeonpea + Rice (2:5) 30-120-30	20	80+66.7	
Pigeonpea + Ragi (2:4) 30-100-30	20	92+62.5	
Pigeonpea + Maize (2:2)	30-90-30	30	100+100
Pigeonpea + Cotton (2:8)	30-810-30	90	14.3+86
Pigeonpea + Niger (2:4)	30-150-30	30	66.7+66.7
Pigeonpea + Rice (Mixed)	-	-	40+60
Rice + Greengram/Blackgram (4:1)	Uniform rows	15	80+40
Rice + Greengram/Blackgram (4:1)	Uniform rows	20	80+30
Rice + Groundnut (4:1) Uniform rows	20	80+30	
Maize + Cowpea (2:2) 30-90-30	30	100+50	
Maize + Cowpea (1:1) Uniform rows	30	100+75	
Maize +Runnerbean (2:1)	30+120+30	60	80+66.7
Groundnut + Greengram (6:2)	Uniform rows	30	75+25
Cotton + Greengram/Blackgram (1:1)	Uniform rows	45	100+33
Cotton + Greengram/Blackgram (1:2)	Uniform rows	30	100+66.6
Castor + Greengram/Blackgram (1:2)	Uniform rows	30	100+66.6

Table 5 : Pulse based Double cropping systems under rainfed conditions

Land type	1st crop	2nd crop
Upland	Rice/Maize/Ragi	Greengram/Blackgram/Cowpea/Horsegram
	Groundnut	Greengram/Blackgram
	Jute	Greengram/Blackgram/Horsegram
Medium land	Rice	Greengram/Blackgram/Gram/Pea/Lentil
Low land	Rice	Greengram/Blackgram/Pea

Table 6 : Pulse based Triple/Quadruple cropping systems under irrigated conditions

Land type	1st crop	2nd crop	3rd crop	4th crop
Upland	Rice	Maize/Wheat	Greengram	
	Rice	Groundnut/Mustard/ Sunflower	Greengram	
	Rice	Greengram	Groundnut	
	Rice	Radish	Maize	Cowpea/ Greengram
	Maize	Wheat/Vegetable	Greengram/ Blackgram/ Cowpea	
	Jute	Potato	Greengram	
	Jute	Blackgram	Vegetables	
Medium land	Rice	Groundnut/ Maize/Wheat	Greengram/ Blackgram	
	Jute	Rice	Greengram/ Blackgram	
Low land	Rice	Groundnut	Cowpea	

Yield Gaps

Pulses are the energy rich crops grown in energy-starved conditions. Over the years, these crops are grown in marginal and sub marginal lands with poor management practices. The state average yield of pulse crops is about 30 to 50% of the demonstrated yield and only 25 to 35% of the potential yields reported from research plots (Table 7).

Table 7 : Yield gap in major pulse crops in Orissa

Crop	Potential yield at national level,t/ha	Demonstrated yield at national level,t/ha	Average yield in Orissa, t/ha
Greengram	1.1-1.2	0.9-1.0	0.452
Blackgram	1.0-1.2	0.8-0.9	0.455
Horsegram	1.0-1.2	0.9-1.0	0.428
Pigeonpea	1.5-1.7	1.2-1.5	0.860
Bengalgram	2.0-2.2	1.5-1.8	0.663
Lentil	1.5-1.8	1.2-1.4	0.535
Fieldpea	2.0-2.2	1.5-1.8	0.715

Orissa is coming under the low productivity region (<600 kg/ha) of the country, where yield can substantially be increased by addressing various production constraints. This has been well illustrated through FLD

Programmes of Orissa University of Agriculture and Technology (OUAT) where yield improvement of greengram, blackgram, pigeonpea and fieldpea was by 26-47, 34-48, 30-63 and 22-35%, respectively (Table 8).

Table 8 : Yield increase (%) of major pulse crops of Orissa in FLD over farmers' practice

Crop	*Kharif*	*Rabi*
Greengram	26-47	24-56
Blackgram	34-48	34-45
Pigeonpea	30-63	-
Field pea	-	22-35

Constraints in Pulse Production

Major constraints in pulse production in Orissa can be grouped into three categories, *viz.*, Agro-ecological, Biological and Socio-economic as discussed below.

1. Agro-ecological constraints

- About 92% area under pulses is rainfed.
- Pulses are sensitive to acidity and about 70% soils in Orissa are acidic in nature.
- Pulses are grown in marginal and sub- marginal land with poor fertility and low moisture availability. High temperatures during summer season and moisture stress during *rabi* season in rice-pulse system greatly increase flower drop and bud abortion.
- Due to erratic, uncertain and uneven distribution of rainfall, the crops suffer from moisture stress in the western tract and water logging conditions in the coastal tract.
- Canal irrigated lands in rice-rice sequence are gradually becoming unsuitable for pulse cultivation due to high water table, water logging and poor moisture regulated conditions. As a result, production base of pulses gets shifted to extremely diverse and unproductive environmental conditions.
- Productivity of pulses in rice fallows is becoming unsustainable due to deteriorated soil structure following puddling.
- Poor land preparation in medium land soils, where pulses are grown after paddy
- Low coverage under *paira* cropping

2. Biological constraints

- Lack of high yielding varieties suited to varied agro-ecological conditions particularly cold tolerant varieties of greengram and blackgram for interior districts of Orissa.
- The improved high yielding pulse varieties so far identified are suitable only for medium input conditions, which neither perform well in farmers' field nor compete with high yielding and high input responsive varieties of cereals.
- Non availability of good quality seeds of YMV resistant and cold tolerant greengram and blackgram varieties for rice-fallows, early maturing pigeonpea for pre- *rabi* sowing, BOAA free variety of *Lathyrus,* wilt resistant thermo-insensitive varieties of chickpea and suitable genotypes for intercropping.
- Pulses being rich in proteins are highly susceptible to insects and pathogens, particularly to YMV, powdery mildew and leaf spot in greengram, blackgram and cowpea; powdery mildew in peas; wilt, blight, sterility mosaic, pod borer and blister beetle in pigeonpea; wilt, blight and pod borer in chick pea.
- More susceptibility to stored grain pests as pulse seeds absorb moisture easily from the surroundings due to high protein content resulting in about 50 % loss during storage.
- Poor *Rhizobium* population in medium and low land paddy fields and non-availability of quality *Rhizobium* cultures.
- Heavy infestation of weeds particularly in the initial growth stages of 20 to 45 days after sowing. Yield reduction ranged between 25 to 50% in greengram, 30 to 50% in blackgram, 20 to 40% in pigeonpea, 20 to 30% in peas and lentil, 15 to 30% in cowpea and 15 to 25% in chickpea.
- Besides, the inherent physiological barriers to increase pulse productivity are : i) more energy is required to synthesize proteins than carbohydrates, ii) high concentration of protein in seeds requires early mobilization of N from leaves, causing heavy leaf fall and reduced source capacity, iii) maintenance of symbiotic N_2 fixation in nodules requires prolonged use of photosynthates, thus reducing the amount for storage in seed, iv) indeterminate growth habit, progressive flowering, fruiting and seed setting, and low harvest index.

3. Socio-economic constraints

- Rice is the staple food and farmers do not give much attention to the pulses, which are given subsidiary status and are reluctant to invest more on costly inputs, particularly phosphatic fertilizers and plant protection chemicals on pulses.
- Mostly grown as mixed crops, intercrops and bund crops
- Low seed multiplication rate. Seed replacement ratio is hardly 2-3% as against the desired level of 10-15%.
- Non-availability of inputs including quality seeds in time.
- Post harvest losses are to the extent of 25-30%.
- Lack of social fencing. Stray cattle menace restricts their cultivation in rice fallows

Constraints in Rice Fallows

- Low coverage
- Improper land levelling for *kharif* paddy
- Delayed sowing
- Poor crop stand due to i) use of poor quality seeds, ii) poor land leveling, iii) poor germination of seeds, iv) improper water management just before the harvest of *kharif* rice, v) sub-optimal seed rate, vi) lack of proper varieties to withstand waterlogged condition in initial stage, including ability to germinate under excess water conditions and resist YMV and powdery mildew, and, vii) non-adoption of gap filling techniques i.e., resowing of soaked seeds after harvest of rice
- Water stress at flowering stage
- Lack of weed control
- Poor nutrition
- High pest and disease incidence
- Cattle grazing

Pulse Production Strategies

The two important approaches to increase pulse production in the state are: i) expansion of area, and ii) improving productivity. It is easily possible through inclusion of short duration pulse crops/varieties into various cropping systems such as:

- Complete and partial substitution for upland rice (7 lakh ha) by pulses like pigeonpea, cowpea, greengram and blackgram as well as pulse based intercropping systems
- Intercropping pulses in interspaces of agro-forestry systems
- Inclusion of pulses such as greengram, blackgram, chickpea, lentil, peas and *Lathyrus* in rice-fallows covering about 14 lakh ha in medium and lowland area through (i) *paira* cropping of pulse crops in standing *kharif* rice, (ii) high density sowing of extra short-duration pigeonpea cultivars maturing in 90-100 days during pre-*rabi* season in command areas taking advantage of agronomic dwarfing, residual moisture and limited irrigation facility, and (iii) introduction of pulses like horsegram into reclaimed and wastelands, as contingent crop following failure of upland rice due to moisture stress and as a second crop following maize or rice in upland.
- Increasing area under summer greengram, blackgram and cowpea, as their water requirement is less and can be easily fitted into limited irrigation water availability situations.
- Inclusion of blackgram and greengram as intercrop with spring planted sugarcane and as cover crops in newly developed orchards.
- Increasing acreage under non-conventional and new pulses like Rajmash and rice bean.

Cropwise Production Technologies

Adoption of high yielding genotypes, quality seeds, optimum plant stand, efficient and timely farm operations, balanced nutrition, timely irrigation, integrated pest management and appropriate post harvest management can bring about substantial increase in pulse production in the state.

Improved Varieties of Pulse Crops

Improved varieties of pulse crops suitable for different agro-climatic zones of Orissa in different seasons, rice-fallows and *paira* cropping having resistance/tolerance to major pests and diseases have been identified (Table 9). Availability of quality seeds of these varieties in time can boost up production by about 15%. Thus, multiplication of breeder, foundation and certified seeds of the recommended varieties need be taken up to ensure high rate of seed replacement.

Table 9 : Suitable varieties of pulse crops for Orissa

S. No.	Variety	Season	Duration (days)	Yield (q/ha)	Year of release	Remarks
	Greengram					
1	PDM- 54	K, PR, S	60-65	8.0	1987	MR to YMV
2	HUM-1	K, R, S	60-65	7.0	1999	MR to YMV
3	Dhauli (TT 9E)	K, PR, S	60-70	7.4	1985	
4	Durga (OBGG-52)	K,PR,R,S	60-65	8.0	2002	MR to YMV
5	Kamdev (OUM-) 11-5	K,PR,S	50-55	7.0	2002	Very early maturity
6	Sujata (Hyb 12-4)	K,PR,R,S	65-70	7.0	1985	Shinning green seeds
7	Jyoti (Hyb 4-3)	PR, R, S	65-70	6.0	1985	MR to PM & YMV, black seed
8	K-851	K, S	65-70	7.0	1982	Semi spreading
9	TARM-1	PR, R	70-75	8.0	1997	Small seed, MR to PM & T to YMV
10	Pusa-9072	PR, R	65-70	8.0	1995	T to PM
11	HUM-12	S	65-70	8.0	2002	
12	LGG-460	K, PR, R	65-70	7.0	2001	Rt to PM, T to YMV, Suitable for *piara* cropping
13	Pusa Vishal	K, S	60-65	9.0	2000	Bold seed, Rt to PM
14	Pusa-9531	S	65-70	7.0	2000	MR to YMV
15	Samrat (PDM 139)	K,PR,S	60-65	9.0	2001	Multiple disease resistance
16	Pusa Baisakhi	K,S	60-65	7.0	1970	Erect type
17	SML 668	S	55-60	8.0	2005	
18	LGG 410	PR, R	65-70	7.0	2005	Rt to PM, T to YMV, Suitable for *paira* cropping
19	Pragnya (RUM 1-20)	R	60-65	8.0	2005	MR to YMV & leaf spot
20	COGG 912	K	60-65	8.0	2005	MR to YMV, *Cercospora* leaf spot
	Blackgram					
1	Sarala (B 12-4)	K, PR	70-75	7.0	1985	MR to YMV
2	T-9	K	75-80	8.0	1975	Susceptible to YMV
3	KU-301	R	75-80	9.0	1998	Rt to YMV, green seeded

Contd.

4	Pant U-19	K,PR,R,S	75-85	8.0	1982	MR to YMV & leaf spot, *paira*
5	Pant U-30	K,PR,R,S	70-80	9.0	1982	MR to YMV, erect
6	TU-94-2	R	70-85	9.0	1998	Rt to PM & YMV, for *paira*
7	LBG-17	R	75-85	9.0	1985	T to PM, bold black seed, *paira*
8	Prava (LBG-402)	R	75-85	9.0	1991	Bold dull black seed, *paira*
9	LBG 685	R	75-80	8.0	1995	Suitable for *paira*
10	WBU-108	K, R	75-85	9.0	1995	Bold black seed
11	WBG-26	R	70-75	8.0	1999	Rt to PM
12	ADT 4	PR,R	75-80	9.0	1991	Suitable for *paira*
13	KU 300	PR,R	70.75	10.0	2001	MR to YMV
14	Ujala (OBG 17)	K,R,S	65-70	9.0	2006	MR to YMV, *paira*
15	Prasad (B-3-8-8)	K,PR	65-70	10.0	2008	T to leaf chlorosis, *paira*
	Pigeonpea					
1	UPAS-120	K, PR	120-130	10.0	1976	Early, intermediate
2	Durga (ICPL-84031)	K	125-130	9.0	1995	Early, determinate
3	Jagruti (ICPL-151)	K, PR	120-130	14.0	1989	Early, determinate, cream colour seed
4	Pragati (ICPL-87)	K, PR	115-125	10.0	1986	Early, determinate, light brown seed
5	Manak	K	120-130	14.0	1985	Early, indeterminate
6	Maruthi (ICPL-8863)	K	140-160	12.0	1986	Medium, semi determinate, R to wilt
7	Asha (ICPL-87119)	K	160-170	15.0	1993	Medium, indeterminate, R to wilt
8	Laxmi (ICPL-85063)	K	170-190	15.0	2000	Semi determinate
9	ICPH-8	K, PR	120-130	14.0	1991	Early, indeterminate
10	Co 5	K	110-120	10.0	1991	Early, semi determinate
	Horsegram					
1	Urmi (DS-2-2)	PR, R	90-95	8.0	1992	Seeds mottled grey

Contd.

2	Madhu	PR,R	95-100	7.0	1978	Seeds creamy with red spots
3	Deepali (HPK 6)	PR,R	95-100	8.0	1988	Seeds mottled grey
4	Paiyur-2	PR, R	95-100	8.0	1999	
5	PHG 9	R	100-105	8.0	2001	Semi spreading, thick foliage
	***Lathyrus* (Khesari)**					
1	Ratan (Bio-L 212)	R	85-95	9.0	1997	Low BOAA
2	Pusa 24	R	90-100	8.0	1976	Medium BOAA (0.2%)
3	Prateek (LS 157-14)	R	90-100	9.0	–	Low BOAA (<0.07%)
	Field pea					
1	Rachna (KPMR-10)	R	95-105	12.0	1982	Rt to PM, tall, smooth bold round seed
2	HUDP-15	R	95-105	12.0	1999	Dwarf, leafless, Rt to PM
3	DDR-23	R	90-95	10.0	2000	Dwarf, extra early, Rt to PM
4	Aparna (HFP-4)	R	100-110	13.0	1988	Early, dwarf, leafless, Rt to PM
5	Pusa Panna (DDR-27)	R	90-95	12.0	2001	Dwarf, extra early
6	Uttara (HFP 8909)	R	100-110	14.0	1997	Dwarf, Rt to PM
7	Ambika	R	95-105	12.0	2000	Rt to PM
8	Alankar (DMR 7)	R	100-110	12.0	1996	Tall, Rt to PM
	Rajmash					
1	Uday (PDR-14)	R	90-100	13.0	1987	Red variegated seeds
2	Amber (IPR-96-4)	R	90-100	15.0	2002	Determinate, red variegated
3	HUR-137	R	90-100	15.0	1992	Deep pink seeds, erect, semi dwarf
4	HUR-15	R	90-100	14.0	1989	White seeds, erect, determinate
	Lentil					
1	Asha (B 77)	R	90-100	8.0	1980	Rt to rust, small seeds
2	Mallica (K-75)	R	100-105	9.0	1996	Bold seeds
3	Lens-4076	R	100-110	9.0	1993	Bold seeds, Rt to rust

Contd.

4	Sheri (DPL-62)	R	100-110	9.0	1997	Bold seeds
5	Priya (DPL-15)	R	100-110	9.0	1995	Bold seeds
6	Noori (IPL 81)	R	95-105	9.0	2000	T to rust, wilt, bold seeds
	Bengal gram					
1	JG-11	R	85-95	12.0	1990	*Desi*, Early, Bold, Rt to wilt
2	L-5550	R	110-115	11.0	1978	*Kabuli* type, salmon white
3	Pusa-1003	R	110-115	14.0	1999	*Kabuli* type, T to wilt
4	Pusa-256	R	105-115	13.0	1985	Bold seed, light brown
5	Pusa-372	R	105-115	10.0	1993	*Desi*, small seed, suitable for late sowing
6	Bharati (ICCV-10)	R	85-90	12.0	1992	*Desi*, Early, bold white seed, Rt to wilt
7	ICCC-4	R	85-90	12.0	1983	Early
8	Sweta (ICCV-2)	R	80-85	12.0	1993	*Kabuli* type, early, Rt to wilt
9	JG-62	R	110-115	11.0	1982	*Desi*
10	Radhey	R	95-100	11.0	1982	*Desi*, bold seed, suitable for late sowing
11	Annegiri	R	90-100	10.0	1978	*Desi*, Early
12	Vijay (PG 95311)	R	85-95	14.0	2002	*Kabuli*, Rt to wilt
	Cow pea					
1	Swarna (C 152)	K,R	85-95	8.0	–	Grain type
2	FS 68	K,R,S	75-85	9.0	–	Grain type, Bushy
3	Pusa 152	K,R	85-95	9.0	–	Brown seed
4	Amba (V 16)	K,PR,R	85-95	7.0	–	Grain type
5	Co 6	K,PR,R	85-90	10.0	–	Bold seeds
	Rice bean					
1	BRB 1	K,PR	85-95	7.5	2001	Semi erect, light green seeds
2	RBL 6	K,PR	85-95	8.0	2003	Semi erect, yellow green seeds
3	RBL 35	K,PR	95-105	8.5	2003	–
4	RBL 50	K,PR	85-95	8.5	2004	Early

K: *Kharif*, PR: Pre-*rabi*, R: *rabi*, S: Summer, PM: Powdery mildew, MR: Medium Resistant, Rt: Resistant, T: Tolerant

Soil and climatic requirements

Well-drained sandy loam and loam soils with pH range of 5.5-7.5 are ideal for most of the pulse crops. Pulse crops like gram, field pea, rajmash, *Lathyrus* and lentil are grown in relatively heavy soils whereas greengram, blackgram, pigeonpea and horsegram are cultivated in relatively light textured soils in the state. Crops grown in winter and summer seasons with irrigation are grown in soils with high moisture retention capacity.

The pulse crops require well distributed annual rainfall of 600-800 mm. The optimum temperature range is 20-28 °C for pigeonpea, 25-35 °C for greengram, blackgram and cowpea, 25-30 °C for gram, 15-18 °C for field pea and 15-25 °C for lentil. Almost all the pulse crops are resistant to drought but are very much susceptible to heavy rainfall, frost and low temperature.

Land preparation

A well levelled field is a must for raising a successful pulse crop. For *kharif* crops, the land must be well drained and needs to be prepared by one deep ploughing followed by two harrowings with *desi* plough or cultivators and planking. For pre-*rabi* and *rabi* crops grown on residual soil moisture, the field should be ploughed immediately after harvest of the previous crop and seeds should be sown as soon as possible followed by planking to conserve soil moisture and ensure good contact of seeds with soil for better germination. Proper land leveling for *kharif* rice is required if *paira* crop of pulse is to be grown in *rabi* season. For summer season crop, pre-sowing irrigation helps in land preparation and sowing at optimum soil moisture. In acid soils, application of lime @ 0.2 LR (lime requirement) or blanket dose of 500 kg/ha in lines at the time of sowing is recommended to overcome aluminium toxicity and ensure good growth of bacteria and nodulation. Application of lime @ 0.2 LR ameliorates soil acidity and increases crop productivity in pulse crops (Table 10).

Table 10 : Effect of lime application (0.2 LR) on seed yield (kg/ha) of pulses

Pulses	-Lime	+Lime	% increase
Pigeonpea	344	820	138
Blackgram	392	518	32
Soybean	520	700	35
Cowpea	2192	3204	46

Source : Das and Rout (2006)

Seed rate and spacing

Optimum seed rate is essential to get adequate plant stand for higher yield. For *paira* cropping of greengram, blackgram, fieldpea, chickpea, *Lathyrus* and lentil, 25-50% more seeds than the recommended rate are required. Higher seed rate is also recommended for delayed sowing and broadcast sowing. The seed rates and spacing of different pulse crops are given in Tables 11 and 12.

Table 11 : Recommended seed rate (kg/ha) for different pulse crops

Crop	Season		
	Kharif	**Pre-*rabi/rabi***	**Summer**
Greengram	25	30	25
Blackgram	20	30	25
Pigeonpea	15-20	30	-
Cowpea	20	25	25
Horsegram	-	30	-
Fieldpea	-	50-60	-
Chickpea	-	50	-
Lentil	-	40	-
Rajmash	-	50	-
Ricebean	25	30	-

Table 12 : Recommended spacing (cm x cm) for different pulse crops

Crop	Season		
	Kharif	**Pre-*rabi/rabi***	**Summer**
Greengram	30 x 10	25 x 10	30 x 10
Blackgram	30 x 15	25 x 10	30 x 10
Pigeonpea	45 x 20	30 x 15	-
Cowpea	60 x 20	45 x 15	45 x 15
Horsegram	45 x 15	45 x 15	-
Fieldpea	-	30 x 10	-
Chickpea	-	30 x 10	-
Lentil	-	30 x 15	-
Rajmash	-	30 x 15	-
Ricebean	30 x 10	30 x 10	-

Sowing Time

Optimum sowing time for *kharif* pulses is 2nd half of June. Delayed sowing in July and August reduces yield significantly (Table 13). Pre-*rabi* and *rabi* crops should be sown in 2nd half of September and 2nd half of October to 1st half of November, respectively. Infestation of whitefly and yellow vein mosaic disease increases in greengram when sowing is delayed to 1st week of January

(Table 14). Optimum sowing time for spring/summer pulse is 2nd half of February to 1st half of March.

Table 13 : Effect of date of sowing on seed yield of kharif greengram at Berhampur, Orissa

Date of sowing	Grain yield, kg/ha
10 July	854
25 July	831
09 August	754
24 August	626
CD (0.05)	84

Table 14 : Effect of date of sowing in late winter on yield and pest infestation in greengram at Ranital, Orissa

Date of sowing	Days to maturity	Seed yield (kg/ha)	Whitefly/10 plants	YMV(%)
01 January	77.7	675	08.4	04.3
10 January	77.4	640	11.5	07.8
20 January	76.9	591	14.0	10.3
30 January	76.6	538	19.2	14.0
09 February	75.7	485	25.3	18.1
19 February	74.6	416	17.0	25.8
01 March	74.5	362	16.0	31.8
CD (0.05)	NS	13	0.8	1.01

Seed treatment

Seeds are to be treated with seed dressing fungicides like Carboxin 37.5% + Thiram 37.5% (0.15%) or with Carbendazim 0.15% + Thiram 0.25% to control seed borne diseases such as damping off of seedlings, seedling blight and wilting. For controlling root knot and reniform nematodes, particularly in greengram, blackgram and cowpea under intensive cropping, seeds are to be treated with Carbosulphan dispersible solid (DS) 3%. Seed inoculation with *Rhizobium* culture @ 20 g/ kg seeds improves yield by 10 to 13% in different pulse crops. Seed treatment with Mo @ 160 mg Mo as Sodium molybdate and 8 mg Co as Cobaltous chloride/kg seeds enhances nodulation and seed yield of greengram (Table 15). Seed treatment with insecticides/ fungicides should be done one week before *Rhizobium* treatment. *Rhizobium* inoculation should be done every year when pulses are grown after rice, since bacterial population is reduced drastically under waterlogged conditions. Combined seed inoculation of *Rhizobium,* PSB and VAM enhances yield of greengram at Nayagarh over their sole application (Table 16).

Table 15 : Effect of Mo and Co on nodulation and yield of greengram

Treatment	Yield, kg/ha	% increase	Nodules/ plant	Leghaemoglobin, mg/g nodule
Control	342	-	9	0.69
Rhizobium	430	26	15	0.76
Rhizobium+Mo	568	66	22	0.79
Rhizobium+Co	560	64	20	1.32
Rhizobium+Mo+Co	610	78	27	1.34
CD 5%	52	-	5	-

Source : Pattanayak *et al.* (2000)

Table 16 : Effect of biofertilizers on nodulation and seed yield of greengram

Treatment	Yield, kg/ha	Times increase	Nodules/plant
Control	319	-	20
VAM	832	1.61	33
Rhizobium	876	1.75	47
VAM+*Rhizobium*	1040	2.26	45

Source : Das *et al.* (1997)

Fertilizer Application including Micro-Nutrients

At final land preparation, well decomposed FYM or compost @ 2.5 t/ha is incorporated into the soil. Besides this the recommended dose of fertilizer is applied as basal in furrows at the time of sowing. Use of mixed source of soluble and insoluble (rock phosphate) phosphates in 1:1 ratio enhances yield of pulse based cropping sequence in acid soils (Table 17). Lime in combination with FYM further increases the yield (Table 18). Application of 20 kg S/ha through gypsum, SSP or elemental S increases yield of pulse crops in the range of 22 to 46% (Table 19). Pulses also responded positively to micronutrients like Zinc and Boron (Table 20). Recommended fertilizer doses for different pulse crops are given in Table 21.

Table 17 : Effect of rock phosphate and single super phosphate on grain yield of greengram-rice sequence in sandy clay loam soil (pH 5.9)

Treatment	P_2O_5, kg/ha		Yield, kg/ha	
	Greengram	Rice	Greengram	Rice
Control	-	-	400	3100
SSP	40	40	590	3960
URP+SSP(1:1)	40	40	580	3530
URP+SSP(3:1)	40	40	580	4320
URP	40	40	500	4060
CD 5%	-	-	NS	222

Source : Das and Rout (2006). Udaipur Rock Phosphate (URP) contains 14.6% P_2O_5, 1.2% Citrate soluble P_2O_5, 37% CaO and 13.1% MgO

Table 18 : Yield response (%) of pulses to graded doses of lime with or without FYM

Lime requirement	Pigeonpea		Cowpea	
	-FYM	+FYM	-FYM	+FYM
0.0	-	46	-	32
0.1	42	49	51	78
0.2	39	56	52	78
0.3	46	56	52	78

Source : Mishra (2002)

Table 19 : Effect of different sources and levels of S on seed yield (kg/ha) of pulses

Treatment	Greengram	Blackgram	Pigeonpea	Gram	Lentil
S source					
Elemental S	804	742	1108	1800	1450
Gypsum	861	721	1092	1911	1601
SSP	820	699	1119	1984	1504
S level, kg/ha					
0	705	503	915	1468	1087
20	861	690	1163	1884	1584
40	879	751	1084	1989	1538

Source : AICRP on Pulses (1994)

Table 20 : Response of greengram (kg/ha) crop to secondary and micronutrients

Treatment	Site I	Site II	Site III
Control	483	448	494
Zn 5.5 kg/ha	560	497	603
(Zn 5.5 kg +S 20) kg/ha	563	-	-
(Zn 5.5 +S 20 +B 1.0) kg/ha	607	-	-
(Zn 5 +S 20 +Mo 0.5) kg/ha	-	610	642
CD 5%	36	45	57
Soil status (ppm):			
Zn	0.80	0.75	0.45
B	0.55	0.55	0.39
S	12.0	9.0	8.5

Source : AICRP on Micro and Secondary nutrients (1999-2000, 2000-01, 2002)

Table 21 : Recommended fertilizer dose (kg/ha) for different pulse crops

Crop	N	P_2O_5	K_2O
Greengram	20	40	20
Blackgram	20	40	20
Pigeonpea	20-30	40-60	20
Cowpea	25	50	25
Horsegram	10-20	20-40	20
Fieldpea	25	50	25
Chick pea	20	40	20
Lentil	20	40	20
Rajmash	80	60	40
Ricebean	40	40	20

Nutrient Management in Pulse Based Cropping Systems

Application of 50% N from organic source and the remaining 50% from chemical sources sustained higher productivity of pigeonpea + rice (2:5) and rice-horsegram cropping systems in acid soils of Orissa. Combined application of recommended dose of chemical fertilizers + lime (0.2 LR) + FYM 5 t/ha increased system yield of pigeonpea + groundnut (2:8) intercropping system and groundnut - horsegram sequence by 233 and 186 % over farmers' practice and 119 and 109% over recommended dose of chemical fertilizers, respectively in acid lateritic soils of Orissa (Table 22). In P deficient soils, greengram yields in rice - greengram sequence were higher with phosphatic fertilizers applied to greengram as compared to yields of the crop grown on residual P applied to rice crop.

Table 22 : Response of pulse based cropping systems to integrated use of fertilizers, liming and FYM

Treatment	Pigeonpea + groundnut		Groundnut - horsegram	
	Groundnut equivalent yield, kg/ha	Yield response, %	Groundnut equivalent yield, kg/ha	Yield response, %
Farmers' practice	898	-	1471	-
100% RDF	1365	52	2011	37
100% RDF + Lime (0.2 LR)	2355	162	3496	138
100% RDF+Lime (0.2 LR) + FYM (5 t/ha)	2986	133	4208	186

Weed management

Yield reduction due to weeds ranges from 15 to 50% in different pulse crops and the critical period of crop weed competition varies from 15 to 60 days after sowing (Table 23).

Table 23 : Crop weed competition and yield reduction in pulse crops due to weeds

Crop	Critical period of crop-weed competition, DAS	Yield reduction (%)
Greengram/Blackgram	15-30	25-50
Pigeonpea	15-60	20-40
Gram	30-60	15-25
Fieldpea/Lentil	15-45	20-30
Cowpea	15-30	15-30

The pulse crops are dominated by following categories of weeds in *kharif* and *rabi* seasons:

Grasses

Kharif : *Dactyloctenium aegyptium, Digitaria sanguinalis, Echinochloa colonum, Panicum maximum, Isachene despare, Eleusine indica*

Rabi : *Echinochloa colonum, Digitaria sanguinalis, Eleusine indica, Dactyloctenium aegyptium*

Sedges

Kharif: *Cyperus rotundus, Cyperus iria*

Rabi: *Cyperus iria, Cyperus rotundus*

Broad Leaved

Kharif: *Celosia argentea, Ageratum conizoides, Croton sparciflorus, Commelina bengalensis, Oldenlandia corymbosa, Physalis minima, Euphorbia hirta, Phyllanthus niruri*

Rabi : *Cleome viscose, Physalis minima, Amaranthus viridis, Heliotropium indicum, Leucas aspera*

Control measures : Irrespective of the pulse crops, first hoeing and weeding at 3-week stage followed by second at 15 days later, controls weeds effectively, conserves soil moisture and improves plant growth. Light hoeing creates dust mulch and conserves soil moisture during stress in *kharif* season and for crops grown on residual soil moisture. Pre plant incorporation of Fluchloralin or Trifluralin @ 0.75 kg/ha at one day before sowing (DBS), pre-emergence application of Alachlor or Butachlor or Pendimethalin @ 1.0 kg a.i. /ha or Oxadiazon or metolachor @ 0.75 kg a.i./ha with 500 litres of water one DAS followed by one hand weeding at 50 DAS for pigeonpea and at 35 DAS for other pulse crops is equivalent to two hand weedings. Post-emergence spray of Quizalofop-p- ethyl @ 0.05 kg a.i. /ha or Imazathapyr @ 0.5 kg/ha at 15-20 DAS (2-3 leaf stage of weeds) is also recommended for effective weed control.

Weed management in the early stages of crop growth is essential. Since much progress has not yet been made on the use of herbicides in pulse crops, an integrated approach comprising of mechanical, cultural and chemical measures can ensure higher yields. Some useful steps are as follows:

- Prepare stale seed bed
- Select quick growing and high yielding varieties
- Adopt pre monsoon sowing to ensure successful establishment of the crop stand and to minimize the weed competition
- Ensure suitable crop husbandry (seed treatment, fertilization, post planting plant protection measures) at the right stages to facilitate a faster crop growth and rapid canopy development to enhance competitive ability of crops
- Pre planting or pre emergence herbicides help in reducing weed intensity in the early stages of the crop growth
- Timely intercultural operations
- Use labour saving weeders
- Intercrop with weed smoothering crops to save cost of at least one hand weeding

Water management

Crops during *kharif*, pre-*rabi* and *rabi* seasons are usually grown as rainfed or on residual soil moisture. Drainage rather than irrigation is more important in *kharif* season. During summer season, light irrigations are required for germination, 20 to 25 DAS, flowering and pod-development stages for higher yield. Flowering and pod development stages are the critical periods for irrigation in most of the pulse crops. The water requirement of different pulse crops varies from 150 to 250 mm for greengram, blackgram, horsegram and field peas to 350 to 400 mm for pigeonpea and 500 mm for rajmash. Almost all the pulse crops require 3 to 4 irrigations scheduled at 25 to 50% available soil moisture or at 0.6-0.8 IW (Irrigation Water)/CPE (Cumulative Pan Evaporation) ratio. But Rajmash requires about six irrigations applied at IW/CPE of 1.2 to 1.4 in sandy clay loam soil of North Central Plateau Zone (Table 24).

Table 24 : Effect of irrigation and nitrogen on seed yield (kg/ha) and water requirement of Rajmash

Irrigation	N (kg/ha)				Mean yield	WR, cm
	0	40	80	120		
IW/CPE=1.0 (5)	983	1340	1608	1600	1383	42.9
IW/CPE=1.2 (6)	1285	1759	2025	2026	1774	48.9
IW/CPE=1.4 (6)	1359	1836	2104	2117	1854	49.5
Mean	1200	1645	1912	1916		
CD 5%	Irrigation	Nitrogen	Interaction			
	48	56	63			

Plant Protection Measures

Insect pests

Infestation of insect pests results in 20 to 80% loss in pulse crops. The important pests and their management strategies are briefly described below.

A. Vegetative stage

1) **Flea beetle** (*Phyllotreta* spp.): Attacks at 2-6 leaf stage, making numerous small round holes in leaves during early morning and evening hours.

   - Dusting Chlorpyriphos 1.0% D or Malathion 5% D @ 25 kg/ha in morning dew hours.

2) **Hairy caterpillar:** Red hairy (*Amsacta albistriga*) and Bihar-hairy caterpillars (*Spilosoma obliqua*) attack during monsoon season and completely defoliate the crop at young stage.

- Clean cultivation including summer ploughing and bond trimming
- Putting a band of insecticide (Chlorpyriphos 1% or Malathion 5% D) around the field and dusting the field with the same insecticide. Alternatively, spraying Chlorpyriphos 20 EC or Monocrotophos 36 SL or Triazophos 40 SL @ 1000 ml/ha is effective.

3) **Aphid** (*Aphis craccivora*), **Jassid** (*Amrasca biguttula biguttula*), **Thrips** (*Megalurothrips usitatus*) and **White fly** *(Bemisia tabaci)*: Nymphs and adults suck the cell sap and weaken the plants. Aphids mostly attack greengram, cowpea and blackgram. Jassids are seen throughout the season. Whitefly is a vector of YMV of greengram, blackgram and cowpea.

- Spraying Imidachloprid 200 SL or Spinosad 45 SC @ 125 ml/ha or Thiomethoxam 25 WG or Acetamiprid 50 SP @ 125 g/ha.

4) **Cut worm** (*Spodoptera* spp.): The larvae attack the crop enmass or individually consuming the leaves and cutting the plants during night.

- Control as suggested for pod borer given below.
- In case of enmass attack, control measures suggested for hairy caterpillars are effective.

5) **Mites** (*Schizotetranychus cajani*): Both adults and nymphs feed on plant sap producing yellow spots on the upper surface of the leaves. The infected plants do not bear flowers.

- Dusting with micronized Sulphur dust 5% or spraying with Wettable Sulphur 85% WDP @ 2.5 kg/ha or Dicofol 18.5 EC @ 2.5 l/ha or Propargite 57 EC @ 1000 ml/ha or Fenpyroximate 5%. EC @ 500 ml/ha.

B. Reproductive stage

1) **Pod borer complex:** *Helicoverpa armigera* attacks Bengalgram, pigeonpea, greengram, blackgram, fieldpeas and horsegram. *Maruca testulalis* and plume moth (*Exelastis atomosa*) attack almost all the pulse crops.

- Summer ploughing to destroy the hibernating pupae,
- Setting of pheromone trap for *Helicoverpa* (15-20 /ha),
- Spraying of Nuclear Polyhedrosis Virus (NPV) against *Helicoverpa* and *Spodoptera* @ 250 LE/ha, or

- Two sprays with Cypermethrin 10 EC @ 1000 ml/ha or Indoxacarb 14.5 EC @ 300 ml/ha or Lambda Cyhalothrin 5 EC @ 500 ml/ha or Spinosad 45 SC @ 125 ml/ha at 50% flowering and 15 to 20 days later, or
- Spraying of *Btk* formulations @ 1 kg/ha to control *Spodoptera* and *Helicoverpa* in cowpea, peas and chickpea
- Alternatively, spray neem formulations at 300 ppm @ 2.5 lit/ha or 1500 ppm @ 1.5 lit/ha.

2) **Blister beetle** (*Mylabris pustulatus*)**:** The adults damage the flowers in pigeonpea

- Hand collection of the adults and putting them in kerosinised water is recommended.

3) **Pod bug** (*Clavigralla gibbosa*)**:** The nymphs and adults suck the sap from pods and grains resulting in shrivelled seeds.

- Spraying of Dimethoate 30 EC @ 1000 ml/ha or imidacloprid 200 SL @ 125 ml/ha or Thiamethoxam 25 WG @ 125 g/ha
- Alternatively, spray neem formulations 10,000 ppm @ 500 ml/ha.

C. Storage pests

Pulse beetle (*Callosobruchus* spp.) The grubs destroy the seeds leaving only the outer pericarp. The adults start infesting the matured pods in the fields, thus the infestation is carried from field to storage godown. It can be managed by-

- Sun-drying to maintain the moisture below 10 % in seeds
- Do not use large containers to store small quantity of grains
- Before storing, stitch/repair all torn bags and after filling, stitch the bags to make them tight.
- Impregnate gunny bags by soaking in 0.1% Malathion solution for 10 minutes and dry them before using for grain storage
- Treat the walls, dunnage materials and ceilings of empty godown with malathion 50 EC @ 10 ml/lit or Dichlorvos (DDVP) 75% SC @ 7ml/lit at 3 litre spray fluid/100 m^2 surface.
- The airtight empty godown can be fumigated using 120 Aluminium phosphide tablets/100 m^3 for 5-7 days.

- Smearing with vegetable oil 0.02% (mustard, neem, karanj, palmolene) and ensure thorough mixing of oil.
- Mixing of *begunia* (*Vitex negundo*) or neem (*Azadirachta indica*)/ curry leaf (*Murraya koenigi*) powder, banana (*Musa paradisica*) leaf ash or cowdung ash, extracts of aswagandha (*Withania somnifera*) or kalmegh (*Andragraphis paniculata*) @ 3-5 % with sun dried seeds and storing in air tight containers.
- Surface sprays of bags in storage with Malathion 10 ml/litre or DDVP 7 ml/litre when pest infestation is observed.
- Infestated grains can be fumigated with 2-3 Aluminium phosphide tablets of 3 g each per tonne for 5-7 days with care.
- Bulk grains should be covered with a layer (2.5 cm) of sand or saw dust or dung ash.
- For home use, pulses can be milled after harvest to avoid pulse beetle attack

Diseases

The yield is reduced by 10 to 20% by diseases. The important diseases are:

1. **Leaf spot and blight** (*Cercospora* spp): Occurs in blackgram, greengram, pigeonpea, horsegram and cowpea during vegetative stage. Round to oval or irregular dark brown spots appear on both sides of leaves resulting in yellowing, drying and defoliation.
    - Seed treatment with Carbendazim 0.2% + Thiram 0.3% or Thiram+Carboxin (0.2%)
    - Foliar spraying with Mancozeb (0.3%) or Carbendazim 12% + Mancozeb 63% (0.2%)
2. **Dry root rot and steam rot:** (*Macrophomina phaseolina*): Almost all the pulse crops are affected; they are more prevalent in monsoon season when soil becomes too wet. Dark brown to black deep seated linear lesions appear on the tap root, tertiary roots and stem at collar region resulting in poor root growth and leading finally to the death of the plant.
    - Seed treatment with carbendazim 0.2% + Thiram 0.3% or Thiram+Carboxin (0.2%)
    - Spray the basal portion of the plants with Metalaxyl 8% + Mancozeb 64% @ 0.2%

3. **Yellow vein Mosaic Virus:** Seen in almost all pulse crops. It is transmitted by whitefly. Veinal chlorosis and yellow green mottling of leaf occurs.
   - Use tolerant variety
   - Prophylactic spray to control insect vector with imidachloprid @ 125 ml/ha or methyl demeton @ 1000 ml/ha
4. **Rust** (*Uromyces appendiculatus*): Reddish brown to orange brown rusty pistules/blisters appear on leaves in patches resulting in yellowing, drying and defoliation
   - Seed treatment as mentioned above
   - Spraying with Karthane (0.2%) or Calixin (0.1%)
5. **Fungal wilt** *(Fusarium oxysporum)*: Infects pigeonpea, pea, *Lathyrus*, lentil and bengalgram at various stages of crop growth resulting in partial or complete wilting.
   - Spray the basal portion of plants with Carbendazim (0.2%)/ Carbendazim 12% + Mancozeb 63% (0.2%)
6. **Anthracnose disease** (*Colletotrichum lindemuthianum*): Dark brown to black necrotic patches appear on leaves, twigs, branches and pods.
   - Foliar spray with Carbendazim (0.15%)/Thiophanatamethyl 70% WP (0.2%).
7. **Powdery mildew** (*Leveillula taurica*): Grey white powdery mass covering leaves and entire plant arresting the vegetative growth.
   - Dusting sulphur powder @ 25 kg/ha or spraying with wettable sulphur @ 0.4%
   - Foliar spray with Cabendazim (0.15%), karathane (0.2%) or Calixin (0.1%)
8. **Root knot nematode** (*Meloidogyne incognita*): Incipient wilting, stunted growth, nodules are parasitized by nematode forming galls in greengram, blackgram and pigeonpea.
   - 2-3 deep summer ploughings
   - Crop rotation with rice and oilseeds
   - Seed dressing with Carbosulfan 25% DS @ 3% (w/w)
   - Soil application of Carbofuran 1 kg a.i/ha

Harvesting and storage

Pods of pulse crops are dehiscent in nature. Crops are harvested before splitting of pods and when 80 to 85% of pods are dry. It is better to harvest by cutting the plants just above the ground surface, leaving the roots for incorporation in soil which improves soil fertility as compared to uprooting of the whole plants. Incorporating the green plants of greengram, blackgram or cowpea after one or two pluckings is still better for the subsequent crops (Table 25). Matured pods of greengram and blackgram during *kharif* season are harvested by plucking to avoid viviparous germination. Seeds are dried to proper moisture content (8-10%) for safe storage. Seeds should be treated with Carbendazim 0.2% and Thiram 0.3% for safe storage.

Table 25 : Effect of harvesting management of pulse crops on soil fertility

Treatment	N contribution (kg/ha)	Yield of succeeding *ragi* (kg/ha)	Soil organic carbon (%) after *ragi* crop
Control	-	559	3.8
Incorporation at flowering stage	38.4	1120	5.5
Incorporation after one plucking	31.8	1027	5.3
Harvesting by cutting at ground level	26.5	941	5.0
Harvesting by uprooting	19.3	848	4.2

Source : Das *et al.* (2001)

Cultivation of Pulse Crops in Non Conventional Seasons/Areas

Pulses like greengram, blackgram, pea, gram, lentil and *Lathyrus* are grown as *Paira/Uttera/Amuhan* crop under medium and low land ecosystems in standing crop of rice (2 weeks before harvesting). The advantages are: better utilization of residual soil moisture, saving energy, money and time due to zero till cropping and coverage of more area in short time.

Management of *Paira* Pulse Crop

- Cultivation of stiff straw, non-lodging, medium duration rice varieties maturing in 135-140 days.
- Ideal land type is medium to low land with medium to heavy soil texture.
- Proper land levelling for *kharif* rice.
- Harvesting rice crop with 10 cm stubble height and immediately removing the harvested plants from the field.

- Sowing *paira* crop in the 1st week of November.
- Keeping thin film of water at the time of sowing of *paira* crop.
- Using 25-50% higher seed rate.
- Seed treatment and seed inoculation with *Rhizobium* and PSB @ 200 g each for 10 kg seeds.
- Application of phosphorous dose of *paira* crop to *kharif* rice in addition to the dose of rice crop or Application of 100 kg DAP and 33 kg Muriate of Potash/ha at panicle initiation stage of rice crop.
- Foliar spray of DAP or urea 2% at flowering of *paira* crop and repeated after 15 days.
- Suitable varieties are:

Greengram	:	Pusa-9072, TARM-1& 2, OBGG 52, OUM-11-5, LGG-460,
Blackgram	:	Ujala, Prasad, TU-94-2, LBG-17, LBG-402, LBG-485, ADT4
Fieldpea	:	Rachna, HFP-4 (Aparna), DDR-27 (Pusa Pana), HFP-8909 (Uttara), DMR 7 (Alankar)
Gram	:	JG 11, Radhey, L 550
Lathyrus	:	Ratan, Prateek.

Future Thrust

- Pulse area at present is 20.00 lakh ha which is more or less equal to what it was in 1983-84. The plateau can be broken by crop substitution, intercropping and expansion of area under bund crop, *paira* crop, catch crop and summer pulse.
- Improvement of local germplasm for development of cold tolerant variety of greengram and blackgram.
- Seed multiplication and distribution to increase Seed Replacement Rate (SRR) to 10 to 15 %.
- Promoting lentil cultivation in a big way.
- Soil amelioration to reduce acidity.
- Line sowing to ensure plant population, early interculture and mulching.

- Integrated weed management, especially for *kharif* pulse.
- Irrigation at critical stages (flowering and pod development).
- Integrated nutrient management (seed inoculation, balanced fertilization, organic + inorganic).
- Integrated pest management
    - Insect pest: Pod borer complex – monitoring, need based chemical application.
    - Major diseases: PMD, YMV, leaf curl, *Cercospora*, sterility and wilt. Summer ploughing, resistant variety and seed treatment will reduce intensity of infestation.

References

AICRP on Micro and Secondary Nutrients (1999-2000, 2000-01, 2002) Annual Report, All India Coordinated Research Project on Micro and Secondary Nutrients OUAT, Bhubaneswar.

AICRP on Pulses (1994) Annual Report, All India Coordinated Research Project on Pulses, Berhampur, OUAT, Bhubaneswar.

Das, P.K and Rout, K.K. (2006). Nutrient management of pulses. In. Nutrient Management of Crops in Soils of Orissa (Mitra, G.N. Ed.), pp. 75-89, OUAT, Bhubaneswar.

Das, P.K., Jena, M.K., Sahu, D., Pradhan, N.K. and Nanda, S.K. (2001). Response of ragi crop nitrogen contribution of upland rainfed legumes. Environment and Ecology 19 (1):52-55.

Das, P.K., Sahoo, P.N. and Jena, M.K. (1997). Effect of VA-Mycorrhiza and Rhizobium inoculation on nutrient uptake, growth attributes and yield of greengram (*Vigna radiata*). Environment and Ecology 15 (4):830-833.

Economic Survey. (2009-10). Govt. of Orissa.

Lenka, D. (2000). Agriculture in Orissa..480 p, Kalyani Pub., New Delhi.

Mishra, M. (2002). Response of crops to graded doses of lime amended with or without FYM. M.Sc (Ag.) Thesis submitted to OUAT. Bhubaneshwar.

Orissa Agricultural Statistics, (2008-09). Directorate of Agriculture & Food Production, Orissa.

Pattanayak, S.K., Dash, D., Jena, M.K., and Nayak, R.K. (2000). Seed treatment of greengram with Molybdenum and Cobalt. I. Effect on nodulation, biomass production and N-uptake in an acid soil. Journal of the Indian Society of Soil Science 48 (4): 769-773.

❑❑❑

18

Punjab

S.S. Walia and Roopeinder Singh
Department of Agronomy, PAU, Ludhiana

Pulses play an important role in improving the soil health through biological nitrogen fixation. These crops also help to conserve natural resources and therefore help to conserve natural resources and sustain the agricultural productivity. With almost one-fourth population below the poverty line, the recent steep hike in prices has made the availability of pulses for this section of society even more difficult, thus, further jeopardizing both food and nutritional security.

An adult needs about 85 g of pulses per day in a balanced diet which also includes proteins from animal sources. In purely vegetarian diets, the Food and Agriculture Organization and the Nutritional Advisory Committee of India have recommended 104 g of pulses per capita under the minimum nutritional programme. The present production of pulses in Punjab, the per capita availability and total requirements are 4 g/capita/day and 909 thousand tonnes, respectively. The availability of pulses in 2007 was only 34 g in India and it was only 2 g in Punjab in 2009. Thus, there is a huge gap between the minimum requirement and availability at present levels of production. To meet the increasing demand of the burgeoning population, pulse production has to be increased substantially. In Punjab, the production has to be increased nearly twenty folds.

Trends of Pulses Production

The Punjab state with 1.5 percent area has played a very critical role in ensuring food security. The state with 2.4% of the nation's population, contributes towards 12 % of the cereals to the national food basket. But, the success of the green revolution has negatively impacted pulses cultivation in

Punjab state. The area under pulses has declined from 903 thousand hectares in 1960-61 to 22 thousand hectares in 2008-09. Similarly, the production has declined from 308 thousand tonnes to 19 thousand tonnes. Now, the Punjab Govt. has put a ban on rice transplanting before June 10. So, there is a scope for increasing area under summer moongbean.

Existing Cropping Systems

Rice-wheat is the predominant cropping system in central Punjab and occupies 72 per cent of the total cultivated area. The area under rice has increased tremendously at the cost of area under maize, groundnut and moong, and presently rice is cultivated on about 26.42 million hectares. The rapid spread of rice-wheat system has been attributed mainly to its better adaptability, availability of high yielding varieties and mechanization of the two crops. The farmers are further encouraged by the favourable Government policies of minimum support price (MSP) and committed procurement and free electricity for tubewells. Cultivation of rice crop has caused severe over exploitation of the underground water resources in order to meet its water requirement and resulted in decline of water table (74 cm/year in central Punjab) which has now attained very serious dimensions.

On account of irrigation facilities and better adaptability, its cultivation has extended to sandy soils having low water retentivity, which has led to the irrational use of irrigation water and deficiency of micronutrients *viz.*, iron, zinc, has become very common. Similarly, cotton-wheat is a predominant cropping system in south-western districts and maize-wheat in the sub mountainous districts. Both the crops in these crop rotations are exhaustive and deplete soil nutrients to a great extent. Therefore, to maintain the soil health and to sustain the production of these cereal crops on long term basis and at the same time, to increase the pulse production in the Punjab, it is essential to include a non-exhaustive leguminous pulse crop in these cropping systems.

Efficient Alternative Systems

Each plant of a pulse crop is virtually nature's mini nitrogen fertilizer factory, which enables the pulse crops to meet their own nutritional requirements and have additional benefits as residual effects on succeeding cereal crops. Pulses are also excellent feed and fodder for livestock. Therefore, they are considered to be the best option for intensification of predominant cereal based cropping system in Punjab. The major options of crop intensification through legumes are inclusion of short duration varieties of legumes as catch crop such as moongbean e.g. SML 668 and SML 832. The short duration moongbean varieties require 60-65 days and can easily fit in the cereal based cropping system.

The total numbers of tubewells operating in the State are around 11.9 lakhs irrigating about 72 per cent of the cropped area. Such a high intensity of tubewells aggravated the situation of declining water tables which is further coupled with the non-adoption of irrigation schedules and improved practices ultimately affecting productivity. The total water requirement of rice-wheat system is estimated to be 212 cm, with more than 75 per cent being used by the rice crop alone. Thus, to reduce water consumption, effective savings must be made in rice cultivation in the rice-wheat system. The major cropping systems were developed as per different agro-climatic zones with rational use of the resources.

The data given in Table 1 reveals that there is sufficient scope to replace rice-wheat cropping system with other cropping systems along with substantial increase in economic yields. The maize-potato-onion; summer groundnut-potato-bajra(f), maize-potato-summer moong and maize-wheat-summer moong gave 2.1, 1.8, 1.7 and 1.3 times more productivity over rice-wheat system which clearly elucidates the superiority of these systems over rice-wheat system. These systems also helped to save 82-144 cm of irrigation water (Table 2).

Table 1 : Rice equivalent yield of different cropping systems

Cropping system	Economic yield (q/ha)			Rice equivalent yield (q/ha)	Total calories (Kcal × 1000)
	I	II	III		
Maize-Potato-Onion	55.3	306.0	230.9	278.6	23035
G.nut-Potato-Bajra (F)	29.0	333.0	356.5	233.0	19673
Maize-Potato-Moongbean	54.5	326.7	13.2	191.0	26217
Maize-Wheat-Moongbean	54.5	61.2	12.3	161.8	43922
Rice-Wheat	63.5	51.10	-	132.4	39652

Price (Rs/q)/K.Cal. (per 100 gram) : Wheat, 1005(346); Rice, 745(346); Maize, 700(342); Potato, 250(97); Onion, 400(50); Oilseed, 2500; Mungbean, 1700(334).

The maize-potato-onion system gave the highest productivity (278.6 q/ha/annum) and used 82 cm less water than rice-wheat system with a productivity margin of 132.4 q/ha/annum. The summer groundnut-potato-bajra (fodder) system gave 233.0 q/ha/annum productivity with 103 cm irrigation water (Table 2) leading to 109 cm saving of water.

Maize-potato-summer moong cropping system gave 191.0 q/ha/annum productivity with a total irrigation water use of 92 cm, thereby, resulting in a net saving of irrigation water to the extent of 120 cm. The maize-wheat-summer moong produced 161.8 q/ha/annum productivity and used only 68 cm irrigation water which was 68 per cent less than irrigation water used by the rice-wheat system.

The net returns were maximum at Rs. 1,25,023/ha/annum in maize-potato-onion system which were 2.09 times more over rice-wheat system (Table 2). The net returns in the other cropping systems like maize-wheat-moong, maize-potato-moong and summer groundnut-potato-bajra (fodder) were Rs. 72,797, 78,588 and 1,11,839, respectively.

Table 2 : Net returns, irrigation water applied and production efficiency of different cropping systems

Cropping system	Total variable cost(Rs/ha)	Gross return (Rs/ ha)	Net return (Rs/ ha)	Irrigation water applied (cm)
Maize-Potato-Onion	83383	208406	125023	130
G.nut-Potato-Bajra (F)	62435	174274	111839	103
Maize-Potato-Moong	64250	142838	78588	105
Maize-Wheat-Moong	48255	121052	72797	92
Rice-Wheat	39318	99060	59742	212

The quantity of water used in the maize-potato-onion; summer groundnut-potato-bajra(f), maize-potato-summer moong and maize-wheat-summer moong was 38.7, 51.4, 50.5 and 56.6 per cent less, respectively, than quantity of water used for rice-wheat system (2120 X 10^4 litres). The corresponding value in terms of saving of electricity consumption (per ha basis) was 758, 1008, 990 and 1110 electricity units with electricity bill amounting Rs 2280, 3030, 2978 and 3343 per ha over rice-wheat system for the four cropping systems respectively (Table 3). The groundnut-potato-pearlmillet (fodder) system showed the highest water productivity of 2.262 kg grain/m^3 irrigation water followed by maize-potato-onion (2.138 kg grain/m^3 irrigation water). The least water productivity of 0.625 kg grain/m^3 irrigation water was observed in rice-wheat cropping system.

Table 3 : Energy consumption for irrigation in different cropping systems

Cropping system	Irrigation water use efficiency (kg grain/ cu m water used)	Time (Hours / ha/ annum) of 5 HP motor*	Electric consumption (units/ha/ annum) of 5 HP motor	Cost of electricity consumption (Rs/ha/ annum)
Maize-Potato-Onion	2.138	325	1205	3615
G.nut-Potato-Bajra (F)	2.262	258	955	2865
Maize-Potato-Moong	1.819	263	973	2918
Maize-Wheat-Moong	1.759	230	853	2553
Rice-Wheat	0.625	530	1963	5895

* with discharge of 15 litres/second with 74 % irrigation efficiency

Replacement of 5 per cent rice-wheat area (2.6 million ha) by these alternative crop rotations will amount to the saving of about 135850 ha-m irrigation water with additional net return of Rs. 4650 million to the farmers and saving of 162 million electricity units amounting to Rs 378 million.

Maize-*gobhi sarson*-summer moongbean on an average of five years gave 33.0 q/ha more rice equivalent yield than the maize-wheat system in loamy sand soil. Similarly, rice-*gobhi sarson*-summer moongbean also produced 14.5 q/ha more rice equivalent yield than rice-wheat system. The results further revealed that maize-*gobhi sarson*-summer moongbean gave a net return of Rs 14915/ha over maize-wheat. The corresponding increase in net return under rice-*gobhi sarson*-summer moongbean was Rs 6228/ha, respectively (Table 4).

Table 4 : Grain yield and rice equivalent yield (q/ acre) of different cropping systems

Cropping system	Rice Equivalent Yield (q/ha)	Gross returns (Rs/ ha)	Variable cost (Rs/ ha)	Net returns (Rs/ha)	Increase over maize -wheat/rice -wheat (Rs/ ha)
Maize-wheat	84.0	47845	33928	13918	
Maize-*gobhi sarson*-summer moongbean	117.0	66703	37870	28833	14915
Rice-wheat	120.5	66600	35298	31303	
Rice-*gobhi sarson*-summer moongbean	135.0	76870	39240	37630	6228

Production Potential and Yield Gaps

The average yield of rice in Punjab state was 58.0 q/ha. The potential yield achievable in Punjab is 75.0 q/ha. The yield gap between actual present yield and potential yield is 17.0 q/ha. This shows that there exists scope for improvement in management strategies. Similarly, the system production in other pulse crops can be increased upto a level ranging from 3.0 to 8.4 q/ha in different crops as shown in Table 5 so there is scope for further yield enhancement on farmers fields.

Saving of Virtual Water

From the above discussions, it is evident that pulses are low yielders but if this reduction in pulse grains production, when weighed against the over exploitation of resources such as irrigation water alone, shows its significance nearly 4000 litres of water is required to produce one kg of rice in Panjab.

Table 5 : Individual crop information with respect to variety, duration, average yield of the state and man-days used

S.No.	Crop	Variety	Duration (days	Max. yield of crops harvested (q/ha)	Av. yield of the state (q/ha)	Yield gap (q/ha)	Man days used (No./ha)
1	Rice	PR 116	119	75.0	58.0	17.0	80
2	Wheat	PBW 343	155	65.0	45.1	19.9	23
	Mungbean	PAU 911	75	12.3	8.9	3.4	62
3	Summer Mungbean	SML-668	60	13.2	6.4	6.8	50
4	Mash	Mash 114	83	9.0	4.8	4.2	30
5	Summer mash	Mash 1008	72	11.3	4.8	6.5	30
6	Arhar	AL 201	140	15.5	9.7	5.8	62
7	Rice bean	RBL 6	125	15.0	11.0	4.0	62
8	Soybean	SL 744	139	18.3	13.8	4.5	47
9	Desi gram	PGF 2	170	19.0	11.6	7.4	71
10	Kabuli gram	BG 1053	155	20.0	11.6	8.4	71
11	Lentil	LL 699	145	12.5	6.5	6.0	49
12	Field pea	Field pea 48	135	20.0	11.8	8.2	53

Source : Anonymous (2010); Anonymous (2010-11)

Water cost per hectare of paddy (5t/ha) yield average

Quantity of water used/ ha : 2600X5X1000= 13 million litres of water/ ha/season of 110 days active field life from transplanting (T.P.) to last irrigation (cut off at about 15 days of T.P.)

Cost of electricity in pumping 13 million litres of water

Average electric motor of 5 HP with average discharge of 15 litres/second and irrigation application efficiency of 74 %

$$\text{Time (hours)} = \frac{13\times10^6}{15\times60\times60} = 241 \text{ hours} \times \frac{100}{74} = 325 \text{ hours/season}$$

$$\text{Electric consumption of 5 HP} = \frac{740 \text{ Watts / HP}\times5\text{HP}}{1000} = 3.7 \text{ units/hours}$$

This, infact, is the cost of virtual water equivalent and the paddy exported in the World market carries the subsidy from the Punjab Government (at the cost of general public) an equivalent of nearly 25% (Rs. 722/ton of paddy)

Therefore, states like Punjab which export paddy at MSP of Rs 5900/ tonne (Rs 590/q) are virtually providing electricity equivalent to Rs. 722/ tonne of paddy and taxing indirectly the residents' equivalent to this amount.

Such high use of virtual water is not a viable preposition for such crops which are using resources irrationally. The gram crop requires one and rape seed and mustard two post sowing irrigations as compared to five irrigations in wheat. If 10 % of total wheat area (34 lakh ha) is diverted to gram, it would help to save 765000 ha-m water (7.5 cm of water draw down over 3.4 lakh ha area). Likewise, 10 % area under *gobhi sarson* which is water loving crop among rapeseed and mustard, would save 25500 ha-m irrigation water as compared to wheat. This water can be used for raising other crops or improve the aquifer recharge. Similarly, replacing of one ha rice cultivation will enable to raise four ha of maize/cotton or six ha of groundnut/pigeonpea crops with the same quantity of water.

Total cost of working of motor over the paddy season at prevailing rate of per KWH of Rs 3.00

Cost per hour = 3.7 X 3.0= Rs 11.1 per hour

Cost for 325 hours = 325 X 11.1= Rs. 3608/ha/season

Cost for 5 t paddy= Rs 3608

$$\text{Cost per tonne paddy} = \frac{3608}{5} = \text{Rs. 722/tonne}$$

Favourable Conditions to Increase in area under Pulses in Punjab

- **Punjab Govt. puts a ban on rice transplanting before June 10:** The Punjab Govt. has put a ban on transplanting of rice nursery before June 10. If any farmers transplant the rice nursery then the state Department of Agriculture ploughs up the field. So, the majority of farmers transplant their rice nursery after June 10. After harvesting of wheat and before transplanting of rice there is sufficient time for raising a good crop of summer moongbean.
- **SML 832:** The Punjab Agricultural University has released a new high quality summer moongbean variety which is having medium sized seeds. It has erect and determinate plant type and medium stature. It bears pods in clusters and possesses early and synchronous maturity (61 days). Pods have a thick coat of blackish brown colour at maturity. Each pod contains about 10 grains. Grains are green, medium seized and very shiny with good culinary properties. Its average yield is 11.5 q/ha.
- **Short duration rice/wheat varieties:** Short duration rice varieties are available (PR 115), so these varieties can be transplanted after harvest of summer moongbean. Even after harvesting of rice, the wheat variety, PBW 550, can be sown which takes 10-12 days less than other timely sown varieties (DBW 17 and PBW 621) and produces yield at par with these varieties.
- **Mechanization for pulses:** The pulse crops can be sown and harvested mechanically.
  - **(i) Sowing with happy seeder:** To save time the summer moongbean can be sown with happy seeder without preparing the field. The sowing can be done with zero–till drill if there is no wheat straw in the field. In case of combine harvested wheat crop, summer moongbean can be sown with happy seeder in the presence of wheat straw.
  - **(ii) Combine harvesting of pulses:** When 80 per cent of pods are mature then apply Gramoxone (paraquat) @ 2 L/ha in 375-500 litres of water so that leaves and stem dries up in 5-6 days. Then the moongbean can be harvested with combine harvester to save the crop from untimely rains and save labour. The combine harvesting of pulses will help in leaving the moongbean residues in the field, which upon decomposition will help in improving the soil health.

Area to be put under pulses for self sufficiency: In Punjab rice is grown on an area of about 27 lakh ha during *kharif* season and wheat occupies 35.3 lakh ha during *rabi* season. During *kharif* season 8-10 per cent area is occupied by basmati rice and transplanting is done during first fortnight of July. So, there is scope to increase pulse area before rice transplanting in Punjab. The area under pulse production is 22000 ha. If fifty times area is increased then summer moong would be sown on 11.0 lakh ha area. By taking the average summer moongbean yield of 11.5 q/ha, a total of 1265, 000 tonnes of moong grains amounting to Rs. 633,000 crore will be produced and per capita availability of pulse will be enhanced from 2 g/day to 100 g/day. To achieve this 41,250 tonnes moongbean seed will be required costing about Rs. 248 crores. It is also worth while to mention here that the pulses fix atmospheric nitrogen and the farmers can reduce their fertilizer dose of nitrogen by 1/3 to the following rice crop ($44x10^4$ quintal nitrogen or $96x10^4$ quintal urea can be saved) amounting to Rs. 460 crores which is almost double the seed cost for mungbean.

Improved Varieties

Moong

PAU 911- Plants are erect, compact, determinate and medium statured. The average height of plants is about 70 cm. The variety shows good resistance against mungbean yellow mosaic virus (MYMV), cercospora leaf spot (CLS) and bacterial leaf spot (BLS). It takes about 75 days to mature. Average grain yield is 12.3 q/ha.

ML 818- Plants are erect, determinate and medium in stature. The average height of plants is about 75 cm. The variety shows reasonable resistance against mungbean yellow mosaic virus (MYMV) and fairly resistant to cercospora leaf spot (CLS) and bacterial leaf spot (BLS). It takes about 80 days to mature. Average grain yield is 12.3 q/ha.

PBM 1- Plants are erect, compact, determinate and medium statured. The average height of plants is about 67 cm. The variety is tolerant to mungbean yellow mosaic virus (MYMV) and reasonably resistant to bacterial leaf spot (BLS) and anthracnose diseases. It takes about 75 days to mature. Average grain yield is 10.5 q/ha.

ML 613- Plants are determinate in growth and medium statured. The average height of plants is about 85 cm. The variety is tolerant to mungbean yellow mosaic virus (MYMV), cercospora leaf spot (CLS) and bacterial leaf spot (BLS). This variety is also fairly resistant against white fly and jassids. It takes about 85 days to mature. Average grain yield is 10.8 q/ha.

ML 267- Plants are erect, determinate in growth and medium statured. The average height of plants is about 80 cm. The hairy stem and branches are with single tinge of purple colour. The variety shows considerable resistance to mungbean yellow mosaic virus (MYMV). It takes about 85 days to mature. Average grain yield is 10.5 q/ha.

SML 668- Plants are erect, determinate in growth and short statured. It is an early maturing variety which matures in about 60 days. Average yield is 11.3 q/ha. It has very good cooking quality. It is resistant against yellow mosaic virus and thrip attack. Most suited for summer season.

SML 832- Plants are of erect and determinate type and medium in stature. It bears pods in clusters and possesses synchronous maturity (61 days). Each pod contains about 10 grains. Grains are green medium sized, very shining with good culinary properties. Its average yield is 11.5 q/ha. Most suited for summer season.

Mash (Blackgram)

Mash 114- The variety is recommended for the whole of Punjab. Plants are dwarf, erect and determinate. It takes about 83 days to mature. Its seeds are very good for culinary purposes. The variety is appreciably resistant to mungbean yellow mosaic virus (MYMV), cercospora leaf spot (CLS) and bacterial leaf spot (BLS). Average yield is 9 q/ha.

Mash 1-1 - The variety is also recommended for the whole of Punjab. Plants are dwarf, erect and determinate. It takes about 90 days to mature. Its seeds are very good for culinary purposes. The variety is appreciably resistant to mungbean yellow mosaic virus (MYMV), *cercospora* leaf spot (CLS) and bacterial leaf spot (BLS). Besides this it also shows resistance against jassids and white fly. Average yield is 8.8 q/ha.

Mash 338- The variety is recommended mainly for humid areas of Punjab. Plants are tall and indeterminate. It takes about 115 days to mature. Its seeds are also very good for culinary purposes. The variety is appreciably resistant to mungbean yellow mosaic virus (MYMV), *cercospora* leaf spot (CLS) and bacterial leaf spot (BLS). Besides this it also shows resistance against jassids and white fly. Average yield is 8.75 q/ha.

Mash 1008- Plants are dwarf and erect with average height of 25 cm. It matures in about 72 days. Yellow mosaic virus attack is low. Its average yield is 11.3 q/ha. Seed has about 24 % protein and has good culinary properties. Suitable for summer.

Mash 414- Plants are dwarf and erect with average height of 30 cm. It matures in about 72 days. Its average yield is 10.5 q/ha. Seed has about 23 % protein and has good culinary properties. Most suited for summer.

Mash 218- Plants are dwarf, compact and erect with average height of 30 cm. It matures in about 75 days. Its average yield is 10.0 q/ha. Seed are bold and dull black in colour. Seed has about 23 % protein and has good culinary properties. Suited for summer season.

Pigeonpea

PAU 881- It is an early maturing variety with indeterminate growth habit. It takes about 132 days to mature. Plant height is about 2 metres. Average yield is 14.0 q/ha.

AL 201 - It is an early maturing variety with indeterminate growth habit. It takes about 140 days to mature. Erect plants are about 2 metres in height. The main stem is much longer than side branches. The standard petal in flower has prominent red streak. Average yield is 15.5 q/ha.

AL 15 - It is short duration variety with determinate growth habit. It takes about 135 days to mature. Short statured plants are erect and compact with a height of about 1.5 to 1.8 metres. Plants have synchronous maturity. Average yield is 13.8 q/ha.

Ricebean

RBL 6- The variety is photosensitive and is resistant to bacterial, fungus and, in particular, to viral diseases. Growth is spreading type and maturity is synchronous. Seeds are not affected by storage pests. Time taken for maturity is about 125 days under irrigated conditions. Its average yield is 15 q/ha.

Cowpea

CL-367 - Dual purpose variety i.e. used as both fodder and pulse crop. Plants are erect and leaves are dark green in colour. Resistant against Anthracnose and *YMV* diseases. It has good fodder qualities. Variety is good for human consumption as it has good cooking quality. Average yield is 270 q/ha of green fodder and 12.3 q/ha of grains.

Cowpea 88 - Dual purpose variety i.e. used both as fodder and pulse crop, recommended for the entire state. Seed colour due to its chocolate brown colour, is very attractive for culinary purposes. Resistant against Anthracnose and *YMV* diseases. Average yield is 250 q/ha green fodder and 10.8 q/ha of grains.

Gram- *DESI*

PBG 5- Stem is sturdy and leaves are dark green. It takes about 165 days to mature. Seeds are medium bold and dark brown in colour. 100-seed weight is 18 g. It can tolerate blight and root diseases. Average yield is 17q/ha.

PGF 2- Plants are semi-erect with long branches. It takes about 170 days to mature. Seeds are bold. It can tolerate blight and root diseases. Average yield is 19q/ha.

PBG 1- Plants are tall and erect in growth with long branches. It takes about 160 days to mature. Seeds are bold and brown in colour. It can tolerate blight and wilt diseases. Average yield is 16q/ha.

PDG 4- Erect plants with dark green foliage and long branches. Seeds are bold and crop takes about 160 days to mature. It can tolerate root rot and blight. Average yield is 19.5q/ha. Suited for rainfed conditions too.

PDG 3- Semi-erect plants with dark green foliage and long branches. Seeds are bold and crop takes about 160 days to mature. Average yield is 18.0q/ha. Suited for rainfed conditions too.

Kabuli Gram

BG 1053- Early flowering variety with bold seeds and semi-erect growth. It tolerates root rot and foot rot diseases. Matures in about 155 days. Average yield is 20 q/ha.

L 550- Early flowering variety with bold seeds and bushy growth habit. 100-seed weight is about 21.3 g. Matures in about 160 days. Average yield is 15 q/ha.

Lentil

LL 931- Plants are dwarf and erect with profuse branching. Flowers are pink in colour. Crop matures in about 146 days. Pod borer attack is low. It has good culinary properties. Average yield is 12 q/ha.

LL 699- Plants are dwarf and erect with profuse branching. Foliage is dark green. Variety is early flowering type with high number of pods. Crop matures in about 145 days. Pod borer and blight attack is low. It has good culinary properties. Average yield is 12.5 q/ha.

LL 147- Plants are dwarf and erect with profuse branching. Variety is early flowering type with high number of pods. Crop matures in about 150 days. Seeds are with black spots and have good culinary properties. Average yield is 11.3 q/ha.

Field pea

Field pea 48- Early maturing variety, plants are erect and dwarf. Pods are longer. Seeds are attractive with slight wrinkles and have very good culinary properties. Due to early maturity it escapes pod borer attack and fungal disease. Crop matures in about 135 days. It has higher protein content. Average yield is 20 q/ha.

PG 3- It is dwarf variety with white flowers. It is early flowering and maturing variety. Seeds are attractive with slight wrinkles and have very good culinary properties. Due to early maturity, it escapes pod borer attack and fungal disease. Crop matures in about 135 days. Its yield is 17.5 q/ha

Package of Practices

Chickpea

Land Preparation: The soil for gram cultivation should be loose and well aerated as it reduces the chance of wilt attack and results in increased grain yield. Deep tillage is very beneficial to increase yield. July and August are suitable months for deep tillage. Deep tillage has multiple benefits as it reduces weed infestation, loosens the soil, promotes rain water seepage and also encourages development of deep roots.

Seed Rate: The optimum seed rate for *desi* and *kabuli* gram are 37.5-45.0 and 92.5 kg ha^{-1}, respectively. For PBG 5 variety of *desi* gram, a higher seed rate of 60 kg ha^{-1} should be used. Seed rate of *desi* gram increases with the delay in sowing time. Seed rate of 67.5 kg ha^{-1} should be used in case of sowing during second fortnight of November while 90 kg ha^{-1} is recommended for sowing during first fortnight of December.

Sowing Management: Sowing should be accomplished by *pora* method. The proper row spacing is 30 cm. The sowing depth should be 10-12.5 cm. After rice, particularly on heavy soils it is better to sow two rows of gram on raised beds.

Fertilizer Application including Micro-Nutrients : In *desi* gram sown under irrigated and unirrigated situations, 15 kg nitrogen ha^{-1} and 20 kg phosphorus ha^{-1} should be drilled at sowing. In *Kabuli* gram sown under irrigated conditions, apply 15 kg nitrogen ha^{-1} and 40 kg phosphorus ha^{-1} in the same manner as in case of *desi* gram. The nitrogen and phosphorus should be supplied from urea and super phosphate, respectively or any other fertilizer available in the market.

Weed control and inter culture: One or two hand- hoeings at 30 and 60 days after sowing should be done. Otherwise, pre-plant application of Treflan 48 EC (Trifluralin) at 2.5 litre or Stomp 30 EC applied as pre-emergence at 2.5 litre ha^{-1} can be an effective measure for controlling annual weeds.

Water management: A heavy pre-sowing irrigation ensures deep rooting for proper soil moisture utilization. Later on, one more irrigation should be given between mid-December and end-January.

Plant Protection Measures

a) Insect Pests-

1. **Termites-** Crops at seedling stage and near maturity are more prone to its attack. Termites can be observed feeding on roots or near the root zone of a damaged plant. Its attack is more in light soils. The damaged plants dry up and can be easily pulled out of soil. Seed treatment with Chloropyriphos 20 EC per kg seed should be done as a preventive measure. Recommended quantity of both *desi* and *kabuli* seed spread as a thin layer on *pucca* floor should be treated by calculated dose of any of the above insecticides diluted in 1.25 litre water for one hectare.

2. **Gram caterpillar (*Helicoverpa armigera*) -** The larvae feed on leaves, flower buds, flowers, pods and grains. Spray 250 ml Sumicidin/ Fenlik/Agrofen 20 EC (Fenvalerate) or 400 ml Decis 2.8 EC (Deltamethrin) or 200 ml Cymbush 25 EC (Cypermethrin) in 200-250 litres of water per hectare or dust 25 kg Malathion 5% dust per hectare at the start of pod formation and repeat after two weeks, if required.

b) Diseases

1. **Blight (*Ascochyta rabiei*)** – It is characterised by dark brown spots with dot like bodies on stem, branches, leaflets and pods. The seed may also be infected. The dot like bodies may be concentrically arranged on pods and leaves. Excessive rains further aggravate the situation.

   Seed treatment with Bavistin+Thiram (1.5+1.5g) or Captan or Captaf 3 g or Rovral 2.5 g/kg of seed should be done as a preventive measure. If infestation is noticed afterwards, go for 3-5 sprays of Hexacap or Indofil M-45 or Captan or Kavach at 900g/ha in 250 litres of water at 15 days interval. First spray should be given as soon as disease appears. Always use disease free seed. Varieties such as PBG1, PBG5 and C235 which are known to be more resistant should be grown. Diseased plants should be destroyed after harvest.

2. **Grey mould** (*Botrytis cineria*)- Leaflets are characterised by small water soaked spots. Infected leaves turn dark brown. Erect sporophores of fungus are produced on flowers, leaves, branches and pods. Soft rotting of infected portion occurs.

   Seed treatment with a combination of 3g Bavistin+Thiram (1:1) or Indofil M-45 3 g or Ronilan 4 g or Bavistin 3 g/kg seed should be done before sowing. If disease appears particularly in February/March go for one spray of Indofil M-45 at 875 g or Baytan 500 g or Thiram 875 g in 250-300 litres of water/hectare.

3. **Wilt** (*Fusarium oxysporum f. Sp.ciceri*)- Affected plants have dull green colour and drooping petioles in initial stages. Further, the leaves turn yellow and become straw coloured. Dark-brown to black vascular dis-colouration is a very characteristic symptom. Partial plant wilting may occur.

   Desi gram varieties PBG1, PBG5, PDG3, PDG4 and GPF2 and *kabuli* variety BG 1053 which are wilt tolerant, should be grown. Moisture conserving measures should be adopted in rainfed areas.

4. **Stem rot** (*Sclerotina sclerotium*)- All the above ground plant parts are susceptible. Main stem is infected at ground level. The affected parts are covered with white mycelium and black sclerotia. The seed should be free from sclerotia. October 25 to November 10 sown crop escapes the disease. The debris should be destroyed in the field after harvesting. Chickpea should be rotated with crops such as wheat and barley which are non-susceptible.

5. **Foot rot** (*Opercullela padwickii*)- Collar region of plants show light brown to dark brown lesions. Further, these lesions become black and affect the basal tap root. Collar region shows complete girdling. Sowing should be done from 25 October to 10 November. Non-susceptible crops should be rotated with gram.

Lentil

Land Preparation: Two to three ploughings should be done for proper pulverisation of soil.

Seed Rate: 30-37.5 kg seed rate is optimum for sowing one hectare.

Sowing Management: Sowing should be accomplished from end-October to mid-November. However, it can be done even upto end-November without much decline in yields.

Fertilizer application including micro-nutrients: Basal doses of 12.5 kg nitrogen and 20 kg P_2O_5 (if seed is inoculated with *Rhizobium* culture) or 40 kg phosphorus (if seed is not inoculated) per hectare should be applied to crop. Phosphorus should be applied through drilling.

Weed control and inter culture: One/two hand hoeings at 30 and 60 days after sowing are sufficient to control weeds.

Water management: This crop requires only one to two irrigations during growing season depending upon rains. In case of single irrigation application, apply at 6 weeks after sowing. If two irrigations are planned then the first one should be applied at 4 weeks after sowing and the second at flowering or pod formation stage.

Plant Protection Measures

a) **Insect Pests**

1. **Lentil pod borer-** It feeds on leaves, flower buds, pods and grains. Spraying Sevin/Hexavin 50 WP (carbaryl) at 2.25 kg/ha in 200 litres of water at flower initiation provides an effective control. Repeat spray at three weeks, if the need arises.

b) **Diseases**

1. **Blight** (*Ascochyta fabae* f.sp. lentis)- Stem, branches, leaves and pods show dark brown spots. Branched and elongated lesions may be noticed on the stem. Black pycnidial fungal bodies in the form of concentric rings occur on infection spots. These spots girdle the tender branches and cause the above part to wilt and dry up. Seed treatment with 2 g/kg Captan should be done. Diseased plant debris should also be destroyed as a measure to reduce inoculum.

2. **Rust** (*Uromyces fabae*)- Stem, branches, leaves and pods show development of yellowish white pycnial and aecial pustules. These may occur singly or in groups. At later stage, brown uredia may form. Further, dark brown black telia are formed. Plants under severe infestation dry up and give burning appearance. Destroy plant debris, use clean seed and grow tolerant variety LL931.

Field pea

Land Preparation: Two to three times ploughing should be done for proper pulverisation of soil to obtain a fine seed bed.

Seed Rate: 62.5-75 kg/ha seed is optimum.

Sowing Management: Sowing should be done in rows spaced 30 cm apart using either *kera* or *pora* method.

Fertilizer application including micro-nutrients: 30 kg nitrogen and 40 kg P_2O_5 should be applied to crop by drilling along rows.

Weed control and inter culture: One/two hand hoeings at 30 and 60 days after sowing, preferably with wheel hand hoe to control weeds.

For chemical weed control, pre-emergence spray of Stomp 30 EC (Pendimethalin) 2.5 litres dissolved in 375-500 litres of water per hectare is recommended. This application is especially effective against *Phalaris minor* and broad leaf weeds.

Water management: Generally a pre-sowing irrigation should be applied. In case the preceding crop was paddy, it can be sown without irrigation. During the growing season, it requires one irrigation at pre-flowering stage around end of December and another one at pod-formation stage. In sub-montaneous regions, it can be grown as rainfed crop.

Plant Protection Measures

a) Insect Pests

1. **Pea stem fly** (*Ophiomyia phaseoli*)- Seedlings of crop sown earlier i.e. before mid-October are more seriously damaged by its attack. So, avoid earlier sowing.
2. **Pea thrips** (*Thrips indicus*)- Young crop is severely damaged due to cell sap sucking. As the attack starts, the crop should be sprayed with Rogor 30 EC (Dimethoate) at 1 litre dissolved in 200-250 litres of water per hectare. If required, the spray may be repeated at an interval of 15 days.
3. **Pea leaf minor** (*Chromatomyia horticola*)- The larvae of the insect while feeding, make tunnels in the leaves. The damage is severe especially during December-March. As the infestation starts, the crop should be sprayed with Rogor 30 EC (Dimethoate) at 1 litre dissolved in 200-250 litres of water per hectare. If required, the spray may be repeated at an interval of 15 days.
4. **Pea aphid** (*Acyrthosiphon pisum*) Yellowing and drying of leaves due to sap sucking can be noticed. Following the infection, the crop should be sprayed with Rogor 30 EC (Dimethoate) at 1 litre dissolved in 200-250 litres of water per hectare. If required, the spray may be repeated at an interval of 15 days.
5. **Pod borers** (*Etiella zinckenella, Lampides boeticus* and *Helicoverpa armigera*)- The larvae feed on flowers and pods. As soon as the attack starts, spray the crop with 750 ml Thiodan 35 EC or 900 g Sevin/Hexavin 50 WP (Carbaryl) in 200-250 litres of water per hectare. This spray may be repeated at an interval of 15 days if the need arises.
6. **American bollworm** (*Helicoverpa armigera*) Spray 5 litres of Durmet 20 EC (Chloropyriphos) or 2 kg Asataf 75 SP (Acephate) in 250 litres of water per hectare.

b) **Diseases**

1. **Powdery mildew** (*Erysiphe polygoni*)- Stems, branches, leaves and pods show white floury paches covering large areas. Karathane 40 EC spray at 200 ml in 250 litres of water per hectare should be applied as a control measure and may be repeated at an interval of 10 days, if needed.
2. **Wilt** (*Fusarium oxysporum f. sp. pisi*)- Root rotting and yellowing of lower leaves is followed by wilting. Seed treatment with Captan at 2 g/kg seed before sowing should be done. Sowing should not be done earlier than recommended date.
3. **Rust** (*Uromyces fabae*)- Stem, branches, leaves and pods show development of yellowish white pycnial and aecial pustules. These may occur singly or in groups. At a later stage, brown uredia may form. Further, dark brown black telia are formed. Plants under severe infestation dry up and give burning appearance. Destroy plant debris, use clean seed and grow tolerant variety LL931.

Pigeonpea

Land Preparation: Each ploughing should be followed by planking to make field clod and weed free and prepare a fine seed bed.

Seed Rate: Sowing with recommended seed @ 15 kg/ha should be done.

Sowing Management: The crop should be sown during the first fortnight of June at 50 cm row spacing and 25 cm plant spacing.

Fertilizer application including micro-nutrients: At sowing 15 kg nitrogen and 40 kg P_2O_5 per hectare should be applied to crop by drilling along rows. If the soil shows deficiency of K_2O then it may be applied at the rate of 30 kg/ha. These nutrients can be supplied through any fertilizer available in the market.

Weed control and inter culture: Two hand hoeings at 21 and 42 days after sowing, preferably with wheel hand hoe, are needed to control weeds. For chemical weed control, pre-emergence spray of Stomp 30 EC (pendimethalin) 2.5 litres per hectare is recommended. Otherwise, pre-emergence spray of Stomp 30 EC (Pendimethalin) at 1.5 litres per hectare followed by a hand weeding six/seven weeks after this is recommended. Use 375-500 litres of water per hectare to prepare the spray solution.

Water management: First irrigation is given 3-4 weeks after sowing. Another irrigation is needed only if there is no adequate rainfall. After mid-September, irrigation is not required otherwise crop maturity may be delayed.

Plant Protection Measures

a) Insect Pests

Pod borers- Pod borers, in general, and Gram pod borer (*Helicoverpa armigera*) in particular, are the most important insect pests reported for this crop. Their larvae have various colours and sizes. They feed on leaves, flowers, flower buds, pods and seeds in pods, thereby, causing huge losses in grain yield. Damaged plant parts and dark-green faecal matter at base of plants is an important sign of their infestation. At pod initiation stage, Thiodan 35 EC at 2.5 litres in 250-300 litres of water per hectare should be as soon as larvae appear. If need arises, another spray can be done at 10 days interval. It is advisable to spray this chemical during evening as the pollinators' population remains low at this time.

b) Diseases

1. **Cercospora leaf spot** (*Cercospora cajani*)- Under surface of leaves shows greyish-brown to dark spots. These may coalesce to form large blotches. Lesions may also occur on petioles and stems. Premature defoliation occurs. Always use disease free seed and treat it with Captan or Thiram at 3 g per kg seed.

2. **Bacterial leaf spot (***Xanthomonas campestris pv. cajani***)-** Angular dark spots usually concentrated on one side of mid-rib appear on the leaf. Spots may develop on veins, petioles, main stems and branches. Seed used for sowing should be disease free.

3. **Phytophthora stem blight** (*Phytophthora drechslera f. sp. cajani*)- Young seedlings are attacked and killed as soon as they emerge from soil. Slightly depressed brown to black necrotic lesions with definite margins are produced on stem. Sometimes, stem which has swollen into a cancareous structure at the edge of lesion may break. Lesions on leaflets may either be circular or irregular and whole foliage appears blightened. Soils with poor drainage should not be selected for pigeonpea growing and some other crop should be grown in badly infested areas.

4. **Sterility mosaic-** Mild-mosaic, little or no flowering and pale-green leaves are typical symptoms. Affected plants have bushy appearance. Primary source of infection should be destroyed by not allowing *arhar* and its wild relatives such as *Cajanus platycarpa* to grow around sugarcane, cotton and other fields during winter season.

Ricebean

Land Preparation: A fine seed bed should be prepared.

Seed Rate: 25-30 kg/ha seed rate is optimum for its sowing.

Sowing Management: The optimum time for sowing is first to third week of July. Row spacing should be 30 cm while plant spacing should be kept 10-12 cm. Sowing should be done at optimum soil moisture.

Fertilizer application including micro-nutrients: At sowing 15 kg nitrogen and 20 kg P_2O_5 per hectare should be applied to crop by drilling along rows. On soils with poor fertility status, application of 25 to 37.5 tonnes per hectare of well-rotted farmyard manure may be highly beneficial.

Weed control and inter culture: One/two hand hoeings at 30 and 50 days after sowing, are sufficient to control weeds.

Water management: During monsoon period give irrigation if rains fail, whereas, in post monsoon period 2-3 irrigations should be applied.

Plant Protection Measures

a) **Insect Pests**

1 **Pod borer** (*Helicoverpa armigera*)- Pale, brown, black, green or yellow coloured larvae feed on leaves, flower buds, flowers, pods and seeds in pods. Damage to plant as well as presence of faecal matter below the plants are signs of attack. When plants are shaken vigorously, larvae fall on the ground. At pod initiation stage, Thiodan 35 EC at 2.5 litres in 250-300 litres of water per hectare should be sprayed as soon as larvae appear. If need arises, another spray can be done at 10 days interval.

2 **Mite –** Occurence of webbing on underside of leaves which turn pale yellow is an important sign of mite attack. Subsequently, these infected leaves turn light-brown to dark reddish brown. Spraying infested crop with 375 ml Rogor 30 EC (Dimethoate) in 200-250 litres of water per hectare is an effective control measure.

3 **Dhora** - It attacks different pulses. To reduce attack, dry the seeds properly before storage. The gunny bags, godowns and storage bins should be thoroughly disinfected with the recommended fumigants or insecticides. In grains to be used for seed, mix Malathion 5% dust at 250 g/quintal grains. The grains meant for sowing should be treated with 25 ml Cythion 50 EC (Malathion premium grade) or 2 ml Sumicidin 20 EC (Fenvalerate) or 1.5 ml Cymbush 25 EC or any of the other chemicals recommended by PAU.

b) **Diseases- Yellow mosaic virus**

RBL 6 variety is very resistant to its attack.

Cowpea

Land Preparation: Two ploughings followed by planking are sufficient.

Seed Rate and Sowing Management: For fodder, 50-62 kg seed per hectare of Cowpea 88 and 30 kg/ha of CL 367 is optimum for sowing between March to mid-July, in rows 30 cm apart. When it is sown as a mixed crop with maize the seed rate is 15 kg per hectare of CL 367 or 37.5 kg/ha of cowpea 88 and 37.5 kg/ha of maize per hectare.

Fertilizer application including micro-nutrients: At sowing, 19 kg nitrogen per ha (as Urea) and 55 kg phosphorus per ha (as Superphosphate) should be applied to crop by drilling along rows. For cowpea fodder crop, following wheat which had been supplied with required dose of P_2O_5, its application may be omitted.

Weed control and inter culture: Pre-emergence spray of 1.9 litres of Stomp 30 EC (Pendimethalin) dissolved in 500 litres should be done as it provides effective control of annual weeds particularly *itsit/chapatti* (*Trianthema portulacastrum*).

Water management: May sown crop needs irrigation at an interval of 15 days till the arrival of monsoon. 4-5 irrigations are sufficient during whole crop season. Proper drainage ensures further yield increase.

Plant Protection Measures

a) **Insect Pests**

1. **Jassid and Black aphid** – These sucking pests can be controlled by spraying 500 ml of Malathion 50 EC in 200 litres of water per hectare.
2. **Bihar hairy caterpillar** – Commonly called as *bhaboo kuta.* The attack is more in August-November. Planting a row of *til* (Sesamum) around cowpea field at the time of sowing can be an effective measure to control this attack. Female prefers to oviposit on *til* leaves. Thus, larvae in huge numbers can be collected from leaves and destroyed.

b) **Diseases**

1. **Seed Rot and Seedling Mortality-** Many seed borne micro-flora cause this disease. Infected seeds show shrivelling and discolouration. Seedlings are killed even before emerging out of soil. Seed treatment with Emisan @ 2.5 g/kg or Bavistin 2 g/kg

seed before sowing should be done as a preventive measure. Only disease free seed should be used for sowing.

Greengram

Land Preparation: Two/three ploughings followed by planking are sufficient for preparing clod and weed free field. In case of Summer Moong, it can be sown without any preparatory tillage after wheat harvest with zero-till drill if there is no wheat straw in the field. If there is wheat straw, the sowing can be accomplished with happy seeder.

Seed Rate: 20 kg/ha seed rate is optimum for sowing in case of *kharif* sown crop. Whereas, a seed rate of 37.5 kg/ha is recommended in case of summer moong.

Sowing Management: First fortnight of July is the right time for sowing. The row spacing should be kept at 30 cm. The plant to plant distance should be 10 cm and seeds sown at 4-6 cm depth. For higher yields, bi-directional sowing should be done with the same seed rate. Use half seed rate for sowing in one direction while the remaining half in another. In case of *Summer Moong,* sow the crop from 20th March to 10th April. Sowing can be done till 3rd week of April but then there is risk of pre-monsoon showers at maturity. Summer moong can also be sown on raised beds spaced 67.5 cm apart (37.5 cm bed top and 30 cm furrow) on medium to heavy soils with wheat bed planter as per the recommendation of PAU. Sow two rows per bed with row spacing of 20 cm. Use same quantity of seed and fertilizers and follow other cultural practices as in flat sowing. Irrigation is applied in furrows only. The advantages of this method include saving crop from damage by excessive water at emergence, saving about 20-30 percent irrigation water and 10% increase in yield.

Fertilizer application including micro-nutrients: At sowing 15 kg nitrogen and 20 kg P_2O_5 per hectare should be applied to crop by drilling along rows. On soils with poor fertility status, application of 25 to 37.5 tonnes per hectare of well-rotten farmyard manure may be highly beneficial. For Summer Moong, 12.5 kg of nitrogen along with 16 kg of P_2O_5 per hectare should be drilled at sowing. Fertilizer application to crop sown after potato in rice-potato-summer moong and maize-potato-summer moong should be omitted.

Weed control and inter culture: One/two hand hoeings are sufficient to control weeds. Go for first hoeing at four weeks after sowing and the other one two weeks afterwards, if required. In case of zero tillage, pre-sowing spray of 1.25 litres of Gramoxone (Paraquat) in 500 litres of water per hectare should be done to control weeds.

1.5 litres of Basalin 45 EC (Fluchloralin) or 2.0 litres of Treflan 48 EC (Trifluralin) application on a well prepared seed bed on the same day of sowing, but before sowing can effectively control weeds. Otherwise, pre-sowing spray of 2.5 litres of Stomp 30 EC (Pendimethalin) alone or 1.5 litres of Stomp 30 EC (Pendimethalin) followed by hand hoeing two weeks after application can be effective against weeds. The quantity of water required to prepare spraying solution is 375-500 litres. These sprays control most of annual grasses and broad- leaf weeds except perennial weeds. The given measures control weeds in both *kharif* and summer sown crops, effectively.

Water management: In *kharif* sown crop, irrigation is required only if rains fail. In case of summer moong, 3-5 irrigations are needed. The last irrigation should be applied 55 days after sowing of crop. This ensures high yield and synchronous maturity.

Plant Protection Measures

a) Insect Pests

1. **Jassid, Aphid, Whitefly and Flea-beetle**- For their control, spray 938 ml of Malathion 50 EC (Malathion) or 625 ml of either Rogor 30 EC (Dimethoate) or Metasystox 25 EC (Oxydemeton methyl) in 200 litres of water per hectare. Against white fly another option is to spray 100 g Thiamethoxam 25 WG or 1.5 litres of Triazophos 40 EC using 200-250 litres of water per hectare.

2. **Hairy caterpillar –** The caterpillar feeds on green matter of leaves except mid-ribs. The crop may be totally denuded under a severe attack. Bury the infested plant underground to kill young larvae. The grown up larvae can be either crushed under feet or may be killed by putting in kerosene oil. In case of very high population, spray. 1.25 litres of Thiodan 35 EC or 500 ml of Nuvan 100 (Dichlorovos) in 200-250 litres of water per hectare.

3. **Semi-looper**- When touched, the 2-4 cm long larvae form a loop. In severe cases, plants are denuded within few days. Spray 1.25 litres of Thiodan 35 EC (Endosulfan) or 500 ml of Nuvan 100 (Dichlorovos) in 200-250 litres of water per hectare.

4. **Tobacco caterpillar** (*Spodoptera litura*)-Small larvae are black, while grown up larvae are dark green with typical triangular spots on the body. First and second instars of larvae are heavy feeders on foliage. Grown up larvae feed singly on leaves, buds, flowers and pods. Egg masses along with young larvae present on leaves should be destroyed. Spray of 425 ml of novaluron 10 EC or 800 g of Acephate 75 SP or 1.5 litres of Dursban/Durmet (chloropyriphos) 20 EC using 250 litres of water per hectare gives good control.

5. **Pod Borer** (*Helicoverpa armigera*)- Pale, brown, black, green or yellow coloured larvae feed on leaves, flower buds, flowers, pods and seeds in pods. Damage to plant as well as presence of faecal matter below the plants are signs of infection. When plants are shaken vigorously larvae fall on the ground. At pod initiation stage, spray of Thiodan 35 EC at 2.5 litres in 250-300 litres of water per hectare should be done as soon as larvae appear. If need arises, another spray can be done at 10 days interval. In summer moong it has become a limiting factor under Punjab conditions.

6. **Mite-** Webbing on underside of leaves is caused by the mites. Such leaves turn light to reddish brown. Spray of Rogor 30 EC (Dimethoate) at 425 ml per hectare in 200-250 litres of water is very effective.

7. **Dhora** (*Calosobruchus anal*is)- It attacks different pulses. To reduce infection, dry the seeds properly before storage. The gunny bags, godowns and storage bins should be thoroughly disinfected with the recommended fumigants or insecticides. In grains to be used for seed purpose, mix Malathion 5% dust at 250 g/quintal grains. The grains meant for sowing should be treated with 25 ml Cythion 50 EC (Malathion premium grade) or 2 ml Sumicidin 20 EC (Fenvalerate) or 1.5 ml Cymbush 25 EC or any of the other chemicals recommended by PAU.

8. **Thrips-** Thrips also attack summer moong crop severely. These are very small, dark brown insects. They can be found in flowers and cause flower drop, pod deformation and ultimately yield decline both quantitatively as well as qualitatively. Triazophos 40 EC @ 1.5 litres or 250 ml of Rogor in 200-250 litres of water sprayed at bud-initiation stage provides effective control against the thrips attack.

b) Diseases

1. **Yellow mosaic virus-** It is a viral disease transmitted by whitefly. Leaves show irregular yellow and green patches. Pods are either absent or are only few in numbers. Rogue out the affected plants early in the season. Green yellow mosaic virus tolerant varieties such as PAU 911, ML 818, PBM 1, ML 613, ML 267 should be cultivated. Adopt measures to control white fly. The management is same for summer moong also.

2. **Cercospora leaf spot-** (*Cercospora cruenta* and *C. canescens*). The disease spots are circular, brown and necrotic which coalesce to

cover bigger area and cause defoliation. Seed treatment with Captan or Thiram @ 3 g/kg seed should be done to prevent its attack. Grow only disease resistant varieties such as PAU 911, ML 818, ML 613 and ML 267. Spray of Zineb 75 WP (Dithane Z-78) at 400 g in 100 litres of water per hectare at 10 days interval 2-3 times, gives effective control against this disease.

3. **Root rot-** (*Macrophomina phaseolina*)- Dark lesions on leaves, branches, stem and roots are the main symptoms of this disease. Affected tissues become weak and shred easily. Seed treatment with Captan or Thiram @ 3 g/kg seed should be done as a preventive measure. The management is same for summer moong also.
4. **Anthracnose** (*Colletotrichum lindemuthianum*)- Slightly depressed spots with dark centre is the most prominent symptom. Girdling of stems, branches, peduncles and petioles is observed as lesions coalesce. Seed treatment with Captan or Thiram @ 3 g/kg seed should be done to prevent its attack. Grow only disease resistant varieties such as PAU 911, ML 818, ML 613 and ML 267. Spray of Zineb 75 WP (Dithane Z-78) at 400 g in 100 litres of water per hectare at 10 days interval 2-3 times, gives effective control against this disease.
5. **Bacterial leaf spot** (*Xanthomonas compestris* pv. phaseoli)- Its leaf spots are brown in colour and circular to irregular in outline. Disease free seeds only should be used for sowing. Grow only resistant/tolerant varieties such as PAU 911, ML 818, PBM 1or ML 613.
6. **Rhizoctonia blight** (*Rhizoctonia solani*)- It starts initially from petioles, leaf laminae, or young branches. Later on, the top of plants give blightened appearance and they are observed in patches in the field. Web-like growth whitish in colour develops on leaves in humid weather. Keep the field weed free as the infection initially spreads through weeds. The management is same for summer moong also.

Mash (Blackgram)

Land Preparation: Two ploughings followed by planking are sufficient.

Seed Rate: 15-20 kg/ha seed is optimum for sowing. In case of summer mash use 50 kg seed per hectare.

Sowing Management: Use only bold seeds for crop sowing. The optimum time for sowing irrigated crop is 15th to 25th of July in sub-montaneous region and last week of June to the first week of July in other areas. Sowing of rainfed crop should be done with the start of monsoonal rains. Crop should be sown in lines 30 cm apart preferably in N-S direction with sowing depth at 4-6 cm. Maize intercropping at every fifth row makes mash cultivation more profitable. Sowing of summer mash should be done from 15 March to 1st week of April. The row spacing should be 22.5 cm while the plant to plant distance should be at 4-5 cm. The seed should be sown at a depth of 4-6 cm.

Fertilizer application including micro-nutrients: To both *kharif* and summer mash 12.5 kg nitrogen (as Urea) and 25 kg P_2O_5 (as Superphosphate) per hectare should be applied by drilling along rows at sowing. In case of maize intercropping recommended during *kharif*, apply normal recommended fertilizers for moong at the time of sowing. Thereafter, apply fertilizers as per recommendations for maize crop in proportion to the area under it.

Weed control and inter culture: Weed control measures are same for both *kharif* and *rabi* crops. Provide initial protection to crop against weeds by hoeing at one month after sowing. After that, the crop grows vigourously enough so as to smother weeds. For chemical weed control, pre-emergence spray of Stomp 30 EC (Pendimethalin) 2.5 litres per hectare is recommended. Otherwise, pre-emergence spray of Stomp 30 EC (Pendimethalin) at 1.5 litres per hectare followed by a hand weeding six/seven weeks after this is recommended. Use 375-500 litres of water per hectare to prepare the spray solution.

Water management: During *kharif*, generally there is no need to apply irrigation to crop. However, in case of rainfall failure, apply one irrigation. In case of summer mash, 3-4 irrigations are required. The last irrigation should be given about 60 days before harvest of crop.

Plant Protection Measures

a) Insect Pests

1. **Jassid, Aphid, Whitefly and Flea-beetle**- For their control spray 938 ml of Malathion 50 EC (Malathion) or 625 ml of either Rogor 30 EC (Dimethoate) or Metasystox 25 EC (Oxydemeton methyl) in 200 litres of water per hectare. Against white fly, another option is to spray of 100 g Thiamethoxam 25 WG or 1.5 litres of Triazophos 40 EC using 200-250 litres of water per hectare.

2. **Hairy caterpillar –** The caterpillar feeds on green matter of leaves except mid-ribs. The crop may be totally denuded under a severe attack. Bury the infested plant underground to kill young larvae. The grown up larvae can be either crushed under feet or may be killed by putting in kerosene oil. Under very high population, spray 500 ml of Nuvan 100 (Dichlorovos) in 200-250 litres of water per hectare.

3. **Semi-looper**- When touched, the 2-4 cm long larva forms a loop. In severe cases, plants are denuded within a few days. Spray 500 ml of Nuvan 100 (Dichlorovos) in 200-250 litres of water per hectare.

4. **Tobacco caterpillar** (*Spodoptra litura*)-Small larvae are black, while grown up larvae are dark green with typical triangular spots on the body. First and second instars of larvae are heavy feeders on foliage. Grown up larvae feeds singly on leaves, buds, flowers and pods. Egg masses along with young larvae present on leaves should be destroyed. Spray of 425 ml of Novaluron 10 EC or 800 g of Acephate 75 SP or 1.5 litres of Dursban/Durmet (Chloropyriphos) 20 EC using 250 litres of water per hectare gives good control.

5. **Pod Borer** (*Helicoverpa armigera*)- Pale, brown, black, green or yellow coloured larvae feed on leaves, flower buds, flowers, pods and seeds in pods. Damage to plants as well as presence of faecal matter below the plants are signs of attack. The larvae fall on the ground when plants are shaken vigorously. At pod initiation stage, spray of Thiodan 35 EC at 2.5 litres in 250-300 litres of water per hectare should be done as soon as larvae appear. If need arises, another spray can be done at 10 days interval. In summer moong, it has become a limiting factor under Punjab conditions.

6. **Mite**- Webbing on underside of leaves is caused by the mites. Such leaves turn light to reddish brown. Spray of Rogor 30 EC (Dimethoate) at 425 ml per hectare in 200-250 litres of water is quite effective.

7. **Dhora** (*Calosobruchus analis*)- It attacks different pulses. To reduce attack, dry the seeds properly before storage. The gunny bags, godowns and storage bins should be thoroughly disinfected with the recommended fumigants or insecticides. In grains to be used for seed purpose, mix Malathion 5% dust at 250g/quintal grains. The grains meant for sowing should be treated with 25 ml Cythion 50 EC (Malathion premium grade) or 2 ml Sumicidin 20 EC (Fenvalerate) or 1.5 ml Cymbush 25 EC or any of the other chemicals recommended by PAU.

8. **Thrips**- Thrips also attack *summer mash* crop severely. These are very small, dark brown insects. They are found in flowers and cause flower drop, pod deformation and ultimately, the yield declines both quantitatively as well as qualitatively. Triazophos 40 EC @ 1.5 litres or 250 ml of Rogor in 200-250 litres of water per

hectare sprayed at bud-initiation stage provides effective control against the thrips attack.

b) **Diseases**

1. **Yellow mosaic virus**- It is a viral disease transmitted by whitefly. Leaves show irregular yellow and green patches. Pods are either absent or are only few in numbers. Rogue out the affected plants early in the season. Green yellow mosaic virus tolerant varieties such as Mash 114, Mash 1-1 and Mash 338 should be grown. Adopt measures to control white fly. The management is same in summer mash also.

2. **Cercospora leaf spot**- (*Cercospora cruenta* and *C canescens*). The disease spots are circular, brown and necrotic which coalesce to cover bigger area and cause defoliation. Seed treatment with Captan or Thiram @ 3 g/kg seed should be done to prevent its attack. Only disease resistant varieties such as Mash 114, Mash 1-1 and Mash 338 should be grown. Zineb 75 WP (Dithane Z-78) at 400 g in 100 litres of water per hectare sprayed at 10 days interval 2-3 times, gives effective control against this disease.

3. **Root rot**- (*Macrophomina phaseolina*)- Dark lesions on leaves, branches, stem and roots are the main symptoms of this disease. Affected tissues become weak and shred easily. Seed treatment with Captan or Thiram @ 3 g/kg seed should be done as a preventive measure. The management is same for summer mash also.

4. **Anthracnose** (*Colletotrichum lindemuthianum*)- Slightly depressed spots with dark centre is the most prominent symptom. Girdling of stems, branches, peduncles and petioles is observed as lesions coalesce. Seed treatment with Captan or Thiram @ 3 g/kg seed should be done to prevent its attack. Only disease resistant varieties such as Mash 114, Mash 1-1 and Mash 338 should be grown. Spray Zineb 75 WP (Dithane Z-78) at 400 g in 100 litres of water per hectare at 10 days interval for 2-3 times gives effective control against this disease.

5. **Bacterial leaf spot** (*Xanthomonas compestris* pv. phaseoli)- Its leaf spots are brown in colour and circular to irregular in outline. Disease free seeds only should be used for sowing should only be used. Grow only resistant/tolerant varieties such as Mash 114, Mash 1-1 and Mash 338 recommended by PAU.

6. **Rhizoctonia blight** (*Rhizoctonia solani*)- It starts initially from petioles, leaf laminae, or young branches. Later on, the top of

plants give blightened appearance and can be observed in patches in the field. Web-like growth whitish in colour develops on leaves in humid weather. The field should be kept weed free as the infection initially spreads through them. The management is same for summer mash also.

Recommended Sequential Cropping Systems for the State of Punjab

Rice based cropping systems : Rice-wheat-summer moong, Rice-potato -summer moong, Rice-gram -summer moong, Rice-gobhi sarson -summer moong, Rice-pea-wheat, Rice-gram, Rice-lentil-summer moongbean and Rice-Field pea-summer moong/ summer mash.

Maize based cropping systems : Maize-wheat-summer moong, Maize - potato -summer moong, Maize -gobhi sarson -summer moong, Maize-gram and Maize-Field pea-summer moong / summer mash

Soybean based cropping systems : Summer moong-Toria-wheat, Soybean-wheat/ Barley and Soybean-gobhi sarson

Cotton based cropping systems : Desi cotton-lentil, Cotton-summer moong.

Sugarcane based cropping system : Pigeonpea-sugarcane ratoon 1-ratoon2-wheat and Sugarcane-summermoong.

Summer mash-sugarcane cropping systems : Groundnut based, Groundnut-peas, Summer ground nut-pea-wheat and Ground nut-lentil.

Wheat based cropping systems : Moong/ Arhar/ Mash-Wheat and Summer moong-wheat.

Oilseeds based cropping systems : Summer moong-Toria-wheat, Kharif fodder-toria+gobhi sarson-summer moong, Raya-summer moong, Moong/ Arhar-African sarson, Safflower-moong/ mash, Summer moong-kharif moong-raya, Summer moong-kharif moong-wheat and Summer mash-toria/ raya.

Gram based cropping systems : Bajra-gram and Chari -gram

Medicinal /spices crops based cropping systems : Fennel-moong/ mash, Dillseed-moong/ mash, Rice-peas(early)-celery, Honey plant- moong/ mash, Fenugreek- moong/ mash.

Vegetable based cropping systems : Potato -summer moong and Summer mash-potato.

Intercropping Systems

Intercropping of Summer moong and summer Mash with Sugarcane : Intercrop one row of summer moong or summer mash in between two rows of sugarcane to get additional yield of 3.75 to 5.0 q/ha of summer moong or summer mash.

Practice	Summer Moong	Summer Mash
Variety	SML-668 and SML-832	Mash-1008, Mash-414 and Mash-218
Seed rate/ha	10 kg	12.5 kg
Time of sowing	March 20 to April 10	March 15 to April 07

Intercropping of Winter Maize and Peas : Grow one row of pea in between the maize rows raised on flat beds. In this intercrop care should be taken to give frequent irrigations during frost periods. The weed control should be done mechanically.

Intercropping of Soybean with Maize : Grow one row of soybean in between the maize rows to get higher land equivalent ratio.

Cultivation of pulses in non-conventional areas/seasons: The rice is the major crop during *kharif* season and farmers will not divert area from rice to pulses in this season. However, the summer moongbean is gaining popularity among Punjab farmers. The summer moongbean is cultivated on an area of about 20,000 ha in Punjab and there is further scope for increase in area of summer moongbean. The summer mongbean varieties (SML 668 and SML 832) are high yielding with synchronized maturity.

Summer Greengram

Land Preparation: Two/three ploughings followed by planking are sufficient for preparing clod and weed free field. In case of summer moong, when there is no wheat straw in the field it can be sown without any preparatory tillage after wheat harvesting with zero-till drill. If there is wheat straw, the sowing can be accomplished with happy seeder.

Seed Rate: 37.5 kg/ha seed rate is optimum for sowing in case of summer sown crop.

Sowing Management: Sow the crop between 20th March to 10th April. Sowing can be done till 3rd week of April but then there is risk of pre-monsoon showers at maturity. Summer moong can also be sown on raised beds spaced 67.5 cm apart (37.5 cm bed top and 30 cm furrow) on medium to heavy soils with wheat bed planter as per the recommendation of PAU. Sow two rows per bed with a row spacing of 20 cm. Use the same quantity of seed and fertilizers and follow the same cultural practices as in flat sowing. Irrigation is applied in furrows only. The advantages of this method are protection of crop from damage by excessive water at emergence, saving about 20-30 percent irrigation water and 10% increase in yield.

Fertilizer application including micro-nutrients: 12.5 kg of nitrogen along with 16 kg of P_2O_5 per hectare should be drilled at sowing. Fertilizer application to crop sown after potato in rice-potato-summer moong and maize-potato-summer moong should be omitted.

Weed control and inter culture: One/two hand hoeings are sufficient to control weeds. Go for first hoeing at four weeks after sowing and the second one two weeks later, if required. In case of zero tillage, pre-sowing spray of 1.25 litres of Gramoxone (Paraquat) in 500 litres of water per hectare should be carried out to control weeds.

1.5 litres of Basalin 45 EC (Fluchloralin) or 2.0 litres of Treflan 48 EC (Trifluralin) application on a well prepared seed bed on the same day of sowing, but before sowing can effectively control weeds. Otherwise, pre-sowing spray of 2.5 litres of Stomp 30 EC (Pendimethalin) alone or 1.5 litres of Stomp 30 EC (Pendimethalin) followed by hand hoeing two weeks after application can be effective against weeds. The quantity of water required to prepare spraying solution is 375-500 litres. These sprays control most of annual grasses and broad- leaf weeds except perennial weeds.

Water management: 3-5 irrigations are needed. The last irrigation should be applied 55 days after sowing of crop. This ensures high yield and synchronous maturity.

A field experiment was conducted to evaluate summer moong varieties under various planting methods and seed rates. The SML 832 yielded 11.5 per cent more grain yield over existing SML 668 variety (Table 6). The bed planting method of sowing produced 5.9 and 6.6 per cent higher yield in SML 668 and SML 832, respectively over flat planting. The 75 % recommended seed rate produced comparable yield to 100 % recommended seed rate (Table 7).

Table 6 : Plant height, yield contributing characters and grain yield (q/ha) of different summer mungbean genotypes under different planting methods and seed rates

Treatment	Plant height (cm)	Pods/ plant	Seeds/ pod	100-seed wt. (g)	Grain yield (kg/ha)
Genotypes					
SML 668	47.1	22.5	7.8	5.72	1103
SML 832	45.2	24.9	8.4	4.94	1230
CD 5%	NS	0.8	0.4	0.15	53
Method of sowing					
Bed planting	48.0	25.3	9.3	5.56	1202
Flat	44.3	22.1	6.7	5.11	1131
CD 5%	1.3	0.8	0.4	0.2	53
Seed rate (kg/ha)					
75 % of recommended	45.8	24.0	8.1	5.41	1186
100% recommended	46.8	23.5	8.1	5.25	1147
CD 5%	NS	NS	NS	NS	NS

Table 7 : Grain yield (q/ha) of different summer mungbean genotypes under different planting methods and seed rates

Seed rates	SML 668		SML 832		Mean
	Bed Planting	Flat	Bed Planting	Flat	
75 % recommended	1206	1032	1365	1143	1186
100 % recommended	1063	1111	1175	1238	1147
Mean	1135	1072	1270	1191	

CD (0.05): Varieties x Method of sowing x Seed rate/ Varieties x Method of sowing /Varieties x Seed rate/ Method of sowing x Seed rate/NS

Future Thrust

The present day agriculture is changing very rapidly. Understanding soil, crop management and environment is essential to enhance sustain crop productivity. Pulses being low yielders require more advanced technology under the changed situations. In future, there is a dire need to strengthen agronomic work on the following:

- Selection of efficient, short duration, high yielding pulse varieties with synchronized maturity which can fit well in the predominant cropping systems.
- Bed planting technique is becoming popular on heavy soils where water logging is a problem during rainy days. Other agronomic practices need to be worked out under bed planting situations.

- Emphasis needs to be given on integrated nutrient, weed and insect-pest management studies.
- Evolve intercropping systems with pulses for improving soil health and enhancing farmers' income.
- Development of appropriate machinery for various field operations especially in intercropping systems.
- Cropping systems with allelopatihic pulse crops to check weed growth need to be identified.
- Pulse based cropping systems to be developed so that root exudates from pulses may increase the availability of N and P.

References

Anonymous (2010). Packager of Practices for crops of Punjab (Kharif).

Anonymous (2010-11). Packager of Practices for crops of Punjab (Rabi).

Chand, R. and Singh, R. B. (2012). Major Issues in Agriculture Sector for the Twelfth Plan. NAAS News, 12(1): 6-11.

Dutta, S. (2012). Green Revolution Revisited: The Contemporary Agrarian Situation in Punjab, India. Social Change, 42(2): 229-247.

Gangwar, B., Katyal, V. and Anand, K.V. (2004). Stability and efficiency of cropping systems in Chhattisgarh and Madhya Pradesh. Indian J. Agric. Sci., 74(10): 521-528.

Gill, M.S. and Ahlawat, I.P.S. (2006). Crop diversification-its rôle towards sustainability and profitability. Indian J. Fertilizer, 2(9): 125-138.

Gol (2012). Data Base on Indian Agriculture, Department of Agriculture and Coopération. Ministry of Agriculture, Retrieved from http://dacnet.nic.in.

Grover, D.K.. and Kaura, A.S. (2009). Production And Productivity Trends In Major Grain Legumes In Indian Punjab. Int. J. Agric. Appt. Sci., 1(2): 50-58.

Kharub. A.S. , Chauhan, D.S., Sharma, R.K., Chokkar, R.S. and Tripathi, S.C. (2003). Diversifiaction of rice (Oryza sativa)- wheat (Triticum aestivum) system for improving soil fertility and productivity. Indian J. Agron., 48(3): 149-152.

Kumar, S. and Jain, K. K. (2001). Production and consumption pattern of coarse cereals and puises in India with spécial reference to Punjab. J. Agric. Dev, Policy, 15 (1): 73- 83.

Mahal, A.K., Sekhon, M. K. and Kaur, M.(2011). Pulses and Food grains in Punjab: Growth and Instability J. Res. Punjab agric. Univ., 48 (1 & 2) : 58-62.

Roy Bardhan, S. K., Walker, T., Khatana, V.S., Saha, N.K., Verma, V.S., Kadia, M.S., Haverkort, A.J. and Bowen, W. (1999). Intensification of potato production in rice-based cropping systems: a rapid rural appraisal in West Bengal, (In) Impact on a Changing World, International Potato Centre Programme Report 1997-98, pp. 205-212.

❑❑❑

19

Rajasthan

Surendra Singh, O.P. Gill, G.S. Bhatnagar and Ramniwas
All India Coordinated Research Project on Integrated Farming System, AAU, Jorhat-785013

The average productivity of pulses (Table 1) is only 460 kg/ha but it ranges from 199 kg/ha (Moth bean) to 1719 kg/ha (Field pea). Year to year fluctuation in productivity of different pulses was also observed. The highest productivity *i.e.* 1067 kg/ha was observed during the year 2006-07 & the lowest productivity i.e. 497 kg/ha was recorded during the year 2008-09. The trend of area under pulse crop during the last ten years has also shown a wide fluctuation. Year to year and crop to crop variations are most pronounced. The crop wise major pulse growing districts in Rajasthan are as follows.

Gram : Churu, Hanumangarh, Shriganganagar, Jaipur, Ajmer, Sikar, Alwar, Tonk, Bharatpur, Jhalawar, Bhilawara and Chittore

Moongbean : Nagour, Jaipur, Jodhpur, Sikar, Jalore, Jhunjhunu, Ajmer, Barmer, Churu, Tonk

Mothbean : Churu, Bikaner, Nagour, Jodhpur, Barmer

Cowpea : Sikar, Jhunjhunu, Nagour, Jaipur

Urdbean : Chittore, Banswara, Udaipur, Dungarpur, Jhalawar, Bhilawara

Pigeonpea : Banswara, Dholpur, Alwar, Jhalawar, Udaipur, Swaimadhopur

Lentil : Alwar, Ajmer, Bharatpur, Swaimadhopur, Bikaner, Shriganganagar, Pali

Peas : Jaipur, Bharatpur, Dholpur, Bundi, Baran, Kota, Bhilawara, Chittore

Table 1 : Trend of pulses area and productivity in Rajasthan (kg/ha)

S. No	Crop	1999-00	2000-01	2001-02	2002-03	2003-04	2004-05	2005-06	2006-07	2007-08	2008-09
Productivity (kg/ha)											
1.	Chickpea	695	587	759	757	633	746	443	863	446	779
2.	Lentil	1229	1188	907	979	1101	996	-	-	706	866
3.	Field pea	2217	1941	1942	2253	285	2232	-	-	1328	1557
4.	Pigeon pea	609	376	528	236	813	748	649	501	877	770
5.	Greengram	099	173	215	063	620	272	159	361	393	421
6.	Blackgram	283	288	362	182	522	360	240	293	529	296
7.	Cowpea	120	203	346	096	551	312	208	409	358	529
8.	Arid pulses (Moth)	039	144	237	044	530	131	122	166	314	266
	Total Pulses	**797**	**778**	**993**	**744**	**1172**	**917**	**566**	**1067**	**722**	**497**
Area (000, ha)											
1.	Chickpea	975.4	672.6	969.6	449.7	1117.5	1036.8	1081.9	1010.8	1231.3	1259.5
2.	Lentil	41.1	21.7	169.7	5.8	26.1	28.8	-	-	14.8	20.7
3.	Fieldpea	11.3	12.6	125.8	10.6	89.1	12.6	-	-	4.8	3.8
4.	Pigeonpea	25.8	25.5	238.1	16.5	19.6	16.9	20.4	18.6	17.7	19.3
5.	Greengram	411.9	457.9	7202.2	514.6	828.6	757.7	852.3	750.8	1060.2	885.6
6.	Black gram	119.8	112.8	181.5	258.8	239.6	146.2	132.6	107.8	144.8	137.2
7.	Cowpea	105.7	103.9	136.9	110.9	113.4	127.4	126.1	109.8	126.4	117.7
8.	Arid pulses (Moth)	771.6	961.5	1304.1	430.7	1491.8	1436.3	1227.8	1151.2	1351.8	1220.8
	Total Pulses	**3943.2**	**4045.2**	**3347.5**	**1797.6**	**3925.6**	**3562.6**	**3440.9**	**3148.9**	**2600.1**	**3672.5**

The New emerging areas of pulses production are as follows

Chickpea : In recent years area is increasing in Bikaner, Jaiselmer and Barmer districts.

Greengram : New areas for cultivation of greengram fall in Bikaner, Churu, Jodhpur and Jaiselmer districts.

Pigeonpea : New emerging areas for pegeonpea are Jaipur, Jhunjhunu, Sikar and Alwar.

Existing Cropping Systems

Pulse crops are popular for their suitability in different cropping systems. Being leguminous crops, they are capable of fixing atmospheric nitrogen which not only meets the major nitrogen requirement of that crop but enriches the soil for the succeeding crop. Recent advances in the development of a large number of varieties of different pulse crops varying largely in duration to maturity have made it possible to include these crops in different crop sequences. The important crop sequences involving pulse crops followed in different regions of the Rajasthan are pearl millet-chickpea, greengram-mustard, greengram-wheat, cowpea-wheat, moth-fallow, black gram-wheat, sorghum-chickpea, maize-pea, pigeon pea-wheat, black gram-mustard, pearl millet-lentil, cotton-chickpea and fallow-chickpea.

Yield Gaps

In production of pulses a wide gap exists between yield potential and maximum achievable yield ranging from 27.2 to 42.8% as highlighted in table 2.

Table 2 : Yield gaps in pulse crops in Rajasthan.

Crop	Yield potential of most popular variety (q/ha)	Maximum yield achieved under CSR/ECF experiment (q/ha)	Yield gap (%)
Greengram	10.00 (RMG-62)	6.92	30.8
Cowpea	8.00 (RC-19)	5.28	34.0
Moth	7.5 (RMO-40)	4.29	42.8
Black gram (Urd)	9.0 (T-9)	6.20	31.1
Arhar	10.5 (Prabhat)	7.80	25.7
Chickpea	20.00 (RSG-888)	14.56	27.2

Improved Varieties

The improved varieties of important pulses crops along with their salient features are given in table 3.

Table 3 : Recommended varieties of pulse crops for Rajasthan.

S.No.	Variety	Days to maturity	Yield potential (q/ha)	Salient features
Greengram				
1.	RMG-62	65-70	8-10	Synchronous maturity Grain bold & medium sized. Suitable for *kharif* & *zaid,* resistant to leaf blight.
2.	RMG-268	65-70	10-12	Resistant to drought & leaf blight. Suitable for *kharif* & *Zaid* season.
3.	RMG-492	65-70		Suitable for both *kharif* & *zaid* season. Suitable for low rainfall areas. Medium resistant to nematode.
4.	RMG-344	65-70	8-10	Synchronous maturity, resistant to leaf blight, web blight & drought
5.	Ganga-8	65-70	10-12	Tolerant to insects & nematodes & synchronous in maturity
6.	SML-668	65-75	8-10	Medium bold sized grain & resistant to insect pest & diseases.
Cowpea				
1.	C-152	90-95	6-8	Resistant to gall nematode.
2.	RS-9	80-85	6.5-7	Suitable for both seed & grain fodder.
3.	FS-68	65-70	8-10	Suitable for both the *kharif* & *zaid* crops.
4.	RC-19	60-65	7-9	Synchronous maturity.
5.	RC-101	60-65	8-10	White seeded, resistant to stem rot.
Moth				
1.	RMO-40	62-65	6-8	Tolerant to yellow mosaic & drought
2.	IPCMO-912 (Vikash)	75-80	5-6	Resistant to YMV & tolerant to bacterial blight
3.	RMO-257	65	4-5	Early maturing.
Urd				
1.	Krishna	90-110	10-12	Brown seeded.
2.	T-9	80-90	8-10	Grain small size & brown seeded.
3.	RBU-38 (Barkha)	65-70	10-12	Grain bold size & tolerant to YMV.

Contd.

4.	WBU-108 (Sarada)	65-70	8-10	Grain bold size & tolerant to YMV.
Pigeonpea				
1.	T-21	140-180	10-12	High yielding.
2.	Prabhat	115-135	12-15	Yellow seeded.
3.	Gwalior-3	180-250	8-12	-
4.	UPAS -120	130-135	8-12	Early maturing, tolerant to YMV.
5.	Pusa Ageti	140-145	12-15	High yielding.
6.	Pusa-992	140-145	12-15	Tolerant to wilt.
Gram (Chickpea)				
1.	RSG-44	130-135	20-25	Double podded & tolerant to drought.
2.	RSG-888 (Anubhav)	130-135	20-25	Double podded, tolerant to drought, resistant to wilt, dry root rot & nematode.
3.	RSG-931 (Anvita)	125-130	20-25	Double podded, white flowered.
4.	RSG-963 (Aadhar)	130-135	20-25	Suitable for late sowing, tolerant to drought.
5.	RSG-807 (Aabhar)	135-140	20-25	Bold seeded, tolerant to wilt.
6.	RSG-895 (Arpita)	125-130	20-25	Double podded.
7.	RSG-896 (Arpan)	125-130	16-20	Double podded, resistant to dry root rot & wilt.
8.	RSG-902 (Aruna)	130-135	20-25	Double podded, resistant to dry root rot & wilt.
9.	RSG-945 (Asha)	135-140	15-20	Suitable for late sowing, tolerant to wilt.
10.	RSG-973 (Aabha)	135-140	15-20	Tolerant to drought & wilt.
11.	RSG-991 (Arpana)	135-140	20-25	Suitable for late sowing.
12.	RSGK-6 (Aasar)	135-140	18-20	White flowered & bold seeded, tolerant to dry root rot.
13.	CSJD-884 (Aakash)	125-130	20-25	Double podded & tolerant to drought & dry root rot.
14.	Pratap Chana-1	90-95	12-14	Suitable for rainfed areas of south Rajasthan.
French Bean (Rajmash)				
1.	PDR-14 (Uday)	115-120	20-22	White flowered & green coloured pod.
2.	VL-63	110-115	20-22	Green coloured pods.

Contd.

3.	HUR-15	108-110	18-20	White seeded.
4.	HUR-136	105-107	14-16	Red seeded.
Lentil				
1.	LH-84-8 (Sapana)	130-135	12-15	Brown & medium bold.
2.	Pant L-4	130-135	15-20	Resistant to rust & wilt disease.
3.	L-4076	135-140	12-15	Bold seeded & resistant to rust disease.
4.	DPL-15	130-135	12-15	Brown & bold size seed, resistant to wilt.
5.	Pant Masur-406	125-130	15-20	Small sized grains.
Field Pea				
1.	Type-163	135-140	15-18	White & bold size seed.
2.	BR-12	130-135	15-16	Bold & attractive seeds.
3.	HRP-4 (Arpana)	140-145	20-25	Resistant to powdery mildew.
4.	DMR-7 (Alankar)	115-135	20-25	Bold size seed & resistant to powdery mildew.

Constraints in Pulses Production

The major constraints in pulses production in Rajasthan are as follows.

1. The pulses are being cultivated on marginal & sub-marginal lands having poor soil fertility and moisture retention capacity.
2. The inadequate availability of seed of improved varieties continues to be a constraint for increasing pulses production.
3. The pulses are prone to large number of pests & diseases as compared to cereals and other crops.
4. The pulses are treated as secondary in importance even when they are giving the same yield under similar conditions, and in situations where no other crop would survive.
5. They are invariably grown as mixed crop with cereals and, therefore, they don't get a fair chance to express their full yield potential.
6. The technology gap between the research institutions & farmers' level is very wide.
7. Production of efficient Rhizobium strains and their proper distribution are not adequate.

8. Non-availability of short duration & disease resistant varieties of pulses.
9. Inadequate weed control.
10. Most of the pulses are photosensitive so that growth and yield are affected by time of sowing.
11. Fluctuating market prices.
12. Poor post harvest technology.
13. Improper and inadequate use of improved agriculture implements.
14. Frost & low temperatures during the night cause heavy damage to *rabi* pulses particularly in chickpea.

Production Technologies

The detailed improved production technologies of important pulses grown in Rajasthan are as under:

(I) Gram or chickpea

In Rajasthan, chickpea contribute to 40% of total pulse production and is an important pulse crop grown during *rabi* season.

Land preparation: In monocropping system when fields are kept fallow during monsoon, deep ploughing before rains and 2-3 ploughings before withdrawal of rains helps in increasing the profile moisture. If it is cultivated after *kharif* crop, then one deep ploughing by soil turning plough followed by 2-3 harrowings and planking are sufficient. Soil should be friable and weed free. It should be well drained and levelled, so that water does not stagnate in the field. In case of less moisture, pre-sowing irrigation should be given before field preparation. If manure is to be applied, it should be applied before field preparation so that it may be mixed well in the soil.

Seed rate: The optimum seed rate of small seeded chickpea (12-15 g/100 seed) is 50-60 kg/ha, whereas in bold seeded varieties (25 g/100 seeds) the seed rate is 70-80 kg/ha. In *kabuli* type, 100 kg seed/ha is required for sowing.

Sowing management: The optimum time of sowing in rain-fed conditions is the first week of October, whereas in irrigated conditions the optimum time is up to 20th October. Late sown varieties can be planted up to 15th December in irrigated conditions. However, late sowing adversely affects the yield. The spacing between row to row is 30 cm, and plant to plant 10 cm. The depth of sowing depends on soil moisture but in rain-fed conditions, deep sowing (10 cm) proved better and also reduced *Fusarium* wilt. The sowing should be done by the *desi* plough or by seed drill. Sometimes broadcasting method of sowing is also adopted but line sowing is better due to easy intercultural operations. To overcome diseases like root rot and wilt, seeds should be treated with Thiram 2.5 gm + Carbandazim 0.5 gm/kg of seed. The seed should also be treated with *Rhizobium* culture, PSB and PGPR each @ 600 g/ha. *Rhizobium* culture enhances the nitrogen fixation through symbiosis, PSB converts insoluble phosphates into soluble form and makes it available to the plants. PGPR promotes the plant growth, which consequently increases the yield.

Fertilizer management: Being a pulse crop, chickpea fixes atmospheric nitrogen through root nodules which is a symbiotic process. In the later stages, pulses do not require more fertilizers. In rain-fed conditions, 15-20 kg N and 30-40 kg P_2O_5 per hectare, in irrigated conditions with normal sowing, the dose is 20 kg N, 40 kg P_2O_5 and 20 kg K_2O per/ha and in irrigated late sown conditions, 30 kg N, 40 kg P_2O_5 and 20 kg K_2O per/ha as a basal dose lead to better production. Basal application of 25 kg zinc sulphate per/ha is quite effective for good production. Foliar spray of $ZnSO_4$ @ 0.5% or seed coating with zinc is also very much effective in correcting zinc deficiency. In irrigated areas, soil incorporation of 125 kg/ha gypsum is also useful. Foliar spray of DAP @ 2% enhances the yield significantly and foliar application of urea @ 2% also showed positive results. The fertilizers should be applied in furrows. If soil is deficient in organic matter, application of 5-6 tonnes/ha of FYM improves yield significantly.

Weed control and interculture: Farmers do not generally adopt weed control measures in chickpea, which badly affects the yield. For good production the crop should be weed free. However, two hand weedings, at 25-30 DAS and 50-55 DAS are quite effective in controlling the weeds. Pre-emergence application of Pendimethalin 30 EC @ 0.75-1 kg/ha + one hand weeding at the time of 35-40 DAS is also equally effective in controlling weeds.

Water management: Generally chickpea is cultivated in rainfed areas under conserved moisture but in case of irrigated conditions, if there is insufficient moisture at the time of sowing, pre-sowing irrigation should be applied. In case of failure of rains, irrigation at the time of branching and pod formation improves the production. Moisture deficiency at the time of pod filling stage badly affects the chickpea production. If irrigation is to be given through sprinkler then run it for 5 hours and apply irrigation at 40, 80 & 110 DAS.

Protection from frost: Spraying of 0.1% dilute sulphuric acid at pre flowering & flowering stage is effective in mitigating the adverse impact of frost.

Plant protection measures: Summer fallowing and ploughing reduces the incidence of diseases, nematodes and insect-pests. Deep ploughing exposes the insects, their eggs, spores of diseases and nematodes.

(A) Diseases: Series of diseases affects the pulses yield.

I. Wilt: It is a seed-borne as well as soil-borne fungal disease. The pathogen can survive in the soil up to six years even in the absence of the host. It is a typical vascular disease causing xylem necrosis. The disease is systemic and plants may be infected at any stage. It generally appears after 3 weeks of sowing. Internal tissues from the collar region down wards become dark and discolored. The petioles and rachis along with leaflets drop down. Dropping starts from the upper part of the plant but soon the entire plant drops down. Sometimes only partial wilting occurs.

Control Measures: Avoid cultivation in heavily infested fields. Follow inter cropping or crop rotation with wheat, barley, linseed and mustard. Since the pathogen can survive up to six years, normal crop rotation is not sufficient to overcome the disease. Use resistant varieties for sowing. Deep ploughing in summers; soil solarization by covering with transparent polythene sheet for 6-8 weeks during April-May; removal of host plant debris from the field; rouging of diseased plants and seed treatment by Bavistin + Thiram (1:2) @ 3g/kg of seed or Bavistin (Carbendazim) @ 2.5 g/kg of seed are effative measures for its control.

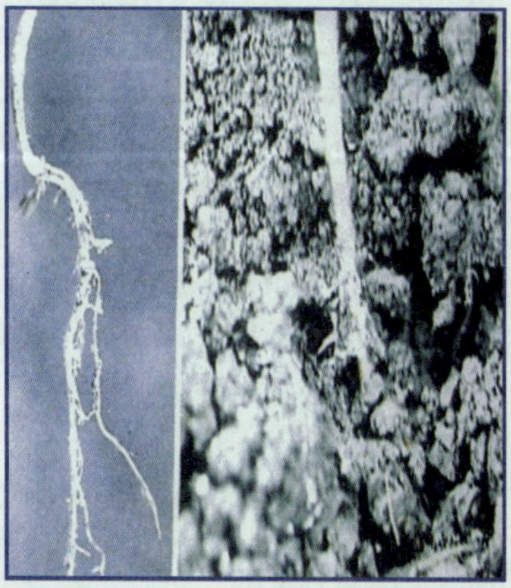

II. Collar rot: It is a soil-borne fungal disease which mostly appears in seedling stage under high moisture conditions. Young seedlings collapse, while in older seedlings, the leaves turn yellow and the plant dries up. Rotting of stem can be seen at the collar region, the rotten portion of the stem covered with whitish mycelial growth of the fungus and the disease may appear in patches.

Control Measures: Seed treatment by Captan @ 2.5 or Thiram @ 3g/kg of seed. Crop rotation, avoiding excess soil moisture and drenching of Demosan @ 15 kg/ha are other measures.

III. Black root rot: It is a seed-borne fungal disease that generally occurs under high moisture conditions at seedling stage. Affected seedlings turn yellow, and petioles and leaflets droop. Generally the seedlings do not collapse; a distinct dark brown lesion appears above the collar portion at the main stem.

Control Measures Seed treatment by Captan @ 2.5 or Thiram @ 3 g/kg of seed. Proper drainage, rouging of infested plants and use of resistant varieties are additional actions to be taken.

IV. Stem rot: The petioles and leaflets droop, the leaves dry up and turn straw coloured prematurely. A web of white mycelial strands appears in the color region and above.

Control Measures: Avoid excess moisture and follow clean cultivation.

V. Ascochyta blight: Small circular brown spots appear on the leaves. These spots have discolored margins. Later black minute dots appear on the necrotic lesion arranged in circles.

Control Measures: Sowing of disease free seed and removal and destruction of diseased plant debris. Deep ploughing, helps in decomposition of fungal straw. Inter cropping with wheat or barley, use of resistant (C 235, GNG 469, GNG 663 etc.) or tolerant (BG 256) varieties for sowing and seed treatment with Bavistin + Thiram (1:2) or Captan @ 3 g/kg of seed are also effective measures.

VI. Botrytis gray mould: Grey and dark brown lesions occur on infected plants. Excessive growth is also conducive for the disease.

Control Measures: Lower seed rate with wider spacing. Late planting in the first fortnight of November, particularly, in irrigated areas. Adopting crop rotation. Seed treatment with Bavistin + Thiram (1:2), or Bavistin @ 3 g/kg of seed. Spray of Captan (0.2%) 3-4 times at an interval of 15 days and use of resistant varieties [{Sweta (ICCV 2) Kabuli}, SAKI 9516 etc.] for sowing can address these problem.

VII. Dry root rot: It is a soil-borne fungal disease that mostly appears at the time of flowering and podding. 30^0C or above temperature is favorable for this disease. Disease appears in scattered patches. The petioles and leaflets droop only at very top. Infected plants dry up. When plants are pulled out root remains in soil. It turns dark showing signs of rotting.

Control Measures: Use of resistant varieties. Adopting of crop rotations along with seed treatment by Carbendazim + Thiram (1:2) 3 g/kg of seed are effective in controlling it.

(B) Nematode: Root-knot nematodes are characterized by the presence of simple multiple knots on the root system and can be seen by naked eyes. The size of knots varies depending up on the genotype reaction and nematode species. Stunting, uneven poor plant growth in patches, plants turn pale yellowish, poor podding, reduced nodulation and reduced yield are common characteristics of root-knot nematode attack. In case of severe infection of disease, blackening of roots takes place in later stages resulting in death of plant.

Control Measures: Soil solarization by covering the soil with transparent polyethylene sheets during the summers (April, May and June) significantly reduces the nematode population. Summer fallowing and deep ploughing (2-3), use of resistant or moderately resistant varieties like RSG 888 for sowing; adoption of crop rotations like Pearl millet/Jowar/maize/paddy/til/onion-chickpea. Apply *Pseudomonas fluorescens* + *Trichoderma viride* @ 2.5 kg/ha are advised seed treatment with Carbosulfan 25 ST @ 3% ww (30 g/kg of seed) are advised. In case of severe infection, soil application of Carbofuran 3G @ 45 kg/ha (1.5 kg ai/ha) or *neem* cake @ 500 kg/ha at the time of field preparation is recommended. Application of organic manure into soil reduces plant parasitic nematode. Application of non-edible deoiled seed cakes of neem, karanj, mahua, castor etc. also reduces nematode population.

(C) Insects-Pests: Pod borer, semilooper, cutworm and termite are the major insect-pests of chickpea. However, pod borer is a key pest causing 90-95% damage alone.

I. **Termites:** These are tiny greyish-white social insects' which live in groups. Termites cause damage right from germination to subsequent growth stages of the crop. They cut the main root of the plant and ultimately the damaged plant dries up. Such plants, if uprooted, show presence of termites and their galleries on damaged portions. Many times, it causes severe damage, resulting in crop failure.

Control Measures: In case of heavy infestation, application of Methyl parathion dust (2%) @ 25 kg/ha in soil at the time of field preparation. Seed treatment with Chloropyriphos @ 10 ml/kg seed is also on effective control measure.

II. **Gram pod borer:** The moth is stoutly built and is yellowish brown in colour. The fore wings are marked with grayish wavy lines, black spots of varying sizes on the upper side, black kidney shaped mark and round spot on the underside. The hind wings are whitish and lighter in colour. The newly hatched larvae are sluggish and whitish-green in colour, become active after 3-4 hours. The caterpillar stage is the damaging stage. Young larvae feed on tender leaves, buds, and flowers and subsequently bore into pods. Normally larvae remain hidden in the foliage of the crop unnoticed till pod formation. It can cause extensive damage. A single larva may destroy 30-40 pods before it reaches maturity.

Control Measures: Timely sowing *i.e.* up to mid October, growing of early cultivars, and complete podding by first week of March in northern region helps in escaping the peak attack of the pest. Spray of 0.1 per cent Monocrotophos (36 WSC) is also recommended which should be repeated at an interval of 10 days if needed. Dusting of Malathion dust (5%) 25 kg/ha is also effective. Mixed cropping with non-preferred host plants like barley, wheat, mustard and linseed is also advised.

III. **Cutworm:** Cutworm is common in the whole country but more serious in northern India, particularly, the flood affected areas and soils rich in organic matter. Adult moths are greyish-brown in colour. The forewings are dirty brown. The hind wings are semi-hyaline white. The larvae are polyphagous feeders infesting wide range of hosts. The moths hide during day times and come out of their hiding places at night. The caterpillar cuts the plant at base, below or just above the soil surface or may even cut the branches. The characteristic feature of damage is that the caterpillar drags the cut parts into the soil for feeding.

Control Measures Application of Methyl parathion dust (2%) @ 25 kg/ha in soil at the time of field preparation and seed treatment with chloropyriphos 20 EC @ 12 ml/kg of seed are recommended.

Harvesting & Threshing: When plants become reddish brown, dry up and start shedding, harvesting should be done. Harvesting should be done preferably in the morning. The harvested plants are to be dried for a week for threshing. Threshing may be done manually or with the help of multi crop thresher with proper adjustment. Seed should be cleaned and dried in the sunlight so that seed is not damaged during storage.

Yield: In rain-fed conditions, the seed yield of chickpea is 8-10 q/ha, whereas in irrigated conditions the yield levels can be up to 25 q/ha.

Pulse based sequential cropping: Pearlmillet-chickpea, paddy-chickpea, cluster bean-chickpea, maize-chickpea, sesame-chickpea, onion-chickpea, groundnut-chickpea and cotton/ rice-chickpea.

Intercropping Systems: Chickpea is also cultivated as an inter crop or mixed with wheat, barley, linseed, safflower, coriander or mustard. In mixed/ inter cropping, chickpea does well when rainfall is low, whereas other crops fare better when rains are adequate or abundant. Intercropping is better than mixed cropping for performing agronomic practices effectively.

Moongbean or Greengram

In Rajasthan, it is grown in an area of 885.6 thousand ha with productivity of 304 kg/ha (2008-09). The average productivity of 278 kg/ha is very low as compared to the genetic potential (1000-1500 kg/ha). Though there are so many reasons of low productivity including cultivation by small and poor farmers on marginal lands, but lack of scientific cultivation is also one of the major factors responsible for low yields, so scientific methods have to adopted for mungbean cultivation.

Land preparation: One deep ploughing by soil turning plough followed by 2-3 harrowings and planking are sufficient. Sandy to black cotton and red laterite soils are best. However, loamy and sandy loam soils are ideal. It is susceptible to alkalinity.

Seed rate: 15-20 kg/ha seed is required for the sowing of moongbean.

Sowing management: *Kharif* crop can be sown at the onset of monsoon in the month of July to first week of August. Summer crop may be sown in the months of March to April. The spacing followed is row to row 30 cm and plant to plant 10 cm. The sowing should be done behind the *desi* plough or by seed drill. Treat seed with Bavistin or Thiram or Captan @ 2, 3, 2.5 g/kg of seed, respectively. The seed should also be treated with *Rhizobium* culture (*Rhizobium leguminosarum*), PSB and PGPR each @ 600 g/ha. For seed treatment, first the seed should be treated with fungicide then with insecticide (if any) and lastly with *Rhizobium (FIR)*. For seed treatment, prepare 10% *gur* solution by boiling it. After cooling *Rhizobium* culture, PSB and PGPR should be mixed well in the *gur* solution and then mixed with the seed. The treated seed should be dried for 6-8 hours in shade and sowing should be done immediately or within 12 hours.

Fertilizer management: 20 kg N and 40 kg P_2O_5 per hectare are recommended as basal application at the time of sowing through drill. Zinc sulphate @ 15-20 kg/ha or 20 kg sulphur through gypsum is also quite effective for good production.

Weed Control and Interculture: The first hand weeding is required to the done at 20-25 DAS and the second at 20 days after first weeding or pre-emergence application of Fluchloralin @ 0.75 kg ai/ha accompanied with one hand weeding at 35-40 DAS.

Water Management: In case of failure of rains during *kharif,* 1–2 life saving irrigations at pre flowering and grain filling stage may be given.

Plant Protection: Summer fallowing and ploughing reduces the incidence of diseases, nematodes and insect-pests. Deep ploughing exposes the insects, their eggs, spores of diseases and nematodes.

(A) Diseases:

I. Root rot & stem rot: The seedlings rot at collar region and plants dry up.

Control Measures: Seed treatment by Bavistin @ 1 g or Thiram 3 g/kg of seed.

II. Bacterial blight: Light yellow circular or irregular raised spots with brown centre appear on the leaves. In the later stage the spots also appears on pods. Pods become yellow and leads to drying and drop.

Control Measures: Use disease free seed, grow resistant varieties like K 851, Seed treatment by Captan @ 2.5 g/kg of seed are advised spray of Streptocycline (100 ppm) + Copper oxychloride (0.3%) are advised on the appearance of the disease.

III. Yellow mosaic virus (YMV): Primary symptoms appear on young leaves in the form of yellow mild scattered spots. The next trifoliate leaf emerging from growing apex shows alternate irregular yellow and green patches. The yellowishness increases and some of the apical leaves turn completely yellow. The diseased plants mature late and bear very few flowers and pods with small seeds.

Control Measures: Clean cultivation is essential. Uproot and destroy diseased plants. Inter/mixed cropping with non host crops like sorghum, pearl millet and maize along with spray of Dimethoate (30 EC) or Metasystox or Melathion (50 EC) @ 0.1% to be repeated at an interval of 15 days is recommended. Cultivation of disease resistant/tolerant varieties for sowing like: MUM 2, , PDM 11, PDM 54.

IV. Powdery mildew: Affected plants show white powdery mass on leaves in the initial stages and later white powder appears on stem and pods also.

Control Measures: Delayed sowing and wider spacing along with field sanitation and clean cultivation. On initiation of symptoms, two sprays of 0.1% Calixin or Karathane are recommended at 10 days interval. Uproot and destroy diseased plants.

V. Nematodes: Stunting, pale yellowish and poor plant growth. Roots swell due to nematode galls. Roots become brittle and is reduced nodulation. Blackening of roots takes place in later stages resulting in plant death.

Control Measures: Soil solarization by covering the soil with transparent polyethylene sheets during the summers (April, May and June). Summer fallowing & deep summer ploughing (2-3) is essential. Avoid monocropping and follow crop rotation. Use resistant or moderately resistant varieties for sowing. Apply *Pseudomonas fluorescens* + *Trichoderma viride* @ 2.5 kg/ha. Seed treatment with carbosulfan 25 ST @ 3% ww (30 g/kg of seed) or TANU 60 EC @ 10 ml/kg of seed is advised. In case of severe infection, soil application of carbofuran @ 45 kg/ha (i.e. 1.5 kg ai/ha) or *neem* cake @ 500 kg/ha at the time of field preparation. Application of organic manure reduces plant parasitic nematode. Application of non-edible deoiled seed cakes of neem, karanj, mahua and castor lower down the nematode population.

VI. Insect-Pests: White grub, Termites, White fly, Aphids, Jassids, caterpillar, semi looper and pod borer are the key pests of mungbean

I. White grub: The beetles appear with the onset of monsoon and may feed on the foliage of a wide variety of host trees and bushes found in the close vicinity like *neem, ber, jamun, sainjana, ashok, guava, khejri* etc. The mated female lays the eggs in soil. The grubs (larvae) cause the damage. These are white with black mouth and are 3-5 cm in length. Being polyphagous in nature, the grubs feed on the living roots of a wide range of cultivated as well as uncultivated plants. The young grubs, after hatching orient towards the roots and start feeding on them. They cut the roots of plants. The affected plants show varying degree of yellowing; some get wilted and ultimately die. They can be easily pulled out.

Control Measures: Adopt proper cultural practices like deep summer ploughing, trapping (light trap and hand collection from host tree) and use of Pheromones and seed treatment with Chlorpyriphos @ 5-6 ml/kg of seed. In the standing crop, apply Imidaclorprid 300 ml, Chlorpyriphos 20 EC @ 4 lit/ha with irrigation water or impregnated in 50 kg soil and broadcast in the affected area followed by light irrigation.

II. **Aphids:** These are small and dark brown or blackish pests. Cause damage by de-sapping. Colonies of nymphs and adults occur on leaves, top shoots and pods. Plants become yellowish. Photosynthesis is interfered with owing to the growth of sooty moulds on the honeydew excreted by the insect. It is also a vector of some viral diseases.

Control Measures: Controlled by spray of Azadirachtin (neem oil and akarin) + Monochrotophos (36 WSC) @ 1 liter + 500 ml/ha, respectively. Spray of Dimethoate (30 EC) or Monocrotophos (36 WSC) @ 0.1% and repeated at an interval of 15 days if needed.

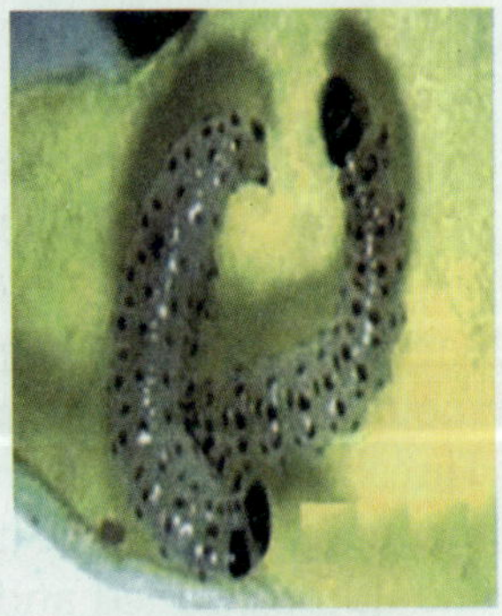

III. **Pod borer:** The newly hatched caterpillars are sluggish and whitish-green in colour, become active after 3-4 hours. The full grown larvae are 3.5-4 mm in length with pale green colour. Dorsal surface bears dark broken stripes, head is reddish brown and legs are brown. The young larvae feed on tender leaves, buds, flowers and subsequently bore into pods and eat up all the grains. It may cause extensive damage.

Control Measures: Spray of 0.1% Monocrotophos (36 WSC) & repeated at an interval of 10 days.

IV. **Red hairy caterpillar:** Also known as *katra,* is about 25 mm in length when fully grown. It is reddish-amber to olive-green in colour and body is covered with numerous long hairs. Being polyphagous in nature, they eat up all the vegetation so the damage proves disastrous. In an epidemic form, there may be a complete failure of the crop.

Control Measures: Spray of Dimethoate (30 EC) @ 0.1% or dusting of Methyl parathion (2%) @ 25 kg/ha.

V. **Green semi looper:** The caterpillar is about 25-35 mm in length. These are pale yellowish green with 5 longitudinal white lines on the dorsal surface. The larva forms a loop during movement. The caterpillars eat the leaves only but some times damage the shoots and buds also, resulting in complete defoliation. It occurs in epidemic form in the years of heavy rainfall.

Control Measures: Spray of 0.1% monocrotophos (36 WSC).

Harvesting & Threshing: When more than 90% pods turn to black harvesting should be done. If labour is available easily and economically, one picking may be done with subsequent harvesting of whole crop on maturity, otherwise, one time harvesting may also be done at the time of full maturity of the crop. Early harvesting results in immature seeds, whereas, delayed harvesting results in shattering of pods and reduction in production. Harvesting should be done preferably in the morning to avoid shuttering due to dew effect. Threshing may be done after sufficient drying in the sunlight. It may be done manually or with the help of multi crop thresher with proper adjustment. Seed should be cleaned and dried in the sunlight so that it is not damaged during storage.

Yield: Mungbean is generally cultivated in poor and marginal lands under poor management. However, under normal conditions, its yield can range from 5-25 q/ha.

Pulse based sequential cropping: Mungbean-mustard-urdbean, Maize-potato/wheat-mungbean, Rice-sugar cane-ratoon-mungbean, Mungbean-potato-onion, Mungbean-wheat/barley, Mungbean-toria-wheat, and Rice-wheat-mungbean.

Intercropping Systems: As mix crop or in a definite ratio (2:1, 1:1 or 1:2) with pearl millet, sorghum, maize, pigeonpea, ragi etc.

Urdbean

Urdbean is an important pulse crop of the Indian subcontinent, cultivated almost through out the country in all the three seasons' i.e *kharif, rabi* and summer. In Rajasthan, it is cultivated in the *kharif* season. India is the largest producer of urdbean in the world. By adopting scientific methods of cultivation and improved varieties the production and productivity of urdbean can be improved further easily.

Land preparation: One deep ploughing by soil turning plough followed by 2-3 harrowings and planking is sufficient. Soil should be friable and weed free. It should be well drained and levelled. Urdbean may be cultivated on a wide range of soils ranging from sandy loam to black soils, however, clay to clay loam soils having good water retention capacity with neutral pH are best suited for urdbean cultivation. It is also susceptible to water logging, so the soils should be well drained. It can not be cultivated on saline or alkaline soils.

Seed rate: For the sowing of urdbean, 15-20 kg/ha seed is required.

Sowing management: *Kharif* crop can be sown on the onset of monsoon in the month of July to first week of August. The spacing to be followed in row to row 30 cm and plant to plant 10 cm. The sowing should be done behind the *desi* plough or by seed drill. Seed treatment with Bavistin or Thiram or Captan @ 2, 3 or 2.5 g/kg of seed, respectively, is recommended. The seed should also be treated with *Rhizobium* culture (*Rhizobium leguminosarum*), PSB and PGPR each @ 600 g/ha. For seed treatment, the seed should be first treated with fungicide then with insecticide (if any) and lastly with *Rhizobium (FIR)* as described for the mungbean.

Fertilizer management: 20 kg N and 40 kg P_2O_5 per hectare as basal application at the time of sowing through drill so that it is available in the root zone of the crop. Application of 200 kg gypsum/ha fulfills the requirement of calcium and sulphur and results in good yields.

Weed Control and Interculture: First hand weeding at 20-25 DAS and second at 20 days after first weeding or pre-emergence application of Fluchloralin @ 0.75 kg ai/ha accompanied with one hand weeding at 35-40 DAS are quite effective.

Water Management: In case of failure of rains during *kharif,* 1–2 life saving irrigations at pre flowering and grain filling stage may be given. In case of heavy rains, proper drainage should be provided in the heavy soils.

Plant Protection: Summer fallowing and ploughing reduces the incidence of diseases, nematodes and insect-pests. Deep ploughing exposes the insects, their eggs, spores of diseases and nematodes.

(A) Diseases:

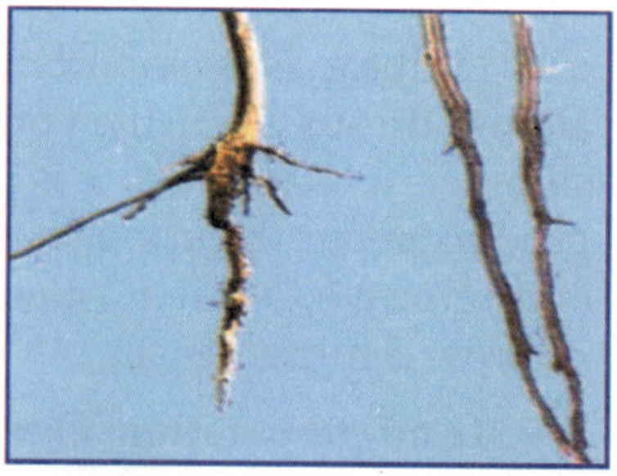

I. **Root rot & stem rot:** The seedlings rot at collar region and plants dry up.

Control Measures: Seed treatment by Bavistin @ 1 g or Thiram 3 g/kg of seed. Adoption of appropriate crop rotation.

II. **Bacterial blight:** Light yellow circular or irregular raised spots with brown centres appear on the leaves. In the later stages, the spots also appear on pods. Pods become yellow and finally dry up and drop.

Control Measures: Use disease free seed and resistant varieties. Seed treatment by Captan @ 2.5 g/kg of seed and spray of Streptocycline (100 ppm) + Copper oxychloride (0.3%) on the appearance of the disease are effective control measures.

III. **Yellow mosaic virus (YMV):** Primary symptoms appear on young leaves in the form of yellow mild scattered spots. The next trifoliate leaf emerging from growing apex shows alternating irregular yellow and green patches. The yellowishness increases and some of the apical leaves turn completely yellow. The diseased plants mature late and bear very few flowers and pods with small seeds.

Control Measures: Clean cultivation. Uprooting and destruction of diseased plants following, Inter/mixed cropping with non host crops like sorghum, pearl millet and maize, spraying of Dimethoate (30 EC) or Metasystox or Melathion (50 EC) @ 0.1% and repeat at an interval of 15 days are recommended.

IV. **Powdery mildew:** Affected plants show white powdery mass on leaves in the initial stages and later on the white powder appears on stem and pods also.

Control Measures: Use disease resistant varieties for sowing. Delayed sowing and wider spacing field sanitation and clean cultivation are other recommendations. On initiation of symptoms two sprays of 0.1% Calixin or Karathane at 10 days interval are recommended. Uproot and destroy diseased plants.

V. **Nematodes:** Stunted, pale yellowish and poor plant growth. Roots swell due to nematode galls and become brittle and nodulation is reduced. Blackening of roots takes place in later stages resulting in death of plants.

Control Measures: The control measures are same as described for moongbean on previous pages.

VI. **Insect-Pests:** White grub, Termites, White fly, Aphids, catter pillar, semi looper and pod borer are the key pests of urdbean.

Control Measures: The control measures for these insect-pests are the same as discussed for mungbean on previous pages.

Harvesting & Threshing: When most of the pods turn black in colour, harvesting may be done. Early harvesting results in immature seeds, whereas, delayed harvesting results in shattering of pods and reduction in production. Harvesting should be done properly in the morning to avoid shattering due to dew effect. Threshing may be done after sufficient drying in the sunlight manually or with the help of multi crop thresher with proper adjustments. Seed should be cleaned and dried in the sunlight so that it is not damaged during storage.

Yield: Generally urdbean is cultivated in poor and marginal lands under poor management and least inputs resulting in yield levels that are definitely low. However, under normal conditions, its yield ranged from 5-15 q/ha, whereas the summer crop may yield 10-20 q/ha.

Pulse based sequential cropping: Pearl millet + urdbean-wheat/barley, Sorghum/maize + urdbean-wheat/barley, Urdbean-toria-wheat-mungbean/urdbean, Maize-urdbean-wheat-green manuring, Mungbean-mustard-urdbean, Rice-urdbean, Pigeonpea + urdbean-wheat, Maize-potato-urdbean.

Intercropping Systems: *Kharif* urdbean is generally cultivated as a mixture with crops like pearl millet, sorghum, maize and pigeonpea. Generally seed of different crops are mixed together and broadcasted or sown in lines but inter cropping is more profitable when cultivated with pearl millet, ragi, maize, pigeonpea, safflower or sugar cane. One or two lines of urdbean may be sown between two rows of above crops eg. Pigeonpea is planted at a distance of 75 cm and a lot of space remains unutilized. If we plant urdbean with pigeonpea, the urdbean production will be additional and weed intensity will be less. In spring, it may be cultivated as an inter crop with sugarcane (1:2 row ratio) and sunflower (2:6 row ratio).

Cowpea

Cowpea is a multi-season, multipurpose leguminous crop cultivated in arid and semi-arid regions throughout the tropics and sub-tropics. It is one of the most productive, genetically versatile and phenotypically hard enough to sustain environmental shocks and stress. In India it is cultivated since ancient times. In general, cowpeas are grown for seed, leaf and pod vegetables, fresh immature seeds, green fodder, cover crop, green manuring etc. It contains 24.1% of protein and is a rich source of calcium and iron. Cowpea also provides small amount of beta carotine, thiamin, riboflavin, vitamin A, Niacin, folic acid and ascorbic acid. It thrives well in dry land conditions and like all other pulses improves soil fertility through nitrogen fixation. It is reported to fix 198 kg nitrogen/ha/year in the soil. However, improvement in production and productivity is only possible through scientific cultivation.

Land preparation: One deep ploughing by soil turning plough followed by 2-3 harrowings and planking is sufficient. Soil should be well drained and levelled, so that water does not stagnate. It can be grown successfully on a wide range of soils like sandy, sandy loam, coarse gravel, red loam, black clayey, loam, rice fields or black cotton soils. However, well-drained loam to slightly heavy soil with neutral pH is best suited for its cultivation. It is not fit for cultivation in saline or alkaline soils.

Seed rate: For the sowing of Cowpea in 20-25 kg/ha seed is required.

Sowing management: *Kharif* crop can be sown on the onset of monsoon in the month of July to the first week of August. Row to row 30-45 cm and plant to plant 10 cm spacing is recommended for this crop. The sowing should be done behind the *desi* plough or by seed drill. Treat seeds with Bavistin or Thiram or Captan @ 2, 3, 2.5 g/kg of seed respectively. The seed should also be treated with *Rhizobium* culture (*Rhizobium leguminosarum*), PSB and PGPR each @ 600 g/ha as discussed/prescribed for moongbean.

Fertilizer management: In poor soils, FYM @ 5-10 tonnes/ha may be applied at the time of 1st ploughing. Similar to other pulses, cowpea requires 15-20 kg N and 40 kg P_2O_5/ha. Application of 20 kg K_2O/ha also increases the yield. Half dose of nitrogen and whole quantity of P_2O_5 and K_2O should be applied at the time of sowing and the remaining nitrogen at 15 to 20 DAS. Foliar application of 2% nitrogen also increases the yield. Application of micronutrients like boron, copper, iron, manganese, molybdenum and zinc also increases the yield.

Weed Control and Interculture: First hand weeding at 20-25 DAS and the second at 20 days after first weeding or pre-emergence application of Fluchloralin @ 0.75 kg ai/ha accompanied with one hand weeding at 35-40 DAS are effective controls.

Water Management: In case of failure of rains during *kharif*, 1–2 life saving irrigations at pre flowering and grain filling stage may be given. In case of heavy rains, proper drainage provision should be made in heavy soils.

Plant Protection: Summer fallowing and ploughing reduces the incidence of diseases, nematodes and insect-pests. Deep ploughing exposes the insects, their eggs, spores of diseases and nematodes.

The major diseases namatodes, insect-pests which affect the crop are same as for moongbean and urdbean. The control measures are also same as described for these crops.

Harvesting & Threshing: When most of the pods mature, harvesting may be done. Early harvesting results in immature seeds, whereas, delayed harvesting results in shattering of pods and reduction in production. Harvesting should be done preferably in the morning to avoid shattering due to dew effect. Threshing may be done after sufficient drying in the sunlight manually or with the help of multi crop thresher with proper adjustments. Seed should be cleaned and dried in the sunlight so that it is not damaged during storage.

Yield: Though synchronous cowpea varieties are available, it is also fact that the whole pods do not mature at a time and there is some difference in maturity. Therefore, if labour is available easily and economical, one picking may be done earlier with subsequent harvesting of whole crop on maturity, alternatively, one time harvesting may also be done at the time of full maturity of crop. Harvesting should be done preferably in the morning to avoid shattering due to dew effect. Threshing may be done after sufficient drying in the sunlight manually or with the help of multi crop thresher with proper adjustments. It is not damaged during storage.

Pulse based sequential cropping: Cowpea-wheat-mungbean, Cowpea-potato-urdbean, Rice-mustard-cowpea, Maize-potato-cowpea, Rice-cowpea, Maize-wheat-cowpea, Rice-wheat-cowpea, Maize-toria-wheat-cowpea, Cowpea-cotton-wheat, Maize-potato-wheat-cowpea.

Intercropping Systems: Cowpea may be cultivated in mixture with *kharif* crops like pearl millet, sorghum, maize, pigeonpea, sesame etc. Sometimes one or two rows of cowpea can be grown in between two rows of pigeonpea, pearl millet, sorghum, maize etc. Inter cropping is more beneficial than mixed cropping because inter-culture operations are easier in inter cropping.

Pigeon pea (Arhar)

Pigeonpea or redgram is the oldest and popular pulse crop. The major pigeonpea growing states are UP, MP, Karnataka, Maharashtra, Gujarat, Rajasthan, Haryana, Punjab, Tamil Nadu, Orissa and Bihar. Though, the area has increased but the productivity has been fluctuating over years at national and state level. In general, the productivity of pigeonpea is low. Though, there are so many reasons of low productivity including cultivation by small and poor farmers on marginal lands but lack of scientific cultivation is one of the major factors responsible for low yields, Scientific methods of pigeonpea cultivation have to adopt for better yield.

Land preparation: It is a very deep-rooted crop and their rooting system is capable of penetrating in fairly hard soils. 1-2 deep ploughings by soil turning plough followed by 2-3 harrowings and planking is sufficient. Soil should be friable and weed free. It should be well drained and levelled, so that water does not stagnate in the field. In areas receiving heavy rainfall, raised bed system with 2.7 m wide beds 0.3 m furrow is ideal for pigeon pea. It can be grown over a wide range of well-drained soils having neutral pH. It can be cultivated in sandy loam, alluvial and deep black soils. Medium heavy loam soils are ideal for its cultivation.

Seed rate: The optimum seed rates for short, medium and late duration varieties are 25-30, 20-22 and 15-18 kg/ha, respectively.

Sowing management: In *kharif* season, pigeonpea is generally sown with the onset of monsoon in the month of June and July, both early and late sowings adversely affect the production. So, optimum sowing time is best for good yields. Row to row spacing of 30-45 cm, and plant to plant of 20 cm is the recommended spacing. The sowing should be done behind the *desi* plough or by seed drill. Sometimes broadcasting method of sowing is also adopted but line sowing is better due to easy inter-culture operation.Treat seeds with Bavistin or Thiram or Captan @ 2, 3, 2.5 g/kg of seed, respectively. The seed should also be treated with *Rhizobium* culture (*Rhizobium leguminosarum*), PSB and PGPR each @ 600 g/ha.

Fertilizer management: Being a pulse crop pigeonpea fixes atmospheric nitrogen through root nodules in a symbiotic process in the later growth stages and does not require much fertilizer but, it does require a starter dose of 20-25 kg N and 50-60 kg P_2O_5 per hectare as basal application at the time of sowing through drill. Application of fertilizer through drill is more beneficial than broadcasting because the nutrients become available in the root zone of the crop.

Weed Control and Interculture: Pigeonpea is slow growing during the first 40-50 days of its growth and very sensitive to weeds during its early growth stage. However, first hand weeding at 25-30 DAS and second at 25 days after first weeding, is quite effective. Pre-emergence application of Fluchloralin @ 0.5 kg/ha accompanied with one hand weeding at 35-40 DAS is also equally good for controlling weeds. It also saves one hand weeding.

Water Management: Pigeonpea is a rainfed crop and it does not require any irrigation unless there is a prolonged drought. In case of failure of rains, irrigation at the time of flowering and pod filling enhances the productivity. In case of heavy rains, proper drainage should be provided in the heavy soils so that excess water is drained out.

Plant Protection: Though, there are several diseases causing damage to the pigeonpea crop, fusarium wilt, sterility mosaic, phytophthora blight and alternaria leaf blight are the most destructive diseases of pigeonpea.

(A) Diseases:

I. **Dry root rot and stem rot:** The seedlings rot at collar region and plants dry up.

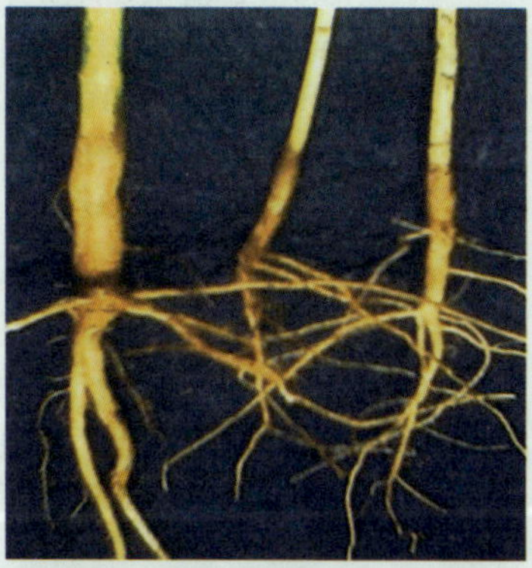

Control Measures: Seed treatment by Bavistin @ 1g or Thiram @ 3 g/kg of seed.

II. **Alternaria blight:** Small circular necrotic spots on leaves develop in concentric rings. These lessons appear on all areal parts of plants including pods.

Control Measures: Use disease free seed for sowing. Use resistant varieties for sowing like Pusa 9, Sharad etc. Spray of Dithane M-45 @ 2.5 g/litre of water is also effective.

III. **Wilt:** Internal tissues from the collar region downwards become dark and discoloured. The petioles and rachis along with leaflets drop down. The entire plant also drops down. Sometimes only partial wilting occurs.

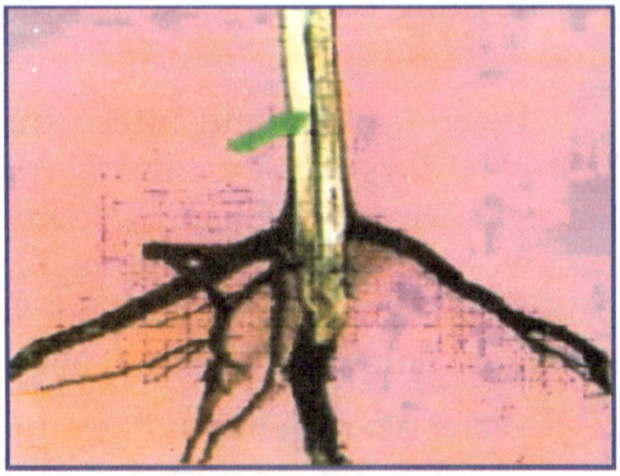

Control Measures: Long crop rotation (with sorghum and other cereals) and inter-cropping with non-host plant (pigeonpea: sorghum in the ratio of 1:1). Deep summer ploughing and field sanitation. Timely sowing. Soil solarization using polythene sheets. Removal of diseased plants and crop residues. Seed treatment by Carbendazim + Thiram (1+2 g/kg of seed) or Carbendazim @ 2 g or Captan @ 2.5 g or Thiram 3 g/kg of seed. Seed treatment with *Trichoderma harzianum* and *T. viride* @ 10 g/kg of seed. Use resistant varieties for sowing like Asha (ICPL 87119), BDN 2, Birsa Arhar, TT 6 etc.

IV. **Cercospora leaf spot:** Irregular yellow and brown margins appear on upper surface of leaves with scattered small spots. Ultimately, the leaves curl and affect the photosynthesis of plant, resulting in reduction of yield and seeds quality.

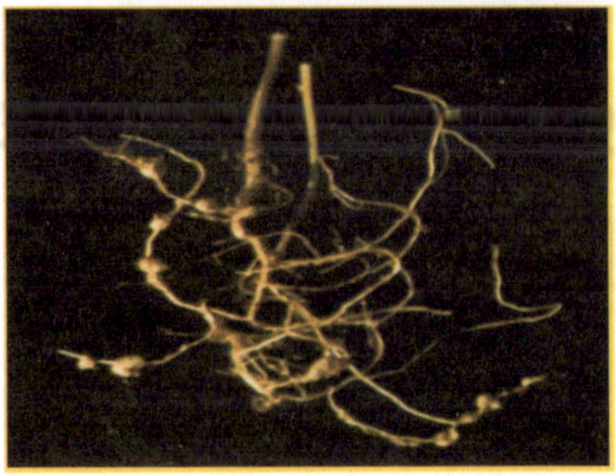

Control Measures: Use of resistant varieties. Seed treatment with Captan @ 2.5 g or Thiram 3 g/kg of seed. Spray of Mancozeb (0.2%) or Carbandazim (0.05%) on appearance of the disease and repeat at an interval of 10 days. Up-root and burn the diseased plants. Use healthy seed, for sowing.

V. **Nematodes:** Stunted, pale yellowish and poor plant growth. Roots swell due to nematode galls and nodulation is reduced. Blackening of roots in later stages, resulting in plant death.

Control Measures: The control measures are same as described for mungbean on previous pages.

VI. **Insect-Pests:** Termites, blister beetle, gram pod borer, plume moth, semi looper and pod borer are the key pests of Pigeon pea.

(i) Blister beetle: Beetles are medium sized and 12-25 mm long. These are conspicuous in appearance and are moderately robust. They are bright metallic blue, green, black and yellow or brown in colour. Their head is joined to the prothorax by a distinct narrow neck. Adults feed on flowers causing flower drop resulting in low yield.

Control Measures: Manual picking and killing. Dusting of Melathion (5%) or Carbaryl (10%) @ 25 kg/ha.

(ii) Gram pod borer: The newly hatched larvae of gram pod borer are sluggish and whitish-green in colour, but become active after 3-4 hours. The full-grown larvae are 3.5-4 mm in length pale green in colour. Dorsal surface bears dark broken stripes, head is reddish brown and legs are brown. The caterpillar stage is the most damaging stage. Young larvae feed on tender leaves, buds, flowers and subsequently bore into pods and feed on seeds. A single larva may destroy 30-40 pods before it reaches maturity. It causes extensive damage.

Control Measures: Spray 0.1 per cent Monocrotophos (36 WSC) and repeat at an interval of 10 days, if needed.

(iii) Plume moth: The adult moths are dark brown in colour, wings are deeply fissured, the forewings longitudinally cleft into two to four divisions and hind-wings into three divisions. The forewings are extremely elongated. The full-grown catterpillar is about 1.3 cm in length, greenish-brown in colour and fringed with hairs and spines. It pupates on the pod surface or in the entrance hole itself. The young larvae bore into the unopened flower buds for consuming the developing anthers. Grown-up larvae first scrape the surface of the pods and then bore into it. The larva never enters the pod completely.

Control Measures: Spray 0.1% Monocrotophos (36 WSC) and repeat at an interval of 10 days if needed.

(iv) Pigeon pea pod fly: It is a small metallic-black fly. Fully-grown larvae are cylindrical in shape and are narrower at the head end which bears black mouth parts. It is creamy white but acquires a yellowish tinge just before pupation. Young larvae bore into the pods and feed on seeds. The maggots eat away a part of the seed and partially damaged seed easily comes under bacterial and fungal attack.

Control Measures: Spray 1% Monocrotophos 36 EC or Dimethoate 30 EC and repeat at an interval of 15 days, if needed.

(v) Termites: These are tiny grayish-white social insects living in groups. Termites cause damage to the crop after its germination and at subsequent growth stages. They cut the main root of the plant and ultimately the damaged plant dries up. Many times it causes severe damage, leading to crop failure.

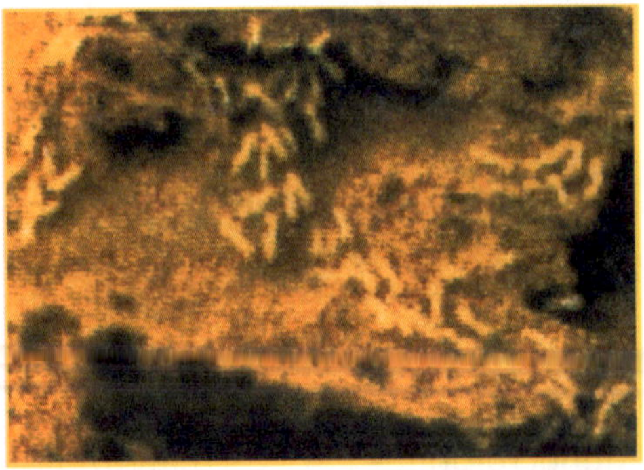

Control Measures: Application of Methyl parathion dust (2%) @ 25 kg/ha in soil at the time of field preparation. Seed treatment with Chloropyriphos @ 5-6 ml/kg of seed.

Harvesting & Threshing: Harvesting should be done at the time of full maturity of the crop. Early harvesting results in immature seeds, whereas, delayed harvesting results in shattering of pods and reduction in production. Harvesting should be done preferably in the morning to avoid shuttering due to dew effect. Threshing may be done after sufficient drying in the sunlight manually or with the help of multi crop thresher with proper adjustments. Seed should be cleaned and dried in the sunlight so that it is not damaged during storage.

Yield: Pigeonpea is cultivated in various land management conditions so the yield levels also vary. However, under normal conditions, its yield ranges from 10-20 q/ha.

Pulse based sequential cropping: Maize-pigeonpea, Pigeonpea-sugarcane, Pigeonpea-wheat, Pigeonpea-potato-urd/mungbean etc.

Intercropping Systems: Pigeonpea is cultivated as an inter crop with cereals, pulses and oil seeds. In cereals, two rows of jowar, maize or pearl millet (at a distance of 30 cm) may be sown with one row of pigeonpea. With pulses, it is cultivated with mung, urd or cowpea. With oil seeds three rows of soyabean or groundnut or *til* (at a distance of 30 cm) may be sown with one row of pigeonpea. It is also cultivated with cotton and sorghum (one row of pigeonpea and sorghum each with six rows of cotton).

Mothbean

Mothbean, also known as dewbean is an important pulse crop cultivated during *kharif* in arid and semi-arid regions throughout the tropics and sub-tropics and even in warm temperate areas of different countries. In India, it is cultivated in north-west parts of sandy areas by poor farmers on marginal lands. Due to its characteristics like extreme tolerance to moisture stress and high temperature, it is the most assured and remunerative arid crop. It is highly salt sensitive so in the areas where there is a problem of salts, its cultivation is not possible. Rajasthan is the largest mothbean producer (near by 80%) in the country. Due to its fast growing and deep root system, it can survive in high temperatures (35-40°C) for 30-40 days without rains and can be cultivated in adverse conditions of Rajasthan successfully. Though it is cultivated in large area in the state, the productivity is very low. The successful cultivation of mothbean, therefore, depends on the scientific cultivation of the crop.

Land preparation: One deep ploughing by soil turning plough followed by harrowing and planking are sufficient. Soil should be friable, weed free, well drained and levelled. Summer ploughing helps to overcome diseases and insect-pests. Mothbean can be grown successfully on a wide range of soils (light to heavy soils). However, well drained sandy or sandy loam soils are best suited for its cultivation.

Seed rate: 10-15 kg seed/ha for early maturing and 10-12 kg/ha for late maturing is sufficient for sowing.

Sowing management: In Rajasthan, sowing of mothbean should be done the day following an effective rainfall event of 30-40 mm. The second fortnight of July is the appropriate time for sowing. Early planting results in more vegetative growth, whereas, late planting invites YMV. Spacing recommended is row to row 30-45 cm and plant, to plant 10-15 cm. The sowing should be done behind the *desi* plough or by seed drill. Treat seeds with Bavistin or Thiram or Captan @ 2, 3 and 2.5 g/kg of seed, respectively.

Fertilizer management: In general, mothbean is cultivated on poor and marginal low fertile soils. To improve organic matter in the soil, 10-15 tonnes of FYM should be added for good crops. Similar to other pulses, mothbean requires 10 kg N and 40 kg P_2O_5/ha. The fertilizer should be applied 2-3 cm below the seed. In potash deficient soils, application of K_2O helps in improving the yields. Seed soaking in 500 ppm thiourea + 100 ppm dimethyl sulphoxide (DMSO) for 4-5 hours and drying for 2-3 hours in shade before sowing, increases yield significantly. In soils with more than 8.0 pH, application of sulphur @ 20 kg/ha through gypsum increases the yield.

Weed Control and Interculture: Mothbean is a rain-fed crop so field should be weed free, otherwise, crop will compete with weeds for moisture content, space, nutrients and sunlight which will ultimately result in low yields. One hoeing and weeding at 30 DAS is sufficient for controlling the weeds effectively. If hoeing and weeding is not possible, Fluchloralin @ 1.0 kg ai/ha should be incorporated in the soil at the time of final field preparation.

Water Management: Mothbean is grown as a rainfed crop where irrigation facilities are not available or limited so irrigation is generally not applied.

Plant Protection: Summer fallowing and ploughing reduces the incidence of diseases, nematodes and insect-pests. Deep ploughing exposes the insects, their eggs, spores of diseases and nematodes.

Major diseases, nematodes, insect-pest affects mothbean are the same as for mungbean and the control measures are also same to be adopted.

Harvesting & Threshing: Harvesting should be done when all the pods turn to yellow and leaves starts drying. Early harvesting results in immature seeds, whereas, delayed harvesting results in shattering of pods and reduction in production. Harvesting should be done preferably in the morning to avoid shuttering due to dew effect. The harvested plants should be put in bulk for drying in the sunlight for 3-5 days. After sufficient drying, threshing may be done manually or with the help of multi crop thresher with proper adjustments. Seeds should be cleaned and dried in the sunlight so that they are not damaged during storage.

Yield: Generally, mothbean is cultivated in poor and marginal lands under poor management and least inputs, so the yield levels are definitely low. However, under normal conditions, its yield ranges from 5-10 q/ha.

Pulse based sequential cropping: Mothbean-wheat-mungbean, Pearl millet + mothbean-oat, Mothbean-fallow-pearl millet, Mothbean-barley-clusterbean.

Intercropping Systems: Mothbean may be cultivated in mixtures with pearl millet or clusterbean. Intercropping of early maturing mothbean varieties with pearl millet in 3:1 ratio is more beneficial. Some times, one or two rows of mothbean are planted in between two rows of pearl millet or clusterbean. Inter cropping is more beneficial than mixed cropping because inter-culture operations and harvesting is easier in intercropping

Lentil (Masur)

Lentil is an important pulse crop cultivated during *rabi* in arid and semi-arid regions throughout the tropics and sub-tropics and even warm temperate areas of different countries.

Land preparation: One deep ploughing by soil turning plough followed by 2-3 harrowings and planking are sufficient. Soil should be friable and weed free. It should be well drained and levelled, so that the water does not stagnate in the field. If manure is to be applied, it should be applied before field preparation so that it is mixed well with the soil.

Seed rate: The optimum seed rate of lentil is 40 kg/ha for small seeded and 60 kg/ha for bold seeded varieties.

Sowing management: 3rd week of October to 1st week of November is the best time for sowing. Delayed planting results in low yields. In lentil cultivation, the row to row distance should be 22.5–30 cm, whereas plant to plant distance should be 10 cm. The sowing should be done behind the *desi* plough or by seed drill. In *utera* cultivation, seed sowing is done by broadcasting. Treat seeds with Bavistin or Thiram or Captan @ 2, 3 or 2.5 g/kg of seed, respectively. The seed should also be treated with *Rhizobium* culture (*Rhizobium leguminosarum*), PSB and PGPR each @ 600 g/ha. as described for moongbean.

Fertilizer management: It only requires a starter dose of 20 kg N, 40 kg P_2O_5 and 20-40 kg K_2O per hectare as basal application. Application of 15-25 kg zinc sulphate and 2 kg molybdenum per hectare increases the yield significantly. Foliar spray of 2% urea is also helpful in increasing the yield.

Weed Control and Interculture: For obtaining good yields, the crop should be weed free. However, two hand weedings, at 25-30 DAS and 50-55 DAS, are quite effective in controlling the weeds. Pre-emergence application of Pendimethalin 30 EC @ 0.75-1 kg/ha + one hand weeding at the time of 35-40 DAS is also equally effective in controlling weeds.

Water Management: Lentil is generally cultivated in rain-fed areas under conserved moisture but in case of irrigated conditions, if there is insufficient moisture at the time of sowing, pre-sowing irrigation should be applied. In case of failure of rains, irrigation at the time of branching and pod formation improves the production. Moisture deficiency at the time of pod filling stage badly affects the production.

Plant Protection: Summer fallowing and ploughing reduces the incidence of diseases, nematodes and insect-pests. Deep ploughing exposes the insects, their eggs, spores of diseases and nematodes.

Diseases and Nematodes : Major diseases affect the lentil are wilt, color rot, black root rot, stem rot and nematodes. The control measures are same as discussed/prescribed for chickpea.

Insects-Pests : The pod borer, Aphids and termites are the major insect-pests of lentil. However, pod borer is a key pest causing 90-95% damage alone. The control measures for termite & pod borer are same as discussed/described for chickpea in previous page.

Aphids: Aphids occur throughout the country but are usually observed in late sown crop. These are shining dark brown, black or greenish black pests. The nymphs and adults infest the leaves, tender shoots and pods. They cause damage by de-sapping of plants which results into drying up of the tender shoots. In later stages, the tender pods may shrivel and dry seeds remain poorly developed.

Control Measures: Spray of 0.1 per cent Dimethoate (30 EC) or Monocrotophos (36 WSC) repeated at an interval of 15 days, if needed.

Harvesting & Threshing: Harvesting should be done when all the pods turn to yellow and leaves starts drying. Early harvesting results in immature seeds, whereas, delayed harvesting results in shattering of pods and reduction in production. Harvesting should be done preferably in the morning to avoid shuttering due to dew effect. The harvested plants should be put in bulk for drying in the sunlight for 3-5 days. After sufficient drying, threshing may be done manually or with the help of multi crop thresher with proper adjustments. Seed should be cleaned and dried in the sunlight so that it is not damaged during storage.

Yield: Generally, lentil is cultivated in poor and marginal lands under poor management in the least inputs, so the yield levels are quite low. However, under normal conditions, the yields range from 15-25 q/ha.

Pulse based sequential cropping: Rice-lentil + mustard-maize (fodder), Maize/ pearl millet/ sorghum-lentil, Cotton/ groundnut/ rice-lentil, Rice-lentil-green manuring, Maize- lentil- green manuring, Maize- lentil + sugar cane.

Intercropping Systems: Lentil is mostly cultivated in mixture with crops like mustard, barley, safflower, linseed and toria. It is also grown as an inter crop with sugarcane. Three rows of lentil may be sown in between two rows of sugar cane.

Field Pea

Field pea or *Matar* is an important pulse crop cultivated during *rabi* in arid and semi-arid regions throughout the tropics and sub-tropics and even in warm temperate areas of different countries.

Land preparation: One deep ploughing by soil turning plough followed by 2-3 harrowings and planking are sufficient. Soil should be friable and weed free. It should be well drained and levelled, so that water does not stagnate in the field. If manure is to be applied, it should be applied before field preparation so that it is mixed well within the soil. Field pea may be cultivated on a wide range of soils from sandy loam to heavy clay soils. However, loam soils with good water retention capacity are best for pea cultivation.

Seed rate: The optimum seed rate of field pea is 40-60 kg/ha in small seeded and 80-100 kg/ha in bold seeded varieties.

Sowing management: 15th October to 15th November is the right time for sowing. Delayed planting results in low yields. Row to row distance should be 30 cms, whereas, plant to plant distance should be 10 cms. The sowing should be done behind the *desi* plough or by seed drill. In *utera* cultivation, sowing is done by broadcasting in the standing previous crop at nearing maturity stage. Treat seeds as described for mungbean.

Fertilizer management: Field pea requires a starter dose of 20 kg N, 40 kg P_2O_5 and 20 kg K_2O per hectare as basal application. Application of 15-20 kg zinc sulphate and 10 kg borax improves the yield considerably.

Weed Control and Interculture: For good production, the crop should be weed free. However, two hand weedings, at 25-30 DAS and 50-55 DAS, are quite effective in controlling the weeds. Pre-emergence application of Pendimethalin 30 EC @ 0.75-1 kg/ha + one hand weeding at the time of 35-40 DAS is also equally effective in controlling weeds.

Water Management: First irrigation at 45 days of sowing & second at pod formation is recommended.

Plant Protection: Summer fallowing and ploughing reduces the incidence of diseases, nematodes and insect-pests. Deep ploughing exposes the insects, their eggs, spores of diseases and nematodes.

(A) Diseases:

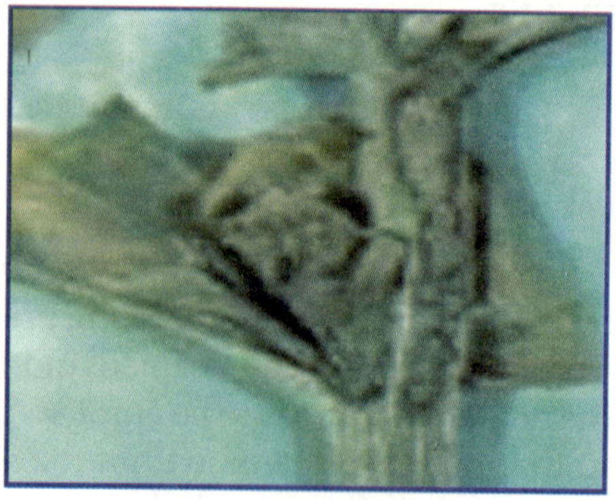

I. **Root rot:** It is a seed-borne fungal disease generally occurring under high moisture condition at seedling stage. Affected seedlings turn yellow and petioles and leaflets droop. Generally, the seedlings do not collapse; a distinct dark brown lesion appears above the collar portion at the main stem.

Control Measures: Seed treatment by Captan @ 2.5 or Thiram @ 3 g/kg of seed. Proper drainage. Rouging of infested plants. Use of resistant varieties is recommended.

II. **Stem rot:** The symptoms and control measures are same as for chickpea described in previous page.

III. **Cercospora leaf spot:** Irregular yellow and brown margins appear on upper surface of leaf with scattered small spots.

Control Measures: Use disease free seed. Clean cultivation. Adopt crop rotations. Uproot and burn the diseased plants. Avoid collateral hosts. Use resistant varieties for sowing. Seed treatment with Captan @ 2.5 g or Thiram @ 3 g/kg of seed. Spray of Mancozeb (0.2%) or Carbandazim (0.05%) on appearance of the disease and repeat at an interval of 10 days.

IV. **Powdery mildew:** Affected plants show white powdery mass on leaves in the initial stages and later, white powder appears on stem and pods also.

Control Measures: Use disease resistant varieties. Delayed sowing and wider spacing also help in control. Field sanitation and clean cultivation are also essential. On initiation of symptoms, two sprays of 0.1% Calixin or Karathane at 10 days interval should be done. Uproot and destroy the diseased plants.

(B) Nematode: The symptoms of incidence and control measures are same as described for chickpea.

(C) Insects-Pests: The major insect-pest attack the crop are pod borer, aphids and termites. The control measures are same as described for Chickpea and Lentil.

Harvesting & Threshing: Harvesting should be done when all the pods turn to yellow and leaves starts drying. Early harvesting results in immature seeds, whereas, delayed harvesting results in shattering of pods and reduction in production. Harvesting should be done preferably in the morning to avoid shuttering due to dew effect. The harvested plants should be put in bulks for drying in the sunlight for 3-5 days. After sufficient drying, threshing may be done manually or with the help of multi crop thresher with proper adjustments. Seeds should be cleaned and dried in the sunlight so that they are not damaged during storage.

Yield: Generally, field pea is cultivated in poor and marginal lands under poor management with least inputs but even then the yield levels are How much bad. However, under normal conditions, its yield ranges from 15-30 q/ha.

Pulse based sequential cropping: Jowar-Field pea, Sesame-Field pea, Pearlmillet- Field pea, Maize- Field pea.

Intercropping Systems: Field pea is mostly cultivated in mixture with crops like mustard, barley, safflower, linseed and *toria*. It is also grown as an inter crop with sugar cane. Three rows of lentil may be sown in between two rows of sugarcane.

French bean

French bean or Rajmash is an important pulse crop cultivated during *rabi* in arid and semi-arid regions throughout the tropics and sub-tropics. Rajmash thrives well in a moderately warm to cool climate. Wet and frosty conditions are unfavorable to the rajmash crop.

Land preparation: One deep ploughing by soil turning plough followed by 2-3 harrowings and planking are sufficient. Soil should be friable and weed free. It should be well drained and levelled, so that water does not stagnate in the field. If manure is to be applied, it should be applied before field preparation so that it may be mixed well with the soil. Rajmash may be cultivated on a wide range of soils from sandy loam to black soils. However, well-drained loam soils rich in organic matter with good water retention capacity are best for rajmash cultivation.

Seed rate: In general the seed requirement is 30-60 kg for small seeded and 60-100 kg for bold seeded varieties/ha.

Sowing management: The optimum sowing time is 15th October to 30th October. In plains, the appropriate time of sowing is the last week of October to Ist week of November. In rajmash cultivation, row to row 25-30 cm and plant to plant 10 cm, spacing is recommended.The sowing should be done behind the *desi* plough or by seed drill. Treat the seeds with Bavistin & Thiram (1:2) @ 3 g/ kg of seed.

Fertilizer management: It requires more nutrients in comparison to other pulses. It requires 5 tonnes FYM, 80-120 kg N, 60-80 kg P_2O_5 and 30-40 kg K_2O per hectare. Full dose of P_2O_5, K_2O and half dose of nitrogen should be applied as basal application and remaining nitrogen should be applied in the form of urea at the time of I, II and III irrigation. Application of 15-20 kg Zinc sulphate and 10 kg Borax per ha at the time of sowing also gives good response.

Weed Control and Interculture: For good production, the crop should be weed free. Two hand weedings, at 25-30 DAS and 50-55 DAS, are quite effective in controlling the weeds. Pre-emergence application of Pendimethalin 30 EC @ 0.75-1 kg/ha + one hand weeding at the time of 35-40 DAS is also equally effective in controlling weeds.

Water Management: Rajmash requires more irrigations than other pulses like lentil and fieldpea. If there is absence of sufficient moisture in the soil at the time of sowing, pre-sowing irrigation should be given before field preparation and in case of absence of winter rains, 3-5 irrigations are sufficient for good production. Irrigation at pre-flowering stage and at the time of pod filling stage is very important for good yields.

Plant Protection: Summer fallowing and ploughing reduces the incidence of diseases, nematodes and insect-pests. Deep ploughing exposes the insects, their eggs, spores of diseases and nematodes.

(A) Diseases:

I. **Anthracnose:** It is caused by *Collectotrichum gloeosporioides.* Reddish-brown scattered spots, which, later coalesce and cover larger portion of leaf as in CLS in mungbean.

Control Measures: Seed treatment with Captan or Thiram (2g/kg of seed). Spray of

Mancozeb (0.2 per cent) or Carbandazim (0.05 per cent) on appearance of the disease and repeated at the interval of 10 days. Uproot and burn the diseased plants.

II. Stem rot: The petioles and leaflets droop, the leaves dry up and turn straw coloured prematurely. A web of white mycelial strands appears in the color region and above.

Control Measures: Avoid excess moisture. Clean cultivation.

III. BCMV (Bean common mosaic virus): It is caused by a virus and transmitted through aphids (*Aphis craccivora*). It is a serious disease and affected plants show mosaic mottling on young leaves and stipules. Plants become dwarf and pale yellow, bear very few flowers and pods with small seeds.

Control Measures: Use disease free seed and disease resistant varieties like- HUR 15, HUR 137 and PDR 14. Spray of 0.1 per cent Rogor (30 EC) or Melathione (50 EC), repeated at an interval of 15 days, if needed. It helps in controlling the carrier insect, which transmits the disease by injecting the virus through de-sapping.

(B) Nematode: The symptoms and control measures are same as described for Chickpea.

(C) Insects-Pests: Pod borer, Stemfly and termites are the major insect-pests of French bean. The symptoms and control measures for termites and pod borer are the same as described for chickpea and lentil.

Stem fly: The stem fly is small in size and maggots are of cream colour and 2-4 mm long. Affected plants show yellowing in initial stages and later on, wilting takes place.

Control Measures: Avoid early planting. Spray of Monochrotophos (36 WSC) @ 0.1 per cent at 7 days after germination.

Harvesting & Threshing: Harvesting should be done when all the pods turn to yellow and leaves starts drying. Early harvesting results in immature seeds, whereas, delayed harvesting results in shattering of pods and reduction in production. Harvesting should be done preferably in the morning to avoid shuttering due to dew effect. The harvested plants should be put in bulk for drying in the sunlight for 3-5 days. After sufficient drying threshing may be done manually or with the help of multi crop thresher with proper adjustments. Seed should be cleaned and dried in the sunlight so that it is not damaged during storage.

Yield: Under normal condition, the yields range from 10-25 q/ha.

Pulse based sequential cropping: Maize/sorghum/pearl millet-rajmash, Follow- rajmash, Mungbean-rajmash-bajara (fodder).

Intercropping Systems: Rajmash may be cultivated as mixed/inter crop with chickpea, lentil, barley, mustard etc. Sowing of two rows of potato at the distance of 40 cm and two rows of rajmash at the distance of 25-30 cm is very much economical.

Future Thrusts

1. Developing varieties resistant to abiotic stresses like cold, heat & salinity.
2. Developing low cost and effective options of seed treatment for disease control.
3. Evolving more economic I.P.M. modules for the protection of crop damage by the insects like pod borer, beetles, thrips etc.
4. Developing varieties suitable for combine harvesting and inter-cropping.
5. Developing economic post harvest technology for storage of seed and grains.
6. Assuring remunerative prices *ni-a-ni* other competing crops.
7. Providing suitable marketing support.

References

Ali, Masood and Kumar Shiv. (2001). An overview of chickpea research in India. Indian J. of Pulses Res. 14: 81-89.

Ali, Masood and Mishra, J.P. (2004). *Rabi* pulses (Chickpea, lentil, lathyrus and frenchbean). Text book of field crops production. Directorate of Information and Publication of Agriculture by Rajendra Kumar. ICAR Krishi Anusandhan Bhawan, Pusa, New Delhi. 317-371.

Atwal, A.S. and Dhaliwal, G.S. (2002). Pests of pulse crops. Agricultural pests of South Asia and their management. Kalyani Publishers, Ludhiana-New Delhi- Noida-Hyderabad-Chennai-Kolkata-Cuttack. 199-203pp.

Bahl, P.N. 1988. Breeding approaches for self pollinated crops "Grain legumes".Plant Breeding by V.L.Chopra. Oxford and IBH Publishing Co. Pvt. Ltd. New Delhi, Bombay and Culcutta. pp 121-137.

Chhabra, K.S. Lal, S.; Kooner, B.S. and Verma, M. M. (1992). Insect pests of pulses. Identification and control manual. Punjab Agricultural University, Ludhiana and Directorate of Pulses Research, Kanpur.

Lal, S.S. and Ahmad, R. (2002). Integrated Insect Pest Management - Present Status and Future Strategies in Pulses. Proceedings of National Symposium on "Pulses for Sustainable Agriculture and Nutritional Security" held at New Delhi, April 17-19, 2001. pp 101-110.

Mishra, S.P. (2004). Project Coordinator's Report. All India Coordinated Research Project on Chickpea. Indian Institute of Pulses Research, Kanpur. Annual Group meet held at JAU, Junagarh, Sept. 15-17, 2004.

Rathore, Y.S. and Lal, Shankar. (1994). Pest management through cultural practices in India. Indian Institute of Pulses Research Kanpur.

Rathore, Y.S. and Sharma, Vivek (2002). Management of Bruchid infestation in pulses. Proceedings of National Symposium on "Pulses for Sustainable Agriculture and Nutritional Security" held at New Delhi, April 17-19, 2001. pp 111-124.

Singh, D.P. (1991). Genetics and Breeding of Pulse Crops. Kalyani publishers, New Delhi – Ludhiana.

Singh, R.A. Dhar, Vishwa and Lal, S. (1994). The Powdery mildew of Pea. Indian Institute of Pulses Research, Kanpur.

Singh, R.A. Gurha, S.N. and Chand, Hari. (2000). Integrated management of fungal foliar diseases in mungbean. Advances in Management of Biotic and Abiotic Stresses in Pulse Crops. Proceedings of National Symposium on "Management of Biotic and Abiotic Stresses in Pulse Crops" held at Kanpur, June 26-28, 1998. pp 99-103.

Sinha, S.N. (1999). Safe storage of seeds: Insect pest Management. Indian Agricultural Research Institute, Regional Station, Karnal (Haryana).

Tripathi,D.P. (2003). Project Coordinator's Report. Indian Institute of Pulses Research, Kanpur. *Kharif* Pulses Group meet held at Gujarat Agricultural University, Model Farm, Vadodara, May 8-10, 2003.

Wanjari, K.B. and Raut, B.T. (2000). Management of Powdery Mildew Disease in Mungbean. Advances in Management of Biotic and Abiotic Stresses in Pulse Crops. Proceedings of National Symposium on "Management of Biotic and Abiotic Stresses in Pulse Crops" held at Kanpur, June 26-28, 1998. pp 93-97.

❑❑❑

20

Tamil Nadu

K. Siddeswaran and P.M. Shanmugam
AICRP on Integrated Farming System, Tamil Nadu Agricultural University
Coimbatore 641 003, Tamil Nadu

Tamil Nadu state is situated at the south-eastern extremity of the Indian peninsula bounded on the north by Karnataka and Andhra Pradesh, in the east by Bay of Bengal, in the south by Indian Ocean and in the west by Kerala state. The state has a coastal line of 922 km and a land boundary of 1200 km. Tamil Nadu lies between 8°5′ and 13° 35′N latitude and 76°15′ and 80°20′E longitude. The total geographical area of Tamil Nadu is 13.03 m ha.

The climate of Tamil Nadu is basically tropical. Due to its proximity to the sea, the *summer* is less hot and winter is mild with mean maximum temperature of 33.1°C and mean minimum temperature of 23.8 °C. The State is benefited by both south west and north east monsoons. The mean annual rainfall of Tamil Nadu is 967 mm out of which, 48 per cent is received during the north east monsoon season (October-December) and 32 per cent during the south west monsoon season (June-September). Among the different soil types prevailing in the state, red soil is the most prominent one accounting to 79.8 lakh ha (62%). The other major types of soils are: river alluvium (16.2%), black soil (11.6%), coastal alluvium (7.6%) and lateritic soil (2.9%). The net area cultivated area is 5.81 m ha with cropping intensity of 114.90%. Nearly 56 per cent of cultivated area is under irrigation. Canals (27.1%), tanks (18.4%) and wells (54.2%) are the major sources of irrigation.

Based on rainfall distribution, irrigation pattern, soil characteristics, cropping pattern, other physical, ecological and social characteristics, Tamil Nadu state has bean classified into seven agro-climatic zones *viz.*, north eastern zone, north western zone, western zone, cauvery delta zone, southern zone, high rainfall zone and high altitude and hilly zone.

Present Scenario of Pulses

Agriculture continues to be the most predominant sector of the state economy, as 70% of the population is engaged in agriculture and allied activities. The predominant crops of Tamil Nadu are paddy, maize, groundnut, sunflower, sugarcane, cotton and vegetables, fruit crops and pulses. Among the pulse crops, blackgram, greengram, horsegram and redgram are grown in major area in Tamil Nadu. Cowpea, lablab, Bengalgram, soybean and horsegram are the other pulses grown to limited extent. There is no perceptible trend in area under pulses but decline in pulses production and productivity has been noticed over the years (2000-2008) in Tamil Nadu (Fig. 1).

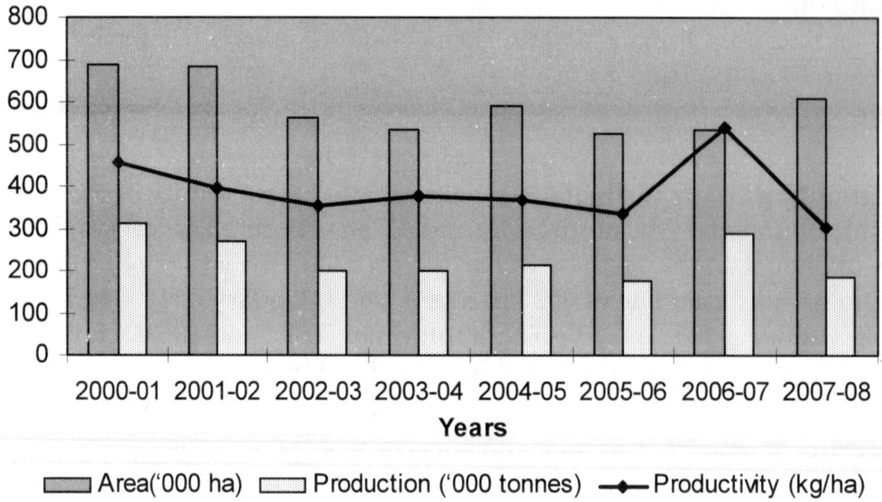

Fig. 1 : Tamil Nadu pulses scenario (2000- 2008)

In Tamil Nadu, pulses are grown in 6.09 lakh ha (10.5% of total cultivable area) with an annual production of 1.85 lakh tonnes and a productivity of around 303 kg per hectare (Table1). Blackgram occupies nearly 50% of total pulsse area in Tamil Nadu. It is mainly grown as rice-fallow pulse (January-February to April-May) utilizing the residual moisture. It is also grown as inter crop in crops like cotton and ground nut. The share of other pulses, area wise are: greengram (26.0%), redgram (4.9%), horse gram (7.9%), Bengalgram (1.1%) and other pulses (9.7%). Considering the consumption of 9 lakh tonnes per annum in the state, there is a deficit of nearly 7.15 lakhs tonnes of pulses/annum which is imported from other states or from other countries.

Table 1 : Area, Production and Productivity of Pulses in Tamil Nadu in 2007-08

S.No.	Crop	Area (lakh.ha)	Production (lakh. t)	Productivity (kg/ha)
1	Blackgram	3.07	0.80	260
2	Greengram	1.59	0.46	291
3	Redgram	0.30	0.21	701
4	Horsegram	0.48	0.21	440
5	Bengalgram	0.07	0.05	675
6	Other Pulses	0.58	0.12	207
	Total	6.09	1.85	303

Source : Season and Crop Report, (2007-08)

Existing pulse based cropping systems: The existing cropping systems in different agro-climatic zones of Tamil Nadu state are given in table-2.

Table 2 : Existing pulse based cropping system in different zones of T.N.

Agro-climatic zone	Cropping systems
1. Command and Tank fed areas	
North Eastern Zone : (Kancheepuram, Thiruvallur, Villupuram, Thiruvannamalai, Vellore, Cuddalore)	• Rice (Aug.-Jan.) - Pulses / groundnut / sesame (Jan.- April) • Rice (June-Sep.) - Rice (Oct.-Jan.) - Pulses / sesame (Feb.-May)
North Western Zone : (Dharmapuri, Krishnagiri, Salem, Namakkal)	• Rice (Aug.-Jan) - Pulses • Tapioca + Pulses
Western Zone : (Erode, Coimbatore, Karur, Dindigul, Theni)	• Rice (Aug-Dec.)- Pulses/Groundnut/ sesame (Dec-Mar) • Rice (Aug.-Jan.) - Pulses /Sesame / cotton / sorghum (Feb.-May)
Cauvery Delta Zone : (Thanjavur, Thiruvarur, Nagapattinam, Trichirapalli, Perambalur, Pudukottai, Cuddalore)	• Rice (June-Sep.) - Rice (Oct.-Jan.) - Pulses / sesame (Feb.-May) • Rice (Aug.-Jan.) - Pulses / sesame / cotton (Jan.-April) • Green manure/ Fallow(May-July) - Rice (Aug.-Jan.)- Pulses (Feb.-April)
Southern Zone : (Madurai, Ramanathapuram, Virudhunagar, Sivagangai, Tirunelveli, Thoothukudi)	• Rice (June-Sep.) - Rice (Oct.-Jan.) - Pulses (Feb.-April) • Geenmanure/fallow(June-Aug.)- Rice(Sep.-Jan.) - pulses /Cotton/ millets / sesame (Feb.-Aug.)
High Rainfall Zone : (Kanyakumari)	• Rice (June-Oct.)-Rice (Oct.-Feb.)-Pulses (Feb.-May)

Contd.

2. Well irrigated areas	
North Eastern Zone : (Kancheepuram, Thiruvallur, Villupuram, Thiruvannamalai, Vellore, Cuddalore)	• Rice (Aug.-Dec.) - Groundnut (Jan.-April) - Pulses / Sesame/ maize (April-June)
North Western Zone : (Dharmapuri, Salem, Namakkal)	• Groundnut (June-Sep.) - Wheat / cole vegetables (Oct.-Feb.) - Pulses / sesame (Feb.-May) • Cotton (Aug.-Feb.) - Pulses/sesame/ groundnut (Feb.- May) • Pulses (April - June) - Onion (Aug.-Dec.) - Sorghum (Jan.-April)
Western Zone : (Erode, Coimbatore, Karur, Dindigul, Theni)	• Cotton (Aug.-Jan.) - Pulses / millets / vegetables (Feb.-May) • Groundnut (June-Sep.) - Green manure (Oct.-Nov.) - Pulses / sesame (Dec.-Mar.) • Pearl millet (March-June) - Maize (July-Oct.) - Pulses (Nov.-Jan.) • Rice (Aug.-Jan.) - Ragi (Feb.-June) - Pulses (June-Aug.) • Rice (June-Sep.) - Rice (Oct.-Jan.) - Green manure / pulses (Feb.-April)
Cauvery Delta Zone : (Thanjavur, Thiruvarur, Nagapattinam, Trichirapalli, Perambalur, Pudukottai, Cuddalore)	• Green manure (May-Aug.)-Rice(Aug.-Jan.)-Pulse (Jan.-Apr.) • Groundnut/ragi (April-July) - Rice/maize (Aug.-Jan.) - pulses (Feb.-April)
Southern Zone : (Madurai, Ramand, Virudhunagar, Sivagangai, Tirunelveli, Thoothukudi)	• Rice (Oct.-Jan.) - pulses (Feb.-April) • Groundnut(Jun-Sep.)- Vegetables/pulses (Oct-Jan)
High Rainfall Zone : (Kanyakumari)	• Tapioca + pulses (Sep.-July) • Rice (Oct.-Feb.) - Pulses (Feb.-May)
3. Rainfed areas	
North Eastern Zone : (Kancheepuram, Thiruvallur, Villupuram, Thiruvannamalai, Vellore, Cuddalore)	• Groundnut (June-Sep.) - Pulses (Dec.-Mar.) • Dry rice (June-Sep.) - Pulses / vegetables / sesame (Dec.-Feb.) • Pulses (June-Sep.) - Groundnut/sesame (Sep.-Feb.)
North Western Zone : (Dharmapuri, Krishnagiri, Salem, Namakkal)	• Groundnut + castor + redgram / pulses/ sunflower/ sorghum / ragi / samai/ Kudiraivali (May-Oct.) - Horse gram (Nov.-Jan.) • Groundnut + castor /pulses (May-Sep.) - Sorghum (fodder) + pulses (Oct.-Feb.) • Rainfed tomato (June-Sep.) - Horsegram (Oct.-Jan.)

Contd.

	• Sorghum/maize/ragi/pearl millet (May-Sep.) - Pulses (Oct.-Jan.)
Western Zone : (Erode, Coimbatore, Karur, Dindigul, Theni)	• Groundnut + castor (May-Jan.) - Horsegram (Oct.-Jan.)
	• Millets (June-Sep.) - Maize / pulses (Oct.-Jan.)
	• Castor / redgram (July-Feb.) - fallow
	• Maize (July-Oct.) - Bengalgram (Nov.-Feb.)
	• Groundnut (June-Sep.)-Maize/pulses (Oct.-Nov.)
	• Millets / maize / groundnut / pulses (June-Oct.) - Horse gram (Nov.-Feb.)
	• Pearlmillet / sorghum / maize + pulses (Aug.-Nov.) - Pulses (Dec.-Feb.)
	• Castor + pulses (July-Jan.)
Cauvery Delta Zone : (Thanjavur, Thiruvarur, Nagapattinam, Trichirapalli, Perambalur, Pudukottai, Cuddalore)	• Groundnut + redgram (Oct.-Feb.) • Sesame (July-Oct.) - Horsegram (Nov.-Feb.)
Southern Zone : (Madurai, Ramanathapuram, Virudhunagar, Sivagangai, Tirunelveli, Thoothukudi)	• Upland rice/millets / pulses/groundnut/ cotton / Chillies (Sep.-Feb.) • Groundnut + pulses (June-Sep.)-Sesame (Oct.-Jan.)
	• Sorghum / groundnut / sunflower / sesame (June-Sep.) - Pulses (Oct.-Jan.)
	• Cotton + blackgram / chillies (Sep.-Feb.)
High Rainfall Zone : (Kanyakumari)	• Tapioca + pulses (April-Dec.) • Groundnut (June-Sep.) - Pulses (Oct.-Jan.)

Yield Gap Analysis

Yield gap in pulses continues to be one of the major constraints in pulse production in Tamil Nadu. A wide gap exists both at progressive and average farm level compared to the experimental station yield. Srigokila (2009) studied the yield gap prevailing in major pulses of Tamil Nadu under different cropping situations. Yield gap I is the difference in the level of productivity of pulses obtained in the experimental station and maximum yield obtained in the farmers field. This gap is due to the existence of obstacles while transferring the technology or partial adoption of technology. This could be eliminated through inter disciplinary research efforts. Yield gap II is the difference between the experimental station yields and average yield of pulse farmer. This happens due to variations in soil fertility, amounts of inputs used and pest management. Yield gap III is the difference between the maximum farmers yield and the actual yield of individual farmers. This is due to inadequate extension activities, which in turn result in inefficient crop management and non optimum use of resources. (Table 3).

Table 3 : Yield gap in major pulses under different cropping situations in Tamil Nadu (kg/ha)

Particulars	Black gram	Green gram	Cowpea	Black gram intercrop in cotton	Black gram intercrop in maize	Rice fallow Black gram
Expt. station yield	750	770	950	700	700	720
Max. yield at farmers level	497	498	562	500	499	410
Av. yield at farmers level	395	397	403	339	335	307
Gap I	253	272	388	200	201	310
Gap II	355	373	547	361	365	413
Gap III	102	101	159	161	164	103

Source: Srigokila,(2009)

Main Reasons for Low Productivity of Pulses

- Around 78% of pulse area in Tamil Nadu is under rainfed conditions.
- The pulse crops are grown mainly in marginal soils poor in soil fertility and moisture retention capacity.
- Mostly grown as mixed crop, intercrop, rice fallow, bund crop, etc.
- Unfavourable weather conditions like erratic and uncertain rainfall, low and high temperature and moisture stress at various crop growth stages.
- Highly susceptible to drought and water logged conditions.
- Soil salinity and alkalinity problems, poor quality of irrigation water.
- Farmers do not devote proper attention to the pulses cultivation.
- Non availability of high yielding varieties/hybrids and quality seed materials.
- Susceptibility to pest and diseases.
- Number of storage grain pests cause considerable losses.
- Poor storability and lack of storage facility.

Improved Varieties of Pulses

Tamil Nadu Agricultural University and State Dept. of Agriculture, Tamil Nadu released 95 varieties in different pulse crops from 1972-2010. Special features of pulse crop varieties released by TNAU are given in table 4.

Table 4 : Recommended pulses varieties for Tamil Nadu.

Crop/ Varieties	Parentage	Duration (days)	Special features
Blackgram			
Co 5	Pure line selection from Musiri type	70-75	Moderately resistant to powdery mildew, leaf crinkle, pod borer and tip blight
ADT 3	Pure line selection from Tirunelveli	70-75	Suitable for rice fallow
ADT 5	Pure line selection from Kanpur variety	70-75	Suitable for rice fallow. Resistant to major diseases like YMV, root rot and leaf crinkle and to stem fly
K1	Co 3 x VS 131	70-75	Suitable for southern districts, YMV resistant, 24.2% protein, suitable for intercropping in cotton
Vamban 1	KM 1 x H 76-1	60-65	High yielding ,suitable for entire state and tolerant to YMV
Vamban 2	Spontaneous mutant selection from Type 9	65	Resistant to Yellow Mosaic Virus.
Vamban 3	LBG 402 x LBG 17	65-70	Resistant to Yellow Mosaic Virus
Vamban (Bg) 4	CO 4 x PDU 102	75-80	Resistant to Yellow Mosaic Virus
Vamban (Bg) 5	Vamban 1 x UK 17	65 - 70	High yield and resistant to yellow mosaic virus disease in all seasons
Greengram			
KM 2	No.122 x S.9	6--65	Tolerant to YMV and pod borer
Co 4	Gamma irradiated mutant of Co 1 (20 kR)	85	High biomass, suitable for rain fed condition
Co 6	WGG 37 x Co 5	62-67	Resistant to yellow mosaic virus, suited to all greengram growing tracts of Tamil Nadu
Co(Gg) 7	MGG 336 x CoGG 902	62	High yielder, short duration, high protein content (25.2%). Seeds are medium bold with good cooking quality

Contd.

ADT 3	H 70 - 16 / Rajendran/G.65	70	Resistant to YMV and stem fly
Paiyur1	Pure line selection form DPT 703	85-90	Higher grain yield, low incidence of YMV, suited for rainfed tracts of Dharmapuri, Madurai, Ramnad, Tirunelveli and Erode Districts
K1	Co 4 x ML 65	70-75	Tolerant to drought. Suitable for pure and cotton based intercropping systems
Vamban 1	Hybrid derivative of (S.8 x PIMS 3)	65-70	High yield, tolerant to YMV
Vamban (Gg) 2	VGG 4 x MH 309	65-70	Shiny grains, lobed leaves and resistant to YMV
VRM (Gg)1	Selection from K851	60-70	Location specific variety. Suited to Vellore and Tiruvannamalai districts
Redgram			
SA 1	Pure line selection from Tiruppatur local	180	A very good long duration strain for rainfed conditions for the entire State
Co 6	Mutant of SA 1 (gamma rays 25 kR)	175-180	Photosensitive type to replace SA 1. Tolerant to pod borer
Co(RG) 7	Selection from PB 9825	120-130	Bold seeded and suitable for late onset of monsoon
COPH 2	ms Co 5 x ICPL 83027	120-130	Suitable as a pure crop in irrigated and rainfed conditions. Also suitable for entire redgram growing tracts of Tamil Nadu
Vamban 1	(Prabath x HY 3A) x (T 21 x 102)	95-100	High yield suitable for intercropping in groundnut
Vamban 2	ICPL 341 x Bhavanisagar local	170 - 180	Suited for entire Tamil Nadu to replace SA 1 & Co 6
VBN(Rg)3	VBN 1 x Gulbarga	100 - 105	Determinate, high yielding, short duration
APK 1	Pureline selection from ICPL 87101	95-105	Resistant to Sterility Mosaic Disease. Short duration and suited for pure crop situation
BSR 1	Selection from Mayiladumparai	Perennial,	Perennial and suitable for entire state. First harvest from 150 days
Cowpea			
Co 2	C521 x C49	90	Vegetable type. Seeds reddish brown with irregular patches
Co 6	Ms 9804 x C152	65-70	Dwarf and suited to rainfed conditions

Contd.

Co(CP) 7	Mutant of CO 4(20 kR)	70-75	Square, dull brown seeds
Vamban 1	Pure line selection from IT 85-F2020	55-65	White grain type and suited to rainfed condition throughout Tamil Nadu
Vamban 2	Selection from IT-81-D-1228-10	75-85	Vegetable type, Ivory coloured seeds and suited for all seasons
Paiyur 1	Selection from VM 16	90	Suited for rainfed tracts of Dharmapuri, Madurai, Ramnad, Tirunelveli and Erode Districts
Horsegram			
Co 1	Pure line selection from Mudukulathur local	110	A good rainfed strain
Paiyur 1	Selection from Mettur local	110	Suited for rainfed tracts of Dharmapuri, Madurai, Ramnad, Tirunelveli and Erode Districts
Paiyur 2	Gamma irradiated mutant of CO 1	105	Suited to rainfed tracts of entire Tamil Nadu. High protein content (19.25%)
Bengalgram			
Co 3	Pure line selection from Maharastra collection	85	Tolerant to root rot, wilt and philology.
Co 4	ICCC 42 x ICC12237	80-85	Attractive desi bold grains (30-32g/100 seeds). Tolerant to root rot
Soybean			
Co 1	Reselection from a Thailand variety	85	Erect, bushy, determinate and photo-insensitive
Co 2	Derivative of UGM 21 x JS 335	75-80	Photo insensitive, moderately resistant to yellow mosaic virus, stem fly and pod borer
Co (Soy) 3	UGM 69 x JS 335	85 - 90	Photo insensitive, high protein and oil content. Semi determinate and bushy type. Resistant to yellow mosaic virus.
ADT 1	Pure line selection from "Hill"	85-90	Short duration, high yield, high per day productivity and tolerant to YMV and leaf miner

Recommended Production Technologies for Pulses

Tamil Nadu Agricultural University and the State Department of Agriculture periodically update and publish the Crop Production Guide for the benefit of researchers, extension functionaries and farmers. The important technologies are briefly provided below.

Seed Rate

Blackgram

Normal condition	-	20 kg /ha
Mixed crop	-	10 kg /ha

Bengalgram

CO 3 : 90 kg/ha and CO 4 : 75 kg/ha

Seed treatment

- Treat the seeds with Carbendazim or Thiram @ 2 g/kg of seed 24 hours before sowing or with talc formulation of *Trichoderma viride* @ 4g/kg of seed (or) *Pseudomonas fluorescens* @ 10 g/kg seed.
- Bio control agents are compatible with biofertilizers. First treat the seeds with biocontrol agents and then with biofertilizers.
- Fungicides and biocontrol agents are incompatible.
- Treat the seeds with 3 packets (600 g/ha) of Rhizobial culture + 3 packets (600 g/ha) of PGPR and 3 packets (600 g/ha) of Phosphobacteria developed at TNAU using rice kanji as binder.
- If the seed treatment is not carried out, apply 10 packets of Rhizobium (2000 g/ha) + 10 packets of PGPR (2000 g/ha) and 10 packets (2000 g) of Phosphobacteria with 25 kg of FYM and 25 kg of soil before sowing.

Fertilizer Application

- Apply fertilizers basally before sowing.
    - Rainfed : 12.5 kg N + 25 kg P_2O_5 + 12.5 kg K_2O +10 kg S*/ha
    - Irrigated : 25 kg N + 50 kg P_2O_5 + 25 kg K_2O + 20 kg S*/ha

Sulpher applied in the form of gypsum if Single Super Phosphate is not applied as a source of phosphorus. Soil application of 25 kg $ZnSO_4$/ha under irrigated conditions.

Sowing of seeds

- For irrigated crop, dibble the seeds adopting 30 x 10 cm cm spacing.
- For rainfed crop, dibble the seeds adopting 25 cm x 10 cm spacing.
- For redgram, dibble the seeds adopting the following spacing.

Variety	Pure crop	Mixed crop
CO (Rg) 7	45 cm x 30 cm	120 cm x 30 cm
COPH 2	5 cm x 15 cm	……..
Vamban 1, APK 1, Vamban (Rg) 3	45 cm x 20 cm	120 cm x 30 cm
CO 6, Vamban 2	90 cm x 30 cm	240 cm x 30 cm
Bund Crop	60 cm for BSR 1 and 30 cm for others	

For Soybean, dibble the seeds at a depth of 2 - 3 cm with a spacing of 30 x 5 cm.

Water Management

- Irrigate immediately after sowing, followed by light irrigation on the third day.
- Irrigate at intervals of 7 to 10 days depending upon soil and climatic conditions.
- Flowering and pod formation stages are critical periods when irrigation is a must.
- Avoid water stagnation at all stages.
- Apply KCl at 0.5 per cent as foliar spray during vegetative stage if there is moisture stress.

Spraying of DAP or Urea, NAA and Salicylic Acid

- Foliar spray of NAA 40 mg/litre and Salicylic acid 100 mg/litre once at pre-flowering and another 15 days later.
- For rice fallow crops, foliar spray of DAP 20 g/litre once at flowering and another 15 days later.
- For irrigated and rainfed crops, foliar spray of DAP 20 g/litre or Urea 20 g/litre once at flowering and another 15 days later.

Weed Management

- Pre emergence application of Pendimethalin 2 litres/ha 3 days after sowing using Backpack/Knapsack/Rocker sprayer fitted with flat fan nozzle using 500 l of water for spraying one ha. After this, one hand weeding 30 days after sowing gives weed free environment throughout the crop period.
- If herbicides are not applied, two hand weedings on 15 and 30 days after sowing are required.

- For soybean, apply Alachlor 4 litres/ha or Pendimethalin 3.3 litre ai/ha after sowing followed by one hand weeding 30 days after sowing.

Harvesting

- Picking the matured pods, drying and processing
- Uprooting or cutting the whole plants, heaping, drying and processing

Inter-Cropping

Redgram

- One row of long duration redgram varieties as inter crop for every six rows of groundnut (6:1) is recommended for rainfed crops.
- One row of short and medium duration redgram as inter crop for every four rows of groundnut (4:1) is recommended for rainfed as well as for irrigated crops.

Bengalgram

- Bengalgram in paired row planting with one or two rows of Coriander as intercrop would give the highest return.
- Wheat can also be intercropped in deep black cotton soil in Coimbatore, Erode, Salem, Namakkal and Dharmapuri districts.

Special situations

- Intercropping of Soybean CO_2 in Sugarcane is recommended for North Western Zone.
- Intercropping of Soybean in coconut gardens of more than 10 years is recommended.
- Vermipelleting (50 g/kg) and adopting spacing of 30 x 10 cm and two foliar sprays of 2% DAP during flowering is recommended to achieve higher yield.

Pulses in Cropping Systems

Field experiments were conducted at Coimbatore to diversify traditional finger millet-cotton-sorghum cropping system of irrigated uplands of western zone of Tamil Nadu. The alternate systems with pulses were compared with the existing sequence based on production efficiency and economic efficiency. The results indicated that beetroot-greengram-maize + cowpea system recorded the higher total productivity of 5753 kg/ha/year (Table 5). Among the different cropping systems, beetroot-greengram-maize + cowpea system registered the highest production efficiency of 18.9 kg/ha/day over the existing system and it also recorded higher economic efficiency of Rs. 264.1/ha/day over the existing system (Shanmugam *et al.,* 2010). Studies conducted in the new Cauvery delta to identify the efficient cropping systems revealed that onion - rice - blackgram recorded higher rice grain equivalent yield of 15305 kg/ha/year compared to the existing system of rice-rice-blackgram (13120 kg/ha/year). The drum seeded rice-rice-maize + blackgram recorded 14732 kg/ha/year. The benefit cost ratio was more (3.29) in drum seeded rice-rice-maize + blackgram (Table 6) followed by maize + blackgram-rice-blackgram cropping sequence (3.21).

Table 5 : Effect of different pulse based crop sequences on productivity, economic and production efficiency (average of 3 years from 2005-08)

Crop sequencse			Cotton equivalent yield (kg/ha)	Production efficiency (kg/ha/day)	Economic Efficiency (Rs/ha/day)
Cotton	Finger millet	Sorghum	3538	10.1	117.6
Beetroot	Green gram	Maize + Cowpea	5753	18.9	264.1
Cowpea	White gingelly	Maize	3165	11.5	134.2
Lablab	Gingelly	Brinjal	4348	12.2	129.7
Maize	Cowpea	Tomato	5470	17.1	243.3

Table 6 : Economic analysis of different pulse based crop sequences for Cauvery delta zone

Cropping sequence			Rice grain equivalent yield (kg/ha/yr	Gross income Rs/ha/yr)	Net income (Rs/ha/yr)	B:C Ratio
Rice	– Rice	– Blackgram(Seed crop & residue incorporation)	13120	137760	89460	2.85
Drum seeded rice	– Rice	– Maize + Blackgram (IC)	14732	154686	107686	3.39
Maize + Blackgram	– Rice	– Blackgram (Seed crop & residue incorporation)	12033	126346	87096	3.21
Onion –	Rice	– Blackgram (Seed crop & residue incorporation)	15305	160702	107903	3.04
Blackgram –	Rice	– Cowpea (Seed & residue incorporation)	8276	86898	49618	2.33

In a cotton based intercropping study conducted at Coimbatore, growing blackgram as intercrop in irrigated cotton did not affect the yield of cotton and provided the highest net returns and B-C ratio (Table 7).

Table 7 : Production potential and economics of cotton based intercropping systems

Treatments	Yield of cotton (kg/ha)	Yield of intercrop (kg/ha)	Net return (Rs/ha/yr)	B:C ratio
Cotton(sole)	1977	-	28196	2.64
Cotton + onion	1750	3398	32025	2.45
Cotton + blackgram	1855	442	40532	2.88
Cotton + greengram	1659	379	28047	2.58
Cotton + lucerne	1559	20300	24592	2.43

(*Source :* Jayakumar, (2004)

Rice Fallow Pulses

Pulses are cultivated in another unique eco-system known as rice fallow conditions in Tamil Nadu. Pulses are sown under this condition in about 2.9 lakh ha in Tamil Nadu which is 30.75 per cent of the total area under pulses in this state. Rice fallow pulses contribute to about 40.5 per cent of the total pulse production. They are cultivated in Thiruvarur, Thanjavur, Nagapattinam Districts and some parts of Pudukottai, Trichy and Cuddalore Districts. Recent statistics (2005-06) show that an area of 1,46,663 ha has been under pulses (blackgram and greengram) in Cauvery Delta Zone with a production of 29,949 tonnes indicating an average productivity of 204 kg/ha.

In the residual soil moisture of the Cauvery old delta soils, pulses particularly, balckgram and greengram, are broadcasted 7 to 10 days before harvest of paddy and allowed to germinate and grow. Since pulses are grown under paddy stubbles, the pulse crops have to survive in the residual moisture of the soil. The average productivity is 300-500 kg/ha in favourable years and 100 kg/ha in unfavourable seasons. Hence, the yield recorded in this ecosystem is highly variable and mostly depends on the management practices followed. Rice fallow black gram responds to foliar nutrition of DAP and KCl (Table 8).

Table 8 : Effect of foliar nutrition on yield and economics of rice-fallow-black gram

Treatments	Grain Yield (kg/ha)	Net return (Rs/Ra/Year)	B-C ratio
Control	679	8578	2.57
Foliar application of 1% urea twice (Flowering, pod filling)	788	10138	2.70
Foliar application of 2% DAP twice (Flowering, pod filling)	914	12608	3.03
Foliar application of 1% KCl twice (Flowering, pod filling)	896	12208	3.02
Foliar application of 2% DAP + 1% KCl twice (Flowering, pod filling)	1007	14380	3.11

(*Source* : Geetha, 2003)

Efficient Zone for Pulses

Among the seven agro climatic zones evaluated, western zone was categorized as the most efficient zone with little year to year variation. North western zone was categorized as the efficient zone. Southern zone and north eastern zone were grouped as efficient zone with little year to year variation. In the Cauvery delta zone, Thiruchirapalli, Perambalur, Pudukottai and Thanjavur districts were categorized as efficient zone with medium to great year to year variation. However, Thiruvarur and Nagapattinam were categorized as not efficient zone with great variation (Latha, 2008). From the analysis, it is interpreted that Coimbatore was the most efficient zone for pulses cultivation wherein hi-tech agriculture can be introduced for pulses. For Cuddalore, Dharmapuri, Krishnagiri, Thiruvarur, Nagapattinam, Dindigul, Viruthunagar, Thirunelveli and Thoothukudi districts there is greater scope to increase the productivity since the area was more. Further, individual pulses wise analysis can be done for precise interpretation since certain pulses are district specific, for example, chickpea in Coimbatore, horsegram in Dharmapuri & Krishnagiri, as well as greengram in Thiruvarur and Nagapattinam districts.

Strategies to Increase the Production and Productivity of Pulse Crops

Blackgram and Greengram

Blackgram and greengram are cultivated in three distinct situations *viz.* rice fallow, dryland and irrigated. Out of the 4.7 lakh ha grown under these crops, nearly 3.0 lakh hectares i.e., 2.2 lakh ha and 0.80 lakh ha are under rice fallow blackgram and rice fallow greengram, respectively. Nearly 2/3 area which is under rice fallow in the entire state faces many constrains like inadequate plant population per unit area due to improper leveling, uncertainty in the time of sowing, lesser availability of quality seeds of suitable varieties, excessive weed growth and moisture stress.

Constraints for Low Productivity in Rice Fallow for Pulses

- Farmers devote very little attention to pulse cultivation.
- Higher degree of susceptibility to both abiotic (soil disorder, nutrient deficiency, moisture stress, water logging, etc.) and biotic stresses (pests, disease and weeds).
- Non-availability of quality seeds of high yielding varieties
- Poor or non adoption of optimum plant population plant protection measures and improved management practices like timely sowing, placement of fertilizers, moisture conservation etc.
- Water stress at flowering and grain filling stages.
- Rat damage and cattle grazing.
- Poor storability and lack of storage facility.
- Exploitation by village traders in marketing the produce and fluctuations in inputs price

Strategies

- Timely supply of improved quality seeds of suitable varieties (ADT 3, VBN(Bg) 3, CO(BG) 653 blackgram and ADT 3 greengram)
- Timely sowing of blackgram and greengram between January 15th and February 15th.
- Designer seed treatment with bio-fertilizer like Rhizobium and Phosphobacteria and with *Trichoderma viride* @ 4 g/kg of seeds or Pseudomonas @ 10g/kg or Carbendazim @ 2 g/kg must be done to prevent root rot disease.
- Optimum plant population @ 33/sq.m should be maintained
- Controlling of weeds by application of weedicides
- 2% DAP+1%KCl+ 40 ppm NAA foliar spray must be given during morning or evening on 25 and 40 days after sowing.
- To overcome moisture stress during critical stages, use of mobile sprinklers is recommended.

The average productivity under rice fallows is only 200 kg/ha and if the productivity could be enhanced to 450 kg/ha through various strategies suggested above, an additional production of 0.75 lakh tonnes per annum is possible.

Redgram

The area under redgram in the year 1989-90 was 1.25 lakh ha which reduced gradually to 0.30 lakh ha in 2008-09 and the production was 0.21 lakh metric tonnes with a productivity of 701 kg/ha. To increase this productivity to a level of 1050 kg per ha, the following strategies are recommended.

Short term

- One of the main reasons for the reduction in production in Tur/ redgram, is the extensive damage by pod borers. Pod borer tolerant redgram varieties are to be grown along with integrated pest management practices (IPM).
- Growing short duration redgram varieties like VBN (Rg) 3, ICPL 87 as pure crop in irrigated conditions in major pigeon pea growing districts which require less amount of water compared to maize.
- The redgram variety Co(Rg) 7 which matures in 120-135 days as an irrigated pure crop in Erode, Coimbatore, Salem and Krishnagiri districts has to be promoted.
- Vamban (Rg) 3 to be grown in southern districts during summer as pure irrigated crop and an additional area of 12,000 ha can be brought under this variety. A horizontal growth will result in an additional production of 0.12 lakh tonnes.
- Cultivating redgram on bunds in low land areas and varieties like LRG 41, Gwalior 3, TTB 7, Hy 3C could be popularized.
- Growing BSR 1 redgram variety on the borders of turmeric fields.
- Provision of mobile sprinklers to alleviate moisture stress.

Long term

- The farmers shall be assured of getting a minimum support price, MSP (i.e. the farm gate price) of Rs. 50/- to 60/- per kg in future for pulse grains (blackgram, greengram and redgram).
- The irrigated area under pulses in the state is ranging from 5-8% at present. Steps to be taken to increase the area under irrigated pulses to a level of 15% at least in future.

- Cultivating perennial redgram (BSR 1) in kitchen gardens and backyards of dwellings to meet the needs of home consumption.
- Cultivating blackgram and greengram in the bunds of low lying lands in the districts of Cauvery delta zone.
- The present Seed Replacement Rate (SRR) in redgram (11%), blackgram (26%) and greengram (20%) is low due to non-availability of quality seeds and most of the seeds used for sowing are farmers seeds which are of poor quality due to genetic deterioration. Hence, demonstration and farmers participatory seed production is also to be undertaken to increase the SRR.
- Inspite of production and supply of huge quantity of breeder seeds, the seed multiplication chain consisting of foundation and certified seed production is not being effective for various reasons. Hence, for strengthening seed multiplication chain, foundation and certified seed production should be intensified. Accordingly, production of required quantity of foundation seeds for production of 50% certified seed requirement by TNAU and production of certified seeds to the tune of 50% of total requirement by the Department of Agriculture is proposed.
- Certified seed production and supply through contract farming involving NGOs, Self Help Groups, Agri Clinic Centres, Seed producers etc., and through public- private partnership mode need to be encouraged.
- Mechanization to be promoted in pulse production.
- To avoid post harvest losses, infrastructure for cold storage has to be developed at the village level.

Popularisation of less known pulses

High yielding varieties of rice bean, lima bean, sword bean, garden lablab (*avarai*) and field lab lab (*mochai*) are to be popularised to achieve more production of pulses in the state.

Cold season pulses

Among the cold season pulses, horse gram is the most important crop which occupies about 0.48 lakh ha of marginal lands with a production of 0.21 lakh tonnes. The productivity is about 440 kg/ha. In certain years, for example in the year 1989-90, the productivity of this crop was very low i.e. 319 kg/ha, therefore, improved varieties like CO 1, Paiyur 1 and Paiyur 2 are to be grown to get higher productivity. Bengal gram is grown only in an area of about 0.07 lakh ha and the productivity of this crop is high i.e. about 675 to 700 kg/ha.

Improved root rot resistant CO 3 and wilt resistant CO 4 bengal gram varities should be popularized to get higher productivity.

Future Strategies

- Bringing more pulse area under irrigation.
- Mitigating measures to deal terminal drought under rice fallow conditions.
- Identification of donors for resistance to YMV, powdery mildew and leaf crinkle virus diseases.
- Intensifying inter varietal and inter specific hybridization.
- Development of fertilizer responsive varieties and adoption of designer seed technology.
- Development of varieties with determinate growth habit for mechanical harvesting.
- To formulate cost effective growth regulator based nutrient mix for rice fallow pulses.
- Evaluation of suitable herbicides for weed control.
- Introduction of mobile sprinklers to overcome drought during critical stages.
- Management of sucking pests, pod borer, YMV, powdery mildew and leaf crinkle diseases.
- Designing of combined harvester for rice and mini combine harvester for pulses.
- Seed production by farmers themselves through seed village concept besides intensive trainings and distance education.
- Dissemination of technologies through demonstrations, OFTs, FFS and village awareness campaigns.

References

Anonymous (2005). Crop Production Guide-2005. Published by Dept. of Agriculture, Govt. of Tamil Nadu and Tamil Nadu Agricultural University, Coimbatore.

Geetha, P. (2003). Refinement of nutrient management techniques for rice fallow Blackgram. M.Sc.(Ag.) thesis, Department of Agronomy, TNAU, Coimbatore.

Jayakumar, M. (2004). Production potential and economics of intensive cotton based intercropping system through nutrient management practices under irrigated condition. Ph.D. thesis, Department of Agronomy, TNAU, Coimbatore.

Latha, K.R. 2008. Delineation of Efficient Crop Zones of Pulses. In CASA Training on Recent advances in the production and utilization of pulses and oilseed crops held at Department of Agronomy, Tamil Nadu Agricultural University, Coimbatore during 10.09.2008 to 30.09.2008. pp-56-59.

Ramasamy,C., Ramanathan,S., Balasubramanian,T.N., Ragupathi,N., Natarajan,S., Devasenapathy,P. and Natarajan,N. (2004). Alternative cropping pattern for Tamil Nadu, Published by Directorate of Research, Tamil Nadu Agricultural University, Coimbatore.

Season and Crop Report - Tamil Nadu 2007-2008 (Fasli-1417). Published by the Dept. of Economics and Statistics, Chennai - 600006.

Shanmugam,P.M., Siddeswaran, K. and Senthilkumar, G. (2010). Alternate cropping systems for increasing productivity and profitability in irrigated uplands of Tamil Nadu. Green Fmg Int.J. 1(5):38-40.

Singh, J.P. , Salaria, A. Singh, K. and Gangwar, B. (2005). Diversification of rice - wheat cropping system through inclusion of basmati rice, potato and sunflower in Trans-Gangetic plains. J. Farming Systems Research & Development.11(1):12-18.

Srigokila, S.T.K. (2009). An analysis of yield gap and technical efficiency in pulses production in Thoothukudi district, Tamil Nadu. M.Sc. (Ag.) thesis, TNAU, Coimbatore.

❑❑❑

21

Uttar Pradesh

H.P.Tripathi, Alok Kumar and M.P.Yadav
Department of Agronomy, N.D. University of Agriculture and Technology, Kumarganj Faizabad, UP.

Pulse crops provide nutritious food, feed and fodder and thrive well in fragile eco-systems where other crops often fail. They have the ability to trap atmospheric nitrogen in their root nodules in association with *Rhizobium*, bringing qualitative improvement in soil health i.e. physical, chemical and biological properties of the soil. These virtues enable pulse crops to remain an integral part of subsistence farming system of semi-arid tropics as well as of Indo-gangatic plains. The role of pulses has also been recognized for sustaining soil productivity under intensive cropping systems in irrigated conditions.

India is the key player in pulse production contributing 28% to the global basket from an area of about 37%. The country contributes more than 75% to the global production of chickpea and pigeonpea, the two major pulse crops of Indian sub continent. At the national level, the most important states for pulses are Madhya Pradesh, Rajasthan, Maharastra, Uttar Pradesh and Andhra Pradesh which together account for more than 80% of the production from 75% of area. The most important pulse crop in the country is chickpea which accounts for 47% of the total pulse production from 33% of the total pulse area. In other states as well as in Uttar Pradesh, the pulses are grown in both the seasons. During summer season, pigeonpea, greengram and blackgram are cultivated while during winter season chickpea, lentil and field peas are grown.

The relative contribution of U.P. in India's pulse production (Table 1) indicates that U.P. contributes 40% area in lentil and field peas and 15.9 and 8.8% in blackgram and pigeonpea, respectively, while the area under chickpea and greengram is only 6.2 and 2.3%, respectively.

Table 1 : Area, production and productivity of different pulse crops in India (2009-10)

Crop	Area (million ha)	Production (million tonnes)	Yield (kg/ha)	Major states	Area in UP m ha (%)
Chickpea	8.75	8.25	943	Madhya Pradesh, Rajasthan, Maharastra, Uttar Pradesh, Karnataka, Haryana	0.55 (6.2)
Pigeonpea	3.53	2.46	697	Madhya Pradesh, Gujrat, Maharastra, Uttar Pradesh, Karnataka,	0.31 (8.8)
Lentil	1.48	1.03	697	Uttar Pradesh, Madhya Pradesh, Bihar	0.59 (40.0)
Greengram	2.84	1.03	386	Rajasthan, Maharastra, Andhra Pradesh, Karnataka, Orrisa	0.065 (2.3)
Blackgram	2.67	1.17	463	Madhya Pradesh, Rajasthan, Maharastra, Uttar Pradesh, Rajsthan and Andhra Pradesh	0.42 (15.9)
Peas	0.76	0.67	886	Uttar Pradesh, Madhya Pradesh	0.31 (40.8)
Total Pulses	23.35	14.66	625		Area (100)

Note: The values in parenthesis indicate the percent share of the state in the country

The Uttar Pradesh has a wide range of agro climatic conditions and have been divided into 9 agro climatic (NARP) zones. The area, production and productivity of these six crops, presented in Table-2, indicate that the area of all pulses except pigeonpea is the highest under Bundelkhand region (Jhansi and Chitrakoot divisions) followed by Central plain zone (Kanpur, Lucknow and Allahabad divisions) for gram, greengram and blackgram, Eastern plain zone (Faizabad, Azamgarh and Varanasi divisions) for pea and North eastern plain zone (Devipatan, Basti & Gorakhpur divisions) for lentil. The major area of pigeonpea lies in Central plain zone followed by Bundelkhand zone. The pulses area in other regions, *viz.*, Western plain zone, central western plain zone, south western semi arid zone and vindhyan zone are very less as compared to other regions (Table 2).

Table 2 : Area of various pulse crops under various agroclimatic zones of U.P. during 2009-10 (000 ha)

Agro-climatic zones	Lentil	Gram	Pea	Pigeon pea	Green gram	Black gram	Total	% share of U.P.
Western plain zone (WPZ)	4.5	0.2	1.0	15.5	1.1	6.6	28.9	1.21
Central western plain zone (CWPZ)	41.4	0.5	5.2	2.9	0.5	66.5	117.0	4.89
South western semi arid zone (SWSAZ)	5.3	4.2	2.3	19.1	1.1	2.3	34.3	1.40
Central plain zone (CPZ)	75.3	145.6	27.5	83.0	6.0	102.8	440.2	18.57
Bundelkhand zone (BZ)	266.7	404.0	197.3	68.6	26.0	298.8	1261.4	53.21
North eastern plain zone (NEPZ)	123.4	5.5	24.7	39.3	-	4.7	197.6	8.33
Eastern plain zone (EPZ)	62.7	34.3	45.1	48.9	0.5	22.2	213.7	9.01
Vindhyan zone (VZ)	13.0	23.8	8.8	28.0	0.1	3.7	77.4	3.27
Total	592.3 (24.9)	618.1 (26.0)	311.9 (13.2)	305.3 (12.9)	35.3 (1.5)	507.6 (21.5)	2370.5	-
Major zones	BZ, NEPZ, CPZ	BZ, CPZ, EPZ	BZ, EPZ, CPZ, NEPZ	CPZ, BZ, EPZ, NEPZ	BZ, CPZ	BZ, CPZ, CWPZ		BZ, CPZ, NEPZ, MEPZ
% share of major zones	89.7	94.5	94.4	78.5	90.6	92.2		89.12

Note: Values in parenthesis are percent of total pulse area of the zone in the state.
Source : Anonymous, (2009) & Anonymous (2011a).

Characteristics of various agro climatic zones of U.P.

1. Tarai zone

This zone comprises the *tarai* part of Shaharanpur, Bijnor, Jyotibaphule nagar, Rampur, Bareilly, Khiri and Bahraich districts which is very meagre. As the area is adjoining hilly areas thus climate is wet and humid. Sugarcane, rice and wheat are the major crops and very less area is under pulses. Since, the major part of all the districts lies in other zones, hence, the data of this zone are being included in other zones (Anonymous, 2009, Anonymous 2010).

2. Western plain zone

This zone covers the districts Shaharanpur, Muzaffarnagar, Meerut, Bagpat, Bulandshahar, Ghaziabad and Gautambuddh Nagar. The geographical and net cultivated area of the zone is 19.37 and 13.42 lakh ha, respectively. The cropping intensity is 152.1% and irrigated area is 98.0% with tubewell as major source of irrigation. The major cropping systems of the zone are rice-wheat-blackgram, rice-wheat, maize-potato-greengram, rice-potato-sugarcane-ratoon-wheat. Pigeonpea is the major pulse crop followed by greengram contributing 5% and 3% of the state share. Though the productivity is nearly of the same level as the state but there is vast scope for its improvement as the soil and climatic conditions are very congenial for pulse production therefore, special care must be taken for enhancing productivity of these two crops in this zone.

3. Central western plain zone

The zone covers Bareilly, Badaun, Shahjahanpur, Pilibhit, Bijnor, Moradabad, Jyotibaphulenagar and Rampur districts. The total geographical and net cultivated area of the zone is 38.06 and 24.36 lakh ha, respectively. The cropping intensity is 161.1% and the irrigated area is 86.2% with tube well as the major source of irrigation. The major cropping systems of the zone are: rice-wheat, rice-wheat-blackgram, rice-potato-sugarcane, rice-lentil/gram, pigeonpea + blackgram. The main pulses are blackgram and lentil covering 13 and 7% share, respectively.

4. South western semi-arid zone

This zone comprises of the districts of Agra, Mathura, Ferozabad, Mainpuri, Aligarh, Hathras, Etah and Kanshiram Nagar. The geographical and net cultivated area of the zone is 22.31 and 16.15 lakh ha, respectively. The cropping intensity in the zone is 160.4%. The irrigated area is 95.1% with tubewells as the major source of irrigation. Fallow-mustard, pigeonpea + greengram, greengram-mustard-sugarcane and maize-potato-blackgram are the major cropping systems. Pigeonpea and greengram are the two major pulse crops grown in the region having shares of 6% and 3% of the state area, respectively.

5. Central plain zone

This zone covers the districts of Farrukhabad, Kannauj, Etawah, Auraiya, Kanpur Dehat, Kanpur city, Fatehpur, Allahabad, Kaushambi, Pratapgarh, Lucknow, Unnao, Raibarelli, Sitapur, Hardoi and Kheri. The geographical and net cultivated area of the zone is 42.51 and 23.54 lakh ha, respectively. The cropping intensity of the zone is 151.6% with 82.6% irrigated area with tubewells as the major source of irrigation. Blackgram—potato-wheat, pigeonpea+ blackgram, rice-lentil/gram, wheat-greengram are the major cropping systems. This is the major zone of pulse production having area under all pulse crops. The pigeonpea, gram, pea, lentil, greengram and blackgram are grown in the area having 27, 23, 9, 13, 17 and 20% of the state share, respectively.

6. Bundelkhand zone

This zone comprises of the districts of Jhansi, Lalitpur, Jalaun, Hamirpur, Mahoba, Banda and Chitrakoot. The total geographical and net cultivated area of the zone is 29.58 and 10.08 lakh ha, respectively. The cropping intensity of the zone is 120.3% and the irrigated area is 50.6% havig canal as a major source of irrigation. The major soil types are *Rakar, Parwa, Kabar* and *Mor* with shallow soil depth. Greengram-wheat, pigeonpea + greengram, Jowar + blackgram-wheat, rice-wheat or rice-gram/lentil are the major cropping systems. This zone is famous for having maximum area under all the pulses *viz.* gram, lentil, pea, blackgram, greengram and pigeonpea contributing about 65% , 45%, 63%, 58%, 74% and 22% area, respectively, of total pulses area in the state.

7. North eastern plain zone

This zone covers Bahraich, Srawasti, Gonda, Balrampur, Basti, Santkabirnagar, Sidharthnagar, Mahrajganj, Gorakhpur, Deoria and Kushinagar districts. The geographical and net cultivated areas are 33.21 and 16.22 lakh ha, respectively. The cropping intensity is 154.9% and irrigated area is 67% source of irrigation. The major cropping systems are rice-lentil/gram, rice-wheat, blackgram-potato-wheat, rice-wheat-blackgram, rice-sugarcane-ratoon-wheat. The major pulses grown in this zone are lentil, pea and pigeonpea occupying 21, 8 and 11% of the state area, respectively, under these crops.

8. Eastern plain zone

This zone covers the districts of Barabanki, Faizabad, Ambedkarnagar, Sultanpur, Chatrapati Shahuji Maharajnagar, Jaunpur, Varanasi, Chandauli, Ghazipur, Mau, Azamagarh, Ballia and Sant Ravidasnagar. The total geographical and net cultivated area of the zone is 45.63 and 25.52 lakh ha, respectively. The cropping intensity is 160.8% and irrigated area is 83.97 % with the major source of irrigation being tube wells. The cropping systems are rice-wheat, rice-wheat-greengram/blackgram, maize-potato-blackgram, pigeonpea + blackgram, pigeonpea + blackgram-wheat, maize - wheat /barley /rai/ mustard, rice-wheat-sugarcane, rice-gram/ pea/ lentil/ mustard. The major pulses grown in area are pigeonpea, pea, lentil and gram having 16, 15, 10 and 5% share of the state area, respectively.

9. Vindhyan zone

This zone covers Mirzapur, Sonbhadra districts and some parts of Chandauli and Sant Ravidasnagar districts. The total geographical and net cultivated area of the zone is 11.33 and 1.46 lakh ha, respectively. The cropping intensity of the zone is 131.1% and net irrigated area is 44%. The major source of irrigation is tube wells. The major cropping systems are: pigeonpea + blackgram, rice-wheat-blackgram, rice-gram/lentil, fallow-gram/pea/mustard. The major pulse grown in the zone is pigeonpea which covers 9% of the total area of the state.

Major Constraints in Pulse Production

The major constraints that limit the realization of potential yield of pulses in the state are given below:

1. Non-unavailability of high yielding disease and insect resistant varieties adapted to diverse growing situations.
2. Large area under rainfed conditions (88%) and moisture stress at reproductive phase of the crop.
3. Biotic and abiotic stresses causes losses upto 30%.
4. Poor plant stand due to broadcasting of seed.
5. Emerging deficiency of secondary and micro nutrients due to use of high analysis fertilizers.
6. Resource poor and low risk bearing farmers.
7. Timely availability of inputs specially phosphatic fertilizers.
8. Limited policy directive and incentives i.e. declaration of minimum support price before sowing of crops.
9. Crop damage due to menace of blue bulls.
10. Inadequate seed replacement rate.
11. Poor storage facilities.

Technologies for Pulse Production

Considering the importance of pulses for sustaining soil health, their suitability to fit in the various cropping systems and their role in household nutrition security, sustained efforts are required to raise the production of pulses substantially. Though, there is a good scope for the expansion of area under pulses, the major thrust required now is on improving the productivity. Even with existing technologies, there is a considerable scope for improvement their productivity of various pulses. Necessary inputs like seeds of improved varieties, use of phosphatic fertilizers, timely irrigation and plant protection umbrella will have to be provided to the farmers in order to realize the potential of the crops. The production of pulses may be increased through both horizontal and vertical expansion.

Horizontal Expansion

1. Intensification of cropping system : The area under pulse crops may be increased by inclusion of short duration pulses like blackgram, greengram in the existing rice-wheat cropping system specially in summer season where assured irrigation facilities are available, *viz.*, rice-potato-greengram, rice-mustard-blackgram (Table 3).

Table 3 : Yield and economics of various rice-based crop sequences (mean of 6 years)

Treatment	Yield (t/ha)			Rice yield equivalent (t/ha/year)	Net returns (x 10^3 Rs/ ha / year)
	Kharif	*Rabi*	*Zaid*		
Rice – wheat-green manuring	7.5	4.5	-	13.0	73.9
Rice – potato-green gram	7.6	25.3	1.06	22.1	125.3
Rice – mustard-black gram	7.4	1.8	0.96	14.7	89.2
CD (0.05)					1.31

Note: Sale price (Rs./kg): Rice- 9.50, wheat grain- 11.50, greengram- 35.00, black gram- 35.00, mustard-20.00, potato-4.00,

Source : Tripathi & Kumar (2010)

2. Diversification of existing system : Diversification of existing cereal-cereal cropping system to cereal-pulses system for higher monetary benefits and improvement in soil health like rice-wheat to rice-gram and rice-lentil have been recommended under irrigated conditions of eastern plain zone (Table-4).

From the experiments conducted under north eastern plain zone in normal irrigated situations, rice-mustard-blackgram and rice-lentil-greengram were found more remunerative than traditional rice-wheat system. While under low lying unirrigated conditions, rice-gram, rice-lentil and rice-linseed were more remunerative than rice-fallow system (Table 4).

Table 4 : Grain yield and economic evaluation of various cropping systems in NARP Zones 7 and 8 (1990-91 to 1994-95)

Crop sequences	Average grain yield (q/ha)			Gross return (Rs/ha/year)
	Kharif	*Rabi*	*Zaid*	
North eastern plain zone (Gonda, Basti, Baharaich)				
Normal irrigated				
Rice-wheat	48.9	41.8	-	33835
Rice-mustard-blackgram	47.7	15.6	5.7	43965
Rice-lentil-greengram	47.4	14.4	6.3	43320
Low lying unirrigated				
Rice-fallow	27.8	-	-	9730
Rice-lentil	29.2	15.0	-	28220
Rice-linseed	28.6	10.8	-	20810
Rice-gram	30.1	18.0	-	32135
Eastern plain zone (Ghazipur, Mau, Varanasi, Jaunpur)				
Normal land irrigated				
Rice-wheat	43.8	43.2	-	32610
Rice-gram	44.0	20.4	-	39880
Rice-lentil	43.5	21.1	-	40545

Note: Sale price (Rs/q) Rice- 350, wheat- 400, maize- 350, gram- 1200, greengram- 1500, blackgram- 1500, lentil- 1200, linseed- 1000, mustard- 1200

Source: Tripathi & Kumar (2010)

3. Introduction of pulses in non traditional areas : The short duration pulses may be substituted in place of upland rice or coarse millets in rain fed areas during *kharif* season for higher returns. The experiments conducted under upland irrigated conditions in Gonda, Basti and Bahraich districts of NARP Zone 7, revealed that three crop sequences *viz. til*-gram-greengram, blackgram-mustard-greengram and maize-gram-greengram were more profitable than traditional maize-wheat system (Table 5).

Similarly, the experiments conducted under upland-irrigated situations in Ghazipur, Mau, Jaunpur and Varanasi districts of the NARP Zone 8 revealed that maize-gram, blackgram-wheat and blackgram-mustard were more remunerative than traditional maize-wheat system.

Table 5 : Grain yield and economic evaluation of various cropping systems under upland irrigated conditions. (1990-91 to 1994-95)

Crop sequences	Average grain yield (q/ha)			Gross return (Rs/ha/year)
	Kharif	*Rabi*	*Zaid*	
North eastern plain zone (Gonda, Basti, Baharaich)				
Upland irrigated				
Maize-wheat	20.4	38.1	-	22380
Maize-gram-greengram	20.7	15.4	7.4	36825
Blackgram –mustard-greengram 38100		8.0	12.5	7.4
Til- gram- greengram	4.3	14.9	6.9	38550
Eastern plain zone (Ghazipur, Mau, Varanasi, Jaunpur)				
Upland irrigated				
Maize-wheat	26.9	41.5	-	26015
Maize-gram	26.5	20.1	-	33395
Blackgram-wheat	9.9	43.2	-	32130
Blackgram-mustard	10.0	13.6	-	31320

Note: Sale price (Rs/q) Wheat- 400, maize- 350, gram- 1200, greengram- 1500, blackgram- 1500, mustard- 1200, til- 2400.

4. Intercropping of pulses : The area of pulses may be increased through intercropping of greengram in spring sugarcane, blackgram in spring sunflower, greengram/blackgram in maize and pigeonpea in Jowar/bajra etc.

The experiments carried out in Gorakhpur, Padrauna and Deoria districts of NARP Zone 7, revealed that intercropping of pigeonpea + maize in *kharif* and intercropping of gram + mustard in *rabi* were more profitable than sole cropping of pigeonpea or gram (Table 6). The use of recommended fertilizers for intercrops further showed an additional income of Rs. 4180/ha/year over unfertilized inter cropping system.

Similarly experiments conducted in the Eastern plain zone showed that intercropping of maize with blackgram in *kharif* and wheat with mustard in *rabi* was more remunerative than sole cropping. The application of recommended dose of fertilizers to intercrops was Rs. 9990/ha/year more remunerative than unfertilized intercropping systems.

Table 6 : Comparison of different intercropping systems with varying levels of fertilizers under upland irrigated conditions (1995-96 to 1999-2000)

Treatment	Mean grain yield (q/ha)				Gross return (Rs/ha)
	Kharif		*Rabi*		
	Main crop	Inter crop	Main crop	Inter crop	
North eastern plain zone (Gorakhpur, Deoria, Padrauna)					
Upland irrigated	**Pigeonpea**	**Maize**	**Gram**	**Mustard**	
Sole main crop	19.9	-	14.9	-	63640
Intercropping with out fertilizers	16.6	14.2	11.3	6.2	67680
Intercropping with fertilizers	16.0	21.4	14.1	4.4	71860
Eastern plain zone (Ghazipur, Mau, Ballia and Jaunpur)					
Upland irrigated	**Maize**	**blackgram**	**Wheat**	**Mustard**	
Sole main crop	23.9	-	45.8	-	39430
Intercropping with out fertilizers	19.0	3.9	39.4	4.6	47840
Intercropping with fertilizers	20.1	5.8	42.3	7.2	57830

Note: Sale price (Rs/q): Wheat- 600, maize- 500, barley- 500, gram- 1600, blackgram- 2000, pegionpea- 2000, mustard- 1500.

Source: Tripathi & Kumar (2010)

Vertical Expansion

Through varietal adoption keeping in view the soil, resources and agro-climatic situations, the productivity of pulses can be enhanced considerably (Table 7). The cultivation of high input responsive varieties having multiple resistances suitable to these areas needs to be encouraged. Similarly, the cultivation of drought tolerant varieties having higher harvest index and hybrids/transgenic also needs due attention.

Table 7 : Improved varieties of pulse crops suitable for Uttar Pradesh

Crop	Region	Varieties	Remarks
Gram	Plain zone	K-850	
	Eastern U.P.	Gujrat Gram-4	Wilt resistant
	Whole U.P.	DCP-92-3*, KGD-1168*, Awrodhi*, Pusa-256, KWR-108*	
	Western U.P.	Adhar (RSG-936)*, WCG-1, WCG-2	
	Bundelkhand	Radhey & JG-16*,	
Late sowing	Whole U.P.	Pusa 372, Udai and Pant G-186	
Kabuli gram	Western U.P.	Chamatkar (BG-1053) & HK-94-134	
	Bundelkhand	JGK-1, Shubhra & Ujjwal	Powdery mildew Powdery resistant mildew and rust resistant
Pea	Whole U.P.	Rachna*, Malviya pea-15 **, Sapana (KPMR-144-1)	
	Plain zones	Pant pea-5*	
	Bundelkhand & Central U.P.	Indra (KPMR-400), Shikha (KFPD-103)	
	Bundelkhand	JP-885, Adrash*, Prakash*, Vikash (IPFD-99-13)*	
	Eastern U.P.	Pusa Prabhat (DDR-23)*, Palthi Pea **	
	Western U.P.	Jai (KPMR 522)** Hariyal*, Pant Pea-42, Aman	
Lentil	Whole U.P.	Narendra Masoor-1*, Pant masoor-5*, DPL-62, L-4076, K-75, HUL-57*, Shekhar-2**, Shekhar-3**	Rust resistant
	Plain zone S	Pant masoor-4*, DPL-15, Pusa Baibhaw	
	Bundelkhand	IPL-81	
	Eastern U.P.	KLS-218*	
	Western U.P.	IPL-406*	
Rajmash	Central & eastern U.P.	Udai (PDR-14), Malviya-13)	
	Eastern U.P.	Amber (IIPR 96-4), Utkarsh (IIPR-98-5)	
	Whole U.P.	Arun	
	Plain zones	VL-63	
Pigeonpea	Whole U.P.	Pusa 992 & T-21	
Short duration	Western U.P.	Paras	
(4.5-5.5 months)	Plain zone	UPS-120	

Contd.

Long duration (8-9 months)	Whole U.P.	Bahar*, Amar*, Narendra Arhar-1**, Azad*, Narendra Arhar-2**, Malviya chamatkar (MAL-13)*, Malviuya vikash (MA 6)**, PDA-11***, Pusa-9***	Sterility mosaic resistant, ** sterility mosaic & wilt resistant, *** sterilty mosaic resistant & suitable for September sowing
Greengram	Whole U.P.	Type 44, Pant moong-3, Narendra moong-1, Pant moong-4*, PDM-11, Malviya Jyoti, Samrat, Malviya Janchetna, Malviya Janpriya, Malviya Jograti, Asha, Meha, TM 9937, Malviya Jankalyani, MH 2-15, IPM 2-3	Yellow mosaic resistant
	Eastern U.P. & plain zone	Pant moong-1	
	Eastern U.P.	PDM-54	
Blackgram	Whole U.P.	Narendra Urd-1, Pant U-30, Azad Urd-2, Shekhar-1, Shekhar-3, Azad urd-3, WB-108, Shekhar-2, Pant urd-31*	Mosaic resistant

Source : Anonymous (2010) & Anonymous (2011b)

Adoption of Proven Production Technologies

The adoption of proven technologies can result in doubling the productivity of pulses and pulse based cropping systems. The proven technologies are detailed in Table 8 for ready reference of farmers and extension workers. The selection of crops and varieties suitable for the area alongwith other management practices are considered important. In addition, to the above the following aspects deserve special attention.

- The raised bed planting in the areas prone to water logging or in poorly drained soils.
- Foliar spray of nutrients, if not applied in sufficient quantity as basal.
- Integrated nutrient management including secondary and micronutrients.
- Integrated weed management specially in *kharif* season where weeds can cause a total loss if not controlled.

- Integrated pest management specially for pod borer (*Helicoverpa*) which is the major insect pest for chickpea and pigeonpea. This can be controlled by summer ploughing.

Table 8 : Production technologies of various pulse crops

Practices	***Rabi* crops**			
	Chickpea	**Pea**	**Lentil**	**Rajmash**
Soil	Loam and clay loam	Loam and sandy loam	Clay loam	Loam & sandy loam
Land preparation	One deep ploughing by soil turning plough and 2 - 3 ploughings by cultivator followed by planking	One deep ploughing by soil turning plough and 2 - 3 ploughings by cultivator followed by planking	One deep ploughing by soil turning plough and 2 - 3 ploughings by cultivator followed by planking	One deep ploughing by soil turning plough and 2 - 3 ploughings by cultivator followed by planking
Sowing time	October to 1st week of Nov.	Mid October to November	Mid Oct. to Mid Nov.	2nd fortnight of October
Seed rate	75-80 kg/ha (small seeded) 90-100 kg/ha (bold seeded)	75-100 kg/ha	Normal sown 40-60 kg/ha Late sown 65-80 kg/ha	120-140 kg/ha
Seed treatment (before rhyzobium inoculation)	By Thiram 2 g + Carbendazim 1 g/kg seed	By Thiram 2g + Carbendazim 1 g/kg seed	By Thiram 2 g + Carbendazim 1 g/kg seed	Carbendazim 2 g/kg seed
Seed inoculation	By specific rhyzobium culture 1 packet (200 g) /10 kg seed and soil inoculation of PSB	By specific rhyzobium culture 1 packet (200 g) /10 kg seed and PSB soil inoculation	By specific rhyzobium culture 1 packet (200 g) /10 kg seed and PSB soil inoculation	-
Method of sowing and spacing	Drilling behind desi plough (6-8 cm depth) at 30 cm apart	Drilling behind desi plough at 30 cm (tall) and 20 cm (dwarf) apart	Drilling behind desi plough at 25 cm (early) and 15-20 cm (late sown) apart	Drilling behind desi plough (6-8 cm depth) at 30-40 cm apart and plant - plant 10 cm
Fertilizer replication	20 kg N + 60 kg P_2O_5 + 20 kg K_2O + 20 kg S/ha	20 kg N + 60 kg P_2O_5 + 40 kg K_2O + 20 kg S + 1 kg Mo/ha	20 kg N + 60 kg P_2O_5 + 20 kg K_2O + 20 kg S/ha	120 kg N + 60 kg P_2O_5 + 30 kg K_2O + 20 kg S/ha

Contd.

Irrigation	1st at preflowering and 2nd at pod development (if needed)	1st at preflowering and 2nd at pod development (if needed)	1st at preflowering and 2nd at pod development (if needed)	3 irrigations at 30 days interval
Weed control	Alachlore 50 EC @ 3.0 l/ha or Pendimethalin 30 EC 3.3 l/ha as Pre-emergence or basaline @ 2l/ha as pre-sowing	Alachlore 50 EC @ 3.0 l/ha or Pendimethalin 30 EC 3.3 l/ha as pre-emergence or Basaline @ 2 l/ha as pre-sowing	Alachlore 50 EC @ 3.0 l/ha or Pendimethalin 30 EC 3.3 l/ha as pre-emergence or Basaline @ 2 l/ha as pre-sowing	Pendimethalin 30 EC 3.3 l/ha as pre-emergence
Plant protection-diseases	**Wilt** Seed treatment and resistant varieties with 3 years crop rotations	**Powdery mildew** 2 sprays of Dinacap 600 ml/ha or Carbendazim 500 g/ha or Tridomarf 500 ml/ha in 600-800 litre water/ha at 10 days interval	**Rust** Sowing of resistant varieties and 2-3 sparying of Carbaxin 0.2% or Propicanazol 0.1% solution at 10 days interval	**Mosaic** Spraying of soulution of Democron or Nuvacron or Rogor @ 1.5 ml/litre of water for the control of white fly
Insect	**Pod borer** Indosulfon 2 l/ha or Melathion 2 l/ha or Cypermethlin 750 ml/ha or Fenvelrate 1 l/ha spray in 700-800 litre of water/ha	**Pod borer/ Shoot Fly** Indosulfon 1.5 l/ha or Fenvelrate 1 l/ha Monocrotophos 1 l/ha spray in 700-800 litre of water/ha	**Pod borer/ Aphids** Indosulfon 1.25 l/ha or Melathion 1 l/ha or Monocrotophos 0.75 l/ha spray in 700-800 litre of water/ha	-

Kharif **crops**

Practices	**Pigeonpea**	**Greengram**	**Blackgram**
Soil	Well drained loam and sandy loam	Well drained loam and sandy loam	Well drained loam and sandy loam
Land preparation	One deep ploughing by soil turning plough and 2 - 3 ploughings by cultivator followed by planking	One deep ploughing by soil turning plough and 2 - 3 ploughings by cultivator followed by planking	One deep ploughing by soil turning plough and 2 - 3 ploughings by cultivator followed by planking
Sowing time	15 June – 10 July (*kharif*)	15 July – 8 Aug. (*kharif*) 15 Feb.-10 April (leaid)	2nd fortnight of July (*kharif*) 15 Feb. – 15 March (leaid)
Seed rate	12-15 kg/ha	12-15 kg/ha (*kharif*) 20-25 kg/ha (*zaid*)	12-15 kg/ha

Contd.

Seed treatment (before rhyzobium inoculation)	By Thiram 2 g + carbendazim 1 g/kg seed	By Thiram 2 g + carbendazim 1 g/kg seed	By Thiram 2 g + carbendazim 1 g/kg seed
Seed inoculation	By specific rhyzobium culture 1 packet (200 g) /10 kg seed	By specific rhyzobium culture 1 packet (200 g) /10 kg seed	By specific rhyzobium culture 1 packet (200 g) /10 kg seed
Method of sowing and spacing	60-75 cm in lines and 20-30 cm plant - plant Ridge sowing is beneficial	Drilling behind *desi* plough at 30 cm (*kharif*) & 25 cm (*zaid*) apart	Drilling behind *desi* plough at 30 cm (*kharif*) & 25 cm (*zaid*) apart
Fertilizer replication	15 kg N + 45 kg P_2O_5 + 20 kg S/ha	15 kg N + 40 kg P_2O_5 + 20 kg S /ha	15 kg N + 40 kg P_2O_5 + 20 kg S/ha
Irrigation	1 irrigation pre monsoon and 1 at pod development (if needed)	1 at pod development (if needed) for *kharif* and 3-4 irrigations for zaid	1 at pod development (if needed) for *kharif* and 3-4 irrigations for *zaid*
Weed control	Alachlore 50 WP @ 2.5 Kg/ha or Pendimethalin 30 EC 3.3 l/ha as pre-emergence	Alachlore 50 WP @ 2.5 Kg/ha or Pendimethalin 30 EC 3.3 l/ha as pre-emergence	Alachlore 50 WP @ 2.5 Kg/ha or Pendimethalin 30 EC 3.3 l/ha as pre-emergence
Plant protection-diseases	**Wilt** seed treatment and resistant varieties with 3 years crop rotations **Sterility mosaic** : use resistant varieties and destroy effected plants.	**Yellow mosaic virus** : 2-3 spraying of Diamethrate 1 l/ha or Methyl-demethian 1 l/ha in 600-800 litre of water at 10 days interval.	**Yellow mosaic virus** : 2-3 spraying of Diamethrate 1 l/ha or Methyl-demethian 1 l/ ha in 600-800 litre of water at 10 days interval
Insect	**Pod borer** : Fenbelrate 0.75 l/ha or Cypermethrin 0.75 l/ha or Decamethrin 0.45 l/ha Quinalfas 1.5 l/ha spraying in800-1000 litres water/ha	**Pod borer** : BT 5% @ 1.5 kg/ ha or Fenbelrate 0.75 l/ha or Cypermethlin 0.75 l/ha or Decamethrin 0.45 l/ha spraying in800-1000 litres water/ha	**Pod borer** : BT 5% @ 1.5 kg/ha or Fenbelrate 0.75 l/ha or Cypermethrin 0.75 l/ha or Decamethrin 0.45 l/ ha spraying in800-1000 litre water/ha

Source : Anonymous (2010) & Anonymous (2011b)

Future Research /Extension Strategies

- Development of high yielding pulse varieties/hybrids/transgenics.
- Development of short duration high yielding drought resistant varieties for Bundelkhand and Vindhyan zones.
- Development of high yielding varieties resistant to water logging situations of North eastern plain zones and tarai zones of U.P.
- Development of fertilizer and other input responsive varieties for plain zones.
- Development and validation of integrated nutrient and plant protection management practices for higher and sustainable productivity.

- Development of extension network for minimizing the gap between potential achievable yields and yield at farmers fields.
- Declaration of government policies at appropriate time e.g. incentives for pulse growers, minimum support price before sowing of crops etc.

References

Ali. M. and Mishra J.P. 2003. Strategies of pulse production in India. IIPR Kanpur.

Anonymous (2009). Uttar Pradesh ke krishi ankare. Directorate of Agriculture, Uttar Pradesh.

Anonymous (2010). Rabi phasalon ki saghan padhatiyan. Krishi Vibhag U.P. pp: 63-78.

Anonymous (2011a). Uttar Pradesh ke krishi ankare Directorate of Agriculture, (Uttar Pradesh).

Anonymous (2011b). Kharif phasalon ki saghan padhatiyan. Krishi vibhag U.P. pp: 139-150.

Tripathi H.P. and Kumar, Alok (2010). Compendium of cropping system for 3 decades. Deptt. of Agronomy, NDUAT Kumarganj, Faizabad.

❑❑❑

22

Uttarakhand

A.K. Bharadwaj, Purushottam Kumar, Sumit Chaturvedi, D.K. Singh, A.P. Singh, Mohit Sharma and Vivek Patrak

Govind Ballabh Pant University of Agriculture and Technology, Pantnagar - 243 145, U.S. Nagar, Uttarakhand, India

Pulses are an essential source of protein in the diet of the vegetarian population. Cultivation of pulses has been a long-standing tradition in Uttarakhand, where the availability of alternative sources of protein such as meat and milk is barely sufficient to fulfill the daily dietary requirements. Pulses assume big importance in these rainfed and environmentally harsh regions by being an ideal crop for the prevalent low-input agricultural conditions. In Uttarakhand, pulses not only form the major constituents of the staple diet but also serve as an important source of feed and fodder for the livestock. Pulses also improve the soil health by fixing atmospheric nitrogen as well as by increasing soil organic matter

In Uttarakhand, the pulse crops constitute a key component of various cropping systems. During the last five years (2002-03 to 2007-08), the area and production of pulse crops in the state has increased by 30.5 and 112.0%, respectively. Uttarakhand accounts for about 52 % of total area and production of the NWHZ, where urdbean, horsegram (*Gahat* or *Kulthi*), rajma, and black soybean (*bhat*) are the main *kharif* pulse crops and lentil is the major *rabi* crop. Blackgram or urdbean (*Vigna mungo,* L.) is grown in low and mid hills. Its cultivation above 1500 m is rare and it is substituted by rice bean (*V. umbellata*), adzuki bean (*V. angularis*) and pseudo-cereals. Horsegram (*Macrotyloma uniflorum*) and black soybean (*Glycine max*), locally known as *gahat* and *bhat,* respectively, are grown in the low and mid hills and are popular for their characteristic flavour and medicinal properties. The grain type of French bean (*Phaseolus vulgaris*) is known as "*rajma*". It is an important crop of rainy season and is grown along with maize and grain amaranth. Fieldpea (*Pisum sativum*) is a minor pulse but has great potential in the mid and high hills. Other pulse crops like mungbean (*V. radiate,* L.), pigeonpea (*Cajanus cajan*) and chickpea (*Cicer arietinum*) have negligible area in the hills of Uttarakhand. The area, production and productivity of pulse crops has increased substantially during the last one decade in Uttarakhand. This is due to the adoption of high yielding and disease resistant varieties of pulses which are adapted to the diverse agro-ecological and cropping systems. Not with standing that, the state reels under a production deficit of 125.8 thousand tonnes in pulse crops (FAI, 2005). The wide gap between the demand and production warrants intensified efforts for enhancing yield levels in the hilly regions of the state that account for 82.4% of the total area and 79.11% of the total pulse production in the state through introduction and adoption of improved pulse production technologies.

Demand Supply Gap

In Uttarakhand, there is good scope of inclusion of grain legumes in different cereal-based cropping systems due to their ability to enhance soil fertility, drought tolerance and better adaptation on marginal lands. However, their yield levels are very low (about 0.8 t/ha), and they cannot compete other crops such as rice and wheat unless market prices are favorable and their indirect benefits to soil fertility and the environment are taken into account.

Pulses are not only a low cost source of protein for vegetarian population, they are also a low cost substitute for vegetables in periods of high prices of vegetables. Improvement in pulses production technology can reduce the production cost and hence, prices and create scope for further increase in demand for pulse crops by replacing some portion of the disproportionately high level of cereals in the consumption basket for a balanced diet.

The demand-supply gap is also reflected in the higher prices in recent years- retail price of pigeonpea reached as high as Rs. 120 per kg and other pulses remained above Rs. 70 per kg for more than six months. The recent price hike is the result of the simultaneous occurrence of lower stock levels and less domestic production both in domestic and global markets, and to some extent speculative activity in the commodity futures markets. Keeping this in view, the National Food Security Mission (NFSM) has targeted an increase in pulses production by 2 mt by 2012. It is important to mention that since last 10 years, the minimum support price (MSP) announced for all pulse crops has been below the market price. For example, the current increased MSP is Rs. 30 per kg, while the market price is hovering around Rs. 100 for pigeonpea. High volatility in prices for long periods, low productivity, and stagnation in production technology has acted as disincentives for pulses production (Reddy 2006).

Scenario in Uttarakhand

Pulses are grown on 0.63 lakh hectares of area with the annual production of 0.50 lakh tonnes. Productivity of pulses is 797 kg/ha which is more than the national average of 612 kg/ha (Anonymous, 2009). The major pulses grown are urdbean, lentil, horse gram, peas, black soybean and fieldpea.

Area under pulse crops has increased progressively from 0.48 lakh ha in 2002-03 to 0.63 lakh ha in 2007-08 (Fig 1). Total pulse production and average productivity has also gone up as compared to 2002-03.

Urdbean and lentil are the major pulse crops of the state, grown widely both in plains and hills. The area, production and productivity of urdbean and lentil are presented in (Table 1). In 2007-08, urdbean accounted for about 35 per cent area with 32 per cent of total production of *kharif* pulses followed by horse gram which occupied 19.2 per cent of the area with 21.4 per cent of production. The share of black soybean (*bhat*) in the total area and production was 9.28 and 14.0 per cent, respectively. However, the contribution of pigeonpea and mungbean were meagre.

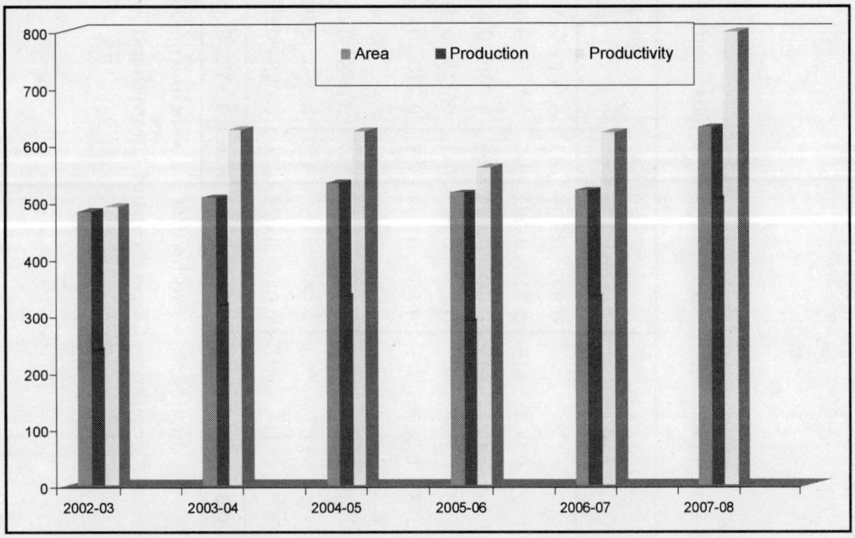

Fig. 1 : Area, Production and Productivity of Pulses in Uttarakhand from 2002-03 to 2007-08

Lentil is the major pulse crop which constitutes 56 and 36 per cent of area and production of all *rabi* pulses, respectively (2007-08). Peas (field and vegetable pea) also contribute significantly to pulse production of the state having 39 per cent of the total area and 59 per cent of the total production.

Table 1 : Crop wise area (lakh ha), production (lakh tonnes) and productivity (kg/ha) of pulses in Uttarakhand (2002-03 to 2007-08)

Crops	2002-03			2003-04			2004-05			2005-06			2006-07			2007-08		
	A	P	Y	A	P	Y	A	P	Y	A	P	Y	A	P	Y	A	P	Y
Chickpea	-	-	-	0.007	0.006	787	0.0[illegible]8	0.006	776	0.009	0.006	698	0.008	0.007	808	0.011	0.007	682
Lentil	0.144	0.064	445	0.116	0.072	617	0.1[illegible]8	0.070	593	0.130	0.064	495	0.135	0.061	453	0.126	0.053	418
Peas*	0.035	0.026	743	0.044	0.034	759	0.0[illegible]9	0.030	775	0.038	0.026	685	0.035	0.033	951	0.088	0.087	990
Pigeonpea	0.018	0.008	480	0.019	0.015	741	0.0[illegible]8	0.012	645	0.016	0.009	546	0.017	0.009	544	0.026	0.02	779
Urdbean	0.098	0.037	375	0.108	0.039	359	0.1[illegible]2	0.050	442	0.107	0.058	540	0.119	0.077	645	0.139	0.111	800
Rajmas	-	-	-	-	-	-	-	-	-	-	-	-	0.037	0.027	735	0.048	0.049	1024
Horsegram	0.076	0.035	461	0.082	0.061	742	0.0[illegible]6	0.061	715	0.086	0.051	597	0.109	0.077	706	0.12	0.107	896
Black soybean	0.034	0.02	595	0.049	0.055	1117	0.0[illegible]3	0.045	848	0.051	0.032	636	0.056	0.038	670	0.058	0.056	960
Others	0.077	0.046	597	0.080	0.034	425	0.0[illegible]8	0.058	592	0.078	0.042	538	0.003	0.002	667	0.013	0.012	923
Total	**0.482**	**0.236**	**491**	**0.505**	**0.316**	**626**	**0.5[illegible]2**	**0.332**	**624**	**0.515**	**0.288**	**560**	**0.519**	**0.331**	**621**	**0.629**	**0.502**	**798**

Note : *A = Area, P = Production, Y = Productivity • Includes both vegetable and field pea

Source : Department of Agriculture, Government of Uttarakhand

Table 2 : District wise pulse area (hectare), production (metric tonnes) and productivity (q/ha), for different districts in Uttarakhand

District	Urd			Gahat			Arhar			Rajma			Lentil		
	Area	Production	Productivity	Area	Production	Productivity	Area	Production	Productivity	Area	Production	Productivity	Area	Production	Productivity
Chamoli	441	369	8.37	325	1048	7.29	187	147	7.86	873	1048	12.00	36	13	3.61
Dehradun (Hills)	661	615	9.30	633	868	9.49	215	204	9.49	723	868	12.01	291	160	5.50
Dehradun (Plains)	212	138	6.51	0	0	0.00	1	1	10.00	0	0	0.00	146	102	6.99
Dehradun	873	753	8.63	633	868	9.49	216	205	9.49	723	868	12.01	437	262	6.00
Haridwar	49	22	4.49	0	0	0.00	0	0	0.00	0	0	0.00	789	357	4.52
Pauri	4573	3713	8.12	2185	9	8.08	206	155	7.52	11	9	8.18	1385	485	3.50
Rudraprayag	356	153	4.30	202	178	3.07	73	66	9.04	223	178	7.98	7	2	2.86
Tehri	1912	1851	9.68	3134	893	12.19	1449	1077	7.43	1076	893	8.30	1161	581	5.00
Uttarkashi	823	644	7.83	909	1782	8.61	457	366	8.01	1697	1782	10.50	379	167	4.41
Garhwal Mandal	**9027**	**7505**	**8.31**	**7388**	**4778**	**9.84**	**2588**	**2016**	**7.79**	**4603**	**4778**	**10.38**	**4194**	**1867**	**4.45**
Almora	986	790	8.01	1510	1	7.80	1	1	10.00	1	1	10.00	1374	481	3.50
Bageshwar	155	123	7.94	230	35	11.61	0	0	0.00	47	35	7.45	1051	473	4.50
Champawat	956	574	6.00	936	43	5.50	0	0	0.00	61	43	7.05	728	254	3.49
Nainital (Hills)	426	341	8.00	823	35	7.51	18	15	8.33	37	35	9.46	267	93	3.48
Nainital (Plains)	1167	817	7.00	0	0	0.00	0	0	0.00	0	0	0.00	287	287	10.00
Nainital	1593	1158	7.27	823	618	7.51	18	15	8.33	37	35	9.46	554	380	6.86
Pithoragarh	1050	913	8.70	1086	881	8.11	0	0	0.00	124	97	7.82	4384	1534	3.50
Udhamsingh Nagar	131	61	4.66	0	0	0.00	0	0	0.00	0	0	0.00	394	315	7.99
Kumaon Mandal	**4871**	**3619**	**7.43**	**4585**	**3459**	**7.54**	**19**	**16**	**8.42**	**270**	**211**	**7.81**	**8485**	**3437**	**4.05**
Uttarakhand (Hills)	**12339**	**10086**	**8.17**	**11973**	**10728**	**8.96**	**2606**	**2031**	**7.79**	**4873**	**4989**	**10.24**	**11063**	**4243**	**3.84**
Uttarakhand (Plains)	**1559**	**1038**	**6.66**	**0**	**0**	**0.00**	**1**	**1**	**10.00**	**0**	**0**	**0.00**	**1616**	**1061**	**6.57**
Uttarakhand	**13898**	**11124**	**8.00**	**11973**	**10728**	**8.96**	**2607**	**2032**	**7.79**	**4873**	**4989**	**10.24**	**12679**	**5304**	**4.18**

Contd.

District	Bhat			Others (rice bean, mung, etc.)			Chickpea			Pea (including vegetable pea)			Total		
	Area	Production	Productivity	Area	production	Productivity	Area	Production	Productivity	Area	Production	Productivity	Area (ha)	Production (met. tonnes)	Productivity (q/ha)
Chamoli	405	306	7.56	30	24	8.00	0	0	0.00	113	45	3.98	2410	2189	9.08
Dehradun (Hills)	6	6	10.00	122	101	8.28	0	0	0.00	284	199	7.01	2935	2754	9.38
Dehradun (Plains)	0	0	0.00	227	193	8.50	81	69	8.52	29	26	8.97	696	529	7.60
Dehradun	6	6	10.00	349	294	8.42	81	69	8.52	313	225	7.19	3631	3283	9.04
Haridwar	0	0	0.00	0	0	0.00	40	28	7.00	47	47	10.00	925	454	4.91
Pauri	443	301	6.79	11	9	8.18	339	119	3.51	82	33	4.02	9235	6590	7.14
Rudraprayag	73	69	9.45	0	0	0.00	2	1	5.00	8	3	3.75	944	534	5.66
Tehri	115	94	8.17	240	188	7.83	5	1	2.00	87	52	5.98	9179	8557	9.32
Uttarkashi	310	264	8.52	85	72	8.47	12	4	3.33	1112	778	7.00	5784	4860	8.40
Garhwal Mandal	**1352**	**1040**	**7.69**	**715**	**587**	**8.21**	**479**	**222**	**4.63**	**1762**	**1183**	**6.71**	**32108**	**26467**	**8.24**
Almora	1853	1915	10.33	0	0	0.00	4	1	2.50	19	8	4.21	5748	4375	7.61
Bageshwar	517	511	9.88	5	4	8.00	7	2	2.86	19	12	6.32	2031	1427	7.03
Champawat	412	392	9.51	1	1	10.00	30	11	3.67	67	27	4.03	3191	1817	5.69
Nainital (Hills)	209	177	8.47	19	14	7.37	16	6	3.75	109	44	4.04	1924	1343	6.98
Nainital (Plains)	0	0	0.00	22	16	7.27	570	513	9.00	146	124	8.49	2192	1757	8.02
Nainital	209	177	8.47	41	30	7.32	586	519	8.86	255	168	6.59	4116	3100	7.53
Pithoragarh	1470	1543	10.50	59	44	7.46	5	2	4.00	57	26	4.56	8235	5040	6.12
Udhamsingh Nagar	0	0	0.00	0	0	0.00	1	1	10.00	6637	7301	11.00	7163	7678	10.72
Kumaon Mandal	**4461**	**4538**	**10.17**	**106**	**79**	**7.45**	**633**	**536**	**8.47**	**7054**	**7542**	**10.69**	**30484**	**23437**	**7.69**
Uttarakhand (Hills)	**5813**	**5578**	**9.60**	**572**	**457**	**7.99**	**420**	**147**	**3.50**	**1957**	**1227**	**6.27**	**51616**	**39486**	**7.65**
Uttarakhand (Plains)	**0**	**0**	**0.00**	**249**	**209**	**8.39**	**692**	**611**	**8.83**	**6859**	**7498**	**10.93**	**10976**	**10418**	**9.49**
Uttarakhand	**5813**	**5578**	**9.60**	**821**	**666**	**8.11**	**1112**	**758**	**6.82**	**8816**	**8725**	**9.90**	**62592**	**49904**	**7.97**

The area under field pea is very small as compared to vegetable pea. The contribution of chickpea is small both in terms of area and production. District wise area, production and productivity have been represented in Table 2.

Constraints in Pulses Production

Even with the best of efforts, production and productivity of pulses has been stagnant. Due to the low productivity-low input nature, pulses are grown as alternate crops on marginal lands after taking input intensive crops like paddy and wheat by most of the farmers. Also, they are grown as rainfed crops with little or no modern yield enhancing inputs. As a crop of secondary importance, pulse crops do not attract much of the farmers' attention. In addition to this, these crops are more susceptible to a number of biotic and abiotic stresses, which are responsible to a large extent for the instability and low yields.

Abiotic Constraints

There has been a high degree of risk associated with pulses production. More than 90% of the area under pulses is rainfed. The mean rainfall of the state is about 1,300 mm and the coefficient of variation of the rainfall is 20-25%. Moisture stress is the often cited reason for crop failures in plains, however; it is also the major reason for crop failure in hills. High heat stress results in forced maturity with low yields. Drought stress alone may reduce grain yields by 50 per cent. A quantum jump in productivity can be achieved by applying life saving irrigation especially in *rabi* pulses grown on residual moisture. This could be tackled by development of varieties tolerant to moisture stress and bringing more area under irrigation.

Irrigated area under pulses is virtually small. Availability of adequate soil moisture for crop growth depends on rainfall, water holding capacity and depth of soil in rainfed areas. Higher evapo-transpiration during the *rabi* season causes severe yield reductions.

Grain yield is mainly influenced by temperature. Terminal drought stress in chickpea and lentil is normally accompanied by high temperatures (more than 30 °C) towards maturity, which may cause poor pod filling. Cold is also an abiotic stress, limiting the grain yield of pulse crops by causing frost injury, mortality or delayed podding (especially in chickpea and lentil). All warm season pulses are sensitive to low temperatures, but generally they are not exposed to low temperatures. On the other hand, cool season pulses are often subjected to chilling temperatures especially in areas of north India. However, not much improvement has taken place in the development of chill and frost tolerant varieties.

Waterlogging damage in submontane soils is also a limiting factor in some of the pulse crops especially pea. *Khesari* and horse gram can establish well in water logged soils, even with relay sowing of rice, and can grow until maturity in water logged soils.

Poor drainage/water stagnation during the rainy season causes heavy losses to pulses on account of low plant stand and increased incidence of phytophthora blight disease. Ridge planting has been found very effective in ensuring optimal plant stand and consequently higher yield. A simple ridger, already available, can effectively be used for this purpose. Since most pulse crops are drought tolerant, most of the research efforts have been confined to developing genotypes and associated production technologies to suit dry land conditions. Consequently, germplasm suited to high rainfall and irrigated conditions are lacking.

Biotic Constraints

More than 250 insect species are reported to affect pulses in India. Among these, nearly one dozen cause heavy crop losses. Pod borer causes maximum yield losses, followed by pod fly, wilt and root rot. Another important pest affecting pulses are nematodes and among them root-knot nematode is more important in terms of spread and damage to crops, which have been effectively controlled by bio-agents.

Among important diseases, wilt, sterility mosaic virus (SMV), yellow mosaic virus (YMV) and powdery mildew (PM) are common and cause considerable damage. A few varieties possessing vertical resistance to one or two diseases such as wilt resistance in chickpea, cercospora leaf blight, web blight and anthracnose in mungbean, sterility mosaic in pigeonpea have been identified at Pantnagar. YMV resistant mungbean (Pant Moong-1, -2 and -4) and urdbean (Pant-U-19, Uttara) and rust resistant lentil (Pant-L-406) varieties have been developed.

Physiological Constraints

Pulses (C-3 plants) suffer from inherently low yield potential and are a physiologically inefficient group of plants compared to cereals (C-4 plants) such as sorghum and maize. The fact that C-3 plants usually do better in cool climates suggests that C-3 plants are better for *rabi* season. Comparatively lower harvest index (HI) than cereal is also a limiting factor in pulses. In pulses it is only 15-20% compared to 45-50% in case of cereals such as wheat and rice.

Pulses in general have a high rate of flower drop. In pigeonpea, over 80% of the flowers produced in a plant are shed; by decreasing flower drop, yield can be increased considerably. This can be done either by breeding lines which retain a large proportion of flowers producing pods or through physiological manipulations, such as spray of hormones which reduce flower drop.

Management and Socioeconomic Constraints

In addition to the usual constraints associated with pulses, production in hills is affected adversely by poor management and socioeconomic factors as listed below:

- In general, the soils are poor in nutrients and uncertainty associated with rainfed conditions does not encourage the farmers to go in for heavy fertilization by high analysis fertilizers.
- Broadcasting is the common method of sowing, which leads to uneven plant stand and difficulties in performing necessary interculture operations.
- There is inadequate availability of efficient Rhizobium, PSB and other cultures for proper inoculation and yield performance.
- There is paucity of improved varieties of black soybean, horsegram and *rajma* suitable for hills.
- There is poor marketing system and inadequate efforts by the governmental agencies to ensure premium on pulse produce from hills through 'organic' branding.

Improved Production and Approaches to Overcome Constraints

The wide gap between the potential and average yields shows that there is great scope for increasing the pulse production in Uttarakhand. The production can be increased by both increasing the productivity and bringing the additional area under the pulses. In view of diverse nature of constraints in pulse production, there is an urgent need for adoption of diverse approaches to effectively address them. Crop wise production technologies for increased productivity of pulses in Uttarakhand are given in table 3.

1. Improved Varieties for Different Agro-climatic Zones of Uttarakhand

Several high yielding varieties of pulse crops suitable for Uttarakhand have been developed in recent years. Some of these varieties have played a significant role in increasing pulse production in the state of Uttarakhand during the last decade.

Table 3 : Crop wise production technology for increased productivity of pulses in Uttarakhand

CROP	LAND PREPARATION	SEED RATE (kg/ha)	SOWING MANAGEMENT	FERTILIZER MANAGEMENT (kg/ha)	WEED CONTROL	WATER MANAGEMENT	RECOMMENDED VARIETIES
	1	2	3	4	5	6	7
KHARIF PULSES							
URDBEAN	Well drained sandy loam	12-15	**SOWING TIME** **Hills-**2nd fortnight of june **Plains and tarai and bhabhar-**3rd week of aug-1st week of sept. **SOWING METHODS** 30-45cm in line at 3-4 cm depth, plant to plant spacing of 10 cm.	**DOSE:** 15:40 N:P_2O_5 Rhizobium culture @ 200g/10 kg seed	1. Two hand weedings at 20 -30 and 40-50 DAS. 2. Fluch lorolin 45 EC @ 2.0 l/ha and Pendimethalin 30 EC @3.3l/ha in 800-1000 l water	As and when needed	**Hills and Tarai bhabhar-** Pant urd 19, 35, 31, 40, Narendra urd 1 **Plains-** Azad 1, Uttara, Shekhar 1 and 3
MUNGBEAN	Well drained sandy loam	12-15	**SOWING TIME** **Hills-**2nd fortnight of June **Plains and tarai and bhabhar-**4th week of July-2nd week of Aug. **SOWING METHODS** 30-45cm in line at 3 -4 cm depth plant to plant spacing of 10 cm	**DOSE:** 15:40 N: P_2O_5 Rhizobium culture @ 200g/10 kg seed	1. Two hand weedings at 20 -30 and 40-50 DAS. 2. Fluch lorolin 45 EC @ 2.0 l/ha and Pendimethalin 30 EC @3.3l/ha in 800-1000 l water	As and when needed	**Hills and bhabhar-** Pant moong 2, 4, 5, 6 Narendra moong 1 **Plains-** PDM 11, Samrat, Malviya Jyoti, Malviya janchetna, Malviya janpriya, Malviya jagriti and Asha
RAJMA	Well drained sandy loam	75-80	**SOWING TIME** **Hills-**2nd fortnight of June. (Kharif) **Tarai and bhabhar-** 1st week of Feb.(Zaid) and 2nd fortnight of Oct.(Rabi) **SOWING METHODS** 30 cm in line and after 15 -20 days plants are maintained at 10 cm spacing, depth. About 5 cm.	**DOSE:** FYM- 4-5 t/ha N- 100 kg/ha P_2O_5- 80 kg/ha K_2O- 40 kg/ha S- 20 kg/ha	Two hand weedings at 15 -20 and 35-40 DAS.	As needed and when needed. Critical Stages are flowering and pod formation	**Tarai and bhabhar-** PDR 14, VL rajma 63 **Hills** – Vl rajma 63
RICE BEAN	Well drained sandy loam	15-20	**SOWING TIME** **Higher hills-**2nd fortnight of May. **Lower hills-**1st fortnight of June **METHODS** 40-50 cm in line , plant to plant/depth spaces of 10 cm and depth of 3 cm.	**DOSE:** N- 20kg/ha P_2O_5- 40 kg/ha	Two hand weedings at 15 -20 and 35-40 DAS.		PRR 1 and PRR 2

HORSEGRAM	Well drained sandy loam	40-50	**<u>SOWING TIME</u>** 1st fortnight of June is best sowing time **<u>METHODS</u>** 30-35 cm in line , plant to pl ant/ depth.	**<u>DOSE</u>:** N- 20 kg/ha P_2O_5- 40 kg/ha K_2O- 20 kg/ha	Two hand weedings at 20 -30 and 40-50 DAS.		VL 1, VL8 and VL10
PIGEONPEA	Well drained sandy loam	12-15 and IPCL @ 25-30 kg /ha	**<u>SOWING TIME</u>** **Hills-**Mid April-mid May **Plains and tarai and bhabhar-** Early maturing var.- June mid Late maturing var.- July **<u>SOWING METHODS</u>** **<u>Line x Line</u>** Late maturing var.- 60-75 cm Early maturing var.- 45-60 cm **<u>Plant x Plant</u>** Late maturing var.- 20-25 cm Early maturing var. 15 cm line at about 5 cm depth.	**<u>DOSE</u>:** N- 15kg/ha P_2O_5- 40-50 kg/ha S- 30 kg/ha	1. Two hand weedings at 20 and 30 DAS. 2. Pendimethalin 30 EC @3.3 l/ha in 800-1000 l water Alachlor 50 EC @4.0 l/ha Fluchlorolin 45 EC @ 2.0 l/ha	One irrigation at the time of pod formation	**Hills** – UPAS 120, Pusa 6, ICPL 151, VL Arhar 1 **Plains-** Ta 21, manak, Paras, UPAS 120 **Late maturing var.-** Type 7 and 17, Bahar, Amar, Narendra Arhar 1 and Malviya chamatkar
GUAR	Well drained sandy loam	Grain purpose - 25 Fodder- 40	**<u>SOWING TIME</u>** **Grains-**1st week to 3rd week of July **Fodder-** April **Irrigated-** March/April (Vegetable) **<u>SOWING METHODS</u>** 30 cm in line with plant to plant/depth spacing of 10-15 cm at a depth of 5 cm.	**<u>DOSE</u>:** N- 20-25kg/ha P_2O_5- 50-60 kg/ha K_2O- 30-40 kg/ha	1. Two hand weedings at 20 and 30 DAS. 2. Fluchlorolin 45 EC @ 2.0 l/ha	In kharif crop 1 or 2 irrigations and in summer crop, irrigation at 12-15 days interval	**Grains purpose-** FS 227, Durgapur Safeda, D 111, D 128 and Guar no. 2 **Vegetable purpose-** Pusa Mausmi, Pusa sadabahar, Pusa navbahar and **Fodder purpose-** HFG 129
RABI PULSES							
CHICKPEA	Well drained sandy loam	Small grain and Medium grain var.-60-80 Large grain var.-100	**<u>SOWING TIME</u>** **Hills-**Mid Oct-1st week of Nov. **Plains and tarai -** Normal sown-1st week of Nov. Late sown- 2nd week of dec. **<u>SOWING METHODS</u>** **<u>Line x Line</u>** 30-45 cm with plant to plant/depth spacing of 10 cm at depth of 5 cm.	**<u>DOSE</u>:** N- 15-20 kg/ha P_2O_5- 40-45 kg/ha K_2O- 20-30 kg/ha Rhizobium culture @ 200g/10 kg seed	1. Two hand weedings at 20 and 30 DAS. 2. Pendimethalin 35 EC @3.3 l/ha in 800-1000 l water	One irrigation before flowering and one at pod formation stage	**Normal sown-** Pusa 256, K-850, Avarodhi, KWR-108 and pant G-114 **Latel sown-** Udai, Pant G 186 **Kabuli-** sadabahar, L550, Pusa-1053, 267 and Pant Kabuli chana 1

PEA	Well drained loam	Normal var.-80-100 Short var.- 125	**<u>SOWING TIME</u>** **Hills**-Mid Oct-1st week of Nov. **Plains and tarai -** Oct mid-mid Nov. **<u>SOWING METHODS</u>** **<u>Line x Line</u>** Normal var.-30 cm Short var.- 20-25 cm, plant to plant with spacing of 10 cm at the depth of 5 cm.	**<u>DOSE</u>:** FYM- 60 Qt/ha N- 20kg/ha P_2O_5- 60 kg/ha K_2O- 40 kg/ha S- 30 kg/ha Rhizobium culture @ 200g/10 kg seed	1. Two hand weeding s at 20 and 30 DAS. 2. Pendimethalin 35 EC @3.3 l/ha in 800 - 1000 l water	One irrigation before flow ering and one at pod formation stage
LENTIL	Well drained sandy loam	Tarai and Bhabhar- 30-40 Hills- 40-50 Late sown- 50-60	**<u>SOWING TIME</u>** **Hills**-Mid Oct-1st week of Nov. **Plains and tarai -** 1st fortnight of Nov. **Late sown-** 2nd week of dec. **<u>SOWING METHODS</u>** **<u>Line x Line</u>** Normal sown-25 cm x10 cm. Late sown.- 15-20 cm x 10 cm , plant to plant depth	**<u>DOSE</u>:** N- 15-20 kg/ha P_2O_5- 40-50 kg/ha K_2O- 20-30 kg/ha Rhizobium culture @ 200g/10 kg seed	1. Two hand weeding s at 20 and 30 DAS. 2. Pendimethalin 35 EC @3.3 l/ha in 800 - 1000 l water	One irrigation before flow ering and one at pod formation stage

a) Blackgram

Pant Urd-31 : This is a dwarf variety of blackgram. It has given an average yield of 10 q/ha. In addition to its early maturity and higher yield, the variety is resistant to MYMV. It has been released for cultivation in plains and lower hills of Uttarakhand for pure cropping in *kharif* season. Pant U-31 was also found suitable under late planting and intercropping with cereals during *kharif* season.

Pant Urd-40 : It is an early maturing variety of black gram with slow senescence and resistance to MYMV, web blight and *Macrophomina* blight. Its average yield in pure cropping is about 10 q/ha. It has also performed better in 1:1 row ratio in sorghum + urdbean, maize + urdbean and ragi + urdbean intercropping. Owing to its yield superiority and resistance to diseases, Pant U-40 was released for pure and intercropping with cereals for the plains and lower hills of Uttarakhand.

b) Lentil

Pant Lentil-4 : It matures in 130-135 days and yields about 18-20 q/ha. It is a small seeded variety with multiple resistances to diseases. It is also suitable for both plains and hills of Uttarakhand.

Pant Lentil-5 : It has large seeds and multiple resistance to rust, wilt and blight diseases. It matures in 125-130 days and yields 18-20 q/ha. Pant Lentil-5 is suitable for both plains and hills of Uttarakhand.

Pant Lentil-6 : It matures in 125-130 days and yields about 12-15 q/ha. It is resistant to rust, Ascochyta blight and wilt diseases. Pant L-6 has light green foliage and stem, white flowers and erect growth habit. It was released for cultivation in the plains of Uttarakhand. This variety is also suitable for the hills.

Pant Lentil-7 : It matures in 120-125 days and yields about 14-15 q/ha. Pant L-7 is resistant to rust disease and tolerant to pod borer pest. It was released for cultivation in the plains of Uttarakhand. This variety is also suitable for the hills.

c) Field pea

Pant Pea-13 : It is a dwarf variety of fieldpea with average plant height from 55 to 60 cm. It matures in about 120 days in plains and 140-145 days in hills and has shown resistance to powdery mildew and tolerance to the rust diseases. It has been released for cultivation in plains and hills of Uttarakhand.

Pant Pea-14 : It is a dwarf variety of field pea and is resistant to powdery mildew and tolerant to the rust diseases. Pant P-14 has an average yield of 15 to 20 q/ha. It matures in 120-125 days in plains and 140-150 days in hills and has been released for cultivation in plains and hills of Uttarakhand for timely sown conditions in the *rabi* season.

d) Ricebean

Pant Ranichauri Ricebean-1 : It is resistant to yellow mosaic virus and has good nodulation even under acidic soils. It is released in the state of Uttarakhand and recommended for low input condition of Uttarakhand. Its average yield is 8-10 q/ha.

Pant Ranichauri Ricebean-2 : It has shown very good nodulation even under acidic soils and is suitable for cultivation under rainfed, low input conditions of Uttarakhand. Its average yield is 8.85 q/ha.

e) Cowpea

Pant Lobia-1 : It is an early maturing (70-75 days in *kharif* and 75-80 days in spring) variety with short (55-60 cm) and compact plant type. It has determinate growth habit. This variety has large (15.5 g/100), white seeds and yields about 10-12 q/ha. It can also be cultivated during spring under irrigated conditions in the mid-hills. It was released in 2008 for cultivation in the hills and plains of Uttarakhand.

f) Gahat

VL Gahat-10 : Its yield potential ranged from 10-12 q/ha and it matures in about 115-135 days. Seeds are large with 100-grain weight of 3.35 g as compared to VL Gahat 1 (2.93g). It has shown resistance against stem rot and anthracnose.

g) Frenchbean

VL Rajma-125 : It is white-seeded, 36-40 cm tall and matures in 73-98 days. Its protein content is 25.74 % and requires very less cooking time (25 min.). It yields 11-13 q/ha and is resistant to bacterial blight and stem rot.

h) Bhat

VLS 65 (*Bhat*) : It It matures in 120 days and yields 12.8 q/ha. It is resistant to frog eye leaf spot, pod blight, Cercospora and bacterial blight.

Frenchbean var. Pant Anupama at Maturity

Rajma crop var. Contendor

Vegetable Pea var. Arkel

Chickpea var. Pant G 186

Lentil var. Pant lentil 4

Lentil var. Pant lentil 5

Fig. 1 : Different pulse crops grown in *tarai* and *bhabhar* region of Uttarakhand

2. Diversification and Intensification

The area under pulse crops in the state has increased considerably during the last decade, but the trend may not remain same owing to availability of more remunerative options, such as vegetable crops, with the farmers. There is, therefore, a need to cultivate large tracts of land which is uncultivated or barren for varying reasons. Bringing such lands under pulse cultivation would increase total pulse production initially and also ameliorate the soil. Cultivation of pulse crops in such lands will also reduce their exposure to soil erosion.

Another approach for enhancing area is through increased use of pulses in mixed and intercropping. Diversification and intensification of major cropping systems involving pulse crops is an approach of immense potential for increasing pulse production through lateral enhancement of pulse area and increased cropping intensity.

In plains of Uttarakhand, the dominant cropping system of rice-wheat can be intensified by inclusion of summer mung or lobia through zero tillage technology. In maize-wheat system, alternative cropping systems including pulses can be adopted i.e., Maize-Early potato-Wheat+Lobia/French Bean (Paired Row), Maize - Lahi-ZT Lentil/French Bean/Rajmash and Maize - Lahi - Wheat+French Bean (Paired Row). Maximum opportunity for intercropping lies in widely spaced crops like sugarcane based systems. Peas, razmah, french bean can be suitable intercrops for October planted sugarcane at the spacing of 120/90/75cm wide, while lentil and chickpea for November planted sugarcane. For spring planting (February/March) scope for moong/ urd/lobia are immense. Lentil, rajma, frenchbean for December harvested ratoon crop and moong/urd/lobia for February harvested ratoon crop are suitable with optimum agronomic management.

Paired row intercropping of mung in spring sugarcane

Alternate row intercropping of mung in spring sugarcane

Lentil intercropping in autumn sugarcane

Lentil intercropping in autumn sugarcane

Fig. 2 : Diversification and intensification of pulses in sugarcane based cropping systems

In the hilly rainfed regions of Uttarakhand, *chaiti dhan* - wheat - finger millet - fallow is the most popular crop rotation, which gives a cropping intensity of 150 per cent. Data collected from farmers' fields have shown that by replacing the late maturing traditional cultivars with improved short-duration varieties, a crop of lentil can be accommodated in the 'fallow' period. This crop rotation increases the cropping intensity from 150 to 200 per cent and has been successfully adopted by farmers in the rainfed areas of Tallighat, Pyura, Fulwari goond and adjoining villages of district Bageshwar (Gupta, 2008).

In addition to incorporation of pulses in dominant cropping systems, there is need to introduce pulses as intercrops in different row ratios or geometry. Sahu *et al.* (2009) reported that in urdbean+ragi intercropping system at GBPUA&T, Pantnagar planting pattern of 1:1 recorded maximum urdbean equivalent yield (1404 kg/ha) followed by 2:1 (1348 kg/ha). Both of these planting patterns i.e., 1:1 and 2:1 recorded 32 and 27 % higher urdbean equivalent yield as compared to sole cropping of urdbean (1066 kg/ha). In maize+urdbean intercropping system with different planting patterns, normal row planting of 1:1 gave higher maize equivalent yield than 2:1 and 2:2 planting patterns (Singh, 2009). At Ranichauri, intercropping in 1:1 ratio of grain amaranth and horse gram produced highest grain amaranth yield (Singh and Ram Bhajan, 2009). Wheat and lentil are generally sown as a mixed crop in the hills. Intercropping experiments conducted by VPKAS, Almora in Garud block of district Bageshwar, revealed higher profits from planting wheat and lentil in 2:1 ratio as compared to the mixed crop. The grain yield of wheat and lentil in 2:1 ratio planting using improved cultivars was 24.0 q/ha and 4.5 q/ha, and the dry fodder yields were 39.8 q/ha and 4.8 q/ha, respectively. The wheat and lentil grain yields under mixed cropping using local cultivars were 20.5 q/ha and 3.8 q/ha, and dry fodder yields were 34.7 q/ha and 3.9 q/ha, respectively (Gupta, 2008).

3. Improved Crop Management Practices

Adoption of proper crop management practices and judicious use of resources enables better performance of crops within limited resources besides conservation of precious natural resources for future.

a) Seed and sowing

Pre sowing moisture is a pre-requisite for growing *rabi* pulses. Most of the *rabi* pulse crops like lentil, chickpea and fieldpea are grown on residual moisture. At times, the moisture is the limiting factor for germination due to early withdrawal of south west monsoon. Under such conditions, seed priming by soaking of seeds for 6-8 h in water has proved beneficial in improving plant stand and early vigour, which resulted in increased production (Singh and Ram Bhajan, 2009).

Timely sowing is one of the most important factors deciding the production potential of pulses. Optimum planting time was found to be last week of July to first week of August for urdbean. Drastic reduction in yield took place when urdbean was planted beyond first week of August. Narrow row spacing (20 cm) out yielded wider spacing (30 cm) under late planting conditions (beyond August 16). Under late sown condition, 50 per cent increase in seed rate of urdbean has been found advantageous. Time of sowing also affected the incidence of insect pest and disease. July sown crop of urdbean has been found to show higher incidence of web blight in comparison to August sown crop. The disease severity is low at wider row spacing as compared to closer row spacing. Early sown lentil (October 5 to November 5) had minimum disease severity of rust and maximum grain yield, while under late sown conditions, (November 19 to December 19) maximum disease severity and low grain yield were obtained (Singh, 2009).

b) Nutrient management

Nutrient management in pulses is an important factor for higher yield particularly in the area of scarce resources like Uttarakhand. Pulses are endowed with the unique ability of biological nitrogen fixation. This phenomenon, however, occurs effectively with adequate population and effective strain of soil rhizobia, which are generally low in stressed soils of rainfed areas. Seed inoculation of pulse crops with biofertilizers like Rhizobium has shown 12-21% increase in productivity. The technology involves inoculation of seeds with *rhizobium* culture @ 25g/kg seed 10-12 h before sowing (Singh and Ram Bhajan, 2009).

In rainfed areas, pulse crops generally suffer from inadequacy of N at the reproductive phase due to moisture stress and degeneration of root nodules. Foliar spray of 2 % urea helps in improving N content of leaves making them photosynthetically more active. Foliar sprays of urea have demonstrated an increase in yield of pulse crops by 13-20%. Sulphur (S) deficiency is widespread in areas under pulse crops. Among these, chickpea, fieldpea and pigeonpea require 40 kg S ha^{-1}, while lentil, greengram and black gram need 20 kg S/ha. Application of S in pulse crops increased yield in the range of 176-592 kg/ha (Singh and Ram Bhajan, 2009).

c) Integrated pest management

The use of chemicals for disease and insect control is rare in the hilly areas of Uttarakhand due to their higher cost and limited availability in the local market. Cultural and biological means of pest control may be a viable option for majority of farmers of Uttarakhand. Adoption of such plant protection technologies would contribute significantly towards increasing production by minimizing losses due to diseases and pests. Neem (*Azadirachta indica*) leaf extract was found most effective against *Cercospora* leaf blight in urdbean followed by kaner (*Nerium indicum*) and eucalyptus (*Eucalyptus sp.*) extracts. Geranium (*Pelargonium graveolens*) oil and peppermint (*Mentha virdis*) oil were found very effective against lentil wilt. Seed treatment with cow urine was also effective. The biocontrol agent, *Gliocladium virense,* used as seed treatment (0.4 %) and spray was effective in urdbean and mungbean crops under glasshouse and field conditions. In lentil, seed treatment with bio-agents and fungicides improved germination and plant stand significantly in comparison to the untreated check. Seed treatment with Thiram + Vitavax, (2:1), bio agent *Trichoderma harzianum* + Vitavax, Gliocladium virense + Vitavax and Pseudomonas fluorescence used as seed dresser @ 4 g/kg decreased the seedling mortality, improved germination, plant stand and yield as compared to the untreated check (Singh, 2009).

White grub is a major insect of hills and is responsible for causing heavy economic losses in a wide range of crops by feeding on the underground parts of the plants. In a bio-assay for 11 local isolates of entomopathogenic bacteria done against the grubs of, *Anomala dimidiata* under laboratory conditions, strain WHPSB 2 of the bacterium, *Bacillus cereus,* was found highly effective with 83.0 % mortality of grubs of Blister beetles which causes considerable damage to *kharif* pulses, particularly pigeonpea and rice bean, by feeding on the floral parts and tender pods. At VPKAS, Almora, neem formulation at 3 ml/l was found better for controlling the insect (Anonymous 2007-08)

4. Availability of quality seed

Timely and adequate availability of quality seed of improved varieties is critical for enhancing production of pulses. The current seed replacement ratio in pulse crops is 2-3 % against the desired level of 10-15 %. In most of the pulse crops, there is enough breeder seed available but its multiplication through foundation and certified seed is essential for timely availability of quality seed to farmers.

5. Extensive extension of improved technology

The extension of improved technologies developed by researchers to the farmers is crucial for bringing out a tangible impact of the technology. For this purpose, Front line Demonstrations have proved to be a very effective. FLDs on performance of new varieties, management of diseases and insects, nutrient management, weed management and effect of bio-fertilizers can be conducted extensively to disseminate the improved pulse production technologies in the state.

6. Strengthening market facilities

One of the chief reasons dissuading farmers from expanding cultivating of pulse crops in the hills are the poor profit margins. Most of the pulses in hills is under low input organic mode with low productivity. Moreover, their produce is also sold at ordinary prices as conventional produce.

The shift in consumer demand from inorganic to organic can be channelized by premium-pricing of the organic produce and passing on the additional gains to the growers. This model needs to be propagated in the state on a wide scale so that farmers engaged in organic farming receive due compensation for their organic produce.

Local cultivars, such as 'Harshil' and 'Munsyari' types, in rajmash and local strains of other pulses such as *gahat* and *bhatt* are preferred owing to their characteristic flavour and nutritive value. The demand of these crops is high, so much so, that the crop is bought by the traders while it still stands in the field. The produce is sold on by the traders at high premium but, unfortunately, only a small portion of the profit reaches the farmers. A dedicated marketing setup will ensure organized collection and disposal of the organic produce to reputed food brands, enabling higher profits to reach the farmers.

Acknowledgement

Authors of the article would like to thank the research undertaken at Gobind Ballabh Pant University of Agriculture and Technology Pantnagar; VPKAS, Almora; and College of Forestry and Horticulture, Ranichauri. Special thanks to the pulse group of GBPUA&T, Pantnagar for providing help and necessary support for compilation of the chapter. Contribution of AICRP-IFS and University Farm is also acknowledged for promoting pulses under different cropping and agroclimatic conditions through research and demonstration.

References

Amarender Reddy A. (2009). Pulses Production Technology: Status and Way Forward. Economic & Political Weekly EPW. December 26, 2009 Vol xliv no 52.

Anonymous. (2008). Annual Report 2007-08. Vivekanand Parvatiya Krishi Anusandhan Sansthan (ICAR), Almora-263 601, Uttarakhand, p 119.

Anonymous (2009). District wise area and production of Principal crops of India. Directorate of agriculture and Cooperation, Ministry of agriculture: Government of India.

Gupta, H.S. (ed.). (2008). Uttar-paschimi parvatiya khsetron mein pramukh dalhani fasalon ki kheti ke liye unnet taknikein. Vivekanand Parvatiya Krishi Anusandhan Sansthan (ICAR), Almora-263 601, Uttarakhand.

Reddy, A.A. (2006). "Impact Assessment of Pulses Production Technology", Research Report No 3, Indian Institute of Pulses Research, Kanpur.

Sahu, J.P., Sharma, B.B., Singh, V.K. and Chandra Bhushan (2009). Pulse Agronomy. In:. Pulse Research at Pantnagar (2000-2009) Ed D.P.Singh, G. B. Pant University of Agriculture & Technology, Pantnagar, pp 42-51.

Singh, D.P. and Khulbe, R.K. (2009). Scenario of pulse production in Uttarakhand. In: Pulse research at Pantnagar (2000-2009) Ed: D.P.Singh, G.B.P.U.A. & T., Pantnagar pp-1-4.

Singh, D. P. (2009). Pulse Research at Pantnagar (2000-2009). G. B. Pant University of Agriculture & Technology, Pantnagar - 263145, Uttarakhand, India, p 100.

Singh, D. P. and Ram Bhajan. (2009). Technologies for augmenting oilseed and pulses production in rainfed agriculture. Convocation Souvenir. G. B. Pant University of Agriculture & Technology, Pantnagar - 263145, Uttarakhand, India, pp 69-78.

❑❑❑

23

West Bengal

Manabendra Ray and Kajal Sengupta
Bidhan Chandra Krishi Viswavidyalaya, Dist. Nadia, West Bengal, 741 252

Pulses are recognized as an integral part of Bengali as well as Indian diet and the ideal supplement of cereals by virtue of their high protein and essential amino acid content. Pulses are inseparable ingredients of vegetarian diet and one of the cheapest sources of dietary nutrition protein in the state. The present emphasis on soil health, environmental quality and economic consideration, has stimulated a paradigm shift in cropping patterns leading to pulse based cropping systems. Any cropping system can become compatible and complete only when a pulse crop is included in it. Pulses won a strategic position in intensive as well as subsistence agriculture, as they are excellent source of dietary protein for millions of people, nutritional feed for livestock and having profound ameliorative effect on soil.

West Bengal has 2% of India's total area with 8% of its population (8.7 crores). Unfortunately, the supply of pulses is not sufficient to meet the full requirement. As such the state suffers from acute shortage of this commodity. During the last 12 years (1996-97 to 2007-08), the production of pulses varied from 141580 metric tonnes to 214110 metric tonne. To alleviate protein malnutrition, a minimum of 50 g pulses per capita per day should be available in addition to other sources of protein such as cereals, milk, meat or eggs. With its current production level, the availability of pulses per head per day in West Bengal will be 4.46 to 6.75 g. To make up this shortfall in supply, production of pulses required meet the deficit has to be increased 10 times. This is a herculean task being faced by agricultural scientists, researchers and planners.

Present Scenario

At present, pulse crops are grown in an area of 200940 ha in West Bengal with an average yield of 786 kg/ha and production about 158020 tonne of pulse crop. Lentil (58740 ha), black gram (52540 ha), grass pea/ lathyrus (33500 ha) and greengram (16740 ha) are the major pulse crop grown in the State (Table 1). For meeting the prescribed availability of 50 g pulse/day/head, the State

Table 1 : Area, production and yield (kg/ha) of major pulse crops in West Bengal

Crop		1998 -99	1999 -00	2000 -01	2001 -02	2002 -03	2003 -04	2004 -05	2005 -06	2006 -07	2007-08
Gram	A	23.29	26.62	54.68	50.83	47.54	46.49	37.99	40.04	31.21	25.12
	Y	815	825	917	851	780	1026	1024	911	768	983
	P	18.97	21.96	50.17	43.24	37.08	47.68	38.59	36.46	23.98	24.69
Pigeonpea	A	2.61	3.08	8.87	3.94	3.08	3.36	1.48	1.83	2.04	1.1
	Y	641	754	669	830	913	1006	710	866	691	810
	P	1.67	2.32	5.93	3.27	2.81	3.38	1.05	1.58	1.41	0.89
Mung bean	A	12.87	10.9	11.24	10.9	9.77	11.12	11.72	11.5	12.64	16.47
	Y	389	522	580	554	517	431	375	396	400	693
	P	5.01	5.69	6.52	6.04	5.05	4.79	4.39	4.55	5.06	11.42
Black gram	A	73.99	78.78	70.09	62.79	62.15	66.37	58.93	57.12	60.82	52.54
	Y	470	590	523	740	651	637	686	717	662	764
	P	34.77	46.51	36.63	46.49	40.47	42.27	40.42	40.93	40.28	40.15
Lentil	A	51.92	58.76	75.99	71.38	68.71	69.38	62.7	61.49	64.24	58.74
	Y	686	581	901	525	629	767	604	764	664	763
	P	35.6	34.17	68.5	37.47	43.25	53.19	37.87	46.96	42.67	44.82
Field Pea	A	9.23	7.04	7.33	8.84	10.76	14.61	14.35	13.13	12.5	9.86
	Y	733	856	833	634	804	1178	655	876	675	727
	P	6.78	6.03	6.11	5.6	8.65	17.21	9.4	11.5	8.44	7.16
Lathyrus	A	23.3	23.09	40.39	34.67	32.8	35.67	35.05	33.25	32.15	33.5
	Y	895	951	1053	828	827	1148	949	913	948	804
	P	20.85	21.96	42.55	28.7	27.12	40.94	33.28	30.38	30.5	29.93
Kulti kalai	A	4.96	4.33	4.31	3.68	5.75	4.09	2.79	2.8	2.41	2.21
	Y	412	482	502	860	464	436	418	461	472	500
	P	2.04	2.09	2.16	3.17	2.67	1.78	1.16	1.29	1.13	1.11
Soybean	A	0.8	0.73	0.54	0.52	0.52	0.02	0.54	0.57	0.53	0.54
	Y	576	604	599	597	598	714	565	572	597	601
	P	0.46	0.44	0.32	0.31	0.31	0.02	0.31	0.33	0.32	0.32
Other pulses	A	0.77	0.78	0.108	1.59	0.71	0.76	0.85	0.85	1.05	0.89
	Y	414	535	613	543	673	511	612	633	534	605
	P	0.32	0.42	0.66	0.87	0.48	0.39	0.52	0.54	0.56	0.54

A= Area (000′ ha), Y= Yield (kg/ha), P= Production (000′ tonnes)

Source : Estimates of area, production and yield rate of major crops in West Bengal, Govt. of West Bengal (2007-08)

requires 15,87,750 tonne of pulses per year, whereas the State is at present producing only 1,58,020 tonne of pulses per year. Thus, it is clear that for self sufficiency, the present pulse production has to be increased by a factor of 10. Most of the demand of pulses are met through imports while causes a huge loss to the exchequer. Among the different districts of West Bengal, the two adjacent districts *viz.*, Murshidabad (43700 tonne) and Nadia (35000 tonne), have major pulse producing areas (Table 3). Other important pulse producing districts are: Malda, Birbhum, Midnapore (E), South 24- parganas and North 24- parganas (Table 2).

There is limited scope of taking pulse crop during *kharif* season, As *Aman* rice (Winter rice) is the chief and most popular crop of the season, in *rabi* season, there is tough competition with other crops like wheat, oilseeds, potato and seasonal vegetables which are more remunerative, but even then the winter season is the major pulse growing season of West Bengal. At present, whatever the production of pulses is achieved in West Bengal, 77% comes from the *rabi* season.

Trends of Pulse Production

The area as well as yield of pulse crops in West Bengal has decreased during the last 12 years (1996-97 to 2007-08) (Fig.1 & 2). In contrast to this, the area under *boro* rice (summer rice), oilseeds, potato and vegetables have increased remarkably (Table 3). It is not only the total pulse area and total pulse production which has decreased, but also the area and production of every individual crop (except *mung* bean) has decreased during this period (Table 2 and Fig.3). During this period, the average yield of the pulse crop has also decreased to some extent. The expansion of area of other crops, which are more remunerative than pulse crops, is one of the major reasons for stagnation/ decline of production of pulses in the State. The additional area brought under *boro* rice, oilseeds, vegetables and potato has been at the cost of area under pulse crops. With the increase in irrigated area, the pulse area has decreased or stagnant, whereas the area of other crops, which compete for coverage with pulse crops, has increased. Additionally, the pulse crops have been pushed to marginal and sub-marginal lands which are reflected in the decreased productivity (Fig.2).

Yield Gaps

The production as well as the yield of pulse crops could be boosted up substantially through technology transfer by front line demonstrations (FLDs). There are huge gaps between research plot/field yield and farmers' field in each and every agro-climatic zone (Table 4). From the table, it is evident that there is huge scope of increase in yield in every zone for all pulse crops. From

Table 2 : District wise area and production of major pulse crops in West Bengal

District	Chickpea		Pigeon pea		Green gram		Black gram		Lentil		Pea		Lathyrus		Total	
	A	P	A	P	A	P	A	P	A	P	A	P	A	P	A	P
Darjeeling			10	9	176	95	429	291	19	14	9	7	16	13	1141	718
Jalpaiguri	61	58	17	15	286	198	1694	740	1383	622	166	128	486	370	4147	2162
Coochbehar			5	4	180	131	4487	2997	608	272			569	250	5899	3686
Uttar Dinajpur	361	131			146	144	1394	801	480	269	2	2	93	51	2694	1489
Dakshin Dinajpur	12	11					659	326	31	9	16	12	5	4	942	444
Malda	3876	3983	34	30	249	113	12759	12635	4010	3971	463	426	2629	2988	24240	24238
Murshidabad	6330	5149	372	392	814	543	9391	9265	19707	15465	3998	2161	10929	8701	51541	41676
Nadia	4162	3588	9	10	527	334	8241	6974	18735	13693	3837	3455	887	732	36398	28786
24 Pargana (N)	510	432	9	10	59	38	836	808	6537	4259	645	459	556	537	9152	6543
24 Pargana (S)	34	29	9	10	9828	6448	58	56	298	187	359	255	4579	3567	15165	10552
Howrah					149	96	5	5			4	3	398	342	556	446
Hoogly					38	29	125	52	16	9	215	154			394	244
Burdwan	54	50	22	10	62	41	257	231	752	357	22	16	284	60	1518	804
Birbhum	9501	11083	26	12	501	457	116	35	5974	5474	106	76	1181	867	17848	18194
Bankura	55	50	15	7	63	37	18	6	125	160	3	2	1	1	319	281
Purulia	156	131	415	308	353	165	10728	3874	16	14	5	4	139	58	12425	4875
Medinipur (W)			154	72	2818	2351	1222	1018	46	40	1	1	108	71	5589	4337
Medinipur (E)					217	198	117	35					10641	8313	10795	8546
Total West Bengal	25112	24695	1097	889	16496	11418	52536	40149	58737	44815	9851	7161	33501	2692	200943	158021

A= Area (000' ha), P= Production (000'tonnes)

Source : Estimates of area, production and yield rate of major crops in West Bengal, Dept. of Agriculture Govt. of West Bengal (2007-08).

the existing acreage, the total pulse production can be more than doubled if this yield gap can be bridged. This improvement is mainly due to improved varieties coupled with improved agro-techniques.

Table 3 : Area, production and yield of major crops in West Bengal

Crop		1996-97	1997-98	1998-99	1999-00	2000-01	2001-02	2002-03	2003-04	2004-05	2005-06	2006-07	2007-08
Total cereals	A	6212.3	6336.05	6334.96	6573.8	5918.39	6557.97	6297.31	6359.87	6268.45	6241.23	6143.81	6169.49
	Y	2187	2241	2248	2237	2297	2490	2438	2484	2543	2486	2575	2578
	P	13584.85	14201.59	14241.01	14703.98	13595.68	16326.09	15354.66	15797.57	15939.99	15513.49	15820.5	15902.56
Boro rice	A	1056.44	1206.83	1450.48	1474.34	1401.84	1454.99	1406.08	1390.15	1376.39	1381.96	1401.23	1511.59
	Y	3119	2958	3393	3031	3240	3034	2986	3086	3093	2928	3226	3259
	P	3294.88	3569.52	4921.98	4468.4	4541.33	4414.88	4198.6	4289.46	4257.11	4047.04	4520.66	4926.11
Total pulses	A	234.66	221.96	203.73	214.11	274.54	249.13	241.77	251.89	226.38	222.59	219.61	200.94
	Y	737	688	621	661	800	703	694	840	739	784	703	786
	P	172.98	152.67	126.47	141.58	219.55	175.14	167.89	211.64	167.3	174.53	154.35	158.02
Total oilseeds	A	509.07	508.08	493.1	502.32	598.55	604.15	568.39	685.05	673.14	643.47	703.41	707.43
	Y	841	761	769	808	953	816	837	951	827	969	917	998
	P	428.11	386.57	379.14	405.9	570.66	493.04	475.83	651.74	556.82	623.3	645.38	705.71
Jute	A	620.14	641.57	612.07	613.92	651.81	636.07	620.42	620.42	569.16	558.94	594.87	609.82
	Y*	12.1	11.77	12.05	12.12	13.56	13.37	13.49	13.49	13.8	14.29	14.14	13.47
	P*	7505.97	7549.7	7373.63	7428.4	8836.17	8505.52	8366.72	8366.72	7853.2	7989.25	8411.54	8216.04
Potato	A	314.3	284.02	318.2	315.85	299.66	299.82	349.3	308.43	320.55	354.46	407.94	400.79
	Y	26956	20947	21023	23689	25606	26090	19761	24711	221.7	21053	12384	24704
	P	8472.27	5949.37	6689.64	7482.32	7673.13	7822.26	6902.47	7621.64	7106.58	7462.46	5051.96	9900
Vegetable	A	513.7	734	700	722.3	780.545	838.79	874.87	859.86	868.41	949.74	904.7	912.3
	Y	12986	10921	13383	11648	11957	12224	12157	12698	12663	12247	13361	13763
	P	6670.80	8016.00	9367.80	8413.00	9332.98	10252.96	10635.95	10918.85	10996.62	11631.89	12088.00	12556.00

A= Area (000′ ha), Y= Yield (kg/ha), P= Production (000′ tonnes), Y*= Yield (bales/ha), [1 bale=170 kg]; P*= Production in '000bales

Source : Estimates of area, production and yield rate of major crops in West Bengal, Dept. of Agriculture, Govt. of West Bengal (2007-08) and Annual report of Dept. of FPI and Horticulture, Govt. of West Bengal (2004-05, 2005-06, 2006-07).

The comparison of achievable yield and present yield obtained in different NARP zones of the state (Table 4), it was observed that maximum yield gap in gram and mungbean exist in hill zone, while for pigeonpea and blackgram it was in the red and laterite zone. Similarly the wide gap exists in *tarai* zone for lentil and lathyrus but for pea it was coastal saline zone.

Table 4 : Yield gap (kg ha^{-1}) of different pulse crops in different agro-climatic zones of W.B

Zone	Gram	Pigeon pea	*Mung* bean	Black gram	Lentil	Pea	Lathyrus
Hill	1117	983	800	661	906	1317	357
Terai	932	915	630	786.5	1211	1362	599
Old Alluvial	1060.6	1209.8	569	793.8	799.4	1345.2	493
New Alluvial	1056.7	986	580.7	840	823.7	1313.7	305
Red &Laterite	1103.8	1563.3	561	881.8	499	1373	564
Coastal Saline	1029	986	523.7	584	856	1378	393.7

Existing Cropping Systems

The existing cropping systems involving pulse crops are given below:

Rainy season/*Kharif*	Winter/*Rabi*	Summer/*Zaid*
Jute/Black gram	-	*Mung* bean
Maize + Black gram (Inter crop)	-	*Mung* bean
Aus Rice + Black gram	Potato/Mustard/Vegetable	-
Kharif Rice	Potato	*Mung* bean
Kharif Rice	-	*Mung* bean
Kharif Rice	Lentil	-
Kharif Rice	Lentil + Mustard (Mixed crop)	-
Kharif Rice	Lentil	Sesame
Kharif Rice	Chickpea	-
Kharif Rice	Chickpea + Mustard (Mixed crop)	-
Kharif Rice	*Khesari/Lathyrus*(as *Paira/Utera* crop)	-
Pigeonpea	-	-
Pigeonpea + *Mung* bean	-	-
Kharif Rice*Kharif* Rice*Aus* Rice	Pea	-
	Rapeseed	*Mung*bean
	Vegetables	*Rabi Arhar* (Pigeon pea)

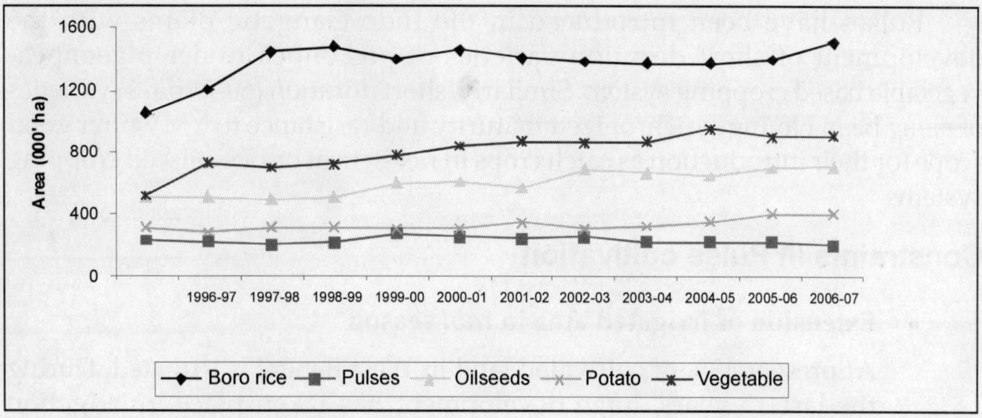

Fig.1 : Trend of area of boro rice, pulses, oilseeds, potato and vegetables in West Bengal (1996-97 to 2007-08)

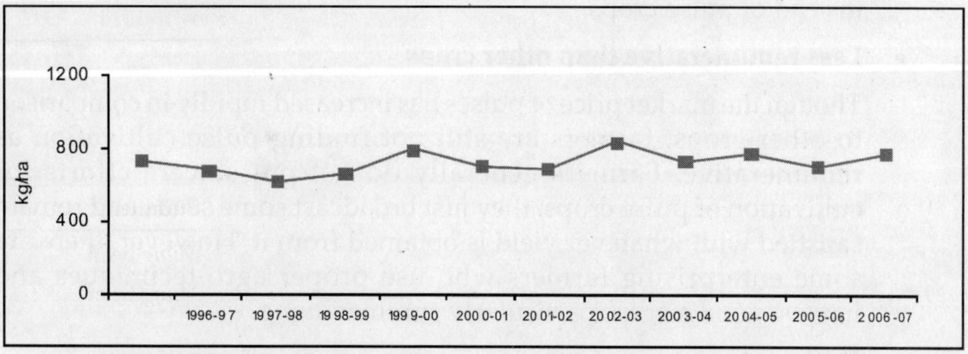

Fig. 2 : Trends of pulse productivity in the West Bengal (1996-97 to 2007-08)

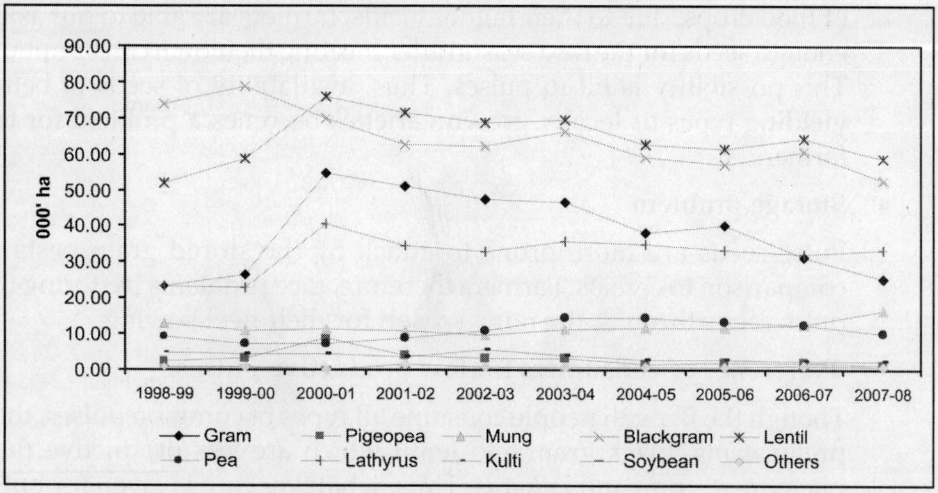

Fig. 3 : Trend of Area under different pulse crops (1998-99 to 2007-08)

Pulses have been introduced in the Indo-Gangetic plains with the development of short duration varieties of pigeonpea under pigeonpea-vegetable based cropping system. Similarly, short duration (60-70 days) varieties of *mung* bean having synchronized maturity and resistance to YMV offer good scope for their introduction as catch crops in rice-wheat or rice-oilseed cropping systems.

Constraints in Pulse cultivation

- **Extension of irrigated area in *rabi* season**

  At present 43% of cultivated land in West Bengal is irrigated. During the last 15 years, huge development has taken place in irrigation resources, particularly ground water irrigation. With the availability of irrigation water, farmers utilized the opportunity to cultivate food crops, cash crops and vegetable crops, which are more remunerative, instead of pulse crops.

- **Less remunerative than other crops**

  Though the market price of pulses has increased rapidly in comparison to other crops, farmers are still not finding pulse cultivation as remunerative. Farmers generally do not put sincere efforts for cultivation of pulse crops, they just broadcast some seeds and remain satisfied with whatever yield is obtained from it. However, there are some enterprising farmers who use proper agro-techniques and harvest good a crop, particularly summer *mung*.

- **Deficient seed supply**

  Interested farmers do not get enough seeds of good genotypes. In case of food crops, due to their higher yields, farmers are able to put aside enough seeds for the next season after meeting their own consumption. This possibility is nil in pulses. Thus, availability of seeds of better yielding types or locally grown varieties becomes a problem for the farmers.

- **Storage problem**

  Pulse seeds are more prone to attack by the stored grain pests in comparison to cereals. Farmers therefore, face problems in storing the pulse seeds through the rainy season for their next sowing.

- **Preference of consumers for low productive pulses.**

  Though the Bengali people consume all types of common pulses, they prefer *mung*, black gram and lentil which are less productive than pigeonpea, gram and *lathyrus*. Thus, whatever care is given for pulse crops, it is for *mung*, black gram and lentil.

- **Abnormal weather condition**

  Delayed precipitation during *kharif* season, or excessive rainfall or prevalence of winter rains is another reason for decrease in area, production and yield as a whole.

- **Incidence of diseases** (particularly viral diseases) or pests is also considered to be an important reason for decrease in area, production and yield of pulses in West Bengal.

Improved Varieties of Pulse Crops Suitable for West Bengal

Gram : Mahamaya-1, Mahamaya-2, Anuradha, etc.

Pigeonpea (*Arhar*) : Bahar, NDA-1, Pusa-9, Azad, Sarada, Manak, UPAS-120, etc.

Mung bean : Sonali (B-1), Panna (B-105), PDM-54, PDM-84-139, PDM-11, PDM-84-143, BSL-1, Pant Mung-4, Sukumar (WBM-29), HUM-16, etc.

Black gram (*Urd* Bean) : Uttara, Kalindi (B-76), T-9, WBU 108, Sujata (WBU 109), WBU 105, Pant U-19, etc.

Lentil : Ranjan (B-256), WBL 77, WBL 58, Asha (B 77), Pant L 639, NDL-1, KLS-218, etc.

Field Pea : G. F.-68, Dhusar (B-22), VL-42, Rachna, HUDP-15, DDR-23 and H. F. P. – 4, etc.

Khesari / Lathyrus : Varieties with low ODPA (βOAA) content (< 0.2%) are Nirmal-1, Nirmal-2, BioL-212, BioR-202, LSD-3, P-24, etc.

Crop-wise Production Technologies

Chickpea

In West Bengal, among all the *rabi* pulses, chickpea is ranked 3rd with respect to area (25.12 thousand ha) and production (24.69 thousand tonne) next to lentil and *lathyrus* (Annonymus, 2009).

Chickpea is mostly consumed in the form of processed whole seed (boiled, roasted, parched, fried, steamed and sprouted, etc.) or *dal* (decorticated split cotyledons boiled and mashed to make a soup) or as *dal* flour (*besan*). Different snacks, sweets and condiments are prepared from chickpea. Fresh green seeds are also consumed as green vegetable. It is valued not only as human food but it is also an important animal feed. Chickpea straw is considered as good forage. This crop produces 30-40 number of nodules plant^{-1} in their roots, where *Rhizobium* bacteria live and fix atmospheric nitrogen. Chickpea, grown in rotation with cereal crops, can contribute to the total pool of N in the soil and improve yields of the cereals. N fixed by legume crops not only meets their own N requirement but also a sizable quantity (30-90 kg ha^{-1}) is left for the succeeding crop (Prasad, 1996).

Rice-based cropping system is the most common system in West Bengal as rice is the staple food crop. After *aman* paddy, farmers have a choice to select their crops according to their need and will. Chickpea is one of the most important pulse crops that farmers grow after *aman* rice. In West Bengal, farmers generally grow chickpea in rice-fallow as rainfed crop utilizing the residual moisture and get a yield of 983 kg ha^{-1}.

Land preparation : After harvesting of *kharif* crop, especially rice, chickpea is sown under West Bengal conditions. After harvesting of previous *kharif* crop or cessation of monsoon rain, land preparation is started during October-November and consists of with 3-4 cross ploughings followed by laddering or levelling.

Seed rate and sowing management : The sole crop requires 45-50 kg ha^{-1} seed. However, seed rate varies with the variety. Yield is higher under line sowing with a spacing of 30 cm x 10 cm. In case of deep sowing (at 7-10 cm depth), the incidence of collar rot is reduced. Chickpea can be grown as mixed or inter crop and requires less seed rate. Under rainfed conditions, pre-soaking of seed with water (4-5 hrs) increases the grain yield of chickpea significantly (Annonymus, 2008). Middle of November to first week December is the ideal time for sowing (Ray *et al.,* 2009 and Sur *et al.,* 1966).

Fertilizer application : Chickpea requires N: P_2O_5: K_2O @ 20: 40: 40 kg ha^{-1} along with 8-10 tonnes of FYM ha^{-1}. Under deficient conditions, Borax spray @ 0.2% at 30 and 50 DAS may increase the yield. If there is any deficiency of molybdenum, spraying of ammonium molibdate @ 0.05% may also increase the seed yield of chickpea. Spraying of 2% urea at flowering and 10 days thereafter increased the grain yield of chickpea under rainfed conditions (Bhowmik, 2006).

Weed control and interculture : This crop quickly covers the field and there is little need for weeding at later stages. However, two hand weedings at 30 and 60 DAS gave higher yield which was at par with the integrated weed management practices through application of pendimethalin 30% EC @ 1.0 kg *a.i./* ha as pre-emergence followed by one hand weeding at 30 DAS (Annonymus, 2004).

Water management : Chickpea is generally grown as rainfed crop. However, two light irrigations at branching and pod formation stage increased the yield (Ray *et al.,* 2009 and Ray *et al.,* 2001).

Plant protection measures : Due to high moisture content in soil, the crop some times suffers at seedling stage from soil borne fungal diseases like collar rot as a result of which young seedlings collapse; in older seedlings the leaves turn yellow and plants dry up. Seed treatment with Thiram or Captan @ 0.2% by weight may check the disease. Gram pod borer is the key insect-pest of chickpea. An integrated management system involving cultural practices like summer ploughing along with chemical control like application of Malathion 5% dust or, Carbaryl 5% dust is found effective.

Lentil

Among the all pulses, lentil ranks first in respect of area and production in West Bengal (Table 1). Like chickpea, this crop is also cultivated during *rabi* season after harvest of *kharif* crop, mainly *aman* paddy. It contains a considerable amount of protein (25.1%) and is easily digestible. People of West Bengal prefer to consume lentil more than any other pulse. The stover serves as a very good fodder.

Land preparation : After cessation of monsoon rain land, preparation is started during October- November, after harvesting of *aus* rice or early *aman* rice (*kharif* paddy), jute, etc. Early November sowing crop gives better yield (Roy *et al.*, 2009). Land preparation is done with 3-4 cross ploughings followed by laddering. Lentil does not require very friable land.

Seed rate and sowing management : Under West Bengal conditions, the crop is generally cultivated as sole or mixed crop with rapeseed/mustard, gram, linseed, wheat or barley. Generally the seed of lentil is broadcast and thereafter, one laddering is done. It can be sown with seed drill with a row spacing of 25-30 cm. When grown as sole crop, it requires 30-35 kg seed ha^{-1} but when cultivated as *paira* crop or in case of late sowing, it requires higher seed rate. When it is sown as *paira* crop in standing paddy, the seed rate of 70 kg ha^{-1} recorded highest yield (Anonymous, 2008). Mixed crop requires less seed rate. Seed treatment is done with Thiram @ 1g kg^{-1} of seed. Seed treatment with *Rhizobium* strain is found beneficial.

Fertilizer application : Lentil requires N: P_2O_5: K_2O @ 20 : 40:40 kg ha^{-1} along with 8-10 tonnes of FYM ha^{-1}. Bhowmik (2008) observed highest grain yield of lentil with 2% urea spray at pre-flowering and 10 days thereafter along with 20-40-20-20 kg N-P_2O_5-K_2O-S ha^{-1}. The highest seed yield of lentil was obtained with combined application of *Rhizobium* + PSB + PGPR (Annonymous, 2006).

Weed control and interculture : The field should be kept weed free for the first 40 days. There is no need of weeding for *paira* crop. However, for the sole crop, one hand weeding is sufficient.

Water management : Lentil requires 15-20 cm of water in its life cycle. It is generally grown as rainfed crop. Excess water or excessive rainfall in its life cycle is detrimental. Under moisture stressed condition, 1-2 light irrigations are sufficient for getting higher yield.

Plant protection measures : The most important disease of lentil is wilt (*Fusarium oxysporum*). Seed treatment with Thiram or Thiram + Carbedazim (1:1) @ 2.5 g / kg of seed controls the disease effectively.

Lathyrus

In West Bengal Grass pea (*Lathyrus sativus* L.) is commonly grown in rainfed rice-fallows during winter season. In this State, grass pea (*Lathyrus*) is mainly cultivated as *utera/ paira* crop (no-tillage relay crop) in the medium/low lying rainfed rice-areas. Due to high ODPA (βOAA) content of local races and with the advent of irrigation facility, farmers of this region shifted from relay cropping of *Lathyrus* to more remunerative rapeseed/mustard, potato and other vegetables. Accordingly, the area under *Lathyrus* reduced drastically from 54,000 ha during 1981-82 to 33,500 ha during 2007-08.

Land preparation : It is general practice of the farmers of this region to sow *Lathyrus* seed in the standing rice crop just before its harvest to ensure its germination using the residual moisture and to avoid tillage operation during its growing period. Das (2000) reported that the *paira* crop, after rainy season rice, gave the best seed yields, compared to two or four tillage operations.

Seed rate and sowing management : *Lathyrus* requires a seed rate of 60 75 kg ha^{-1} under *utera* or *paira* system (Annonymus, 2006). Das and Jha (1997) reported that after receding of soil water during October/November, broadcasting of grasspea seed @ 60 kg ha^{-1} under no-tillage condition recorded the highest seed yield. Gupta and Bhowmick (2005) reported that sowing of *Lathyrus* seed 20 days before harvesting of rice, recorded the highest yield. Preceding rice stubble or straw maintaining a height of 15 cm above ground, was found beneficial for production of both seed and fodder under rainfed situations (Das, 2000).

Fertilizer application : The normal dose of N: P_2O_5 : K_2O is 20 : 40: 20 kg ha^{-1}. A *paira* crop after rainy season transplanted rice, did not respond to phosphate at different levels of NPK- fertilizers in combination (Das and Bhanja, 1993). Gupta and Bhowmick (2005) recorded higher yield with 20-50-20 kg ha^{-1} of N : P_2O_5 : K_2O, however, they got maximum yield with the basal application 10-25-20 kg ha^{-1} of N : P_2O_5 : K_2O + 2% DAP spray at flowering.

Weed control and interculture : *Lathyrus* as *paira* crop after *aman* rice, under rainfed conditions, heavily suppressed several weeds (e.g., *Cyperus rotundus, Cynodon dactylon, Melilotus alba/indica, Chenopodium album, Vicia hirsute)* and grew well (Das *et a.,l* 1997). However, keeping the field weed free with two hand weedings at 45 and 65 DAS was found to be significantly superior to the farmers' practice (not weeding) (Gupta and Bhowmick, 2005).

Water management : The crop is generally grown as rainfed crop. The *paira* grass pea crop in a rainfed 3-crop system (300% intensity), improved the soil moisture condition during its growing period, thus assisting seed germination of the succeeding crop (Das and Das, 1998).

Plant protection measures : Generally *lethyrus* does not require any plant protection measure. Some times it is attacked by borer that can be controlled by spraying of Carbaryl (5% dust) @ 20-25 kg ha^{-1}.

Field pea

Pea is consumed as *dal* (splitted cotyledons) as well as green vegetable. It is very nutritious and contains about 19.7% protein.

Land preparation : Land is prepared with 3-4 ploughings followed by leveling or laddering.

Seed rate and sowing management : After cessation of monsoon rain during end of September or early October, the seeds are sown in well prepared land. A seed rate of 45-50 kg ha^{-1} with 30 cm x 10 cm spacing is normally practiced and helps to get the potential seed yield.

Fertilizer application : A recommended dose of 20-40-40 kg ha^{-1} of N: P_2O_5 : K_2O increases the yield. Seed inoculation with appropriate *Rhizobium* strain also increases the yield of pea.

Weed control and interculture : One to two hand weedings are sufficient to control weeds in pea field.

Water management : It requires 20-25 cm of water in its life cycle. One pre-sowing irrigation is generally given; it may be followed by two light irrigations at 4 and 8 weeks after sowing. Adequate soil moisture increases the seed yield.

Plant protection measures : To control the attack of pod borer pest, spraying of Carbaryl 5% dust @ 20-25 kg ha^{-1} is sufficient.

Black gram (*Urd* bean)

In West Bengal among the all pulses black gram ranks 2nd in respect of area and production (Table 2). It is cultivated mainly as pre-*kharif* or *kharif* crop. It is an important pulse crops with many uses, the stover is used as green manure or fodder.

Land preparation : Black gram is cultivated with 2-3 ploughings. Under abnormal conditions, the crop can be sown as *paira* crop or with zero tillage.

Seed rate and sowing management : The sole crop of black gram requires about 25-35 kg ha^{-1} seed. It can be cultivated as mixed crop or inter crop with upland rice, maize, etc and under this situation the seed rate will be less. Generally black gram seeds are broadcasted but line sowing with 30 cm x 10 cm spacing gives better yield. Sengupta *et al.* (2001) observed highest grain yield with line sowing at 20 cm x 10 cm spacing (plant density 5 lakhs ha^{-1}). Black gram can be sown from early March to September under West Bengal conditions. Generally, the summer crop gives better yield. Pre-sowing seed treatment with polyethylene glycol (PEG, at 10^{-2} M solution) + $Na_2HPO_4.6H_2O$ (5 x 10^{-4} M solution) for 4 hrs gave a better yield than untreated seeds (Bhattacharya *et al.,* 1984). Inter cropping of black gram with upland rice (in between the rows of rice or in paired rows of rice) recorded better results than sole crop (Sengupta *et al.,* 1985).

Fertilizer application : Generally a fertilizer dose of 20-40-40 kg ha^{-1} N: P_2O_5: K_2O is recommended, but in addition to N: P_2O_5 : K_2O, application of S @ 40 kg ha^{-1} recorded higher yield (Annonymus, 2006). Seed inoculation with *Rhizobium* gave better results than uninoculated seeds in black gram (Biswas and Bhowmick, 2009). Foliar application of fertilizers like 2% urea or 2% NH_4HCO_3 at flowering stage also gave beneficial results (Chakravorty *et al.,* 1984).

Weed control and interculture : The critical crop-weed competition is the first 40 days period. One hand weeding at 25-30 DAS is found beneficial. Application of Pendimethalin 30 EC @ 1.0 ka *a.i.* ha^{-1} + one hand weeding recorded the highest yield (Annonymus, 2006).

Water management : The crop is grown under rainfed conditions. Stagnation of water in the field is harmful for the crop, so leveled land or good drainage system is essential for the crop.

Plant protection measures : To control different diseases of black gram, seed treatment with Mancozeb (2 g / kg seed) + Carbendazim (1 g / kg of seed) or Thiram or Captan @ 0.2% is recommended. Black gram mosaic disease is the major disease in West Bengal.

Mung bean

The *mung* bean is a very short duration crop and since it has a deep root system, it can escape drought. This allows the crop to be cultivated during spring or summer. It is an important pulse crop of West Bengal and is mainly grown as a catch crop.

Land preparation : Land is prepared by 2-3 ploughings followed by levelling.

Seed rate and sowing management : Greengram requires a seed rate of 30-40 kg ha^{-1}. Highest grain yield of greengram was recorded with 30 cm x 10 cm spacing (Bhowmick *et al.,* 2006). Before sowing, seeds should be treated with Mancozeb (2 g / kg seed) + Carbendazim (1 g / kg of seed) or Thiram or Captan @ 0.2%.

Fertilizer application : Generally a fertilizer dose of 20-40-40 kg ha^{-1} N:P_2O_5:K_2O is recommended, but an addition applicational of S @ 30 kg ha^{-1} recorded higher yield (Sengupta, 1998), thus, application of SSP is a better option than DAP or any other source of P. Spraying of growth regulator, like **brassinolide** (phenolic compound) increased the seed yield of greengram (Sengupta *et al.,* 2009). Before sowing, inoculation of seeds with *Rhizobium* bacteria increases the yield. Under Zn deficiency,application of $ZnSO_4$ @ 20 kg ha^{-1} once in 3 years is found to be effective for increasing the grain yield.

Weed control and interculture : First 40 days in its life cycle is critical from the crop-weed competition point of view and during this period, the field must be remain clean and weed free; one hand weeding at 25 DAS is found effective.

Water management : The crop is normally grown as rainfed crop. Under moisture stressed condition, one pre-sowing irrigation and one post-sowing irrigation at pod filling stage are found sufficient to obtain higher yields.

Plant protection measures : To control viral disease, spraying of Metasystox 0.1% is advocated.

Pigeonpea

It is a perennial crop, but can be cultivated as an annual *rabi* crop. It is an important pulse crop at the national level but in West Bengal its coverage is very low (Table 2). It contains nearly 22.3% protein and can be used as *dal* or green vegetable. Husk of the grain is used as fodder and the turqs (stems/ branches) are used as fuel in rural Bengal for cooking.

Land preparation : Warm and humid climate is ideal for its growth. On the onset of monsoon rain (early June), the land is prepared with thorough ploughing followed by laddering. Deep ploughing is recommended since it is a deep-rooted crop. The crop does not tolerate water logging, so good drainage facility must be maintained in the field.

Seed rate and sowing management : Normally 15-20 kg ha^{-1} seed rate is required for sole crop with 50 cm x 30 cm spacing. For *rabi Arhar,* seed rate is higher (30 kg/ha) and the spacing is 30 cm x 30 cm. When pigeonpea is cultivated as inter crop or mixed crop, less seed is required. Before sowing seeds are treated with Mancozeb (2 g / kg seed) + Carbendazim (1 g / kg of seed).

Fertilizer application : Generally a fertilizer dose of 30-60-60 kg ha^{-1} $N:P_2O_5:K_2O$ is recommended, but in addition to $N:P_2O_5:K_2O$ application of S @ 30 kg ha^{-1} recorded higher yield. In addition to chemical fertilizers, 8-10 tonnes of FYM ha^{-1} are applied. Pre-sowing inoculation of seeds with *Rhizobium* bacteria increases the yield. Under Zn deficiency application of $ZnSO_4$ @ 20 kg ha^{-1} once in a 3-year rotation increases yield.

Weed control and interculture : First 60 days in its life cycle is critical. Two hand weedings at 25 and 50 DAS are found beneficial for weed management as well as for obtaining higher seed yield.

Water management : The crop is grown under rainfed condition and no irrigation is generally applied. But when grown as a *rabi* crop, 1-2 irrigation may be given as and when required specially at harvesting stage.

Plant protection measures : To control pod borer, application of Metasystox 0.1% starting from flowering is essential.

Future Thrust Areas

- At present very limited area is covered with improved high yielding pulse varieties developed either in West Bengal or in other states but suitable for West Bengal conditions. There are proven experimental results of the trials conducted in the experimental fields over years to show that the adoption of modern improved varieties can enhance the yield 20-25%.
- Information in respect of effect of improved agronomic practices in the pulse crops for improving the crop yield is available in the state. Therefore timely sowing using optimum seed rate, spacing and plant protection management.
- Growing pigeonpea during *rabi* season instead as mixed crop with upland rice.
- In North Bengal, a vast area is acidic in nature and remains fallow during post rainy season, with soil amelioration this area can be brought under pulse cultivation. In some pockets in North Bengal, crops like lentil and black gram, are grown during post-*kharif* season after harvest of jute and upland rice. But there is still ample scope of increasing the acreage under pulses comprising the districts of *Uttar* Dinajpur, *Dakshin* Dinajpur, Jalpaiguri and Cooch Behar, provided the supply of soil ameliorants like dolomite and lime are assured.

- During the last few years it is becoming progressively difficult to grow gram particularly in the Gangetic-Alluvial zone where more than 60% pulse area is located. Gram is one pulse crop through which the total production of pulses can be increased significantly. But, because of too much winter rain, its cultivation is becoming limited bringing down the total area as well as production. The four major districts are Murshidabad, Nadia, North 24 parganas and Malda, where more than 60% pulse area is located. Unlike these gangetic-alluvium districts, the experience is different in red-laterite districts like, Birbhum, where there is good scope of gram cultivation utilizing the residual moisture after harvest of *kharif* rice.
- The sole cropping of pulse in pre- and post-rainy seasons by fitting them into double and multiple cropping systems, has opened out bright prospects in irrigated as well as rainfed areas.
- Inter-cropping of pigeonpea with maize has been identified as profitable. Inter-cropping of soyabean with direct seeded rice or pigeonpea with rice or black gram/greengram with rice under different management and spatial arrangement have shown better results. Inter-cropping of November planted crop like wheat and mustard has opened new vistas in increasing the production of pulses. Similarly, alternate row sowing (30 cm apart rows) of turmeric and greengram has been found to provide 7-10 q/ha of greengram without affecting the yield of turmeric and at the same time minimized the cost of weeding turmeric (Kundu and Chatterjee, 1981, 1982).
- Large field boundaries can be utilized for growing pigeonpea, particularly in watershed areas of Purulia, Birbhum, Bankura, Bankura and Burdwan, and even the field boundaries of rice can be used. In the canal command area or even in the tubewell command area, a large areas along the irrigation channel remain fallow. Pigeonpea may be raised along these irrigation channels, which will boost the pigeonpea production.
- To increase area and production of pulse crops in West Bengal, crop specific approaches are needed, which should be adopted in the overall framework of systems approach. The major thrust areas to be addressed are as follows:
  - Replacement of cereal or other crops in the prevailing rice-wheat or rice-oilseeds cropping systems with high yielding varieties of pulses;
  - Inclusion of short duration varieties of pulses as catch crops;
  - Development of multiple disease and pest resistant varieties;

- Reducing storage loses and improving market information and infrastructure;
- Linking MSP to market prices can bridge the gap between demand and supply;
- Developing high nitrogen fixing varieties or efficient ***Rhizobium*** strains, which will play a crucial role in sustainable agriculture; and,

References

Annonymus (2009). Estimates of area, production and yield rate of major crops in West Bengal (2007-08), Dept. of Agriculture, Govt. of West Bengal.

Annonymus (2008). Annual report pulses and oilseed research station, Berhampore, West Bengal.

Annonymus (2006). Annual report pulses and oilseed research station, Berhampore, West Bengal.

Annonymus (2004). Annual report pulses and oilseed research station, Berhampore, West Bengal.

Bhattacharya, K. K, Sengupta, K.and Chatterjee, B. N. (1984). Seed treatment for increasing black gram [*Vigna mungo* (L.)] yield in rainfed agriculture. Seed Res. 12(1): 40-47.

Bhowmik, M.K. (2008). Effect of foliar nutrition and basal fertilization in lentil under rainfed condition. J.Food Legume 21(2): 115-116.

Bhowmik, M.K. (2006). Foliar nutrition and basal fertilization in chickpea under rainfed condition. Env. and Eco., 24S (4), 1028-1030.

Bhowmick, M.K., Sadhukhan, R. and Gupta, S. (2006). Comparative performance of mungbean genotypes under varying plant populations and fertilizer levels during kharif. Ind. J. Pulses Res. 19 (2): 225-227.

Biswas, P. K. and Bhowmick, M. K. (2009). Evaluation of different Rhizobium strains on growth, nodulationh and seed yield in urdbean. J. Crop and Weed. 5(2): 87-89.

Das, N. R. (2000). Lathyrus stivus in rainfed multiple cropping systems in West Bengal, India – a review. Lathyrus Lathyrism Newsletter 1 : 25-27.

Das, N. R. and Das, D. (1998). Soil moisture retention in rainfed rice-fallows in dry season of West Bengal. Adv. In Agric. Res. In Ind. 9, 133-138.

Das, N. R., Bhattacharya, S.P. and Das, A. K.(1997). Weeds in crop of West Bengal. III Winter season. World Weeds 4, 215-226.

Das, N. R. and Jha., A. (1997). Utiliztion of low wate land by rainfed rabi pulses with no-tillage under different seed rate. J. Interaca. 1: 112-114.

Das, N. R. and Bhanja, N. (1993). Establishing rainfed no-till winter crop under NPK-fertilizer after transplanted wet rice. Int. Rice Res. Newsletter 18, 39.

Gupta, S. and Bhowmick, M. K. (2005). Scope of growing lathyrus and lentil in relay cropping systems after rice in West Bengal, India. Lathyrus Lathyrism Newsletter. 4: 28 -33.

Prasad, R. (1996). Cropping system and sustainability of agriculture. Indian Farming. 46. 39-44.

Ray, M., Chowdhury, S., Nanda, M. K. and Khan, D. K. (2010). Water use in chickpea (Cicer arietinum L.) in the gangetic alluvial zone of West Bengal. J. Agrometrol., 12 (1): 81-84.

Roy, A., Aich, S. S., Bhowmik, M. K., and Biswas, P. K. (2009). Response of lentil varieties to sowing time in the plains of West Bengal. J. Crop and Weed. 5(2): 92-94.

Ray, D., Bandyopadhyay, P., Jana, P.K. and Bhowmick, M.K. (2001). Influence of irrigation schedules on yield, consumptive use and crop coefficient of chickpea (*Cicer arietinum* L.). J. Interacad., 5(1): 33-38.

Sengupta, K., Mitra, S. and Ray, M (2009). Effect of brassinolide on growth and yield of summer greengram crop. Ind. Agric. 53 (3-4): 155-157.

Sengupta, K., Nandi, S. and Baral, Manika (2001). Effect of sowing attributes on grain yield of rainfed black gram [*Vigna mungo* (L.)] crop. Proc. Nat. Sem. of Crop Management. Feb.1-3: 15-18.

Sengupta, K., Sengupta, C. and Chatterjee, B. N. (1985). Intercropping of upland rice with black gram [*Vigna mungo* (L.)]. J. Agric. Sci., Camb. 104, 217-221.

Sengupta, K. (1998). Abst. First . Int. Agron. Cong. Nov.209.

Sur, S.C., Sengupta, K.and Bhowmick, K and Sen, S. (1966). Optimum sowing time of sowing gram (*Cicer arietinum* L.) in West Bengal. Ind.J.Agric.Sci., 36(4):173-179.

❑❑❑

File Edit View History Bookmarks Tools Help
http://www.book....com/index.php?
www.bookfactoryindia.com/index.php?
Feedback
Most Visited Getting Started
PUBLISHERS AND SUPPLIERS OF AGRICULTURAL,HORTICULTURAL,FOOD AND VETERINARY SCIENCES BOOKS
Hello Guest,
Home | About NIPA | Join Mailing List | Contact Us | New Member | Sign In |
Your cart contains 0 item.
NIPA
NEW INDIA PUBLISHING AGENCY
Search
Keywords
GO
Download Catalogues
Invitation To Authors
Featured Authors
Teachers Resources
Students Resources
Booksellers Resources
NIPA's NEW RELEASES
NIPA's FORTHCOMING
NIPA's BESTSELLERS
Browse Books by Subject
Agricultural Sciences
Animal Husbandry
Botanical Sciences
Biotechnology
Climate Change
Engineering
Complete Subject List
Environment
Food Science
Earth Sciences
Horticulture
Pharmaceutical Science
Food and Nutritonal Security by Sustainable Agriculture: Methods to Attain and Sustain
Brijesh Kumar Mishra: Ed.
INR 2205.00
DAIRY TECHNOLOGY
Milk and Milk Processing: Vol.01: Dairy Technology
Singh, Shivashraya
INR 2246.00
DAIRY TECHNOLOGY
Dairy Products and Quality Assurance: Vol.02 : Dairy Technology
Singh, Shivashraya
INR 2246.00
SEED SCIENCE & TECHNOLOGY
Seed Science and Technology
Vanangamudi,K.
INR 1796.00
ADVANCES IN PROTECTED CULTIVATION
Advances in Protected Cultivation
Singh,Brahma, Balraj Singh
INR 1526.00
Bioinformatics in Agriculture
Bioinformatics in Agriculture: Tools and Applications
M.Balakrishnan: Ed.
INR 1526.00
PM 03:43
17-04-2014